Donald S. Maier

What's So Good About Biodiversity?

A Call for Better Reasoning About Nature's Value

 Springer

Donald S. Maier
Middletown Dr. 2251
Campbell, CA, USA

ISBN 978-94-007-3990-1 (hardcover)   ISBN 978-94-007-3991-8 (eBook)
ISBN 978-94-007-7099-7 (softcover)
DOI 10.1007/978-94-007-3991-8
Springer Dordrecht Heidelberg New York London

Library of Congress Control Number: 2012937654

Springer is part of Springer Science+Business Media (www.springer.com)

Contents

Chapter 1
Prologue

Sweet is the lore which Nature brings;
Our meddling intellect
Mis-shapes the beauteous forms of things: —
We murder to dissect.

– from "The Tables Turned" (1798), William Wordsworth

1.1 Why This Book?

I am an environmentalist, trained as a scientist, and a moral philosopher by current profession. Elements from all three components of my background interlink to support an abiding belief: How people behave with regard to natural environments is very important – important enough to rise to the level of major moral consideration. One way to put this is: how we humans treat the natural world can either redound to the credit of our character or derogate it. Conversely, our character is partly expressed in how we treat nature. In other words, how we think about the natural world and how we act on and in it are bound up in issues that are cardinal in leading a good human life. Lately, those who share this feeling about the importance of nature have latched onto one strikingly salient way of expressing it. On this accounting, there is nothing more central to these natural environment-related issues than the diversity of life, or what has come to be known as "biodiversity".

Vocabularies differ. But in their own terms, many scientists, philosophers, economists, and environmentalists express a like view of how important the natural environment is – both in and for the lives of persons. I also know that these like-minded people have labored to understand why biodiversity is preeminent in this picture of natural value. They have done so with markedly accelerating intensity since the invention of that neologism in the mid-1980s. The result is a formidable and fast growing body of work that tries to make the case for nature's value on the back of biodiversity. It might be initially surprising that some substantial amount of this work falls under the label "ecology", which one might suppose to be divorced

D.S. Maier, *What's So Good About Biodiversity?: A Call for Better Reasoning About Nature's Value*, DOI 10.1007/978-94-007-3991-8_1,
© Springer Science+Business Media B.V. 2012

from questions about value. But it is hard to avoid the impression that so many studies investigate biodiversity as an independent variable in "objective" ecological relationships precisely in order to bolster its credibility as the basis for "the good" at which conservation and restoration work then aims. All told, this body of work has grown to the point where it might now constitute the major bulwark holding back those who are skeptical of the value of nature and those whose interests lie elsewhere.

The currently overwhelmingly popular environmentalist view is that we must save biodiversity because biodiversity is too valuable to lose; and that our primary commitment to the natural world is expressed in this act of salvation. This sentiment seems to infiltrate most every argument for why nature is valuable, and it comes from a multitude of the most respected sources. It comes from E.O. Wilson – the most eminent biodiversity advocate of the day. A steady stream of books, lectures, and essays has carried his repeated pleas to head off the Sixth Great Extinction. It is hard to find a biologist who writes about biodiversity and who does not follow Wilson's path. Paul Ehrlich, Osvaldo Sala, Daniel Faith, Sahotra Sarkar, Andrew Dobson, F. Stuart Chapin III, and Gretchen Daily join many other scientists who follow Wilson and follow him into this book. Scientific authorities are not alone. We even have it from the Pope – John Paul II (1999, §25, "Ecological concern") – that "the forests of Amazonia, an immense territory... is one of the world's most precious natural regions because of its bio-diversity which makes it vital for the environmental balance of the entire planet."

The titles trumpet the basic message. Books such as *Sustaining Life: How Human Health Depends on Biodiversity* (Chivian and Bernstein 2008b) are published at a startlingly high rate. Even more frequently published are papers with such titles as "Biodiversity Loss Threatens Human Well-Being". Works such as these are published by the world's most prestigious presses and journals; their authors are among the most distinguished experts in their fields. Moreover, the world's most powerful organizations that identify themselves as "conservation organizations" have made the "save biodiversity" theme their guiding anthem. The World Wildlife Fund, for one, rationalizes actions in terms of "Biodiversity Visions", which identify "priority areas critical to maintaining the biodiversity of the entire ecoregion."[1]

Disturbingly, these prevailing views have barely been examined or questioned.

Some number of months ago, I began to systematically scrutinize these sources and the many arguments that they present. I was looking for a sound foundation on which to build my own contributions to this effort. With the great size of the literature and the capabilities of its authors, I was at first confident that my only significant problem would be one of selecting and integrating an existing and unassailable edifice of argument. Instead of a solid edifice, I found a chimera. I was stunned that I could not find a single argument that does not have serious logical flaws, crippling qualifications, or indefensible assumptions.

My confidence transformed to worry; then to astonishment; and finally to alarm.

---

[1] This phrase appears on the WWF's web site: http://www.worldwildlife.org/science/ecoregions/item1871.html.

As I said, I feel strongly that it is important for people to behave well with respect to our natural environment. A lion's share of the recent rationales marshaled on behalf of this view hinges on the value of biodiversity. However, these rationales are mostly so fragile that they crumble before modest scrutiny. Worse, those who have no true interest in the natural world can easily appropriate much of the reasoning for environmentally questionable and even destructive ends. In fact, this appropriation is already happening, it is accelerating, and it has the imprimatur of the most influential of biodiversity "advocates" and self-identified "conservationists". The problem is that much of the same bad reasoning "for" biodiversity is easily conscripted in defense of the proposition that little in nature, including biodiversity, has much value. But perhaps most disconcerting of all is that the defective rationales, which purport to bolster the case for biodiversity's value, infiltrate and misdirect the "conservation" project goals of the environmentalists and scientists who craft them. When this happens, the most capable would-be friends and protectors of the natural world inadvertently spend their precious time and resources accelerating its demise.

When it comes to biodiversity and the range of arguments that defend and build on its alleged value, it is hard to avoid the impression of culturally conditioned, uncritical acceptance and unhealthy disciplinary inbreeding. Circularity, confusions, missing premises, normative biases, and doubtful empirical claims go unnoted and uncorrected. Worse, these failures of reasoning are often repeated – error for error, detail for detail – by one party after another. It is as if there is a tacit agreement among colleagues not to rock the boat of bad reasoning – perhaps out of fear that there is no other way to defend nature and its value. Confirmation bias – the human tendency to actively seek and interpret information in a way that confirms closely held beliefs and the corresponding tendency to ignore or underweight disconfirming information – could also be at work, blunting the acuity of critical judgment.

But the principal purpose of this book is not to uncover the sociology or psychology behind these problems of reasoning. Rather, the book sets itself first the simpler task of uncovering the problems, and then the more formidable one of using this understanding as a springboard for another way of thinking about natural value. It seems inevitable that, one way or another, the serious defects will eventually emerge. I feel that it is better for someone like myself who deeply believes in natural value to declare the Emperor naked, rather than to leave it for another day, a less scholarly approach, or an outright hostile one.

All of this is cause for a thoroughgoing reassessment of what the environmental community has been saying about biodiversity. This phenomenon will be familiar to the trained philosopher: A plausible thesis is proposed. Here, the thesis is: the signal value of the natural world is ensconced in biodiversity. Sometimes, the thesis is submitted as a kind of warning: that impingements on biodiversity will likely cause the collapse of nature or of the destruction of the planet as a human-inhabitable place. An intense effort to support some form of this thesis ensues. But after decades of effort, no argument can be found that "sticks". This, I believe, is cause to reconsider the thesis. Maybe, just maybe, biodiversity is the wrong hook on which to hang the value of nature.

The intent of this book is to challenge all of us environmentalists to reconsider and reground our arguments for why the natural world is valuable. The book also challenges us to consider whether our actions and projects – which largely have flowed and right now are flowing from faulty logic – might consequently be unreasonable and undesirable. It is a truly sad thought that the greatest of all threats to nature might well come from the misjudged behavior of its would-be advocates.

Finally, the book takes up its own challenge and suggests another way to view and assess the value of nature. Taking up this different viewpoint is at the same time a distancing from currently dominant ones, which largely build on the science and the economics of biodiversity. That is a good thing, for I believe that the principles of current approaches, until now largely unquestioned, promote a way of relating to nature that is disdainful of its core value. A telling symptom of this is that if they were consistently honored, these principles would justify transforming the natural world into something that no lover of nature would recognize as "nature" or "natural".

1.2   Mixing Philosophy with Biology

This book occupies some tricky terrain at the intersection of environmental attitudes, the discipline of philosophy, and the disciplines of several earth sciences, particularly ecology and conservation biology. Much of it deploys the tools of analysis and more or less traditional philosophical moral categories that are stock in the philosophical trade. In an attempt to dig even deeper than these tools allow, the last chapter turns to myths and metaphors expressive of the human condition. All of these devices are brought to bear on a topic that is largely discussed in scientific fora, in which arguments presented by scientists prevail.

What this means, most importantly, is that scientists and philosophers are talking about the same matters of value. Scientists are making moral arguments, whether or not they are aware of this. Like philosophers, they are trying to reason from some moral viewpoint – which means, from some set of moral premises, whether or not these are articulated, acknowledged, or even recognized by their authors. A philosophical consideration of these arguments and their premises is nothing more (nor less) than the project of bringing them to light. It is only by means of this illumination that their moral contours and general soundness can be assessed.

The terrain is tricky partly because scientists in general, and biologists in particular, use a distinctive language and communicate with their fellow scientists in exclusive conferences and journals. Philosophers tend to occupy their own, similarly sequestered disciplinary environment. The assumptions, conventions, traditions, tropes, and values concerning standards for reasoning of one group might initially seem alien to the other. As a consequence, it is difficult to bridge the gap between the two in order to lay claim to the intersecting territory. Perhaps because of the difficulty, it is also rare. But bridging the gap is essential to this discussion and it is one of this book's main goals.

There are other hazards. The signal ones are the strain on the biologist to understand the methods and strictures of philosophy, the strain on the philosopher to understand the biology and other earth sciences, and the strain on everyone else without special training in those disciplines to understand what in the world is going on. To ameliorate these hazards, I have conferred with biologists, philosophers, and non-specialists, asked them to share with me the obstacles to understanding that they encountered, and have incorporated some fair amount of explanatory material with these various backgrounds and perspectives in mind.

For those without specialized training in biology or other earth sciences, I have included a glossary of terms – principally from those disciplines – that might be unfamiliar. For those without specialized training in philosophy and particularly moral philosophy or axiology (the general study of value), I have further populated the glossary with terms from those disciplines; and I have included in the main text a brief introduction to how philosophers think about moral value.

But there is only so much a book such as this can do. The glossary is no substitute for serious study of ecology and other earth sciences. The quick introduction to moral philosophizing cannot replace struggling with the texts of previous thinkers; nor is it a substitute for experience in reflecting on moral conditions and difficulties. The topic of nature's value and how biodiversity fits into it is difficult and complex. There is no getting around the necessity for you, my reader, to stretch your capacities beyond their accustomed bounds.

1.3   The Scope and Chief Goal of This Book

As a philosopher, I could make my life's work the analysis of every biodiversity-based argument ever made on behalf of natural value and responsibility. I have pretty much done that during the extended time I have devoted to writing this book. There seems to be little danger of running out of grist for the mill. At least at first blush, it appears that the supply of arguments is continually refreshed.

This book encompasses an astoundingly large and diverse set of views on biodiversity's value. It is unique in this respect. It scrutinizes most views about why biodiversity is too valuable to lose – and all the most well known views. "New" arguments are continually crossing my desk and computer screen in the form of new journal articles and new books. But the "new" arguments more often than not closely hew to the old ones. Unfortunately, that means that they display the same and often faulty reasoning. The logic and its defects are typically not changed by a slightly altered empirical premise or conclusion.

The book has the goal of touching on all the most promising and credible arguments for biodiversity's value. I hope that it serves as an excellent reference in that regard. On the other hand, the book does not aspire to be so comprehensive as to be able to claim that absolutely no argument for biodiversity's value has escaped its critical attention. Many arguments are simply so flawed that their examination offers no illumination. But also, this has to do with my philosophy of doing philosophy.

I conceive of my job as a moral philosopher to be one of fomenting change for the better. I do this by teasing out the logic in commonly offered lines of reasoning, posing challenging questions, and suggesting ways of thinking that are sufficiently discomfiting for people to feel that they need to reconsider their own lines of thought.

I hope to sometimes provoke people to risk thinking things through on their own – perhaps for the first time – instead of accepting the word of even highly respected colleagues.

My arguments are the burred seeds that get caught in your mind. To accomplish my thorny purpose, an absolutely comprehensive discussion – one that leaves unexposed no detail of any argument anywhere – is not required. I will have done my job if I have sensitized you to the logic of the best arguments now on offer and if I have given you the tools to ask your own questions and to assess similar and even dissimilar arguments on your own.

Also, I know that I have little chance of convincing you of anything directly through my specific arguments. In my experience, the only effective arguments for any of us are the ones that each of us personally constructs. I welcome you to borrow and steal bits and pieces from my efforts. But the most convincing arguments will be your own – from whatever component parts you choose and however you choose to cobble them together.

For my part, I have tried mightily to construct arguments that are sound in fact and logic. At the same time, I know that no attempt to address a topic this complex is without flaws; I will not be surprised if you discover some in this book. But I hope that one or two flaws in my reasoning will not dissuade you from the preponderant impression that the reasoning about biodiversity now before the general public, as well as in scientific, economic, and policy circles, is rife with flaws. I hope that you, like me, will not be able to escape the conclusion that a case for natural value cannot stand on such a flawed foundation. But mostly, I hope that this book prods you into reviewing and even reconstructing your own thinking on these matters. As for my own attempt at such a reconstruction in the last chapter of this book, even if you find it unconvincing, I hope that my example will serve to set you free to find something better.

Of course, you might end up disagreeing with this book's assessment of the case for basing the value of the natural world on biodiversity. I welcome that result, too. However, I would hope that your position is based on reasons that avoid the defects, traps, and recitation of failed arguments that this book might make familiar to you.

Chapter 2
Preliminaries

The days when the value of the natural world was found in the natural value of wilderness are over. And endangered species are just *so* 1970s. In the last two decades, those dated pegs for hanging environmental value have been largely supplanted by biodiversity.

Biodiversity. As an instrument for hitting the value notes of ecology, conservation biology, and popular environmentalism, it has been played like an accordion. With bellows wide open, philosopher Bryan Norton (2006, 57), among others, has played the tune "[biodiversity] must … capture all that we mean by, and value in, nature." Thrilling as this music is, it requires a conception of biodiversity that is at once too broad and too narrow for a usefully focused discussion of environmental value. If biodiversity is what underlies all that is valuable *in nature*, then the issue of its value is so *broad* that it devolves into a biodiversity-encoded way of rejoining the old and ongoing discussion of natural versus artifactual or cultural value – or how those two categories of value intertwine. But this conception is also too *narrow*, for it does not capture how certain artifacts – for example, highly managed and manipulated landscapes or the products of genetic manipulation – might plausibly contribute to the plus column of the biodiversity ledger.

The word "nature", of course, has a broad spectrum of possible meanings, and the various lines of reasoning that this book critically reviews tend to use it in way that relates to some envisioned state of the world, as well as possibly a history of human involvement in bringing about that state. The views about what biodiversity is and why it might be valuable tend to work off a conception of nature of this sort. Instead of trying to explicate this conception up front, I let its various proponents speak for themselves about it in the course of presenting their views about biodiversity. At least, I do that until Chap. 8. There, in Sect. 8.1, I point out some vexing difficulties with conceptions of this general sort; and in Sect. 8.2, I propose another way of thinking about what nature is. That, I argue, is the key to a rather different and more fruitful way of thinking about its value.

D.S. Maier, *What's So Good About Biodiversity?: A Call for Better Reasoning About Nature's Value*, DOI 10.1007/978-94-007-3991-8_2, © Springer Science+Business Media B.V. 2012

The conundrum of Norton's definition partly has to do with how uneasily biodiversity straddles the worlds of nature and human projects – something that this book visits, revisits, and then finally comes to grip with in its final chapter. Unfortunately, this sort of puzzle is the rule rather than the exception in the many dozens of biodiversity definitions and the many more dozens of value attributions that have infused the staggeringly voluminous scientific and environmentalist literature on biodiversity since the neologism's invention in the mid-1980s.

Often, biodiversity is simply *presumed* to have value with no explicit or coherent account of what biodiversity is, the genesis and justification for any value that it might have, and how this value relates to biodiversity itself. We see this in emotional pleas to conserve biodiversity in order to prevent the "impoverishment" – a word that plainly embeds a value judgment – of the earth. These pleas leave the meaning of biodiversity to guesswork. Alongside them are operational definitions of biodiversity "as what is being conserved by the practice of conservation biology" (Sarkar 2002, 132). The value is then directly built into the promotion of this practice to a norm: We are told that it is a practice that we *ought to* adopt.

Sahotra Sarkar (2005, 182), a widely published conservation biologist and philosopher of biology, posits that biodiversity, quite literally, *is* what is conserved (or "optimized", as he says) by the output of the rules and procedures that *he* advocates for conservation:

> … biodiversity should be (implicitly) operationally defined as what is being optimized by the place prioritization procedures [of Sarkar's invention] that prioritize all places [i.e. assign values to them] on the basis of their biodiversity content using true surrogates.[1]

On this approach, the answer to the question "Is biodiversity good?" is an essentially tautologous "yes" – for according to him, good biodiversity is (by operational definition) that which should conserved by virtue of being the output of his algorithm's execution. But Sarkar offers no discernible independent argument for why we should believe in his algorithm's value-revealing powers.[2]

---

[1] Sarkar's operationally defined attribution of value is all the more remarkable for relying on a "greedy" heuristic algorithm (or class of such algorithms) in order to solve his "place prioritization problem". Very roughly, this is the problem of finding the fewest or least costly set of places that satisfy certain diversity requirements. It is a variant of and shares the computational complexity of the well-known set cover problem, which is NP-complete. It is easy to show that the greedy set cover heurisitic can yield not just sub-optimal results, but even worst possible results. That concern would be academic were it not for the fact that there seems to be no credible evidence that real conservation problems generally avoid conditions or problem structures that yield bad or worst-case results. This combination of circumstances seems to eliminate even a "practical" defense of Sarkar's suggestion that biodiversity be valued according to its operationalization, which entails that we ought to conserve that which an execution of his algorithm chooses to conserve. See Note 2 on possible underlying values.

[2] To be fair, Sarkar's operational recommendations can be said to implicitly reveal underlying held values – answers to the question "What, in biodiversity, is valuable?". The revelations come in the form of his justifications for the choice of various stand-ins ("surrogates") for the real (biodiversity) thing (whatever that might be), combined with algorithms for making conservation choices based on the surrogates. Moreover, Sarkar does have a separate account of the value of biodiversity – its

Other accounts of biodiversity and its value might initially appear to be more promising. There are many dozens of them. Unfortunately for biodiversity, quality is not the companion of quantity. Some accounts trip over simple category mistakes in their underlying conception of biodiversity. These include conceptions that conflate the *diversity* of kinds with their particular identities (as discussed in Sect. 4.1.3, Particular species). The archetypal case is when the diversity of species in a place is mistaken for the specific "composition" of the species set – that is, "who" lives there at the moment. Another pervasive error involves confusing biodiversity for one or another of the dozens of "indexes" or measures of biodiversity – Shannon-Wiener entropy, Simpson's index, various kinds of "quadratic" diversity indexes, including ones that "measure" phylogenetic diversity, and many more.[3]

Many attempts to attribute value to biodiversity hinge on tacit and questionable judgments about the value of properties that have no apparent bearing on diversity, as such. Salient examples include endemism or the status of a species as "exotic" (aka "invasive" or "alien" – clearly bad) or native (presumed to be good). Some others (for example, Hooper et al. 2005, 5) hinge on the connection of biodiversity as a means to the end of some one or more ecosystem property or function, such as productivity, carbon storage, hydrology, nutrient cycling, integrity, or stability. In these cases, the empirical causal link is often tenuous and in any case, the general value of at least some *ends* – productivity and integrity stand out – is as elusive as that of biodiversity itself. Yet other theories of value, though promoted by natural scientists, seem to rely on an oddly non-naturalistic theory of natural order (Sect. 6.12, Biodiversity as the natural order) with roots in the philosophy of such thinkers as Thomas Aquinas and Gottfried Leibniz. And there are others. Chapter 6 of this book (Theories of biodiversity value) surveys a broad array of theories that attempt to justify the attributions of value to biodiversity.

Given the highly visible play and cachet of biodiversity in scientific, environmentalist, and political fora; given the apparent abundance of confused and confusing discussion of it; and given that these confusions ground real decisions regarding conservation and development that significantly affect the natural environment, there seems to be a pressing need for careful, philosophical scrutiny. But so far as I can

---

"transformative power" – which this book takes up in Sect. 6.10 (Biodiversity as transformative). However, he offers few clues about why following his operational procedures might have any particular efficacy or efficiency in securing this (transformative) value. This criticism is independent of the question raised in Note 1, which concerns whether or not the algorithm that Sarkar espouses performs a computation that is "correct" according to his description of the (variant of the set cover) problem that it is supposed to solve. It is also independent of whether or not a theory of biodiversity's value as transformative is credible.

[3] See for example Ricotta (2007) and its references, which include Ricotta's go-around with Sarkar on the subject of biodiversity measures. Both scientists lose sight of the distinction between biodiversity (on the one hand) and measures or indexes for it (on the other), despite the fact that Sarkar mentions this distinction in his criticism of Ricotta. See also the seminal paper by Hurlbert (1971), which, despite being commonly cited, suggests questions about diversity measures that are largely ignored or misunderstood in the scientific literature. This topic is discussed further in Sect. 4.1.2 (Measures and indexes).

tell, there is startlingly little such scrutiny. This book is primarily motivated by a desire to take a step towards redressing this deficiency.

While the principal aim of this book – up to the concluding chapter – is to survey, clarify, and scrutinize conceptions of *value* that might be said to attach to biodiversity, this investigation requires some reasonably clear conception of that which is valued. Therefore, the enquiry begins with Chap. 3 (What biodiversity is). In Sect. 3.1, it formulates a core science-based conception of biodiversity as the diversity of kinds in categories that have proven their value in biological science. This offers a sensible basis for the discussion of biodiversity's value. It squeezes in the bellows of the conceptual accordion to make biodiversity less than a nebulous and trivializing "all that we mean by, and value, in nature". Nor does the restriction to the categories of biological sciences excessively compress the concept, for those categories span biology. To allow for the use of multiple categories, Sect. 3.3 arrives at a "multi-dimensional" conception of biodiversity, which, it is argued, remains a sufficiently broad brush for sweeping up most any value in biodiversity, if there is any to be swept up. In truth, this effort in conceptual clarification is not just a preliminary. It is a substantial step toward understanding biodiversity's merits (or lack of them) as a bearer of value.

To help demarcate the boundaries of Chap. 3's definition of biodiversity, Chap. 4 (What biodiversity is not) considers proposals to define biodiversity in ways that head off in various wrong directions. On the one hand (Sect. 4.1) are category mistakes that typically reduce biodiversity to, and confuse it with, the biological identity of some one or another of the entities that contribute to diversity – such as individual organisms or particular species. These reductive slides are evident in such serious and seriously common misstatements as that biodiversity (as opposed to a particular population of flying insects) provides a pollination service. Another kind of slide – of a thing to some measure of it – is ensconced in the commonly used ecological definition of biodiversity as some combination of species richness (by which ecologists mean "the number of species in an area") and the even distribution of individuals amongst existing species. Beyond the category mistakes are various "accretive" conceptions (Sect. 4.2), which try to bulk up biodiversity in order to give it some (more) value muscle. The bulk-producing supplements tend to apply (once again) to the particular identity of the things that are diverse, not diversity itself. They include such properties as rarity (defined either non-relationally, as low abundance within some system or relationally, as uniqueness), distinctiveness (a comparative property), endemism (geographical rarity – a place-related property), the species/area ratio (a kind of biodiversity density or place-related efficiency), endangerment (a risk-related property), and viability (also risk-related). Some of these properties will be familiar from the widely publicized and vigorously promoted concept of "biodiversity hotspot".

At the logical halfway point of the book, Chap. 5 (The calculus of biodiversity value) turns from what biodiversity is to the value proposition for it. Its discussion abstracts from any particular account of the source of biodiversity's value insofar as this is possible. Before looking at specific accounts for how biodiversity might get its value (in Chap. 6), this chapter examines various implicit and explicit assumptions about how, more abstractly, the value of biodiversity relates to biodiversity itself. Section 5.1 asks such questions as: Does biodiversity admit of more and less – that is, increments and decrements – in a way that permits orderings – that is, more or less biodiverse states of a place or of the world? If so, do relative orderings in degrees of biodiversity – for example, one place or state of the world being more biodiverse than another – directly transfer in a parallel ordering of corresponding values? Section 5.2 further explores value orderings, which a richly structured definition of biodiversity along multiple dimensions (as suggested in Chap. 3) might restrict. And finally, Sect. 5.3 asks, what does this imply about possible obligations with regard to biodiversity?

The discussion of the calculus of biodiversity values in Chap. 5 is the final requisite prop for the stage across which Chap. 6 (Theories of biodiversity value) parades accounts of how biodiversity is supposed to be valuable. The selection of these accounts is generous and represents a great majority of discussions on this topic, though it is undoubtedly not all-inclusive. It does include, among other theories, that biodiversity is valuable for "unspecified moral reasons" (Sect. 6.1), that it is valuable as a resource (Sect. 6.2), as a service provider (Sect. 6.3), as a sustainer of human life (Sect. 6.4), as a key to human health, both as pharmaco-poeia and as an inoculation against infection (Sect. 6.5), as the progenitor of human biophilic tendencies (Sect. 6.6), as a generator of (more) value (Sect. 6.7), as font of knowledge (Sect. 6.8), as having option (and quasi-option) value (Sect. 6.9), as transformative (Sect. 6.10), as having experiential value (Sect. 6.11), and as an expression of the natural order (Sect. 6.12). The chapter concludes (Sect. 6.13) with some proposals that are not customarily offered as self-contained theories of biodiversity value, but that are often suggested as still heavily influencing evaluation. These include what conservation biologists call "viability" and "endangerment" as well as conservation "efficiency". Section 6.13 concludes with an extended discussion of the place of the history of human-caused impingements in valuing biodiversity.

Chapter 7 (Some inconvenient implications) steps back to take a broader view of commonly held beliefs about the "good" of biodiversity. It derives from them some implications, which are inconvenient obstacles for those who would hold up biodiversity as a key to nature's value. Section 7.1 explores the discomfiting impli-cation that egalitarianism with respect to biodiversity's categories lets all the bad boys into the club along with the good. Section 7.2 examines how embracing bio-diversity as an unalloyed good authorizes science projects, which seek to build "snow-globe" worlds detached from, and even antithetical to, the natural world. Section 7.3 fills in many considerations that are routinely omitted from, and bias discussions of how the value of biodiversity and the possibility of an ongoing "great extinction" relates to human timeframes. Section 7.4 attempts to get a han-dle on endemic confusions and slides in meaning in discussions about biodiver-sity's value. All of these try to squeeze some value out of biodiversity by conflating biodiversity with what is not biodiversity. Finally, Sect. 7.5 suggests that taken together, the considerations of this chapter should signal caution to philosophers, scientists, and environmentalists who would hold up biodiversity as an emblem of nature's salvation.

The valuation of biodiversity as emblematic of natural value is a product of a more general framework for valuing the natural world, which has acquired a considerable grip on thinking about nature's value. Chapter 8 (Natural value starting from people) starts, in Sect. 8.1, by characterizing this framework, including how, within what I call "the biodiversity project", it serves as a basis for championing biodiversity. The discussion identifies the elements that go into the framework, their salient flaws, and some possible explanations for why the framework can appear to be inevitable, despite these flaws. And it characterizes the biodiversity project's founding normative assumption – namely, the proposition that the value of the natural world supervenes on its instantiation of certain structured properties of biodiversity. This view is both supported by and supports a set of dubious propositions concerning the nature of human responsibility for nature. Joining them are even more problematic propositions concerning the ends of conservation and restoration projects. Most troubling of all, these projects hold up, as an exemplar of right action, "biogeoengineering" nature so as to achieve their problematic ends. Finally, this section shows how this enterprise is aided and abetted by "lesser evil arguments", which foster and are fostered by several moral hazards. Section 8.2 then turns to the task of framing another way to think about nature's value, which is not vulnerable to these many deficiencies. It proposes that the core of natural value resides in a valuing relationship that people can have with respect to the natural world, which requires a stringent kind of "letting it be". The suggestion is novel in a number of significant respects, including the way in which it uses elements that might be familiar from other usages. The implications are wide-ranging and will surprise some – including calling into serious question current-day orthodoxies of conservation, restoration, and environmentalism generally.

The remainder of these Preliminaries arms the reader with some requisite tools for following the many twists and turns of reasoning that this book follows – both in the work of others and later in the book's own (Chap. 8) proposal for why the natural world is so valuable. It first (in Sect. 2.1) provides an overview – aimed primarily at those less conversant with philosophy – of how philosophers think about value in the context of environmental ethics. Then, to give a sense of the state of the art in reasoning about biodiversity, Sect. 2.2 catalogs types of logical lacunae pervading arguments that try to find nature's value in biodiversity. The chapter concludes in Sect. 2.3, with reflections on why it might be difficult to argue for the value of diversity in any domain – not just biological. It traverses a path from the philosophies of Thomas Aquinas and Gottfried Leibniz, on through William Cowper's famous but almost always misinterpreted "Variety's the very spice of life", on its way to a consideration of United States Supreme Court Justice Lewis F. Powell's majority opinion in *Regents of the University of California v. Bakke* about the value of diversity.

2.1 An Environmental Philosopher's Conception of Value[4]

Over the last couple of millennia or so, philosophers have devised categories, concepts, distinctions, questions, and broad theoretical approaches for thinking about value in general, moral value in particular, and how to not just think about but also act on valuations. They are still at it.

This section presents a highly selective and very coarse-grained overview of this slicing and dicing. It also relates some respects in which venerable philosophical axiological and moral instruments have recently been adapted for application to the natural world. The body of current environmental discussion exhibits a very pronounced bias towards one specific toolset, while largely ignoring others. This necessarily influences thinking about what actions are justified with regard to the natural environment in general, and with regard to biodiversity in particular. The bias is also necessarily reflected in which positions and arguments this book most concerns itself to examine.

In its broadest sweep, the theory of value is the theory of what is good and what varieties of goodness there are. Value theory and moral theory intersect in the domain of value that bears on right action. In other words, some value achieves an exalted level or in some way qualifies to enter a domain in which, according to moral theory, a person ought to take it into account in choosing from among alternative actions, different pursuits, various ways of behaving, and more generally, how to live a good life. Such goods are said to command moral respect or consideration. They are said to have moral value.

Of course, very few things are *that* valuable or valuable in such a way that persons *ought* to take them into account – in the moral sense – when deciding how to act and behave. Few people would deny that among the things that hold such compelling value and that are subjects of moral consideration are other persons and their interests. The good for you (an other person) is a good that I have a moral duty to take into account in my treatment of you. But the old computer on which I am typing this manuscript does not command this kind of attention. It would be foolish for me to not treat it well or well enough to finish this book, because I have no reasonable alternative way to accomplish that desired goal. But I have no clear *moral* obligation to afford it good treatment. Moreover, if some new kind of writing device were unexpectedly made available to me – one that permitted a far more efficient recording and transfer of my ideas than this awkward device with a keyboard and misbehaving software – then the treatment of my computer with casual disregard for its continued functioning would not even be foolish, so far as completing this book is concerned.

The value of certain other entities might fall somewhere between the value of you as a person and that of my computer. Paintings and other works of art occupy one special kind of middle ground occupied by aesthetic value. It is possible that aesthetic value sometimes gives rise to moral consideration. But whether or not this

---

[4]A cost of compressing a topic that merits thousands of pages of discussion into a few is the sacrifice of many philosophically worthy distinctions and topics for the sake of a highly relevant few.

is so, and if it is so, how and under what additional conditions, if any, are big and difficult questions.

What about nonhuman living things and the places where they live? Environmental ethicists agonize over where to place their value. These entities might seem different from other persons in some morally significant regard; but they also seem to be morally different from my computer.

The big question lurking in the background of this book's first seven chapters is: "What kind of value, if any, does *biodiversity* have?" Can actions, pursuits, projects, or ways of behaving be morally right or wrong as a result of their regard, disregard, or misjudged regard for biodiversity or for their effects on it? Is biodiversity similar or somewhat similar to other human beings with regard to its moral implications? Or might biodiversity be more like my old computer – something that has served some set of particular human needs well enough, but that is entirely dispensable when better alternatives present themselves or when my needs and desires change? Even if there are no such alternatives right now, might the needs and desires that currently existing biodiversity actually serve be just as well served by some substantially truncated or altered version of it?

After all, I really don't need a Turing machine, or even a general-purpose computer such as my laptop to write this book. If its word-processing software were to migrate into its motherboard so that it could do word processing and only that – no electronic mail, no web surfing, no preparation of taxes owed from this book's undoubtedly fabulous royalties – then it would serve as well for the writing of my book. Might a greatly reduced version of biodiversity be like that? There is also the possibility that biological diversity, as such, really has no particular value at all – not even the dispensable (or substitutable) utilitarian value of my computer. This might be true even if some set of particular things biological do have value – even great value – quite apart from the cardinality of that set or how much each of its elements differs from all of the others.

2.1.1   Concepts and Categories of Value

I begin by introducing some key concepts and categories of axiology – the general study of value. These concepts and categories come to light in sorting through three distinctions that preoccupy environmental philosophy.[5]

The first distinction is between instrumental value and non-instrumental value. "Non-instrumental value" is one of several, oft-conflated meanings of "intrinsic value". We will bump into some of its other meanings shortly.[6] This first meaning

---

[5]This heterogeneous set of distinctions includes elements from both normative ethics and metaethics.

[6]For a seminal discussion of different and often conflated senses of "intrinsic value", see O'Neill (1992).

hinges on the time-honored distinction between an end or final end versus a means to that end. Something is instrumentally good if it is a means to an end whose value is logically prior to that means. There can be a chain or even web of means, all of which are instrumental in achieving an ultimate goal. If people are intrinsically valuable, then it might be thought that biodiversity is instrumentally valuable as one link of several chained or otherwise cooperating means to various human ends – for example, keeping people alive or keeping them healthy.

Whether or not biodiversity is instrumentally valuable, it might also have value as an end in itself and so be intrinsically valuable in this sense of "ultimately valuable". This is a value that would be attributable to it without reference to some other, valuable end that it is a means of achieving. Some call this "a moral reason" for biodiversity's value. But this is a mistake that trades the first meaning of "intrinsic value" – "ultimately valuable" – for another a second meaning – namely, "sufficiently valuable to warrant moral consideration". The two definitions are not congruent because an instrumental value, just as much as an intrinsic one, can command moral attention. For example, if something is a means, or the only means, or the only means at hand of keeping a person (or all members of the human species) alive or healthy, then it almost certainly should feature in moral considerations. According to the second definition, it would then be intrinsically valuable.

Despite the possibly great importance of instrumental goods, the axiological status of something that is *merely* instrumentally valuable does have a certain precarious quality. As a mere means to some other end and valuable for no other reason, it is vulnerable to replacement by something else whose value exceeds it by being a superior means to that end. Even if biodiversity is instrumental in keeping people alive – say, because it ensures a source of food – there is no reason in principle why some other source of food – say, produced via advanced industrialized organic chemistry – might not better serve that end.[7]

Many discussions of instrumental and intrinsic goods convey the impression that these categories are both mutually exclusive and jointly exhaustive. In fact, they are neither. So far as mutual exclusion is concerned, I have already hinted that biodiversity, for example, might have both instrumental and intrinsic value. It might be thought to be a means, and perhaps an indispensable means, to the end of some important aspect of human welfare (for example, good health or survival), while also deserving moral consideration for its own sake, independent of its role in realizing a distinct, separately justified goal.

Nor are the categories instrumental/non-instrumental exhaustive because it is possible for something to be deemed valuable, but neither as a means of achieving a valued end, nor as such an end valued in itself. Rather, it might be valued as a *constituent in* something else independently valued, or *constitutive of* that other

---

[7]One might be warranted in saying that this is a distinct possibility – not just in principle, but in near-term practice. For example (Britten 2009), scientists are well on their way to figuring out how to grow disembodied meat.

thing. The terms "constituent good" and "constitutive good" are sometimes used interchangeably, but it is useful to tease out their distinct meanings.[8]

To be a constituent good is to be a good by virtue of playing a part in the good of a "whole". This does not require being a means, in the sense of a causal factor in bringing about the whole. For example, acquiring knowledge could be said to be a constituent of a good life. However, it doesn't cause a life to be lived well. Nor could the acquisition of knowledge be said (without substantial bending of meaning) to be a means undertaken for the end of a good life. Certainly, we do learn things for other ends – to satisfy our curiosity, to find out how to pursue a project, to get a job. But think of how odd it would sound for someone to say, "I am learning about ecosystems (just) so I can live a good life." Or "I believe that learning about ecosystems will cause my life to be a better life."

It is possible to conceive of biodiversity's value as that of a constituent good. For example, someone who viewed nature as a good and biodiversity as an essential element of nature could conclude that biodiversity is good as a constituent of nature. Or, if diversity of any kind were held up as a good (a precarious proposition, as we shall see), then biological diversity might be valued as a significant constituent of the "whole" of diversity. Biodiversity also might be viewed as a constituent good when considered to be the wellspring of certain kinds of knowledge that is a constituent of a good human life.

Like a constituent good, a constitutive good does not necessarily have a causal or other means-to-end relationship with another good. The relationship of part to a whole undergirds a constituent good. In contrast, a constitutive good is a way of realizing something else whose good is understood prior to and in a broader and more abstract way than the constitutive good's. "Way of realizing" might at first sound much like "cause" or "bringing about". But in this context, "constitutive" has more the sense of "a way of implementing" or "a way of being". I might have the desire to eat healthy food that is produced without the intense expenditure of precious energy, water, and land resources. My choosing to eat a bowl of bulgur wheat instead of a sirloin beefsteak might be constitutive of this dietary or environmental end more than it is a means to it.

The categories of constituent and constitutive value are useful to hold in mind in the context of later discussions in this book. For example, they are often implicit in explications of the "just-so" model of biodiversity value, which is the topic of Sect. 5.1.4 (The just-so model). Also, one might interpret the view, discussed in Sect. 6.8 (Biodiversity as font of knowledge), as an argument for biodiversity as a constituent good in a good human life.[9]

---

[8]Ackrill (1980, 19) introduces both constituent goods and constitutive goods, calling them both "constitutive" goods. His example of the former is the relation of putting to playing golf; his example of the latter is the relation of playing golf to having a good holiday.

[9]It is doubtful that this exhausts the set of value categories. David Schmidtz finds yet another, distinctive category in what he calls "maieutic ends" – ends that are satisfied by engendering other ends. See Schmidtz (2008a, 39).

Perhaps the conception of the good of biodiversity that most dominates discussions of conservation and restoration in what Chap. 8 calls "the biodiversity project" involves both constitutive and constituent goods. On such a conception, a salient good of the world's biota is realized when their numbers and diversity constitute a good biodiverse state of affairs, which in turn, is a constituent good in the greater good of nature. Part of the attraction of this view is that, by its lights, the good of biodiversity is "objective". That is, it is a matter of the actual, biodiverse state of the world, which, in principle, can be determined via careful scientific observation and accounting for the world's biota. The dichotomy of "objective" versus "subjective" value comes up for discussion shortly.

A second distinction – or really, set of distinctions – concerns the genesis of value. It crosses over the boundary from normative ethics into metaethics. I regard this boundary as porous and crossing it as optional – for one can grapple with questions about what kinds of values there are and about which values count most (normative ethics) without delving into others about how values arise in the world (metaethics).[10] Still, some great concern centers on the questions of the extent to which and the respects in which the truth of value assertions[11] depends on the valuing subjects who make these assertions. When it is said that something is "good" – not just "good *for*" something else – is this so only because people "create" or somehow bestow this goodness? Does this value come into being as the result of human acts of attributing value to entities or entering into some other valuing relationship with them? Even if the genesis of value does not require actually existing human valuers going about the business of valuing things, might value still resemble a "secondary" quality of things – conceivable and understandable only by reference to the characteristic experiences and responses of valuers in their valuing acts? On the first view, nothing ever had value, let alone was morally considerable before the arrival of valuing humans on the scene; and all value, including moral value, will expire with the last exhalation of the last of *H. sapiens*. On the second view, value, like redness, could have existed both before the arrival of humankind on the scene; at it might persist after its exit from the universe's stage. In fact, value could exist even in an alternative universe in which *H. sapiens* never evolved – still conceived and understood in terms of a hypothetical valuing and sensing *H. sapiens*-like being.

An alternative view backs more or less completely away from giving valuing subjects a significant role in values' genesis, only allowing that the *attribution* of value is subjective in the very weak sense that valuing subjects alone have the

---

[10]Many philosophers would dispute this, claiming that metaethics is logically prior to normative ethics. But this stance belies the literal translation of τα μετά τα ἠθικά (which would make metaethics "what *follows* ethics"). And I would say that most of us quite capably dig deep into moral questions without much respect for, or even notice of the any boundary between the two lines of inquiry.

[11]I am assuming here that an assertion to the effect that something is valuable can have a truth value. However, this is a matter of lively philosophical debate.

epistemic capacities to find value wherever it might lie "out there" in the world. In other words, it severs any link that value has to valuing subjects except insofar as the latter are fortuitously equipped to perceive the value-bestowing attributes in the entities that, in fact, possess value. This view conjures up the picture of buried value treasures, waiting to be found, even if in fact they forever elude discovery. Finding value requires such creatures as humans who are capable of value sleuthing. Using their moral compass – otherwise known as "moral judgment" – in conjunction with value maps – otherwise known as "a sense of morality" – they poke around the world to find the value that has been there all along. The role of moral philosophy in this enterprise is to help draw those maps and then to construct a theory that explains the contours of the value topography.

This second set of distinctions – concerning whether the key contributor to an object's value is its set of attributes or the inclination of a valuer to value it – is often taken to be central in grounding the metaethical question of whether value is objective or subjective. Out of this dichotomy arise the third and a fourth meanings of "intrinsic value". According to the third, what is intrinsically valuable has "objective value", by which is meant "value apart from human valuers". The fourth meaning of "intrinsic value" is "value that arises from something by virtue of its recognizable properties". When those properties are understood to be natural (in the sense of "not supernatural") properties, then the fourth meaning dovetails into the third – in essence, by explaining both intrinsic value-as-objective value, as well as the human epistemic capacities to recognize that value in the aforementioned value-bestowing attributes.

The question of whether or not value is subjective is much the same as the question of whether or not value is anthropogenic – in some way generated by humans, or at least requiring their actual or possible valuing presence. There is heated debate on this question. But in my judgment, the answer to it is not nearly so important as it might first appear. That is because if the value of anything – for example, biodiversity or nature considered more broadly – is anthropogenic, it is not any less valuable for that. At least, it is not less valuable so far its role in guiding specifically human behavior is concerned. This suffices for normative ethics – insofar as its central concern is to identify and understand the basis for right actions and right behavior of people.

A major motivation for bringing up anthropo*genic* value is to distinguish it from the orthogonal (independent) category of anthropo*centric* – versus non-anthropocentric – value. "Anthropocentrism", despite well exceeding the standard four letters, has become the stigmatizing dirty word of environmentalism. At its core, it is the position that anything that is worth moral consideration is human. But like many derogatory labels, it is indiscriminately pasted onto many positions that connect moral considerations with people in some way that does not (necessarily) make people the sole objects of moral consideration. The fact that values regarded as anthropo*genic* are tied to people as value progenitors has led to the widespread and persistent conflation of that other category – which lurks behind the objective/subjective dichotomy and has to do with the genesis of values – with anthropo*centrism* – which has to do with the objects of value.[12]

If any importance attaches to the distinction between anthropocentric and non-anthropocentric value (properly confined to the domain of objects of value), it mostly seems to reside in the question of whether anything other than humans, human welfare, or human endeavors are objects that can legitimately be held to possess non-instrumental value. This might seem to be of particular interest for questions about the nature of biodiversity's value since at first blush, biodiversity seems to be mostly about *non*humans – no matter if or how it physically and causally connects back to people. After all, on its own, one fur-challenged bipedal mammal can account for only so much biological diversity. So if one insists on occupying an *anthropocentric* viewpoint, then one might be inclined to think that any significant value in biodiversity must largely or entirely lie in its being an instrument for achieving human goods.

But as previously noted, "merely" instrumental value can still rise to the highest level of moral consideration. That would be true of biodiversity if, for example, it promoted or were a necessary condition for the ultimate end of continued human existence (Sect. 6.4, Biodiversity as (human) life sustainer) or of continued healthy existence (Sect. 6.5, Biodiversity as a cornerstone of human health). However, the perceived ante of the anthropocentric/non-anthropocentric divide is upped when "instrumental value" is read as "non-intrinsic value" and the meaning of "intrinsic value" is reinterpreted according to the second definition, as "meriting moral consideration". This illegitimate slide can appear to give license for non-anthropocentrists to tar anthropocentrism as a general stance that holds that, not only is everything of non-instrumental value human, but that everything worthy of moral consideration is human, too. While this particular move cannot fly except on the wings of confusion, it is good to keep in mind that even when some thing's value rises to an extraordinarily high level, its possession of instrumental value *only* creates a tension that arises from its vulnerability to replacement by better means.

However one sorts out the merits of anthropocentric views, it is important to see that the categories of anthropogenic and anthropocentric are orthogonal. In particular, anthropo*genic* values can be *non*-anthropo*centric*. If value is created by human acts of value attribution, by reference to such attributions, or by some other valuing relationship that people have to the things they value, then this says nothing about whether or not the thing centrally valued – the object of value – is human. Nor, by itself, does it say anything about whether or not the thing is valued intrinsically (in any of its meanings), instrumentally, or in some other way. And again, "some other

---

[12]This unfortunate confusion pervades the literature on biodiversity. For example, E.O. Wilson (1996, 175–176) lobbies hard for a non-anthropo*centric* ethic based on species' rights. But he then has second thoughts about whether his proposal is really non-anthropocentric because, he (unsoundly) reasons, the granting of rights is anthropo*genic*.

way" is important to keep in mind because, insofar as the force of "anthropocentric" trades on a false instrumental versus non-instrumental dichotomy that ignores constitutive and constituent values, it can miss a big part of the evaluative point.

The last of the three distinctions that I wish to mention in this brief discussion of value is that between monism and pluralism. This divide is an organizing principle for two distinct but interrelated domains – of value and of objects that are valuable. As the dichotomy customarily applies to the domain of value, there is an implicit restriction to intrinsic – in its first sense of "ultimate" – value. The question then is whether there is just one kind of ultimate value (value monism), or more than one kind of ultimate value (value pluralism).[13] As the dichotomy applies to value-holding objects, the question is whether only one class of entities or more than one class of entities is capable of holding the one (according to value monism) or more (according to value pluralism) kinds of ultimate value.

This last question – regarding objects of value – is the one that tends to receive principal or at least, initial attention under the rubric of "anthropocentrism", characterized as the belief that only one kind of entity – namely humans – are legitimate holders of intrinsic (ultimate) value.[14] On the other hand, a theory that holds biodiversity to be intrinsically valuable is likely be pluralistic with regard to what is intrinsically valuable. That is because it is hard to conceive of a credible theory that views all other values – including those that attach to people – as "not ultimate", as mere means for enhancing biodiversity, or as reducible to biodiversity's value.

But at some point, the question arises of whether or not different kinds of intrinsic value might be possessed by either the unity or plurality of things that are intrinsically valuable. This creates a divide between *value* monism and *value* pluralism, which parallels the monism/pluralism divide that arises from the classification of value-holding things. A theory that embraces value monism might nominate a plurality of kinds of entities as holders of the one and only kind of value. For example, if sentience is held to confer intrinsic (in the sense of "ultimate") value on a creature by virtue of giving it the capacity to both suffer and to experience pleasure, then one must recognize many different kinds of creatures to possess such value. In fact, this is precisely what Jeremy Bentham proposed, as discussed in the next subsection.

---

[13] This question loses much of its interest when applied to other categories of value – most notably, when it comes to instrumental value. On the one hand, there is obviously a plurality of means corresponding to a plurality of ends served; there is often even a plurality of means to any *one* end, since only human capabilities and physical law limit the kinds, numbers, and arrangements of things that can be means to an end. On the other hand, the primary considerations for the value of means – their instrumental effectiveness and efficiency – is quite unidimensional. But these contrasting observations make a decidedly underwhelming contribution to a basic understanding of value.

[14] A monism/pluralism distinction can be applied to moral principles of right action as well as to moral values. Monistic moral theories seek to show that all moral principles are ultimately reducible to a single one. The monism/pluralism distinction regarding principles of right action is orthogonal to that for values.

Conversely, a pluralistic theory of value might find one or more of the plurality of values in just one kind of entity. In fact, this is a salient complaint of anti-anthropocentrists, who charge that traditional moral theorists tailor each and every kind of ultimate value so as to be possessed by precisely one kind of creature – namely *H. sapiens*. That challenge must be taken seriously by a theory that, for example, maintains the sole bases for ultimate value are human self-consciousness and a rational pursuit of a human life. On such a theory, even such goods as health and freedom, which one might initially suppose to be valuable to a wide variety of creatures, would only have intrinsic value in connection with human lives. This, the anti-anthropocentrists point out, is precisely the strategy of Jim Crow laws in the post-Civil War United States, which disenfranchised blacks. The disenfranchising forces largely avoided a test of racial identity by relying on tests of literacy and property ownership coupled with institutions for applying these tests and interpreting their results. These tests excluded blacks with nearly the assurance of a racial test. In short, the "what kinds of value" question is distinct from the "who/what holds the value" question, even if this distinction is not made explicit.

One major, non-anthropocentric and value-monist view is the one that intrinsic (as ultimate) value is found and only found in the condition of being a teleological center of life. This theory rests on the proposition that each living organism has a good of its own and that this good for the organism is not just good *for* that organism (an instrument for its ends), but (and perhaps because of that) is good in itself. That "good for" does not by itself imply "good in itself" is easy to see by considering that what is good for a sadist is access to persons on whom she can inflict pain. In any case, while this theory is value monist, it self-consciously includes all kinds of living organisms as appropriate subjects of its one and only kind of ultimate value.

Another significant non-anthropocentric, value-monist view is the previously mentioned one that finds ultimate good in the experience of pleasure and the avoidance of suffering. This theory takes the set of living organisms – all of whom are morally valuable according to the previous view – and lops out all the non-sentient ones. According to it, only the tiny minority of sentient creatures – particularly ones capable of suffering – are morally considerable. Yet another non-anthropocentric view finds value in conscious striving. This view does yet more pruning of the tree of moral worthiness, lopping away any sentient creature that is not also conscious.

Contemplation of nonhuman subjects of moral consideration commonly leads to the complaint that all or almost all nonhuman organisms, including nonhuman animals, lack the capacities that make most human individuals moral *agents* who can do right and wrong, be worthy of praise and subject to blame, and have duties. However, this objection does not really militate against the moral respect-worthiness of nonhumans. It is irrelevant to the proposition that nonhuman organisms are moral *subjects* or, as we might say, moral *patients*. In other words, a being deserving of moral respect from moral agents need not itself be a moral agent. This circumstance is not unfamiliar – for example, in children too young to have any developed moral judgment and in adults whose judgment has been seriously impaired by injury or illness.

Individuals are not the only entities that are candidate value-holders. Environmental ethicists have tried to open out the set of candidates to include various sorts of collectives, such as ecosystems, species, and even the entire biosphere. But if these collective sorts of entities are loci of intrinsic value, it is less clear than it is for individual organisms what is the basis for any value that they might hold. One suggestion is to redeploy the teleological argument that has been proposed for individual organisms. According to this suggestion, a collective of one or another of these collective kinds has a good of its own that is sufficiently like an individual organism's good of *its* own to make the collective good in itself. I come back to this reasoning in the next subsection's discussion of deontology. And it crops up again and again later in the book under such rubrics as "ecosystem health", "ecosystem integrity", and "balance" in nature.

How about biodiversity as a locus of intrinsic (ultimate) value? As we remark at greater length in Sect. 2.3 (Cautionary signs), biodiversity is neither an individual living thing nor a collective. It is most simply conceived as a collection of collections. Or more precisely, it is a characterization of a state of the world in which such a collection of collections exists. This makes biodiversity a significant extension to the range of traditional and even non-traditional candidates for repositories of value. It is surprising that the radical nature of this extension is rarely, if ever, acknowledged in discussions about its value.

2.1.2   Approaches and Key Questions of Moral Theory

Axiology, the subject of the previous subsection, says nothing directly about how values should guide human behavior. This subsection sketches what customarily are regarded as the three major philosophical traditions for connecting value, considered abstractly, to the evaluation of actions, projects, and ways of living. These approaches are: consequentialism, deontology, and virtue ethics.

2.1.2.1   Consequentialism

In the public sphere, most current discussion about the norms for right action, projects, and living well is centered on the question: "What are the consequences?"[15] Moral theories that make this question central are consequentialist. Answers to it leave open another question, which is equally important to consequentialists. The second question asks *which* consequences, exactly, count? That is, what are the constituent goods in the kind of world that a moral agent should strive to create? The answer to this question is a consequentialist axiology or its theory of intrinsic (ultimate) value. Different answers define different brands of consequentialism.

---

[15] This contrasts with the cloistered world of tradition-bound philosophy, in which consequentialism does not hold such dominant sway and might not be even the majority opinion.

For a consequentialist, biodiversity guides right action insofar as certain states of the world characterized by their biodiversity are among the consequences that persons' actions should aim to create or maintain. Of course, something must be said about why those biodiverse states are good ones. One cannot just presume, for example, that the current, biodiverse state of the world – no matter how it compares in degree or detail to the biodiverse state as it existed at any other time – is good just because it is the one that now happens to obtain. Its good must be justified and defended – either on the grounds that some special quality makes this state good in itself, or on the grounds that it entails some other good that flows from it.

Looking at more traditional answers to the question of what constitutes a good state of the world, there are, as I said, diverse answers. The classical utilitarian Jeremy Bentham (1789) had a refreshingly simple answer. It is an answer that Peter Singer revived two centuries later (in many places, including Singer 1979) with equally refreshing vigor. According to it, what is intrinsically (ultimately) good is the experience of pleasure – broadly conceived – and the absence of suffering in individual creatures. A good state of the world is one in which pleasure overall outweighs suffering. A better state is one in which there is a yet greater preponderance of pleasure relative to pain, taking into account all individuals who can experience pleasure or suffer. Both Bentham and Singer emphasize the egalitarian implications of this conception, which extends to a certain class of *nonhuman* individuals. Roughly, insofar as a creature is sentient – most notably, capable of suffering – it qualifies for inclusion in moral calculus of pleasure and pain.

A rather different brand of consequentialism predominates in scientific and public policy circles these days. According to it, what counts is specifically human, not nonhuman welfare.[16] More precisely, what counts is the well-being of *human* individuals, which is aggregated as "human welfare" – the well-being of all humanity taken together. I will say a bit about the aggregation part, shortly. Here, I wish to emphasize that the dominant welfarist view cuts nonhumans out of the calculus from the start – except insofar as they are instrumentally good as resources for human welfare or provide services for mankind's benefit.

This still leaves open the critical question of what constitutes a person's "well-being". It is not surprising that this question also admits multiple answers. Bentham and Singer stake out a classical utilitarian position. As we have seen, this is to say that, for any sentient organism, its well-being consists in a preponderance of pleasure, as opposed to suffering in its life. However, the view that dominates public discourse and largely determines public policy is the rather different one due to

---

[16] This assessment might be disputed. The phrase "the good of the environment" often turns up alongside the phrase "human welfare" or "what is good for people". This juxtaposition often leaves the impression that these are one and the same thing under different descriptions. At the very least, it suggests a presumed concordance of environmental and human interests. Unfortunately, to my knowledge, no one has offered an account of what the environment is interested in, let alone given reasons for why environmental interests (taken literally) should matter to people. Instead, discussions invariably turn quickly to the question of why certain environmental conditions benefit people.

neoclassical economics. According to it, only a human life is good in itself, and it is good to the extent that a person's preferences are satisfied. The more preferences that are satisfied and the greater degree to which they are satisfied, the better. What a person desires determines the range of possible goods for that person. Human preferences in aggregate determine the range of possible goods in the world.

On the face of it, the proposition that the good for each person and for people collectively is founded in unrelenting preference satisfaction is very odd. There are few boundaries on what a person can desire. The object of a desire might be practically unattainable for circumstantial reasons. It even might be a physical impossibility. More odd from an axiological perspective is that fact that it is possible, and even common, for a person to desire things that are bad for other persons or even bad for herself. The desiring person can even be well aware of this.

These and other reasons motivate other brands of consequentialism, which give preferences a more subservient role – sometimes by focusing instead on the satisfaction of needs, which are thought to be more objectively ascertainable. However, it is difficult for a consequentialist to ignore preferences entirely. That is because most people would say that a person's life does not go unequivocally better if she herself does not endorse as valuable (and therefore prefer) the things that others "objectively" believe she needs.

Whether the well-being of an individual is conceived in classical utilitarian terms, in terms of neoclassical welfare economics, or in some other way, the consequentialist must take into account the consequences for multiple individuals of human actions and behavior. It is the overall consequent state of the world that is the moral measure of an act, which is right to the extent that it betters this overall state. To meet this evaluative need, these consequentialist theories incorporate a theory of aggregating individual goods. This theory of aggregation necessarily rests on a general presumption that individual goods are, in some general and abstract way, comparable.

The presumption of commensurable values is the mount for the welfare-calculating engine of neoclassical economics. Its theory encapsulates this presumption in an especially neat way, by means of a concrete theory of preference satisfaction. According to it, the value of any thing is its value in a marketplace transaction.[17] If a person is willing to trade Y for X, then she prefers X to Y and therefore, X has greater value to her.

Whether measured according to the market paradigm or in some other way, preferences can be stronger or weaker and their satisfaction can add correspondingly more or less to overall human welfare. The salient point is that, to a first order approximation, the strength of a preference is what counts quite aside from what it is a preference for. Therefore, the preference calculus can abstract from the *locus* of a preference. However, discomfort with a pure egalitarianism with respect to preferences leads some to place them in several different categories, which are then treated

---

[17] As I point out elsewhere, the neoclassical economic conception of value does not require a real, existing marketplace; an imaginary or hypothetical one suffices.

as non-substitutable "ingredients" that go into the baking of the human welfare cake. I will shortly have something to say about those ingredients.

But the purist position is attractively simple and clear. Consider a minimally nutritious meal, a car with a multi-thousand-dollar sound system, clean air to breathe, an iPhone, drinking water that does not make you sick, a satisfying job, a second multi-million dollar home in Vail, the most basic of shelters near the landfill, fulfilling friendships, and a certain biodiverse state of affairs. From the viewpoint of economic evaluation, these multifarious things are stripped of all distinguishing characteristics, except insofar as they are objects of desire and distinguishable in value by dint of the strength with which individuals express their preference for them. They are stripped, too, of the particular circumstances in which the lives of the individuals whose welfare is supposed to be at stake are lived. Any one of the above-mentioned things *can* be traded for any other. The theory of right action that builds on these premises presumes that a trade *ought* to be pursued when it increases the overall preference satisfaction of the trading parties and that of any other persons who might also be affected by their trade. The economic value calculus grinds out the answer, which therefore cannot be "objectively" disputed, though of course, it all revolves around the unexamined desires of human subjects.

I have already suggested the answer to the question: Where would something such as biodiversity fit into this picture? As one among many preferred things, biodiversity, no matter how distinctive an instrument of human well-being, is a cipher indistinguishable from any other in the comprehensive accounting of all the ingredients that constitute or facilitate human well-being.

This leads to the conundrum that the sacrifice of biodiversity could facilitate the development of another instrument of human welfare, which more than makes up for biodiversity's contribution. Whether or not this is true, or more generally, what the limits of "substitutability" in welfare calculations are, is a matter of intense debate. Maybe, some say, something like biodiversity is or should be exempt from substitution – even if the substitute clearly outdoes biodiversity in the welfare bidding sweepstakes.

But what does "something like biodiversity" mean in the context of an economic welfare ingredient's qualifying for a substitution deferment? At least two types of substitution can be distinguished. The first is the substitution for one ingredient another that serves the same end and in a recognizably similar way – just as one might substitute polyunsaturated oil for butter in a low-cholesterol version of "the same" cake. The result of this sort of substitution is a state of human welfare that is essentially unchanged in composition. It exemplifies what the previous subsection characterized as the parlous evaluative status of anything whose value is chiefly instrumental: A newly discovered or invented and better means to a given end is almost always conceivable and often, in due time, practically possible. Some authors call this "technological substitution", though neither new nor advanced technology need be involved. As one example in which technology is involved, if biodiversity were a resource for pharmaceuticals that promote human welfare in the form of healthy lives, then another means of finding drugs – such as computer model-generated substitutions on chemicals already known to be bioactive – might well constitute such a substitution.

The second type of substitutability, sometimes called "economic substitutability", is a direct consequence of the possibility that one can sacrifice the ingredients for one welfare cake in the interest of securing a substantially different set of ingredients that bake into an entirely different kind of welfare cake, albeit one suffused with more welfare goodness overall. If overall human welfare is the ultimate good, and more is better, then it makes sense to regard the new cake as a more than an adequate substitute for the old. Thus, for example, biodiversity might rationally be fingered as the sacrificial victim of the agricultural development of land. The latter might reasonably be thought to increase human welfare by virtue of feeding the world's hungry, and do this to a degree outstripping all conceivable contributions to human welfare by biodiversity.[18]

Those who defend biodiversity, among other "natural goods", mostly propose that biodiversity merits a substitutability exemption of the second, economic kind. This proposal still leaves biodiversity vulnerable to the first, technological kind of substitution. But no matter which type of exemption might be proposed, it must be justified by legitimate and compelling reasons. The reasons cannot be arbitrary or question begging.

As will become more clear in Chap. 6 (Theories of biodiversity value), I am not aware of any justification for exempting biodiversity that is not both arbitrary and question begging. The typical argument comprises a mere recounting of the properties of biodiversity – properties that might indeed be unique to a particular biodiverse state of the world. But no convincing evidence is presented that these properties are of consequence to biodiversity's role as a unique and economically non-substitutable instrument of human welfare. Nor am I aware of any evidence that some other, perhaps drastically altered or reduced state of biodiversity might be at least as good, so far as economic welfare is concerned.

Finally, with my apologies to those who know better, it is important to address a serious and widespread conceptual malady that infects economic arguments for valuing biodiversity. I have in mind the conflation of "existence value" with the intrinsic value of the thing valued – in any of its four senses. In economics, the existence value of some thing X has no implication whatever that X has intrinsic value. In particular, if biodiversity is claimed to have existence value in an economic framework, then that means – and only means – that the existence or contemplation of the existence of a biodiverse state of the world satisfies some human preferences. It is these *preferences* (or satisfying them) that the axiology of economics holds to be ultimately valuable, not an entity such as biodiversity that satisfies the preferences.

---

[18] One might think that a sufficiently diverse agricultural enterprise might serve biodiversity as well as food to the hungry. But (as discussed in Sect. 6.2, Biodiversity as resource) the number of species of plants and animals that figure in agriculture is vanishingly small. So the suggestion that one might achieve the best of both worlds with diversified agriculture is not credible.

2.1.2.2   Deontology

The second main philosophical approach to right action and right living is the deontological one. According to this view, right actions and behavior are not just instrumental for bringing about a good or better state of the world. Their rightness does not derive even primarily from the state of the world that they bring about. The world's state might even sometimes suffer as a consequence of morally justified and even morally obligatory behavior. On the most radical of deontological positions, the consequent state of the world would be held to be totally irrelevant to the moral evaluation of actions.[19] Instead of focusing on consequences, deontology holds that actions and behavior are first and foremost good as fulfilling certain duties and obligations that a moral agent has towards entities that command such moral respect.

According to one form of deontological moral theory, what ultimately counts for leading a moral life is conforming to moral norms that determine whether an action or behavior is morally required, forbidden, or permitted with respect to morally considerable entities. And a first step in determining these norms is to identify these entities, which one can think of as possessing "moral mass".

Moral mass, as a deontological model for intrinsic value (in a sense that combines ultimate and objective value with moral considerability), behaves in a strikingly similar way to material mass. Picture a chunk of physical matter – say, a celestial body – that radiates a gravitational field. This field distorts space and time so as to accelerate, decelerate, and otherwise influence the trajectory of any particle in the massive body's neighborhood. So too, a "moral body" – a morally considerable entity – can be thought of as generating a moral force field. That field alters the "space" in which moral judgments are made so as to influence the moral trajectory of a moral agent – or at least, a morally sensitive agent – in the vicinity of the morally massive body. This model is the starting point for a deontological theory whose principles attempt to explain how moral mass is distributed throughout the world. In the context of environmental ethics, a salient question is how morally massive are nonhuman individuals and nonhuman-containing collectives.

As it happens, there is a time-honored tradition of deontological principles that attribute moral mass to individual nonhuman organisms. I mentioned some of these principles in the preceding subsection on axiology. Paul Taylor is a venerable exponent of this tradition. A keystone of his "biocentric outlook on nature" (Taylor 1981, 210–211; 1989, 119–129) is the vital, worth-imbuing and duty-engendering principle of being a "teleological center of life" – that is, being some thing – and specifically some organism – that has a good of its own. According to Taylor, having a *telos* – a purpose – commands moral respect. In terms of the mass model, it generates a force field that, impinging on a morally sensitive moral agent in a living organism's neighborhood, moves that agent to understand her duty to

---

[19] Immanuel Kant famously seems to have come close to such a position. But only a tiny number of other thinkers have defended it. In mentioning it, I am merely suggesting the range of thinking that can fall under the deontological umbrella.

view the world from the perspective of that organism's life and its striving to live
its life well in a sense appropriate to its kind. Most of us would grant this when
that "other living organism" is another person. We humans owe other humans
moral respect – at least partly for the reason that Taylor suggests. But Taylor
insists that human organisms are not special with respect to his principle. His
biocentrism is founded on the supposition that this basis for moral respect does
not privilege the good of human individuals over the individuals of other species.
In other words, Taylor says that there is no non-arbitrary reason to think that
members of *H. sapiens* – that odd, lately-evolved, bipedal mammal among the
millions of species all told – have any special status in the natural world insofar as
their intrinsic value is bound up with their *telos*.

Suppose that Taylor's argument succeeds and individuals of all kinds (even non-
conscious and *pace* Bentham and Singer, non-sentient ones) are found to merit
moral respect. On the face of it, this value of *individual* organisms has little bearing
on the question of whether a *collection* of individual organisms such as a species
independently commands similar respect. Needless to say, the existence of the species is requisite for the existence of its individual members. Conversely, individuals
could not exist but for the species to which they belong. But without considerable
refinement of the sense in which species are necessary for their individual members,
this observation does not earn the species independent moral consideration. Nor
does it justify their intrinsic (ultimate) value. Consider another case of necessity:
The element carbon is just as much requisite (in some sense) for the existence of
individual organisms (or at least the ones we are concerned about). But carbon does
not consequently command the relevant kind of independent respect. Compared to
a simple collective such as species, the independent moral relevance of the *diversity*
of a *collection* of species – a simplified view of biodiversity – seems even more
remote and elusive as a candidate bearer of intrinsic value.

How, then, can a deontologist pull collectives – species, "biological communities", ecosystems, the planet, all of nature – into the sphere of moral consideration? Proponents of several different brands of deontology develop narratives,
many of which can be understood as attempts to rationalize Aldo Leopold's
(Leopold 1949, "The Land Ethic") land ethic. Leopold (1949, 223) proposes that
it is possible to have "an ethical relation to land" involving "love, respect, and
admiration… and a high regard for its value". To this ethical attitude he conjoins
a suggestion for how it might be understood as springing from a concern for the
well-being of the land. That well-being, in turn, is understood (Leopold 1949,
224–225) in terms of "the integrity, stability, and beauty of the biotic community".[20] Therein lies the seed of an idea of how the worth of individuals – as members of "the biotic community" – might extend to "the community" (understood at
any of several levels of organization) itself.

---

[20] My brief treatment avoids the arm-wrestling matches within the environmental philosophy community, which hotly contests which ideology "owns" Leopold's ideas about the land.

Taylor has a notion of "respect for nature", which might initially sound like the fruit of Leopold's seed. His argument for the moral worth of individual organisms is one pillar of several that, in his view, supports a view (Taylor 1981, 209; 1989, 116–119) of the natural world as "a community of life". Though Taylor sees the value of this community as rooted in the valuable lives of individual community members, he regards the community as separately valuable for the web of physical interdependencies and other relationships between these members, both as individuals and as representatives of interacting species. However, Taylor's own comments on Leopold (Taylor 1989, 118–119) shows him to be notably reluctant to stray far from the good of individuals as the foundation for his own biocentric outlook. According to him (Taylor, 1981, 209), "a 'holistic' view of the Earth's ecological systems does not itself constitute a moral norm". Rather, the moral importance of (as he says) "ecological equilibrium" stems from how it serves the good of individuals – human and nonhuman alike.

Whatever the other merits of Taylor's argument for the moral consideration of nonhuman individuals, it is simple, elegant, and direct. A similar maneuver for collectives has none of these desirable attributes – for the simple reason that Taylor's argument for the intrinsic worth of individuals hinges on their having a *telos*. The task of finding a morally compelling version of this element in species, ecosystems, "biological communities", the entire biosphere, or indeed, any biological collective appears to be enormously difficult. None of these collectives pursue a "life of their own" in any obvious or morally leverageable way.

Perhaps the boldest attempt to pull off the move from the moral worth of individuals to the worth of collections of individuals comes from the eminent environmental philosopher Holmes Rolston III. By linking together a chain of arguments that first bridges the gap from individuals to the lowest-level collective of species, and then goes up the collective scale from the less to the more inclusive and from the homogeneous to the heterogeneous, he moves from species to ecosystems to the earth and finally to nature as a whole. He starts by leveraging Taylor's basic argument from the *telos* of individual organisms to construct a more general notion of "value-able", whose meaning again morphs to "systemic value" as he goes up the collective ladder to ecosystems and beyond. To species, Rolston (1994, 20) still attributes a *telos*, according to which it "defends a particular form of life" through the lives of its individuals. But according to him, species are also value-able. At the level of species, he (1994, 21) takes this to mean "able to conserve a biological identity".

In Rolston's narrative, the value of collectives at higher levels of organization depends on the value of species understood in this way. At the level of ecosystems, Rolston leaves *telos* behind, it having served its purpose as a first stage booster engine to get from individual organisms to species. At this point Rolston's story (1994, 22) is propelled by his notion of "value-able", whose meaning, in application to ecosystems, morphs (Rolston 1994, 24) into the capability of "[selecting]… for individuality, for diversity, for adapted fitness, for quantity and quality of life". According to him (1994, 25), ecosystem processes are value-able by virtue of being "able to create value". In particular, they are capable of engendering value-able species. He calls this the "systemic value" of ecosystems, which they have by virtue of their "productivity". This value is, as he says, "interwoven" with both instrumental and intrinsic value, but according to Rolston (1994, 25), "There are no intrinsic values, nor instrumental ones either, without the encompassing systemic creativity".[21] Creativity and the production of values also figure in the final two stages of Rolston's ascent to the pinnacle of collective inclusiveness, which summits at Valuable Nature by way of Valuable Earth.[22]

At each stage in his narrative, Rolston's reasoning becomes more precarious and open to challenge. When he starts talking about productivity, for example, he seems to implicitly rely on an unmentioned and therefore undefended normative gloss that separates good productivity from bad productivity.[23] So I wish to focus on the level of species, for which he defends a fairly straightforward extension of *telos* – the "[defense] of a particular form of life" – that directly ties it to one of the most compelling bases for the moral consideration of a species.

Rolston is in the company of a number of other philosophers who stake out similar claims regarding species. James Sterba (2001, 30), for example, evidently views the extension of Taylor's position on individuals extends to species as so straightforward that he need only observe that

> … species are unlike abstract classes in that they evolve, split, bud off new species, become endangered, go extinct, and have interests distinct from the interests of their members. For example, a particular species of deer… can have an interest in being preyed upon. Hence, species can be benefited and harmed and have a good of their own and so should qualify in Taylor's view as moral subjects.

Unfortunately, being "unlike an abstract class" does not make an entity capable of enduring harm in a morally relevant sense. Ultimately, the first link in Rolston's chain – the one that attempts to establish the intrinsic value of species – is only as strong as the proposition that the sense in which a species "defends" itself or (with Sterba) "has interests" is sufficiently like the way that an individual organism defends itself and has interests so as to merit similar, moral consideration.

Other deontologists shy away from this risky "direct assault" strategy and try routes that seem to them less hazardous, if also more circuitous. We already have had a whiff of "communitarian" thinking from Paul Taylor's channeling of Aldo Leopold. More generally, "communitarians" (some of whom might not welcome being thrown into the deontological bucket) develop a notion of community that extends beyond purely human societies and that is ultimately rooted in the natural world. According to them, communities, so understood, are the subjects of ultimate moral significance. The intrinsic value that affords them this status derives from such attributes as integrity, stability, resilience, health, and autonomy, which are familiar from modern "scientific" accounts of the properties of ecosystems.

---

[21] The more specific contention that *biodiversity* generates value is examined in Sect. 6.7 (Biodiversity as value generator).

[22] This exposition cannot do justice to the depth, breadth, and subtle sophistication of Rolston's unique approach to environmental value.

[23] This sort of problematic gesture – the utilization of a normatively loaded term as the primary means of "establishing" a norm – crops up repeatedly in this book, including just below.

However, there is no simple and, some would say, no compelling argument for why the integrity, stability, resilience, health, or autonomy of an ecosystem should be thought to command moral respect. But that might be a moot point because, so far as I can tell, there is no non-circular definition of these ecosystem properties that does beg the normative question of their value – either by the definition's use of such other normatively laden terms as "violation" and "harm", or by appealing to how conducive one of these ecosystem properties might be to human well-being. It is perfectly fine to say that ecosystems can be "healthy", when a state of ecosystem health is defined in terms that are appropriate for ecosystems. But it is not at all fine to say that a healthy ecosystem is valuable, when that claim is supported only by assuming, rather than giving reasons for why we should think that the normative weight of the term "health" as it applies to humans (for example) transfers to ecosystems. Nor is it fine to merely transfer the normative weight of such a term as "health" to yet another normatively loaded term such as "harm". Both moves amount to moral evaluation by stipulative definition – an issue that this book revisits a number of times, most saliently in Sect. 8.2.6 (What appropriate fit is not).

Sterba (2001, 30) hits on the crux of the matter in his approving interpretation of Paul Taylor, to whom he attributes the position that

> … ecosystems should qualify as moral subjects since they can be benefited and harmed and have a good of their own, having features and interests not shared by their components.

In this passage, he appears to promote – directly and without ado – the notion of "a good of its own" to the ecosystem level. Granting that promotion pretty much settles the moral score, granting as it does that an ecosystem could be harmed or benefited according as its good is thwarted or promoted. But why should we grant it?

There are plausible reasons to think that an individual organism has a *telos* towards which it strives and that moral agents can harm or benefit the organism by obstructing or promoting its striving. But the extension of this line of thought to an ecosystem or other collective seems problematic. One must keep in mind that, for moral assessment at least, the "striving" cannot be merely an epiphenomenon of the chance interactions and interrelationships of the collective's component parts. Nor in the end, can it be a consequence of the independent, uncoordinated, purposive striving of each individual without regard to any *telos* other than its own. Nor can the good of a collective be merely a function – for example, an aggregation – of the good of each of its parts. In sum, the requirement for having a principled way to assess harm and benefit seems to spawn the additional requirement for some kind of tight organizing principle. For a collection, the organizing principle must be attributable to it independent of its parts. This requirement appears very difficult to satisfy for such a complex assemblage of biotic and abiotic parts as an ecosystem.

The collectives mentioned to this point are ones that have received considerable attention from deontologists in the field of environmental ethics. Biodiversity has not received such great attention, and perhaps for good reason. Global biodiversity is so abstract and so loosely organized that even Sterba might be reluctant to assert that it has a good of its own. Local biodiversity, or the diversity of an ecosystem might initially seem to be a better candidate for intrinsic value because the "components" of local biodiversity are together in one place. That circumstance gives rise to at least the possibility of interaction, and hence, coordination under some principle of organization. But this possibility does not seem to be realized. What is needed for intrinsic value is a purposive coordination – an organized and cohesive attempt to reach a goal. This kind of coherent striving does not seem to be present in even local biodiversity.

Section 2.3.1 (Abstraction) includes some further reflections on this topic.

2.1.2.3 Virtue Ethics

This brings us to virtue ethics, which for many non-philosophers will be the least familiar of the three major moral traditions. This tradition does not view right action and right living as necessarily a matter of bringing about better states of the world, as would a consequentialist. Nor does it view them as primarily a matter of fulfilling duties and obligations to entities whose intrinsic value commands moral regard, as would a deontologist. Instead, virtue ethics focuses on the question, "What kind of person should I be?" According to it, leading a good life is a matter of the continual development of person's character, which can be described in terms of the person's "virtues". This most notably involves developing the attitudes and dispositions that enable a person to respond appropriately to the circumstances in which she finds herself – given her own natural, developed, and developing repertoire of capabilities. On this view, the human virtues are the things of ultimate value. Action and behavior are right to the extent to which they exercise the virtues that apply in the circumstances, which include the strengths and limitations of the actor.

Two interrelated considerations are striking about virtue ethics. The first is that it is unabashedly anthropocentric – *not* necessarily in its axiology, but in the special sense that it quite fundamentally proceeds from consideration of how people should be, live, and behave, and how, generally, they can flourish. This most dramatically contrasts with the deontological predilection for starting with value conceived in terms of force fields that emanate from entities possessing moral mass independent of moral agents. It also contrasts with consequentialism, insofar as it fixates on states of the world before considering how a person can act to bring about states that are particularly good.

Virtue ethics' distinctive starting question – regarding human character – also naturally tends to tip this tradition towards the view that value "in the world" is anthropogenic. But even if all moral value is anthropogenic, and even though all or most forms of virtue ethics are anthropocentric in the sense of making consideration of human character their starting point, neither of these conditions imply anything whatever about the range of things valued (the loci of value) and how valuable things are appropriately valued. In particular, persons need not be the only valuable entities. By themselves, virtue ethics' framing assumptions do not entail that a person has special moral standing merely by dint of membership in *H. sapiens* rather than some other species. Nor does virtue ethics require that the value of a nonhuman entity directly derive from human value – except in the sense that it is something to which a virtuous person should respond in certain ways, according to the virtues.

The second consideration, which closely relates to the first, is that a theory of virtue ethics stands or falls on which human character traits it holds to be virtuous. This leads to the critical question of how the virtues qualify as virtues. Different brands of virtue ethics give different answers to this question. One possible answer nods back towards consequentialism. It gives the answer that virtues are those traits that tend to promote "a preponderance of good" (Driver 2001, 48). This "good", in turn, can be taken to include a flourishing human life. Human flourishing has many ingredients that go well beyond biological success. The "good" can extend being "mere" human flourishing to some more generally good state of the world whose characterization might look a lot like a consequentialist's.

The environmental virtue ethicist Ronald Sandler (2007, 28) focuses on human flourishing in proposing that

> A human being is ethically good (i.e. virtuous) insofar as she is well fitted with respect to her (i) emotions, (ii) desires, and (iii) actions (from reason and inclination); whether she is thus well fitted is determined by whether these aspects well serve (1) her survival, (2) the continuance of the species, (3) her characteristic freedom from pain and characteristic enjoyment, (4) the good functioning of her social group, (5) her autonomy, (6) the accumulation of knowledge, (7) a meaningful life, and (8) the realization of any noneudaimonistic ends (grounded in noneudaimonistic goods or values) – in the way characteristic of human beings (i.e., in a way that can rightly be seen as good).

In considering this account of how a character trait qualifies as a virtue, care must be taken to avoid confusing the qualifying list with what initially might appear to be a similar, but actually quite different, consequentialist position. These qualifying conditions are not, as a consequentialist might say, categories or ingredients of human welfare whose aggregate value, insofar as they are a consequence of human actions and behaviors, is a measure of the rightness of those actions and behaviors. Rather, for a virtue ethicist such as Sandler, these are the ingredients of a good human life, which can be drawn upon to justify the various virtues. The moral measure of an *action* is assessed in a separate and second step – as the degree to which it accords with virtues.

In contrast with deontological views, a virtue ethics view regards the notion of moral considerability as derivative – which is not at all to say that it is unimportant. Here is one possible virtue ethics definition of "morally considerable": An entity is morally considerable just in case, for some justified virtue, a virtuous agent who possesses that particular virtue responds to the given entity in a certain way; and that response consists in acting and reacting in ways that are appropriate for expressing that virtue in the presence of that entity. This understanding of moral consideration leads directly to the possibility of nonhuman loci of moral value.

Sandler, for one, finds virtue in care and respect for individual living, nonhuman organisms. Following the broad outline of Taylor's basic argument from the *telos* of all organisms, he argues that for these organisms, the appropriate kind of respect generally takes the form of a broadly conceived restraint and nonmaleficence. This contrasts with the kind of active compassion that is more generally appropriate with regard to other *persons*. He finds justification for extending the kind of moral respect due to nonhuman individuals to such cohesive collectives as ant colonies. There, he feels that Taylor's basic principle of "a good of its own" still applies. But there it also stops. Focusing on that principle, he balks at finding virtues that apply specifically to species. He balks even more strongly at the notion of specifically ecosystem-related virtues.

There are other ways in which a virtue ethics approach can help to identify the natural value of collectives and even specifically of biodiversity. For example, it might be thought that biodiversity engenders knowledge and that that justifies a virtuous regard for it on the basis of a virtue in turn justified by Sandler's "accumulation of knowledge". This thesis, though mostly stripped of virtue ethics dress, is discussed in Sect. 6.8 (Biodiversity as font of knowledge). More generally, one might think that biodiversity is somehow connected to "a meaningful life". But this proposition is also far more vague and far more difficult to defend.

2.1.3   *Where Biodiversity Fits in the Philosophical Picture*

Several junctures in the two preceding subsections touch on how biodiversity, among other parts of the nature, fits into the philosophically conceived world of value and how the major moral traditions might accommodate this value in their respective theories of right action and behavior. The current subsection returns to this topic for a more focused overview. The emphasis here is on identifying and characterizing what pieces of the philosophical landscape current discussions of biodiversity preponderantly occupy, for this necessarily shapes the lines of reasoning as well as their strengths and weaknesses.

I have already suggested that an aggregating, welfarist version of consequentialism predominates nowadays as the moral-theoretical backdrop for discussing biodiversity's value. Much of what is written on this topic is couched strictly in the terms of neoclassical economics' peculiar brand of market-framed, preference-based consequentialism. Sometimes an effort is made to press beyond preferences and alongside them bring human needs and even distributional fairness into the bundle of human welfare goods that makes up the evaluatively relevant consequent state of affairs. But it is rare for any of this to be made clear. In fact, it is hard to find a discussion that evidences even minimal awareness of the concepts and principles that philosophers have found handy in constructing axiological and moral theories. As a result, this most critical part of the discussion tends to remain almost entirely implicit. It mostly requires reconstruction with educated guesswork.[24]

---

[24] Exceptions do exist to the generally dismal state of axiological discussion connected with economics. A shining example is the book by Debra Satz (2010) on *Why Some Things Should Not Be for Sale*.

Over the past two decades, the economic approach to valuing everything in nature – including biodiversity – has gained considerable, indeed, dominant currency under the rubric of "Natural Capitalism". E.O. Wilson is one of the very few authors who write about biodiversity while not entirely confining himself to this approach. Yet, he does not shun it, either. His discussion (Wilson 2002, Chapter 5) of "How Much is the Biosphere Worth?" is an exemplar. When Wilson asks this question, he is asking a question that does not differ in any significant axiological or moral respect from the question "How much is your car worth?" Both have a value and a price tag. And the price tag might or might not exceed the (economic) value.

Typical of many similar discussions, Wilson tries to suggest nature's great worth with a very selective inventory of natural goods. Included are the cost of "ecosystem services",[25] the annual value of a year's ocean catch, the annual cost of water purification New York City's water,[26] and the source "of new pharmaceuticals, crops, fibers, pulp, petroleum substitutes, and agents for the restoration of soil and water" (Wilson 1996, 174). Wilson is also the creative force behind the "biophilia hypothesis", which posits a human affinity for natural things – an affinity that apparently cannot be exercised in their absence. This – the human pleasure that is supposed to derive from satisfying the desire for contact or interaction with nature – is what a neoclassical economist would call a "non-use" value of nature.

The driving principle behind Natural Capitalism is that the environment, particularly the natural environment, and for our purposes most particularly biodiversity, are quite literally capital assets. That is, for purposes of valuation, nature is equivalent to a collection of already-produced goods whose value lies in their suitability for use in producing other goods and services that people are willing to pay for.[27] Just like any other form of capital, they are investments in and constitutive of

---

[25] Ecosystem services are defined and discussed in Sect. 3.3.2.3 (Functions) and in Sect. 6.3 (Biodiversity as service provider). The estimation of the total value that Wilson uses is due to Constanza et al. (1997). This work is cited ubiquitously, with apparent disregard for the fact that it is irrelevant from the standpoint of establishing nature's market price: What counts for that is the value at the margin, not the total cost of replacement, which is what Constanza purports to estimate.

[26] The example of the New York City water supply is also ubiquitous in the literature of Natural Capitalism. Whether or not this case illustrates anything about the value of natural ecosystems – in contrast to systems that are engineered to serve some narrow (if important) human desire – is an open question. On the other hand, the bearing of this case on biodiversity, as discussed in Sect. 2.2.1 (The bare assertion fallacy), is entirely clear. The purification of NYC's water has much to do with reducing pollution. On the other hand, it has little or nothing to do with increasing biodiversity or any attempt to refrain from decreasing it.

[27] This value is a market value; that necessarily depends on market conditions; and therefore this view of nature's value is committed to the position that it depends on market conditions. Of course, as E.O. Wilson's views illustrate, a Natural Capitalist need not be committed to saying that nature's value lies solely in its value as natural capital. But she is committed to providing a *principled* basis for saying when and why a zero or negative valuation does not, on that account, justify regarding nature's value as non-positive, too. I argue in Sect. 8.2.3 ("Living from" nature, uniqueness, and modal robustness) that principles serving this purpose are elusive, given the nature of economic valuation as it applies to nature.

economic welfare. According to the well-known biologist Paul Ehrlich and the environmental economist Lawrence Goulder (Ehrlich and Goulder 2007, 1147), a "direct line" connects capital assets, which include "natural capital", and what they call "well-being". Their definition as "the enjoyment of… goods and services", makes clear that for them, "well-being" is "getting what people want"; in other words, it merely relabels "economic welfare":

> *Well-being* defies easy definition or measurement. We define it as the enjoyment of various goods and services, both material and intangible. Figure 1 illustrates connections among capital stocks, goods and services, and well-being. The designation *capital* applies to the various stocks because they are durable assets that generate streams of goods and services. Natural capital (KN) includes both renewable resources such as forests, pollinators, natural enemies of pests, and fresh water, as well as nonrenewable resources such as reserves of petroleum and other minerals. Reproducible (or built) capital (KR) encompasses factories, roads, vehicles, machinery, and so on. Human capital (KH) is the productive capacity of human beings and reflects the skills acquired by people.
> These capital assets contribute to well-being in two ways (Fig. 1). First, they produce goods and services for consumption. For example, reproducible capital in the form of a hospital building and the medical equipment within it, plus human capital in the form of skilled medical-care personnel, plus some natural capital in the form of medicines derived from plants yield medical-care services, a contributor to well-being. Second, some capital assets contribute directly to well-being when they are enjoyed for their own sake, as natural capital does when people view wildlife. Hence the line in Fig. 1 showing a direct connection between capital assets and well-being. [italics in the original]

Ehrlich and Goulder join other Natural Capitalists in taking the appealingly simple view of assets as the "stuff" of human welfare. Their account also re-presents the Natural Capitalist broad division of assets into two forms – either a resource that people can use to satisfy their desires or needs, or a provider of a desired service. Either way – as natural resource or as natural service – a chunk of the natural world is good insofar as it "pays its own way" and increases the size of humanity's total pile of capital. Its continued existence is endorsable insofar as it enhances economic welfare in its intact state more than spending it or its constituent parts to "purchase" another asset, which constitutes another, worthy capital investment. On the other hand, if the facts indicate that the sacrifice of a natural asset opens up an opportunity to secure or develop another asset that more surely or greatly increases economic wealth, then by Natural Capitalist precepts, that chunk of natural capital *should* be sacrificed or (as an economist would say) "foregone" for the overall good.

This view is the nucleus of the broadly consequentialist approaches that dominate or lurk in the background of much discussion of biodiversity and its value. This dominance necessarily influences the orientation of the critical part of this book. The calculus of biodiversity value (Chap. 5) is almost entirely is framed in consequentialist terms. And among Theories of biodiversity value (Chap. 6), predominating are ones devoted to the proposition that biodiversity either enhances human health and economic welfare, or is essential to them. This is most obvious in Sect. 6.2 (Biodiversity as resource), Sect. 6.3 (Biodiversity as service provider), Sect. 6.4 (Biodiversity as (human) life sustainer), Sect. 6.5 (Biodiversity as a cornerstone of human health), Sect. 6.6 (Biodiversity as progenitor of biophilia), and Sect. 6.11 (The experiential value of biodiversity). Section 6.9 (Biodiversity options) focuses on the notion that some component of biodiversity's value might be folded away in what neoclassical economists call its "option or quasi-option value". The influence of economic thinking is only initially less obvious in Sect. 6.10 (Biodiversity as transformative), which nevertheless reveals itself to be perhaps the argument most clearly and explicitly rooted in economics – to the effect that biodiversity has value as a marketing force that, like any advertising campaign, transforms human preferences. On the other hand, in staking biodiversity's value on its ability to generate yet more biodiversity, the theory of Biodiversity as value generator (Sect. 6.7) avoids explicit commitment to any particular theory of the value that biodiversity is supposed to generate – for its standard version supposes that what it begets is more biodiversity, not anything else with an independently established value. On the face of it, Biodiversity as font of knowledge (Sect. 6.8), seems to lend itself to consideration from within any of the three broad moral frameworks sketched out in Sect. 2.1.2. But even this theory of why biodiversity might be valuable is customarily framed in the way that Ehrlich and Goulder do in the passage cited above – as part of "natural capital".

Few writers on biodiversity evidence awareness of other ways to frame natural value. Some refer to "other" reasons without elaboration. Others refer to "moral reasons", again without elaborating the moral grounding. Reference to otherwise unspecified "spiritual reasons" seems to join "moral reasons" as an unarticulated stand-in for "a strong belief that I don't know how to support". Section 6.1 (Unspecified "moral reasons") touches on this sort of gesture.

A related kind of unawareness routinely plagues discussions that are entrenched in the economic framework. It manifests itself in the mistaken presumption that entities with "non-use" economic value have some special moral standing – that "non-use" value constitutes "moral reasons" because (one must presume) the kind of preference satisfaction that gives rise to that kind of economic value is for some, never-specified reason, more morally compelling than other kinds of preference satisfaction. For example, this is often how discussions of biophilia come off. It is true that there is no real economic marketplace in which to trade the good (and bad) feelings that derive from being in contact with or engaging with nature. But this is no obstacle to economists, who invent imaginary markets for the imaginary trade of items that are assigned imaginary prices, which they nevertheless hold to represent real value.[28]

---

[28] In saying this, I mostly have in mind "contingent evaluation". But also, economists conjure up imaginary prices via "hedonic proxy" – something whose market price is claimed to accurately reflect the price of something else that does not have a market to determine its price. Other prices, known as "shadow prices" (see, for example, Ehrlich and Goulder 2007, 1148ff.), are fabricated by economists from theoretical whole cloth.

To his great credit, E.O. Wilson is one of the few writers on biodiversity and its value who explicitly and at some length venture outside the economic framework and even beyond that – outside any consequentialist framework. Unfortunately, his discussion suffers from several missteps before it gets anywhere. First, much of his writing leaves the impression that he regards the biophilia hypothesis as biodiversity's main ticket of entry into the realm of moral considerability. For the many reasons detailed in Sect. 6.6 (Biodiversity as progenitor of biophilia) this move is ill judged. More centrally for the current discussion, although Wilson proclaims biophilia to be the connecting glue between "the human spirit" and the environment, his discussion does not really escape the gravity of consequentialist or more narrow economic thinking. For according to Wilson, the good in nurturing biophilic tendencies finds a comfortable home in the category of economic "non-use" values. Affixing the additional label "spiritual" to this value does not change this. Nor is this changed by Wilson's evidence-free speculation about biophilia's evolutionary origins.[29]

Second, Wilson loses sight of the two central questions of moral theory – first, the question of what is valuable, and second, the question of what this implies for how people ought to act. Instead, he (Wilson 1996, 175) gets hung up on the question of whether moral value is anthropogenic or not, and tortures himself with the question of

> … whether moral values exist apart from humanity, in the same manner as mathematical laws, or whether they are idiosyncratic constructs that evolved in the human mind through natural selection, and thus of the spirit.

As I have suggested, the metaethical debate on whether moral value is objective or subjective is at once one of the most vociferous and one of the least consequential for normative ethics. In any case, Wilson's characterization of moral values as inhabiting a Platonic realm of *a priori*, eternal mathematical truths is one that would be alien to most moral objectivists, who mostly regard the value "out there" as matters of fact – that is, contingent on the identities of the bearers of those values, their histories, their properties, and the happenstance of their very existence.

Lastly, even when Wilson pushes outside the narrow confines of economics, his biophilia argument remains within the broader bounds of consequentialism. He pushes outside those bounds with just three limited ways. The first is a deontologically flavored move to extend the concept of the rights of individuals to species. He (Wilson 1996, 175) proposes "that species have universal and independent rights, regardless of how else human beings feel about the matter." But it is a long way

---

[29] The fact that biophilic feelings might be construed as having "non-use" value in an economic framework does not preclude another interpretation whereby they play a role in moral psychology as a means of sensing something of moral significance. Though he has no evidence to ground his speculation (he seems unaware that "compatibility with modern evolutionary theory" does not constitute evidence), Wilson supposes that such feelings might have been shaped by the adaptive advantages that they once conferred. The fact that some entity elicits positive "moral feelings" – however humans evolved to have such feelings – is sometimes a sign that it might have moral value. But that value must still be justified independently of evolved feelings, which might well turn out to be morally inappropriate.

from Wilson's limp, biophilia-tinged pep talk (in Wilson 2002, 134), which proclaims, "It is not so difficult to love nonhuman life", to "we *ought* to love nonhuman life", and from there, on to "nonhuman life forms (that is, nonhuman species) have (minimally) a right to exist". It is a longer stretch still to the ultimate touchstone of biodiversity. Wilson certainly does not supply the required bridge for that last stretch, which would require the proposition that "we ought to love the diversity that results from there being many different nonhuman forms of life".

Wilson offers a second line of argument that might be counted in his repertoire of deontological musings, though he himself does not present it in this way. This is his "we will be reviled if we don't" argument for preserving nature in general, wilderness especially, and biodiversity in particular.[30] In making this argument, one can suppose Wilson to presume some set of obligations and responsibilities whose shunning will blacken the reputation of current-day persons in the eyes of future ones.

His representation of this argument has many flaws. First is its failure to identify some neglected duty that would justify future persons' severe disapprobation of currently respiring persons. Second, the projection onto future persons of strong disapproval by some of us now is pure speculation, and at least on the evidence of attitudes that we now have towards persons of previous generations, highly implausible speculation. Do we now revile the coastal peoples who (according to Rick and Erlandson 2009) dramatically reduced shellfish populations and shellfish size starting some 23,000 years ago; or whose enormously successful predation on sea otters generated sea urchin "barrens" starting some 9,000 years ago? More generally, do we revile almost all pre-historical peoples whose predation cut huge swaths out of the megafauna that they encountered?[31] Of course, duty implies some kinds of knowledge that these earlier peoples cannot be presumed to have had or be held responsible for not having. But from a broader temporal perspective, it appears that current-day persons resemble our non-reviled early ancestors by merely pursuing the best possible way of life as we now conceive it.[32]

A third attempt to venture outside the bounds of consequentialist thinking only circles back to another claim akin to the biophilia hypothesis. This is his claim (made in Wilson 1996, 176) that "the diversity of life has immense aesthetic and spiritual value", which serve the "hereditary needs of our own species". Even granting that the diversity of life does has aesthetic value, Wilson offers little reason to think that this aesthetic value is morally compelling – that it has serious implications for how persons should behave, especially in the face of conflicting interests. The use of the word "spiritual" – floating detached from any compelling argument or theory – does no real work.

On their own terms, I believe that Wilson's forays outside the confines of consequentialist thinking don't get far. The critical part of this book, which is chiefly

---

[30] This argument is implicit in Wilson (2002, 150), among other places in his writings.

[31] On the human diaspora wiping clean the megafaunal slate, see for example, Wilson (1996, 173). See also the discussion in Sect. 6.12 (Biodiversity as the natural order).

[32] Section 7.3 (Biodiversity value in human timeframes) examines the future persons' retrospective blame gambit at greater length.

devoted to an examination of arguments for biodiversity's value that are at least modestly well developed, does not return to them except briefly, in Sect. 6.1 (Unspecified "moral reasons").

This is not to say that these and other approaches to ground biodiversity's value are not worth further exploration. In fact, I would say that other approaches are needed and that the need is extraordinarily urgent. Ronald Sandler's list of teleological justifications for virtues, for example, provides some tantalizing directional hints. Human knowledge need not be viewed as "human capital" – though I believe that the specific theory of Biodiversity as font of knowledge fairs no better when couched in terms of virtue ethics. The meaning of a human life need not be shoehorned into the paradigm of preference satisfaction. Aesthetic value is probably not very fruitfully viewed as a kind economic non-use value. But a tantalizing direction is not a well-defended theory. The extreme difficulty of finding such a theory in any of these directions likely accounts for a notable absence of them to date.

In sum, questions relating to what one ought to do with regard to the environment in general, and biodiversity in particular need not remain marooned on the shoals of a consequentialist or more narrowly economic conception of human or economic welfare. More generally, "the best possible way to conceive and live a human life" invites a fresh reassessment of which elements are truly important to it. But that is a topic whose exploration must wait for Chap. 8.

2.2 Reasoning About Biodiversity – A Catalog of Fallacies[33]

I already have alluded to bad forms of reasoning that pervade arguments for biodiversity's environmental value. This section is a partial catalog. It differs from Chap. 6 (Theories of biodiversity value) by focusing on *forms* of reasoning rather than on the persuasiveness of specific lines of argument or on the truth values of their premises.

I focus here on informal fallacies. One caveat for this discussion is that the fact that a conclusion is reached by way of fallacious reasoning does *not* determine whether that conclusion is true or false. Even if reached by way of faulty reasoning, a conclusion might be true. To think otherwise is to commit the fallacy fallacy. But as a matter of fact, fallacious reasoning most often leads to incorrect results. And fallacious reasoning on behalf of biodiversity is consistently marshaled in support of conclusions that are, in fact, false.

As I say, this is not a comprehensive catalog. Its primary purpose is to show just how worrisome is the state of reasoning about biodiversity. The fallacies described here are encountered with dismaying frequency, despite the fact that they are often

---

[33] This section grew from seeds planted by discussion with Darren Domsky about "correlates", his term for the fallacy whereby causation is illegitimately inferred from correlation (the fallacy of correlation).

not at all subtle. This might help to convey why, as I say, my initial, high confidence in the case for biodiversity "transformed to worry; then to astonishment; and finally to alarm". A secondary objective is to provide a kind of field guide to species of bad form. Armed with a guide that pastes a label on each species and points up its distinguishing marks, one is better prepared to recognize an individual of any species in the field.

There is no agreed-upon taxonomy for informal fallacies (reasoning that is incorrect by virtue a number of different possible flaws, including ones in its premises and justificatory structure), though they are typically treated as inclusive of formal fallacies (patterns of reasoning that are incorrect by virtue of their logical structure alone). The formal/informal distinction is irrelevant for my purposes and so is the specific taxonomy. Mine is an unapologetically *ad hoc* hybrid of my own invention. Also, the boundaries between different kinds of fallacies are porous. This is why many pieces of bad biodiversity logic can be viewed as simultaneously committing multiple fallacies.

But if that is a reason for simultaneously committing multiple fallacies, it is not the only one. Sometimes it is clear that multiple bad forms of reasoning are combined and intertwined in a single thread of misjudged logic. While my field guide clinically isolates the bad forms, it also tries to convey something of the real conditions out in the field, where there is the further intertwining of multiple arguments made in close proximity – each with its own flaws of logic and questionable premises. Chap. 6 gives a more complete sense of this real-life complexity in teasing out the details of various theories of biodiversity's value.

2.2.1 The Bare Assertion Fallacy

A bare assertion fallacy is committed – in its barest form – when a conclusion is asserted with not a single evident premise. A more general and certainly more annoying form is the repeated assertion of the conclusion – *sans* accompanying premises. Using the bare assertion fallacy, an argument for the proposition 'A' as conclusion 'C' has the form:

C: A
C: A
C: A

Sometimes it is hard to distinguish an instance of bare assertion from a Chewbacca defense (see below). To some degree, the distinction is somewhat arbitrary based on whether the conclusion A is asserted before (bare assertion) or after (Chewbacca defense) some discussion with no apparent relevance to A. Almost always, it seems, a bare assertion is also quite obviously false.

Examples abound.

In discussing the role of biodiversity in New York City's famous water supply, restoration ecologist Laura Meyerson and her colleagues seem to rely on a tacit and therefore unsubstantiated symbiosis between "Biodiversity and Ecosystem Services" – the title that heads a section (Meyerson et al. 2009, 89–90) that repeatedly asserts that biodiversity has kept the water clean:

> Freshwater is essential for the maintenance of life... [It] is a provisioning service affected by local biodiversity... New York City's drinking water has greatly benefited from local biological diversity in the Catskill Mountains watershed... A suite of conservation acts... and implementation of land use restrictions and septic system improvements... is protecting natural ecosystems and biodiversity and guaranteeing New Yorkers a continued supply of chemically untreated clean drinking water.

Here, Meyerson asserts that biodiversity affects the "provisioning service" of supplying potable water. Among the other verbiage, one cannot find a single consideration that is a premise from which one can draw this conclusion.

This logic-defying passage is valuable in showing how a single argument can exemplify multiple fallacies. This one does triple duty – simultaneously serving as exemplar of Chewbacca defense and the fallacy of correlation (each of which is discussed in its own subsection), as well as the bare assertion fallacy. Teasing out its form as a fallacy of correlation, the argument runs (where 'Pn' is the nth premise and 'C' is the conclusion):

P1: If you stop dumping sewage into the water, then it will be better to drink.
P2: If you stop dumping sewage into the water, then more creatures will live in and around it: That is, local biodiversity will increase.
C: Therefore, biodiversity makes water better to drink.

Stripped of content, the logical form (where 'X', 'B', and 'A' stand for arbitrary propositions, and the symbol '=> ' stands for causal entailment)[34]:

P1: X => B
P2: X => A
C: Therefore, A => B

In the second sentence of a paper that accompanies Meyerson's, biologist Andrew Wilby and his colleagues (Wilby et al. 2009, 13) assert:

> In its broadest sense, biodiversity is the source of our current food and will be the source of novel foods in the future.

Neither the sentence that precedes the quoted one (the paper's second sentence) nor anything in the remainder of the paper offers any reason to persuade the reader that this assertion is true. It is just asserted. The authors spend a lot of space discussing "dietary diversity", and this might explain their comfort with their unsupported assertion. Adopting their "diversity speak", one might say that dietary diversity is "a

---

[34] Of course, looking back at the argument from which this form is abstracted, one can make these substitutions:

X ← The dumping of sewage into the water is stopped

B ← The water is better to drink

A ← Biodiversity increases

major component" of biodiversity. This arrangement of meanings might suffice to clinch the truth of their bare assertion: If biodiversity largely comprises dietary diversity, then biodiversity would not be just a *source* of food; it would *be* our food. But then the proposition that the loss of biodiversity implies fewer kinds of food teeters on the brink of uninteresting tautology.

The fallacy fallacy says that a conclusion drawn from fallacious reasoning might yet be true. But if "biodiversity" is understood in the way that includes the diversity of species, then, on the evidence presented in Sect. 6.2 (Biodiversity as resource), Wilby et al.'s assertion is quite unequivocally false. That discussion makes clear that a tiny number of organisms supply almost all food for people. It also observes that maintaining biodiversity is inherently at odds with food production, which requires enormous swathes of land that banish the far greater diversity of organisms that would otherwise live there. The flourishing of one inevitably requires the sacrifice of the other – even given enormous and continuing efforts to ameliorate this conflict. On the other hand, if "biodiversity" is understood along the lines of "organic material", then Wilby et al.'s statement is again trivially true – and again uninteresting.

Psychologist Suzanne Skevington (2009, 130) also avails herself of a bare assertion when she maintains, without argument:

> Loss to biodiversity … is known to weaken life support systems and pose serious risks to health.

It is possible that Skevington regards this "known" proposition to be a premise for her main thesis, which is that the loss of biodiversity entails a corresponding loss in "quality of life". This is her central concept – important enough to merit the acronym "QOL" – to which she attaches an idiosyncratic and not intuitively obvious meaning. So it is noteworthy that, at first, she (Skevington 2009, 129) expresses considerable doubt about her QOL thesis:

> Perhaps the simplest model is that in which biodiversity is viewed as having a direct effect on quality of life (QOL), although the empirical evidence is generally poor and sparse beyond anecdotal and political levels.

This skepticism seems to motivate her move to an "indirect" model that connects biodiversity to QOL via human health. But she then launches into a Chewebacca defense, which loses any connection – direct or indirect – in a morass of irrelevant detail. She says: Climate change can cause illness; extreme weather conditions result in ozone depletion; and therefore (apparently), if biodiversity is reduced, then we should expect the quality of life to be reduced too; but not necessarily, because (according to her definition of QOL), better health does *not* imply better QOL. With that, she rests her case.[35]

We take up the Chewbacca defense next. But before we do, it is worth noting that Skevington's bare assertion also happens to be false. Many losses of biodiversity enormously *enhance* human health – as when competent (in the epidemiological sense), nonhuman reservoirs for zoonotic pathogens are eliminated.

---

[35] This is a re-creation of just a portion of Skevington's "argument" in Skevington (2009, 130).

2.2.2 *Red Herring or Chewbacca Defense*

The red herring, which has lately attained unprecedented fame as the "Chewbacca defense"[36] is a catch-all for fallacies of relevance that don't fall neatly into another subcategory of irrelevant reasoning. There are many such subcategories – including the gambler's fallacy, appeals to pity, and personal attacks (*ad hominem* arguments). The fallacy of equivocation, which hinges on ambiguous language, can also be placed into the category of fallacies of relevance. We visit that fallacy shortly in another subsection.

In the realm of biodiversity, an often-encountered variant of the Chewbacca defense is proof by verbosity.[37] One can think of this as a maze of reasoning that is so meandering, so discursive, so complex, and so full of difficult-to-follow and irrelevant detail that it is impossible to know where one is. And if, against all odds, one figures that out, then it is impossible to figure out how one arrived at that point. The relief of reaching the conclusion leaves little inclination to go back in order to reconstruct the tortuous route from its beginning.

In the domain of biodiversity, perhaps the finest exemplars of overwhelming verbosity come from discussions that throw the term "ecosystem service" at the term "biodiversity" as if by their mere proximity on the written page, the positive evaluative force of "ecosystem service" confers value on biodiversity. This, of course, does not qualify as "compelling logic".

Unfortunately, the argument due to Myerson et al., discussed in the preceding subsection, is not unusual in its use of confused and confusing verbosity in lieu of good logic. Ecologists Jerry Melillo and Osvaldo Sala are among those who join Meyerson and her colleagues in this. They contribute a chapter entitled "Ecosystem Services" to a large volume devoted to the titular topic of *How Human Health Depends on Biodiversity*. They, too, cannot resist the apparently overwhelming temptation to discuss the New York City water supply system. But there is more to notice about their approach.

In their introductory remarks, Melillo and Sala (2008, 75–76) set up the target at which they propose to aim:

> The ecosystems of the world deliver their life-sustaining services for free, and in many cases, they involve such complexity and are on a scale so vast that humanity would find it impossible to substitute for them. In addition, we often do not know what species are necessary for these services to work, or in what numbers and proportions they must be present.

It is useful to note that the proposition "ecosystem services are good" is true insofar as benefits to people are good – by virtue of its standard definition as

---

[36] This latter-day moniker for "red herring" was immortalized by the "Chef Aid" episode of the animated sitcom *South Park*, which forever changed the study of informal logic.

Latin-loving philosophers know this fallacy as *ignoratio elenchi* – literally, "ignorance of refutation". This conveys the sense that there is no obvious connection between the premises and the conclusion. As a result, there can be no understanding of what would constitute a refutation.

[37] For those who like to Latinize bad reasoning, this is *argumentum verbosium*.

(Hooper et al. 2005, 7) "ecosystem properties of benefit to people". For now, I wish to put aside the contestable merits of the substantially stronger claims that ecosystem services are "life-sustaining", that they are "for free", and even that they are irreplaceable or have no substitutes. Instead, I wish to draw attention to something more remarkable. This is a paper that purports to help make the case for *How Human Health Depends on Biodiversity*. Despite that, it has nothing at all to say about biodiversity.

Instead, Melillo and Sala warn us that any change in species *status quo* might be dangerous. Then, over the course of more than forty pages, they take us on a whirlwind tour of ecological science and ecosystem services with a running normatively tinged commentary that has little evident connection with biological diversity, as such. We are told that microbes have their own ecosystems; trees absorb NO_x gases; methanogenic microbes eat methane; and that is good because methane is a greenhouse gas; bivalves eat algae and that is good because people dislike the algae; floods are good because they leave fertile soil for crops; on the other hand, they are not so good when there is no flood plain to flood; if you cut down trees where there can be mudslides, that is very bad because that makes mudslides happen when it rains; hurricanes cause mudslides; the biodiversity of crop pests is bad; introduced species can be good when they eat pests; exotics can be good, too, when they eat toxins; bees pollinate some plants that people like to eat; frugivores transport the seeds of some lucky, fruit-producing plants; ecosystem services are worth a whole lot of money but it's hard to figure out how much; the climate is changing and that is bad for ecosystem services; deforestation is bad for the same reason; so is desertification; it's also bad to urbanize and to drain a wetland and to pollute air and to pollute water and to dam rivers.

At the very end, Melillo and Sala (2008, 114) conclude that:

> All of us, regardless of where we live on this planet, depend completely on its ecosystems and on the services they provide… While… changes have clearly contributed to substantial gains in human well-being and economic development for some, many others have benefited little. In addition, the changes have resulted in a substantial and largely irreversible loss in *the diversity of life on Earth*… [italics added]

In other words, from the human point of view, ecosystems have changed both for the better and for the worse; and by the way, biodiversity has diminished while this has happened. The authors evidently miss their own, self-defined target – an argument lending credibility to their thesis that messing with biodiversity is hazardous to human health. That is partly because they give many examples of how, according to them, messing with biodiversity is often quite *beneficial* to human health.[38] But also, while Melillo and Sala do not explicitly commit Meyerson et al.'s fallacy of correlation (for a separate discussion of which, see Sect. 2.2.4), their conclusion is essentially the first two premises of such a fallacy:

P1: Humans have caused ecosystems and their services to change.
P2: Humans have caused changes in the mix of species.

---

[38] For example, Melillo and Sala (2008, 94) laud the introduction of brake fern and water hyacinth, species that "eat" arsenic.

As I have related, the authors don't restrict their attention to ecosystem changes that are the unintentional side effect of anthropogenic forces among possibly others. They also approvingly acknowledge that some alterations of the mix of species are quite intentional human manipulations that quite self-consciously aim to alter "ecosystem services". Phytoremediation of toxin accumulations and the import of "biological pest control" species are examples. Sometimes these species-manipulation-induced changes end up changing things for the better; many other times for the worse. But this proposition is light years away from the "life-sustaining" necessity suggested in their introductory remarks – namely that changes in or reductions to biodiversity could mean the death of us all.

Melillo and Sala simultaneously commit a fallacy of accident, a type of fallacy that I take up next.

2.2.3 Fallacies of Accident

Fallacies of accident[39] come in two flavors. The first moves fallaciously from a general maxim to a particular case. Such a move is often a completely legitimate application of the general maxim. But the move is fallacious when it ignores one or more factors that disqualify the particular case as one to which the general maxim legitimately applies. The second flavor, sometimes called "converse accident" or "hasty generalization", moves fallaciously in the opposite direction – from a particular case to a general maxim. The fallacy consists in ignoring, simplifying, or otherwise manipulating evidence that is legitimate and that undercuts the generalization.[40]

It serves my discussion to lump these two flavors together for what they have in common. The result is a single class of fallacy that encompasses all arguments that appear to work because they arbitrarily exclude inconvenient and contravening evidence, exclude crucial qualifications, or hinge on evidence that is biased or *ad hoc*. A fallacy of accident is a kind of overreaching that merits special attention in the context of science, which strives to account for *all* evidence, including that which frustrates an initially hypothesized and favored conclusion.

---

[39] In the Latin taxonomy of fallacies, fallacies of accident are called "*a dicto simpliciter*", which means "from the unqualified statement".

[40] In the Latin zoo of fallacies, the first flavor of *a dicto simpliciter* is *a dicto simpliciter ad dictum secundum quid*, or "from the unqualified to the qualified statement". This label relates to arguing from a general maxim to a particular case. As suggested in the main text, applying a general proposition to a particular case is entirely valid – provided that the particular case actually falls within the scope of the general proposition. But it is not valid when conditions disqualify the applicability of the maxim to it.

The second flavor of *a dicto simpliciter* is *a dicto secundum quid ad dictum simpliciter*, or "from the qualified statement to the unqualified". As discussed in the main text, this relates to an induction, whereby particular instances are held to validate a general rule. *A dicto secundum quid ad dictum simpliciter* designates the informal fallacy of making an inductive leap while ignoring cases that show it to be generally unwarranted.

Fallacies of accident are particularly pervasive in discussions of the biophilia hypothesis (discussed at length in Sect. 6.6, Biodiversity as progenitor of biophilia). This is the speculation (Kellert 2009, 114) "that human beings possess innate tendencies to affiliate with nature, as manifest in… inherent values that people hold in relation to the natural environment, each conferring adaptive advantages that foster human physical and mental well-being". The real contents of this theory are anything but clear. To get the best sense of that opaqueness one must plunge into Sect. 6.6's detailed examination of biophilia-based arguments that try to connect biodiversity to "quality of life". My narrower concern here is to point out how those arguments commit fallacies of accident – for example, by making speculative extrapolations from observations that have at best a tenuous connection to biodiversity.

Social ecologist Stephen Kellert offers one of many hasty generalizations made under the biophilia umbrella:

P1: Language depends on… refined distinctions and taxonomic classifications (Kellert 2009, 116).[41]

P2: The development of language is facilitated by references to natural objects, which exemplify a set of such distinctions.

C: Therefore, it is an "adaptive advantage"[42] to have a diversity of natural objects available for reference.

This argument has many flaws. Here I wish to focus on its exclusion of evidence that many other, non-natural objects might serve as well as natural ones – and, perhaps even better, conferring greater "adaptive advantage" – for language development. Also excluded is any evidence that justifies discounting the possibility that even among natural objects, a small number of different kinds might suffice. This possibility is suggested by Kellert himself in several ways – including by his claim (Kellert 2009, 116) that the animals that appear in children's books constitute evidence for the importance of natural objects in language development. Putting aside the fact that the books (and computer images) can (and do) exist without the animals (a matter of increasing concern to environmentalists and apparently, Kellert (2009, 113) himself), it is challenging to think of more than perhaps two dozen kinds of organisms – mainly animals – that feature in that literature. This argument points towards how much diversity we can do without, rather than how little diversity we can afford to lose.

Kellert commits another fallacy of accident in trying to discount or dismiss altogether the countervailing results of his own (Kellert 2005, 30ff.), "Greater New Haven Watershed, or Mastodon Study". This study found (Kellert 2009, 111) that

---

[41] The second premise and conclusion attempt to summarize Kellert's extended discussion, taking into account that he is "Explaining the Relationship between Biodiversity and Quality of Life".

[42] "Adaptive advantage" is Kellert's term, ill-chosen for a context that has nothing to do with evolution. If it did, then the argument would also be an instance of a genetic fallacy wherein the benefit for the species *H. sapiens* of a condition over evolutionary time is mistakenly taken to be a reason for why people ought to realize that condition now.

... ecosystem health indicators such as species diversity, chemical pollution, and nutrient cycling were rarely directly related to human socioeconomic factors such as neighborhood quality, environmental affinity, environmental knowledge, or quality of life.

He acknowledges that, of course, species diversity and the like don't affect people's quality of life – because they are *unaware* of how much the quality of their life, unbeknownst to them, really does depend on biological diversity. This rationale for dismissing his own inconvenient evidence simultaneously commits a fallacy of equivocation – for the definition of "quality of life" that he explicitly espouses is Susan Skevington's (mentioned above). Skevington's definition is emphatically based on subjective preferences – largely a matter of self-perceived personal satisfaction. But Kellert cannot legitimately have it both ways. If quality of life is a matter of people getting what they perceive themselves to want, then they cannot be largely unaware of its basis. The two views cannot be reconciled except by equivocating on the meaning of "quality of life".

The cowardly way to commit a fallacy of accident is to suppress a known inconvenient premise by purposefully neglecting to disclose it. But that kind of sneakiness is rarely evident. With striking frequency in discussions about biodiversity and its value, one or more inconvenient premises are presented. The authors don't attempt to refute them; they simply ignore their invalidating force when drawing their conclusion. This surprising variant of fallacy of accident generally involves drawing an intermediate disjunctive conclusion, from which one of the disjuncts is then asserted as conclusion. An argument for the proposition A takes the form:

P1 ... Pn (assorted premises)
C1: Therefore, (A or B)
C2: Therefore, A

In a common and easy-to-spot variant, the logic simplifies to:

P1: Sometimes A
P2: (present but ignored): Sometimes not-A
C: Therefore, generally A

Here is an example from the emotionally charged topic of "alien invasions"[43]:

P1: It is bad for exotic species to "damage" or reduce the diversity of "native" species.
P2: (present but ignored): It is good for exotic species to "damage" or reduce biodiversity when the exotics supply food or "reduce the diversity" of (possibly "native") "pest" species.
C: Therefore, exotics are bad because they reduce or "damage" biodiversity.

---

[43] As I reiterate in Note 61 of Chap. 3, the pervasively used terms "invasive" and "alien" are emotionally and normatively loaded. For my own discussion, I prefer terms that are emotionally and normatively more neutral, such as "exotic", "newcomer", or "recently naturalized". The issue of "who came first" weaves in and out of the book, ending with Sect. 8.2.7.1 (Implications for natural value generally and biodiversity in particular).

In an alternative presentation of this argument, the class of exotics that P1 has in mind are relabeled "invasive aliens" in the conclusion. As any science fiction fan knows, invading aliens – whether from another planet, from the depths of the sea, or from across the ocean – are generally up to no good and are therefore emphatically unwelcome. The variant of the conclusion that results from this relabeling:

C: Invading aliens are bad because they "damage" biodiversity.

seems to need no supporting premises because the very presence of "invading aliens" is, by the negative normative implication of that phrase, already damaging.

Many arguments use specific exotic species to leverage a general case against "invasive aliens". With respect to *Pueraria lobata*[44] – the infamous plant kudzu, which was intentionally imported into the southeastern United States to control erosion along highways – we have:

P1: *P. lobata* reduces the diversity of "native" plants (Chivian and Bernstein 2008a, 48).
P2: (present but ignored): *P. lobata* controls erosion and absorbs harmful toxins (Melillo and Sala 2008, 95).
P3: *P. lobata* is an "invasive alien".
C: Therefore, invasive aliens are harmful.

An essentially identical argument involves *Eichhomia crassipes* (water hyacinth). The fact (Chivian and Bernstein 2008a, 49–50) that this plant makes life difficult for some other aquatic species in (among other places) east Africa's Lake Victoria is more ammunition hurled at "invasive aliens". Ignored in this argument is the fact that, like *P. lobata*, *E. crassipes* is a toxin sink that is particularly good at binding arsenic – a function that might be regarded as the opposite of harmful in arsenic-laden habitats.

In discussing these arguments about exotics, it is important to keep in mind that, insofar as a premise implies that exotics generally reduce biodiversity in their neighborhood, it is false. While some small minority of exotic species does significantly change ecosystems, the case for their having the general effect of *reducing* biodiversity in their new homes is tenuous, at best.[45]

From the topic of infectious diseases (Rapport et al. 2009, 51):

P1: Sometimes disease-carrying snails thrive in conditions of reduced biodiversity.
P2: (present but ignored): Eliminating the snails (with molluscicides) further reduces biodiversity.
C: Decreases in biodiversity are bad for human health.

---

[44] I generally follow the convention of using scientific name for organisms first, followed parenthetically by the (or a) popular name. My aim in doing this is not to present a façade of scientific wisdom that I do not possess. Rather, I believe that scientific names tend to be not so laden with preconceived value as the popular ones – at least for some well-known and "charismatic" species, as well as for species at the other end of the charisma spectrum. I believe that this helps to promote what I regard as the axiologically defensible attitude of *prima facie* species egalitarianism.

[45] I take up this tendency to equate change with reduction of biodiversity later in Sect. 5.1.4 (The just-so model) and Sect. 6.3 (Biodiversity as service provider).

In analyzing this argument as a fallacy of accident I do not wish to ignore what is perhaps the more obvious fallacy of *correlation* implicit in P1: Clearly, it is not the reduction of biodiversity that sickens people. It is the snail-hosted *Schistosoma* spp. (trematode) organisms.

Because, as I have previously observed, a conclusion fallaciously derived is not necessarily false, it is worth remarking separately on the truth value of these general conclusions about the benefits of biodiversity for human health. They are, quite simply, false. Most obviously, the ignored premise (or disjunct in an intermediate conclusion) disproves the generalization that the final conclusion asserts.

Of course, sometimes countervailing evidence that bears on a conclusion never finds its way into the discussion leading up to it. With regard to the conclusion of the last argument, some studies of the Great Lakes and smaller temperate lakes (Mazmunder 2009, 152) show that increased local aquatic diversity leads to a lengthening of the food chain and also increases the tissue concentrations of methyl mercury in top predators, which are the fish that people prefer to eat. This seems to show that at least sometimes, increased diversity is bad – even very bad – for human health.[46]

Some other very general and substantial pieces of evidence bear on the truth value of the "reduced biodiversity makes people sick" conclusion. First, it is well known (Thomas et al. 2009, 231–232) that "the diversity of human infectious diseases is highest in regions with high biodiversity". Also, as global biodiversity has undoubtedly decreased since 1990, life expectancy has increased for all age groups, and mortality rates – a measure of the burden of disease – have decreased in every global region (World Health Organization 2009, Part II, Table 1). However, it would be a fallacy of correlation to infer that this elevation of human health is an effect caused by the decrease in biodiversity.

Finally (for now), conditions unfavorable to human health are not infrequently favorable to increased biodiversity. For example, *Glossina* (tsetse fly) spp. in some regions of Africa keep out people and cattle. This enables the locally diverse flora and fauna to thrive undisturbed (Rapport et al., 43). Of course, it is again a fallacy of correlation to surmise that this is evidence that human sickness helps to preserve biodiversity.

---

[46] While this observation is a quite straightforward one, some find it "strange" or "misleading". But the justifications for these sentiments are themselves strange. For example, one objection is that food chains often get shorter when creatures at the top trophic level are exterminated. That is true. But it is also completely irrelevant to the point – that extending the food chain upwards results in a greater abundance of the top predators most coveted as food, whose high toxin levels affect the health of the people who consume it. It is important to note that my one example here is not isolated nor even the strongest one that shows that sometimes, biodiversity is quite bad for human health. See Sect. 6.5.2 (Biodiversity as safeguard against infection) for more examples.

2.2.4   The Fallacy of Correlation

A fallacy of correlation[47] is a fallacy of false cause, wherein the correlation of two events or conditions is mistakenly taken to entail that one causes the other. It might be the most pervasive fallacy in the world of biodiversity.

The general form of the fallacy of correlation is:

P: A and B
C: Therefore, A => B

But it often assumes the special form already mentioned in connection with reinterpreting the bare assertion argument concerning New York City's water source. This form results when both A and B are expected in the presence (because they are expected consequences) of separate condition X:

P1: X => A
P2: X => B
C: Therefore, A => B

We have observed how Meyerson et al.'s bare assertion regarding the "effect" of biodiversity on NYC water might also be viewed as a fallacy of correlation. Other examples abound. A number involve zoonoses – diseases caused by pathogens or parasites that bridge nonhuman and human hosts. Zoonoses have become a hot topic in "conservation medicine". By definition, at least two collections of species other than *H. sapiens* – the disease pathogens and their nonhuman reservoirs (hosts) – figure in zoonotic disease. In many interesting cases, a third collection of species – vectors, which communicate the disease organisms from one host to another – admixes with the other two. Those with biodiversity on their mind characterize this complex of relationships as a matter of the diversity of each of three separate collections of organisms defined by their respective functional roles. This proves to be a fertile breeding ground for diseases of logic generally and fallacies of correlation in particular, which infect reasoning about biodiversity.

A typical argument concerning zoonoses assumes the just-mentioned special form of the fallacy of correlation. For example:

P1: Deforestation reduces biodiversity.
P2: Deforestation leads to outbreaks of zoonotic diseases in humans.
C: Therefore, reducing biodiversity causes human diseases to "emerge".

Symptoms of the fallacy of correlation do not always present so explicitly. Sometimes they are more implicit – as in one paper by entomologist and disease

---

[47] In the Latin catalog of fallacies, this is *cum hoc ergo propter hoc*, meaning "with this, therefore on account of this". This fallacy is the twin of *post hoc ergo propter hoc*, which infers causation from mere sequence and which I do not take up in my discussion of fallacies.

ecologist Matthew Thomas and his colleagues (2009) and in another by tropical disease expert David Molyneux and his (2008). In these papers, the fallacious reasoning arises by way of an insistence on characterizing the epidemiological discussion in terms of the *diversity* of pathogens, the diversity of vectors, and the diversity of hosts – instead of saying what zoonoses are really about – not the diversity, but the complex dynamics of interacting pathogens, vectors, and hosts. The opening paragraph of Thomas et al.'s paper (Thomas et al. 2009, 229) makes clear the authors' resolve to use the 'b-'word, no matter what:

> Human activity is degrading biodiversity in many ecosystems across the earth… We consider biodiversity within an infectious disease system at a range of levels, including pathogen/parasite diversity, host species richness and species composition, and habitat and ecosystem diversity… The links back to human health implications are sometimes direct (e.g., some human infections), sometimes indirect (e.g., changes in biodiversity affecting diseases of crops)…

It quickly becomes clear that by "degraded biodiversity" these authors really mean "changes in the vector or host populations that result in more human sickness". This makes the correlate (P2) nothing more than a restatement of the damning premise (P1):

P1: Human activity changes the dynamic balance of interacting hosts, vectors and pathogens so as to make people sick.
P2: Human activity degrades biodiversity.
C: Therefore, degrading biodiversity makes people sick.

From the point of view of the scientific enterprise, something even more troubling than fallacious reasoning is going on here. Hosts, pathogens and parasites, and vectors are, as Thomas et al. (2009, 230) say, "interacting biodiversity elements". In fact, according to them (Thomas et al. 2009, 229), "Infectious diseases [themselves] are [a]… component of biodiversity". This "biodiversity"-infused Newspeak is more than just an open door to fallacious reasoning. It is more than just unhelpful. Rather, it obfuscates some terribly complex causal interactions and dynamics that do not fall neatly – or even sloppily – into the pattern required by claims about the (literally) ill effects of "degrading biodiversity". The crux of the real science therefore tends to be swept into the background.

To the great credit of Thomas et al., their paper does present much of the real science of zoonoses, despite the cognitive dissonance of their running normative commentary about how bad "degrading biodiversity" is for human health. Some snippets of science even come through relatively unalloyed with biodiversity norms – as when they present other scientists' work (Thomas et al. 2009, 230), under the heading "Parasite and Pathogen Diversity":

> … Tang et al. … demonstrated that preinfection with infectious hematopoietic and hypodermal necrosis virus (IHHMV) reduced viral load and mortality of shrimps subsequently exposed to white spot syndrome virus. Similarly, preexposure of rainbow trout to a nonpathogenic virus has been shown to reduce impact of a pathogenic virus… The authors suggested that nonspecific or specific humoral factors (host-derived immune response) may play a role in protection, or that one virus is able to reduce the ability of the second to infect host cells.

But they still insist on reframing this refreshing bit of science in terms of "pathogen diversity". This kind of irrelevance moves in the direction of a Chewbacca defense.

Given all this, it is surprising that the authors finally appear to acknowledge the irrelevance of biodiversity to their discussion – although they do this by way of an awkward fallacy of equivocation involving a slide in the meaning of the term "biodiversity":

> … changes in the various components of biodiversity may be even more influential to disease incidence than those in the absolute amount of biodiversity, either up or down, in a given ecosystem.

If it is stipulated that infectious diseases are "a component of biodiversity", then this statement is trivially true. But it is probably better understood as conflating "diversity" with "the particular, diverse entities in an ecosystem that can change " – an equivocation that I discuss at length in Chap. 4 (What biodiversity is not). This makes sense of the statement as saying that, when an ecosystem changes, the changed interactions of the organisms it contains might make people sicker; or the changed interactions might make them healthier. Of course, this proposition has little to do with biodiversity, properly understood. It has nothing to do with any general conclusion concerning the bad effects of "damaging biodiversity".

Sometimes the term "biodiversity" in zoonotic fallacies of correlation is intended to refer, not to the diversity of various functional groups of species (hosts, pathogens and parasites, and vectors), but to the diversity of habitats. In this version, the term "deforestation" in the example presented above is replaced by a more encompassing notion of "habitat destruction". According to this interpretation, one is supposed to think of "habitat destruction" as reducing the diversity of the entire collection of *habitats*. But this is a dubious premise, as the different rubric of "habitat conversion" makes clear: When one habit is destroyed, it is replaced by another, which might well add to the total diversity of habitats. A preference that one might have for having more instances of the one just replaced is not relevant to the question of diversity. For example, farms might be said to differ from each other and from patches of forest they replace in a way that different patches of forest do not.

This section is mainly concerned with forms of argument rather the truth value of premises. So I put the latter consideration aside in this case to observe that the switch from species diversity to habitat diversity does not rid the argument of its fallacious reasoning. Even if the conversion of forest to farmland is correlated with a higher incidence of disease, causation cannot legitimately be inferred. The cause could (and more likely is) a matter of closer contact of people (the farmers) and their minimally diverse farm animals and crops with their new, forest-dwelling neighbors. According to Molyneux et al. (2008, 302–303), exactly this sort of dynamic is the most likely cause of Nipah virus outbreaks in Malaysia, where the virus jumped from *Pteropus* spp. – *P. vampyrus* (Malayan flying fox) and *P. hypomelanus* (variable flying fox) – to pigs and then to people. The Pteropid fruit bats, whose diet comprises figs and other fruit, sought fruit in the commercial orchards of newly arrived pig farmers.[48]

Concerning the truth value of the conclusion – certainly, the destruction of certain habitats causes some human diseases to subside, as when marshes are drained to rid an area of disease-carrying mosquitoes. This is a case of altering a habitat to eliminate a disease vector. Similar considerations even more directly apply to eliminating host reservoirs of zoonotic pathogens. That is precisely the thinking of villagers in areas of the Congo already touched by outbreaks of Ebola, and of villagers in still-untouched areas of Cameroon who are aware of the risk of such outbreaks in Gabon to their south. To them, the prophylactic slaughter of the disease-carrying Hominidae *Pan paniscus* (bonobo), *P. troglodytes* (common chimpanzee), and *Gorilla gorilla* (western gorilla) – makes a lot of sense (Froment 2009, 222).

Finally, to give a sense of how widespread is the fallacy of correlation, here are a few more arguments that instantiate it in the special form on which we have focused:

P1: Dams can diminish biodiversity.
P2: Dams can cause downstream pollution.
C: Therefore, the reduction of biodiversity causes downstream pollution (Meyerson et al. 2009, 90).

In a very similar vein:

P1: Persistent organic pollutants (POP's) can negatively affect the diversity of aquatic species.
P2: The presence of POP's in food sources can negatively affect human health.
C: Negative effects on species diversity negatively affect human health (Mazmunder 2009, 152).

Also:

P1: Greenhouse gas (GHG) emissions "harm" biodiversity.
P2: GHG emissions are cooking our crops.
C: "Harming" biodiversity harms our crops.

Lastly, a variation involving a slightly more complex causal chain:

P1: Increased levels of nitrogen can reduce plant diversity in a marsh.
P2: The (reduced) set of species can shift towards plants preferred by mosquito vectors of various diseases.
P3: When this happens, the incidence of these diseases increases.
C: Changes in biodiversity are harmful (because they cause disease) (Townsend et al. 2009, 169).

---

[48] Specifically, Nipah-infected fruit bats dropped partially eaten fruit and feces, which sufficed to infect the pigs, which in turn, infected people who came in contact with the pigs' feces and nasal mucous secretions, aerosolized by the pigs' coughing.

I should note that these are distillations of arguments that never occur with the clarity in which I present them here. There is almost always a significant admixture of elements of a Chewbacca's defense and proof-by-verbosity, which obscures everything, including logical defects.

2.2.5   Circularity Fallacies or Begging the Question

Circularity in arguments consists in doubling back to a premise, which is then asserted as conclusion.[49] This form of fallacy can be completely unobvious when the argument follows a long and tortuous path, or if the conclusion is stated in a way that obscures the fact that it is equivalent to a premise. But some arguments are circular in the tightest possible way – by building the conclusion directly into a definition. Specifically, if the "good" of biodiversity is built into its definition, then the conclusion that it is, indeed a good, is clinched – albeit in the most uninteresting and unpersuasive way.

Here I draw examples of both types of circularity from later in the book. In Sect. 6.7 (Biodiversity as value generator), I explore the claim that biodiversity is valuable as a fountainhead of things valuable. But when one follows the thread of the argument from biodiversity as source to the value that it generates, what one encounters is nothing other than biodiversity itself – again.

The tighter, definitional type of circularity is encountered in such definitions of biodiversity as Sarkar's operational definition (mentioned at the beginning of this chapter), which builds the value of biodiversity right into itself as "those properties that we *should* conserve via application of a certain algorithm".

2.2.6   The Fallacy of Modality or Speculation Posed as Fact

A fallacy of modality illicitly slides from the logical modality of possibility to an assertion of contingent fact. An argument for the proposition A takes the form:

P: It is possible that A
C: Therefore, A

Some caution is in order in diagnosing this fallacy. A scientist who responsibly expresses customary scientific caution about the possible limits of the conclusions that her research supports does not thereby commit it. But a scientist who lets speculative explanation push aside the actually supported conclusion, or who, less subtly, allows "is" and "will be" supplant "conceivably might be" and "could be" in sweeping generalizations, does.

---

[49] In the Latin catalog of fallacies, this "*petitio principii*" – literally, "request for the premise". This conveys the sense of "the proposition taken to be proved is assumed" – in other words, circularity.

Entomologist Matthew Thomas and his colleagues (2009, 234), for example, appear to be self-consciously inferring *is* from *could* when they say:

> Many zoonotic reservoir species are rodents…, whose dynamics can be strongly affected by predators. While these dynamic effects have *rarely* been linked to increases in zoonotic disease, *in general terms*, if rodents are maintained at low density away from humans, disease transmission *will be* reduced. [italics added]

The added italics serve to emphasize the particularly dissonant special form that the fallacy assumes in this argument:

P1: It is possible that A
P2: A is almost never true
C: Therefore, generally A

This reasoning might also be taken as an instance of the fallacy of converse accident. And when the discussion of Thomas et al. then devolves into a discussion of how grouse populations fluctuate – more or less depending on the greater or lesser presence of predators – it begins to resemble a Chewbacca defense.

2.2.7 The Fallacy of Equivocation

A fallacy of equivocation arises when the meaning of a term in the conclusion differs from its meaning in a premise. I have already remarked on how much poor reasoning about biodiversity, which can be viewed as fallacious in some other way, can also be viewed as hinging on slides in the meaning of that term. This is unsurprising, given the dearth of attention to careful and consistent definition of the concept.

While biological diversity must have something to do with the diversity of such biological entities as species, the slide from this meaning to "a particular species" is so pervasive that it is even built into phrases such as "biodiversity-derived pharmaceuticals" (Cox 2009, 274). So far as I know, not a single pharmaceutical has been derived from biodiversity, though some have come from one or another from a small set of specific organisms. One might suppose that, taken as a group, the organisms that have yielded beneficial drugs are diverse.[50] But that proposition is rather different from the claim that biodiversity in general is needed to deliver the drug you need to cure what ails you.

2.3 Cautionary Signs

Even before setting out on the biodiversity road to natural value, there are reasons to expect significant trouble along the way. One reason has to do with biodiversity's abstraction. The concept is perched atop several layers of collective abstraction; and

---

[50] Though, that supposition happens to be false, as I show in Sect. 6.5.1 (Biodiversity as pharmacopoeia).

it is couched in terms of properties that are several levels removed from what most palpably and forcefully registers in a human life, such as the individual plants and animals that might figure in our life's experience. Another cautionary lesson derives from obstacles that have confronted attempts to marshal the value of diversity in domains other than biology.

These warning signs in themselves are not reasons to think that biodiversity is not valuable. But I think that they haunt many attempts to establish biodiversity's value. The difficulties that they evidence push those attempts towards the set of views that coalesce into what Chap. 8 calls "the biodiversity project". As I argue there, that project is saddled with the apparently unsolvable problem of finding a coherent and credible set of generally endorsable principles that underwrite its claims that certain biodiverse states of the world are better than others. But I am getting well ahead of myself.

2.3.1 Abstraction

The task of finding good reasons to afford moral consideration to individual nonhuman organisms is already a philosophically challenging one. Attempts at it have already generated a long and robust debate, by the standards of the merely decades-old field of environmental ethics (in contrast to the millennia-old field of general moral philosophy). In this vein, one will recall from Sect. 2.1.2.2 (Deontology) Paul Taylor's well-known "egalitarian individualist" position that each and every organism is a "teleological center of life". According to him (Taylor 1981, 210; 1989, 119), each individual organism's pursuit of its own good in its own way entitles it to the same kind moral respect (on this basis) as every other.

Other individualists are not so generous in spreading the mantle of moral considerability. Most famously, as Sect. 2.1.2.1 (Consequentialism) notes, the utilitarian Peter Singer draws the line at sentience. More accurately, he (Singer 1979, 194–196) draws a sliding scale for meriting moral respect based on the degree to which a sentient creature has an interest in experiencing pleasure and avoiding pain. If you are a creature at the far edge of the sentience scale – lacking all but the most remote semblance of my well-developed nervous system – then you are not much above the chair on which I sit, morally speaking. On that principle, the vast majority of individuals of the vast majority of species – arthropods and microbes of various sorts – are effectively excluded from the morally respectable "in" crowd.

It also bears noting that mainstream ethicists remain nervous about *any* puncture in the moral respect barrier between humans and nonhumans, for fear that even a chink will propagate in fissures that ultimately bring down the entire wall. They are guided by a gnawing fear of making a moral issue out of where a person treads when she walks down a path or out of building even the simplest abode. More generally, the specter seems to lurk of becoming embroiled in the question of how and to what degree we might be obligated to trade off the good of a person for the good of a mouse, the good of a mouse for that of a stalk of grass, and the good of all these kinds of individuals for the good of the vast majority of organisms, comprising arthropods and bacteria. Perhaps its exclusion of the preponderance of organisms as objects of primary moral regard is part of the attraction of Singer's sentiencism.

As discussed in Sect. 2.1.1 on Concepts and categories of value, other environmental ethicists are reluctant to suppose that the primary locus of moral respect for the natural world lies in individual organisms of any kind. The heart of the value matter, they think, might be in "collectives" such as ecosystems or, more relevant for most conceptions of biodiversity, species. For those who would invest moral value in species – as that particular sort of collective of conspecific individuals through time – the sledding is much tougher than for those content to focus on the individual members (of at least certain species), for this is an abstraction of at least the first order. As we noted in the earlier subsection, a species lacks the sort of weighty anchor that might plausibly hold fast the moral value of an individual organism.

Expanding on the earlier discussion reveals some considerations that loom large throughout this book. Certainly, a species is not sentient, even if its members are. Nor does a species appear to have interests, even if its members do. As the political philosopher Joel Feinberg (1974, 55–56) remarks,

> A whole collection, as such, cannot have beliefs, expectations, wants, or desires… Individual elephants can have interests, but the species elephants cannot.

This is not exactly an argument. But the oddness of attributing interests to collectives does seem to place the burden on those who would argue *for* the claim that, at least in this crucial respect, a species is a kind of individual. Holmes Rolston, whose thinking the Sect. 2.1.2.2 on Deontology has already visited, does as well as anyone, I think. He starts by arguing (Rolston 1994, 21) *against* the

> …claim that species-level phenomena (vitality in a population, danger to a species, reproduction of a life form, tracking a changing environment) are only epiphenomena, byproducts of aggregated individuals in their interrelationships.

According to him (Rolston 1994, 20),

> A species is another level of biological identity reasserted genetically over time. Identity need not attach solely to the centred or modular organism; it can persist as a discrete pattern over time.

Finally, Rolston (1994, 21) concludes that

> The species line is the vital living system, the whole, of which individual organisms are the essential parts. The species defends a particular form of life, pursuing a pathway through the world, resisting death (extinction), by regeneration maintaining a normative identity over time.

Ultimately, the ontological flattening of the collective species into an honorary individual hinges on metaphors and analogies rooted in what it is to be an individual. Metaphors and analogies are legitimate tools of persuasion. But their application to this case seems not so much an actual argument – that species, despite being a collective, are morally on a par with individuals in some critical respects – as a restatement of what such an argument must accomplish.

Specifically, such an argument must show how these analogies and metaphors have sufficient binding power – relative to contravening disanalogies – to justify the moral (and ontological) flattening.

Unfortunately, Rolston's analogies do not have the requisite power. The sense in which a species is biologically identifiable as "the same entity" over time is far more fuzzy than the way in which an individual organism is identifiable as "that same entity", tracing out its life's trajectory. Even the concept of "species" is very fuzzy. There are at least a dozen definitions[51]; none apply universally; and all are "leaky" in the sense that there is significant uncertainty about their boundaries. This is true even, or especially – in the context of identifying a species as "the same entity over time" – for chronospecies, for which the boundary is in equal measures fuzzy and arbitrary.

No more promising is the metaphor that reimagines a species as itself an organism, whose working parts – its vital organs or appendages, so to speak – are the more familiar individual plants and animals that we observe in the world. The metaphor works at one level precisely because the organs of real individual organisms have no interest of their own, but nonetheless, metaphorically speaking, "conspire" to sustain their individual owner. But at another level, it works directly against Rolston's purpose because his metaphorical organs – the real individual organisms that sink their roots in the soil or scurry about – are quite unlike real organs. The basis for this radical dissimilarity is that organisms, unlike organs, apparently do have their own interests. This only serves to remind us that it is precisely in individual organisms where the clearly recognizable interests lie. It reminds us that these interests do not reside at the level of an organism's morphological or cellular parts. Nor do they reside at the level of the collections that organisms forms with others of their kind. With rare possible exceptions (for example, ants and some other social insects) these are the inadvertent side-effect of each individual pursuing its own, non-collective interests – interests that often conflict with the interests of many other individual organisms, including those of its own kind.

Despite Rolston's admirable attempt to argue to the contrary (and I have devoted this much space to it because it *is* so admirable), a species does not have a good of its own in any but a sense derived by fragile analogy from that of individuals, or in a sense that isn't in some other way largely or completely dependent on the good of

---

[51] The concept of species that biologists most frequently use and cite is the "biological species" concept. Wilson (1992, 35–50) provides an accessible discussion of it and some of its difficulties. The biological species concept relies on sexual reproduction combined with a tricky – because often unverified and even unverifiable – condition having to do with the potential ability to mate and produce viable offspring should the opportunity arise. The limitations of this concept – even with sexually reproducing organisms has motivated a raft of other concepts, including the concepts of phylogenetic species (sharing an ancestor), chronospecies (an infusion of phenotype similarity requirements into another species concept), genetic species (a requirement for genomic similarity, which figures prominently in the definition for bacteria), and phenotypical species (the much-derogated but handy concept of "if it looks like a duck and quacks like a duck, then it is a duck"), among others. No two of these concepts defines the boundaries between species in the same way.

its individual members or their individual interests. Most often, a species is thought to be "doing well" if it manages to sustain some "viable", reproducing population of its members through time. But in and by itself, that does not appear to command any moral respect. Over that time, as a species changes in various characteristics – including the distribution of traits among its members and the size and distribution of its populations – it is anything but clear whether any of these changes are "good" or "bad" for it, so long as the species continues its run. And what are we to make of a species evolving into a new chronospecies? At the point at which the ancestral species is declared to have completed its course, is it thereby harmed?[52]

More difficulties lurk behind modest reflection on what a species is. A species is a "collective" because it is indeed a collection – of individuals that bear certain biological relationships to each other. The relational focus of the modern biological species concept is on an ability to exchange genetic material in order to produce offspring with a like capability.[53] For some organisms – especially asexual haploids such as prokaryotes – this definition does not apply; so other relational properties are required. These, in turn, tend to "leak" back into a more generalized version of the biological species concept. The additional properties typically revolve around a shared phylogeny (they are monophyletic), a creative patchwork of genomic similarities, and some identifiable loci of phenotypical similarity – in some artfully balanced (but as in all art, ill-defined) combination. What could possibly give special moral status to this fuzzy patchwork convention?

A collection of individuals lacks most (though not necessarily all) of the properties attributable to its component individuals. As I have just noted, in the case of species, those missing properties include ones (such as "having a good of one's own" or "capable of experiencing pleasure and pain") that seem to be the most morally relevant. However, the reverse is also true. A collective has other properties that do not apply to any of the individuals in the collection. Might these offer a handle capable of carrying moral weight?

Taking a look at this, one finds properties such as a species' population; its longevity (which can depend on how one defines chronospecies); its trophic relationships – what it likes to eat and what likes to eat it; some range of phenotypes that are "similar enough" – including traits of morphology, biochemistry, physiology, behavior, and development cycles; its suitability to live in certain environments; and its

---

[52] On one account of humankind's prehistory, *Homo erectus* eventually became *Homo sapiens*, at which point the species *H. erectus* ceased to be. One should be reluctant to refer to this extended event as "the demise of *H. erectus*". However, its fuzzy point in time marks the end of the earlier species' lifetime.

[53] I ignore the fact that the usual biological species concept falls apart for perhaps the majority of organisms on the planet, which prefer agamogenesis. The hordes of organisms that spurn sex include the single-celled archaea, bacteria, and protists, as well as a large number of plants (for example, liverworts) and fungi. The prokaryotic bacteria are equally disruptive of the biological species concept from the other, gene-exchanging side of it. They routinely and promiscuously exchange genes with relative disregard for the exchanging partners' genetic similarity or dissimilarity.

I touch on these difficulties again in the context of discussing the species diversity part of biodiversity.

resilience and ability to adapt, including its ability to transform itself or spin off another species. But it is hard to see in any of these properties some obvious grounding for moral respect.

The case for moral consideration of species is further vexed by the observation that the good of a species (whether or not that good has any moral weight) is often at odds with the good of its members. The conflict is built right into the "need" for a species to avoid going extinct by adapting to changing conditions. The evolutionary process of adaptation requires that those individuals that are dealt genes making them less fit to deal with new conditions succumb before they have a chance to pass them on. Insofar as a species has an interest in continuing, it is in its interest that these individuals die an early death.

The death of individuals often serves the "interest" of their species in other important ways. It might be good for a population that an individual weakened by infection die before the infection spreads to others. It is possible for populations to explode – a phenomenon often followed by a population-threatening, if not species-threatening collapse, which can be avoided by having a sufficiently high and steady death rate to avoid the initial explosion. Put less abstractly, individuals must die off at a good and steady pace to ensure the stability and longevity of a population. Consequently, to the extent that individual organisms are the locus of moral worth, the value of species can be (and has been) called into question, and conversely. At the very least, there are some difficult-to-resolve conflicts between the two.

Gaia theorists attempt to banish these difficulties concerning the moral differences between individuals and the collectives to which they belong. Their suggestion is to banish the ontological distinction between them. Banishing that distinction concomitantly banishes any significant distinction in moral standing. From the viewpoint of Gaia theory, as its originator James Lovelock would express it, the entire planet is a single, discrete, living individual – a superorganism with its own ends.[54] Its component parts – such as the various species – are likewise living super-individuals. Erasing the boundary between real, discrete individuals and their environment removes the barrier to applying individual morality to collectives such as species.

Unfortunately, this seeming gain comes at the great cost of having to defend a neo-Platonic ontology that makes individuals out of kinds (substantial universals) and other sorts of collectives – at least so far as ethical standing is concerned. This ontological and metaethical flattening of the visible and moral world is singularly unappealing to those of us for whom the differences between an individual organism and its species are visible, interesting, and even critical for a nuanced understanding of moral relationships.

---

[54] In fact, Lovelock weaves back and forth across the teleological line. At times and especially in response to criticism, he seems to deny attributing purposive behavior to the Gaian superorganism. But to do this, he falls back to some definition of "life" that makes it uncertain why we should care about such a life, except for purely prudential reasons.

I can think of one route to the interests or rights of a species (as a group) that, to my knowledge, has gone largely or perhaps entirely unexplored. That is the route that, *pace* Gaia theory, takes seriously the fact that a species *is* a group, and that leverages some theory of "group rights". A theory of group rights must include an account of what kinds of groups qualify as rights holders, what considerations ground those rights, and exactly how and by whom those rights are held: One might posit that the group itself is the holder of rights; or alternatively, that real, discrete individuals hold them, albeit as (and only as) members of the group.

The influential political philosopher Joseph Raz has a theory of the latter kind, wherein group rights are held collectively by the group's individual members, in their identity as such.[55] That is, Raz grounds group rights as closely as one might think possible to the interests of individuals. In this respect, he departs from the more traditional view, according to which group rights are somehow held by the group as a unitary entity (which need not require that a group be regarded as an individual or superindividual). But independent of one's view of who holds group rights and how they are held (if at all), a theory of group rights must supply some qualifying condition(s) for ascribing rights to a group. Mere existence or survival of the group does not suffice, since we all can think of groups whose disappearance would leave all (humans and nonhumans alike) better off.

As I said, this area appears to be undeveloped and ripe for development in environmental ethics. Before leaving it, I should observe that, even if a theory of group rights could justify ascribing rights to a species, this would fall far short of ascribing rights or any other kind of moral standing to the *diversity* of species.

Much more can and has been said about the moral status of species. The purpose of this brief survey is served if it conveys the accurate impression of how extraordinarily tricky it is to gauge this status. This, I think, is cause to anticipate even greater difficulties in coming to terms with biodiversity's value.

In Chap. 3, I propose that biodiversity is a kind of "super-collective". Insofar as biodiversity is species diversity, it is essentially the collection of the collections known as "species". But even if it were established that each and every species is valuable as a species, it is not clear that this would have any bearing on the question of whether or not any particular collection of species – or species diversity – is valuable as a collection. It is a fallacy of composition to infer that a collection has a property because its individual members have it. In the context of axiology, this formal hazard cautions against concluding something about the good of a collection of entities from the good of each entity in the collection.

On the other hand, insofar as the difficulty of finding credible reasons for why a (or any) species might have moral standing has to do with its being a collective, one should expect an amplification of this difficulty with the biodiversity's "squaring" of this attribute. No fallacy of division is involved in arriving at this

---

[55] This brief description cannot come close to conveying the sophistication of Raz's views (Raz 1986).

observation, for in this case, the relevant attribute – "is a collective" – is shared by both part and whole.[56]

Avoiding the fallacy of composition in the context of axiology requires caution in concluding something about the good of a collection of entities from the good of each entity in the collection. In the current context, this means that even if there were reasons for regarding species as morally considerable after all, this would not suggest much at all about whether or not species diversity had similar moral status. On the other hand, though it might seem odd at first, starting from the premise that any particular species is morally *dis*regardable as a species might well *facilitate* achieving the most valuable biodiverse state of the world. Whether or not this premise is defensible, it seems to be more and more frequently adopted by those proposing conservations plans, which assert that the best biodiverse mix can be achieved by extirpations and replacements at the species level.

This suggests the possibility of conflicts between values at the two collective levels of species and species diversity. This situation mirrors the more widely discussed possibility of conflict between a species and its individuals – as when culling of the weak individuals serves the continuation of the species. One should expect this potential for conflict to modulate any stance on the value of the conflicting entities, and in the latter case, on the value of biodiversity.

Sometimes, of course, the fallacy of composition be damned and one can see how the value of a collection as a collection does, in fact, at least partly depend on each of the collection's elements having their own separate, individual value. That might not be true of the paintings composed of van Gogh's individual brushstrokes but it might be true of stamp collections and even a collection of van Gogh's paintings. So it is not entirely unreasonable to think that the value of biodiversity might work this way, too – or at least to think that this is one ingredient in the value mix. That is, it is not unreasonable to suppose that biodiversity's value derives from the value of each and every component species, considered as such.

And in fact, much of what is written about biodiversity suggests that this is a large part of what many biodiversity advocates have in mind. This is especially evident in the kind of additive calculus that biodiversity advocates often employ in decrying reductions in biodiversity (and discussed at length in Chap. 5, The calculus of biodiversity value). It is also true to the extent that advocates regularly present a case for the value of a particular species, but mistakenly label this as an argument for biodiversity. I have more to say about this later, in Sect. 4.1.3 (Particular species).

---

[56] Of course, in making the claim that a human brain is capable of consciousness, one need not assume the burden of defending the position that each of the brain's neural cells is likewise conscious. The height of a brick wall does not determine (except as an upper bound) the height of any one or more of its component bricks. Even in the realm of axiology, it is clear that a collection of entities can be valuable despite the relative worthlessness of each individual component. Considered in isolation, no single brushstroke of van Gogh's *Starry Night over the Rhone* has anything like the aesthetic value of the painting. However, none of these are examples of collections of collections.

2.3.2   The Value of Diversity in General

Another major set of considerations, concerning the "diversity" part of "biodiversity", presage difficulties in justifying its value. One might suppose that the value of biological diversity is bound up, not so much with the specific and sometimes epistemically challenging domain of biology as with the property of diversity, variety, or heterogeneity. The property of being diverse applies to a great number of domains – not just the "biotic or biota-encompassing" ones (according to the definition developed in Chap. 3) relevant to biodiversity.

This suggests that it might be worthwhile to glance at how diversity or variety affects value in other domains – as a clue for what domain-spanning difficulties might arise in examining this question with respect to biodiversity.[57] At the same time, one might think that this exercise offers promise of shedding light on the dual question of whether or not variety might be generally valuable – again, not just within some one particular domain or other, but across multiple or even all domains. The very challenging nature of this topic in axiology has not deterred some great thinkers from mulling it over in these most general terms.

Among the most famous is Gottfried Leibniz. He is perhaps best known for his thesis that we live in the best of "an infinitude of possible worlds".[58] For Leibniz, "best" first means a kind of metaphysical perfection or, as he would say, "degree of reality". That, in turn, is bundled up with "physical" and "moral" perfection. Leibniz presumes a perfect God who is the author of all that exists. His thesis that "ours is the best of all possible worlds" is the cornerstone of Leibniz's attempt to dispel an apparent paradox: How could such a perfect God author a world that contains horrible suffering and injustice – a world that, by all reasonable human accounts, is considerably less than perfect in many respects?

In pressing his case, Leibniz is handicapped by his reticence on the subject of what constitutes a possible world. Fortunately and important for the current discussion, he is not quite so reticent on criteria according to which a possible world or the actually existing one might be judged good or best. He provides at least one important criterion. As you might have guessed, it has to do with variety (Leibniz 1714 §10):

> God is supremely perfect, from which it follows that in producing the universe he chose the best possible design – a design in which there was
>
> • the greatest variety along with the greatest order,
> • …

---

[57] I am indebted to Neil Manson for suggesting this as a worthy topic of reflection. In fact, I think it worthy of a more extended treatment than space allows here.

[58] Leibniz (1710), §8:

> …there is an infinitude of possible worlds among which God must needs have chosen the best, since he does nothing without acting in accordance with supreme reason.

Perhaps even more famous than the thesis itself is the merciless mocking it receives in Voltaire's *Candide*, by way of the inveterately optimistic personage of Dr. Pangloss.

This passage haunts some hallowed philosophical ground, which was perhaps most elegantly cultivated by Thomas Aquinas. That earlier philosopher and theologian finds fault with "equality" and a lack of distinctness because, according to him, distinct things are required to occupy distinct levels of goodness. But also, for there to be order in the universe, he maintains (Aquinas 1264, Book III, par. LXXI) there must be something distinct and non-equal to order:

> … perfect goodness would not be found in created things unless there were an order of goodness in them, in the sense that some of them are better than others. Otherwise, all possible grades of goodness would not be realized, nor would any creature be like God by virtue of holding a higher place than another. The highest beauty would be taken away from things, too, if the order of distinct and unequal things were removed. And what is more, multiplicity would be taken away from things if inequality of goodness were removed, since through the differences by which things are distinguished from each other one thing stands out as better than another; for instance, the animate in relation to the inanimate, and the rational in regard to the irrational. And so, if complete equality were present in things, there would be but one created good, which clearly disparages the perfection of the [creation].

We can imagine that Leibniz has this kind of deep-rooted metaphysical principle in mind when he says (Leibniz 1710, §118) that

> It is certain that God sets greater store by a man than a lion; nevertheless it can hardly be said with certainty that God prefers a single man in all respects to the whole of lion-kind.

Later, in a letter to his protégé, Christian Wolff, Leibniz writes (Leibniz 1715, 233–234):

> Perfection is the harmony of things, or the state where everything is worthy of being observed, that is, the state of agreement [consensus] or identity in variety; …

To make sense of this, one must assume that the "identity", as Leibniz uses it in the last passage, is not the self-sameness of particulars. Rather, it must mean something along the lines of "ordered or explained by the same general laws".

In sum, according to Leibniz, the good of the world is its perfection. And perfection, he stipulates, consists of lots of different things – in fact, as many different things as possible – behaving and interacting according to the simplest and most general laws. This is harmony. Without variety, no harmony – *ergo* no perfection – would be evident. In fact, harmony (and perfection) would not even be possible. For a state of affairs comprising one or just a handful of things easily conforms to any number of general patterns. Such a state of affairs does not thereby require or demonstrate any impressive degree of harmony.

Leibniz's very general value proposition is worthy of attention because it takes a general form of strikingly similar, though more specialized, claims made on behalf of biodiversity. Replace the "harmony of the world" with the "harmony of ecosystems" – or their "health" or "integrity" or "stability" or "resilience". Replace "the variety of things" with the "diversity of species, diversity of genotypes, diversity of habitats, and all or any other forms of biological diversity". What you get is a perfectionist theory of the biological world. In truth, such a theory seems to be the implicit foundation of much of the scientific and environmentalist literature on biodiversity, as comes out in the course of this book.

Would that we could infer the value of biodiversity as a special case of the perfection of the world considered *in toto*. Unfortunately, there are stubborn difficulties with Leibniz's metaphysical thesis about the role of a grand, all-encompassing diversity in harmony and perfection. The problems do not stem from the theological motivation for Leibniz's metaphysics. Those can be detached and ignored.[59] But even after extirpating any controversial theology, what remains is quite arbitrary and quite vulnerable to challenge. Why is perfection, conceived as harmonious diversity, so good? If this is *not* the best of all possible worlds and the diversity in it, although generous, is not very harmonious, what reason is there to think that this state is better than a state of lesser but more harmonious diversity?

Antithetical maxims that nevertheless have equal allure provide a perspective on how far out of reach is an answer to that question. Writing about the design of the airplane in a way that could just as well be about the design of nature, Antoine de Saint-Exupéry (1939, 60, Chapter III, "L'Avion") tells us that:

> It seems that perfection is attained not when no more can be added, but when no more can be removed. At the end of its evolution, the machine escapes notice.[60]

Saint-Exupéry here suggests that the essence of perfection in design is an utter simplicity – a simplicity that makes the inner workings of that which is designed transparent to the casual onlooker. The messy chaos of diversity – superfluous diversity, at that – is the enemy of perfection, so conceived. This reflection offers up the maxim: Lop out any variety that does not contribute to the end.

The point of juxtaposing these opposing views is not to try to adjudicate the debate. Rather, it is to ask: What would even constitute a good argument for de Saint-Exupéry's position versus that of Aquinas or of Leibniz?

Despite the difficulties in addressing such challenges, the case that Leibniz and other philosophers press – for some positive value attaching to diversity in general – has a comfortable familiarity. To some extent, this might be due to the power of a common misinterpretation of two lines from the pen of the 18th century English poet William Cowper, which famously urge (Bailey 1905, 279, "The Time-Piece", ll. 606–607) that

> Variety's the very spice of life,
> That gives it all its flavour…

This oft-quoted pair of lines is rarely put in the context of the poem in which they appear. In truth, Cowper is railing against *frivolous* novelty in fashion and how its

---

[59] One can also ignore Leibniz's theory of the good for humans, which hinges on a certain knowledge or apprehension of the universe's harmony.

[60] The translation is my own – more literal than the standard but overly imaginative English re-creation by Lewis Galantière. The original French text is:

> Il semble que la perfection soit atteinte non quand il n'y a plus rien à ajouter, mais quand il n'y a plus rien à retrancher. Au terme de son évolution, la machine se dissimule.

pursuit is the ruination of English households. Just a few lines down, he (Bailey 1905, "The Time-Piece", ll. 610–617) does not mince words:

> And, studious of mutation still, discard
> A real elegance a little used,
> For monstrous novelty and strange disguise.
> We sacrifice to dress till household joys
> And comforts cease. Dress drains our cellar dry,
> And keeps our larder lean; puts out our fires,
> And introduces hunger frost, and woe,
> Where peace and hospitality might reign.

This is hardly a blanket endorsement of variety. To the contrary, Cowper puts his finger on a salient problem with it: There is good variety; but there is also bad variety. To know that a state of affairs or a collection of things is diverse in itself does little or nothing to determine whether it belongs to the good or the bad. This theme figures importantly throughout this book.

I do not wish to claim that there *is* not, let alone *could* not be a cogent argument for Leibniz's general thesis on the value of variety (or even for Cowper's as it is commonly misconstrued). I do wish to claim that the task of finding and legitimating a theory on the value of diversity in general is extraordinarily daunting.

*Bio*diversity is not the diversity of *all* things. So the difficulty of Leibniz's grand project does not necessarily reflect a similar level of difficulty for one confined to the biological world. It is conceivable that something specific to this more narrowly defined and better circumscribed domain could facilitate finding value in the diversity of entities within it. An example might be helpful in getting a sense for how things might go with such a project. Once again, my object is not to "prove" that an argument for the value of diversity in any chosen domain is doomed to fail. Rather, it is the much more modest one of suggesting how difficult it might be to find such an argument.

Outside the domain of biological diversity, social justice is also one that frequently holds diversity up front as a central consideration. Curiously, it does this by devaluing certain kinds of diversity, which are considered frivolous for social justice: The focus is on a social ideal wherein equal consideration of persons is not denied on grounds of certain socially and morally *irrelevant* differences between them. This ideal *removes* from consideration for social or moral standing acknowledged differences between persons. Insofar as it does this, it asks us to see (and value) a core moral sameness in all of humanity – a sameness that demands egalitarian respect, even if not egalitarian treatment.

But this is not the whole story. While social justice devalues the diversity of persons in the sense that certain kinds of differences are on principle removed from the set of morally relevant considerations, it sometimes claims that these same differences are a positive good. This kind of claim is a mainstay in support of affirmative action, where the diverse characteristics are typified by (though not restricted to) race and ethnicity.[61]

---

[61] I deliberately restrict my attention to the argument for affirmative action that bears simply and directly on "justice as fairness". In doing so, I leave aside a complex web of considerations that deal with compensatory justice (righting wrongs) and just desert. These fall outside my limited purpose of teasing out the role of diversity as a good in the affirmative action debate.

The argument from diversity on behalf of the social justice of affirmative action is worthy of attention. I approach this topic through the well-known lens of *Regents of the University of California v. Bakke* – the famous case brought before the 1977 United States Supreme Court concerning Allan Bakke, a white applicant who was twice denied admission into the Medical School of the University of California at Davis, despite having test scores and grades that exceeded those of most non-white applicants admitted through an affirmative action program. The case is a reference because of the brilliance (in the true sense of providing great illumination) of the arguments that Justices William J. Brennan and Lewis F. Powell, Jr. proffered. Powell ultimately wrote the majority opinion for a ruling that ordered Bakke's admission.

Keep in mind that the context for this particular diversity argument is that of a public educational institution. The question (or one of the questions) before the court was whether or not a program of affirmative action in such an institution serves a compelling government interest. The Medical School offered four arguments for denying admission to Bakke. The last was its claim that diversity is a good in such a context. Specifically, the good comprises (Powell 1978, Part IV) "the educational benefits that flow from an ethnically diverse student body."

This is the only argument in which Justice Powell found substantial merit – at least, in principle. However, he found that the Medical School's implementation was flawed. According to Powell (1978, Part V-A),

> The diversity that furthers a compelling state interest encompasses a far broader array of qualifications and characteristics of which racial or ethnic origin is but a single though important element. Petitioner's [i.e., the Medical School's] special admissions program, focused solely on ethnic diversity, would hinder rather than further attainment of genuine diversity.

Powell (1978, Part IV-D) is clear that the state specifically has an interest in facilitating a university's ability "to select those students who will contribute the most to the robust exchange of ideas".

In his opinion, Powell disputes an earlier California Supreme Court ruling, which held that race and ethnicity should be banished completely in deciding whom to admit. According to him, race and ethnicity are legitimate considerations, but only within a broader conception of *diversity* – one in which race and ethnicity join such other considerations as age, work experience, family background, special talents, foreign language fluency, athletic prowess, military service, unusual accomplishments, and aspirations.

Two points merit special attention.[62] First, Powell's conception of "genuine diversity" pries loose the connection of diversity to social justice. Any special or tight binding of race and ethnicity to social justice is all but lost in a flood of other characteristics that are mostly or entirely irrelevant to that particular nexus. Second, the "state interest" at stake – the end which "genuine diversity" is said to serve – is

---

[62] My discussion owes much to that of Anderson (2002) though it does not follow hers in any of its details.

confined to a relatively narrow conception of educational aims – namely, "the robust exchange of ideas". These two points can be seen as representing differing but complementary characterizations of the same broad disjunction between diversity and its role in education on the one hand, and the role of education in serving social justice on the other.

This disjunction inevitably leads to the conclusion that the components (in the sense of "categories") of diversity (involving race and ethnicity) that legitimately concern social justice advocates cannot support the weight of the proposition that diversity more generally – Powell's "genuine diversity" – is a means to the end of social justice. Some components of diversity are good from a social justice viewpoint. Others have little or no bearing on it. These "other components of diversity" are not necessarily unwelcome. But their dominating presence make it nearly impossible to think that human diversity, as such, and ranging over all human characteristics, is a cornerstone of a socially just educational institution.[63]

To put this in a slightly different way: The failure to find a positive connection between human diversity and social justice in the context of public education is consequent on the facts that diversity itself is socially neutral and that components of diversity that have little or no bearing on justice greatly dilute those that might have some bearing. This result is quite striking. But the analysis need not stop there. Consideration of other components of diversity – ones not taken up in *Regents of the University of California v. Bakke* – can take diversity-as-an-educational-good from non-positive to strictly negative territory.

Most obviously, universities do not seek the widest possible diversity of academic competence because this type of diversity is generally not desirable. It is educationally counterproductive to have students who cannot hope to interact effectively with the material and with other students because their capabilities fall short. Yet the presence of such students undeniably would make the student body more diverse. Similar considerations apply to the diversity of characteristics that disposes potential students to behave in ways that span the spectrum from the socially constructive to the socially disruptive. No one thinks that universities should take pains to diversify their student bodies by including students whose presence merely adds to the educational difficulties of most students when this inclusion contributes neither to this centrally important education end nor to any other respectable end.[64]

In the context of educational institutions, some components of diversity serve the purposes of social justice and consequently might be good on those grounds. Other diversity is morally neutral. Yet other diversity is justifiably eschewed. The value of the diversity of students in an educational institution, or more generally, the diversity of persons or entities of any kind in any socially valuable institution might not

---

[63] Anderson (2002, 1220–1222, and elsewhere in her paper) presents a similarly unenthusiastic view of social justice arguments for diversity. She goes on to rework them in a broader and, according to her, more promising "integrative" framework.

[64] Exceptions for important or respectable ends are required to override such difficulties as those that stem from the discomfiture of white students at the admission of qualified non-whites into their school.

descend all the way to the nadir of Cowper's frivolous variety. But there clearly are different ways in which persons can vary. These ways are equivalent to the "components" or categories of biological diversity that I take up in the next chapter. With respect to a social institution, some categories of diversity are appropriate and beneficial for its working; others not. It is hard not to conclude that diversity, in itself, does not confer any special evaluative standing – moral or otherwise – on a diverse state of such an institution.

Neither this nor the preceding subsection presumes to argue that clearly defined principles of axiology automatically and without further discussion deflate any possible case for the value of the diversity, in any possible domain. Nor does this discussion pretend to make the weaker version of this claim that restricts itself to the domain of biological diversity. However, the details of the arguments for the value of biodiversity tend to get bound up in details that are peculiar to the biological domain. These can quickly overwhelm any more general perspective. This short discursion into diversity outside the domain of biology is a way of stepping outside and above these details to get a broader sense of the difficulties that face such an axiological task.

My only claim relating to this discussion is this: On the evidence of attempts to build a value case on diversity in more familiar domains, one should be less than sanguine about the prospect of building a value case in the biological domain. And even greater caution seems advisable, taking into account the inherent slipperiness of value in collection – particularly collections of collections.

Nevertheless, onward to biodiversity and its value.

Chapter 3
What Biodiversity Is

The first of the formidable obstacles to a clear and coherent discussion of biodiversity is the astonishing diversity of existing conceptions – including, sometimes, multiple ones used by the same authors in the same piece of writing. This situation unfortunately results in lots of false agreement, as well as false disagreement on propositions about biodiversity.

I seek a coherent characterization that is both sufficiently broad and sufficiently narrow. The goal is a conception that is *broad* enough to capture the core notion of diversity associated with things biological. It should also offer a sensible and sympathetic interpretation or explicatory reinterpretation of usage in existing discussion of biodiversity. An integrated understanding of the main corpus of writing about biodiversity is the ideal – though in the circumstances, a reasonable understanding will not be kind to conceptions of biodiversity that are not clear, not coherent, or in some other way an inadequate basis for moral consideration.

At the same time, as something of special value, I assume that biodiversity carves something somewhat specialized out of the larger biological world. At least, it is not equivalent the entire realm of the natural versus the realm of the supernatural. Nor even is it equivalent to (though it might be emblematic of) "all of nature" versus the human-constructed world.[1] It requires a characterization that is *narrow* enough to avoid falling into those massive black holes.

Finally, as I have already hinted, I seek a characterization that is still sufficiently broad to include most, if not all conceptions worth axiological consideration.[2] I don't know how to argue that my characterization satisfies this claim. Some of the

---

[1] In fact, it will become clear that neither does all of biodiversity fit comfortably as a proper subset of the natural world, and this is a source of great tension.

[2] It is entirely possible that some or many types of biological diversity are scientifically interesting as possible correlates, causes, or effects of other biological constructs or phenomena. My interest is in types of diversity (that is, the diversity of kinds in appropriate categories) that are candidate bearers of value. Of course, scientific interest in biodiversity might in itself confer value on it, and Section 6.8 (Biodiversity as font of knowledge) explores this possibility.

D.S. Maier, *What's So Good About Biodiversity?: A Call for Better Reasoning About Nature's Value*, DOI 10.1007/978-94-007-3991-8_3, © Springer Science+Business Media B.V. 2012

axiological difficulties that we encounter may cast doubt on this. But it certainly is true that getting clear on what biodiversity is takes us a long way towards understanding what limitations that notion has as a bearer of values.

My strategy in developing a definition of "biodiversity" is to start with something fairly broad, which I then narrow and refine both by imposing constraints on it and by elaborating its properties. Chapter 4 (What biodiversity is not) offers further refining constraints.

3.1 The Core Concept

When one talks informally about diversity, one typically has in mind the presence of characteristics that distinguish the things in a set, which nevertheless share some core resemblance by virtue of qualifying for set membership. For example, one might talk about the "diversity of computers", meaning that this diversity is displayed among multiple things, all of which are computers and none of which are chairs, for example. In terms of a simple ontology, there is a *category* – computers – to which each *kind* of computer belongs. The diversity of computers – the Macs versus the various brands of Windows machine, for example – arises from varying characteristics outside of those that make them all computers. Note that the varying "things" in question are not *particular* things – not my well-worn Model X with the scratch on its case versus your pristine-looking Model X. Rather, they are *kinds* of thing or, in my example, *kinds* of computer – Model X's versus Model Y's. Both Model X and Model Y instantiate the category "computer"; while my computer and yours apparently both instantiate Model X. The concept of biodiversity, I think, is best understood along these lines – as a variety of kinds in some category (or as I shall urge, categories). I develop this idea abstractly and then illustrate how it works via the category "species".

I take the contents of the portmanteau seriously: Biodiversity has *something* to do with the biotic world, and specifically with diversity, variety, variation, or differentness within it. Taking my cue from the organizing concepts that are the mainstays of modern biological science, as a first-order approximation, I presume that this differentness is a matter of, and is defined in terms of, the existence of distinguishable *kinds* in one or more biotic or biota-encompassing *categories*, which define what these kinds have in common. Taking another cue from my computer example, I would suggest that diversity, generally (not just for computers), is best understood as a complex state of affairs structured by a multiplicity of kinds, each of which is instantiated by one or more instances, and each of which in turn instantiates one or more qualified categories. In the case of *bio*diversity, I take the primary qualifying authority for selecting the structuring categories to be modern biology's best descriptions and explanations of the biological world, in which these categories play a framing role. I also assume that the instantiation of the kinds in at least one category is grounded in instances of some kind of real, perceivable biotic object in the world.

The definition of a category is a suitable bootstrap for diversity because it includes criteria for distinguishing or individuating the kinds that are its instances. Distinguishable kinds must have the requisite category-specific differentiating characteristics. The "mere" fact that two or more kinds of a biological category exist (that is, separately and independently instantiate the category) is a matter of diversity relative to that category's individuating criteria. In standard ontological parlance, the term "kind" is reserved for a "substantial universal". I use the term more promiscuously than is standard to refer to non-substantial universals, too – when these are instances of a category that might contribute to biodiversity. Non-substantial universals are usually known as "properties" and I also use the term "characteristics" because they characterize substantial universals. Their (non-substantial universal) instances are traditionally known as "modes".[3] Finally, properties (or characteristics) can be exemplified by (not instantiated by) substantial particulars (objects). This simple ontology is illustrated in Fig. 3.1. Modes are omitted because my discussion has no need for them.[4]

Sorting this out with the archetypal, taxonomic category "species": This is a category of (multiple) substantial universals, which are the various and sundry *kinds* of species that exist, have existed, or might exist. The category species provides a way to (not surprisingly) categorize the biotic world – in terms of the various species that inhabit it. This is of interest in my discussion because it also provides a reasonable basis for characterizing the world's biological diversity as species diversity – by virtue of the fact that within this biotic category (of species) one can distinguish multiple kinds (species), which one or more individual organisms (objects) might instantiate. A certain condition characterized by its species diversity occurs or, as one might alternatively say, a certain species-diverse state of affairs obtains when some number of actual, existing organisms concurrently instantiate some number of species-kinds.

I elaborate – though modestly and not in the terms of philosophical ontology – the notion of "category" in Sect. 3.3 (Biological categories and kinds) below. Almost certainly, there are opportunities for ontological mischief here. With a minor exception in Sect. 3.2 (Characteristics), I choose to sidestep these "opportunities" in the belief that this does not compromise the chief objectives of this book. For now, it suffices to think of the categories in terms of what, in Sect. 3.3.1, I call "*ta legomena*" – literally "what is said". In the case of biodiversity, my suggestion is to boil this down to "what is said by biologists". These amount to the natural biological kinds in their natural biological categories. This is not to say that some categories and kinds that enter the discussion by way of ecology and conservation biology might not seem particularly "natural" in the sense of "intuitively obvious" – at least, to those of us who lack an advanced degree in biology. Features, functions, genes, habitats, and other categories are among the ones discussed in Sect. 3.3.2 (Which categories and kinds qualify) and mentioned in Sect. 3.3.3 (Multiple dimensions).

---

[3] "Trope" supplanted "mode" in mid-twentieth century philosophy, though its ontological meaning bears no relationship to its more common meaning of "an expression used figuratively".

[4] This ontology is loosely based on that of E.J. Lowe (2006).

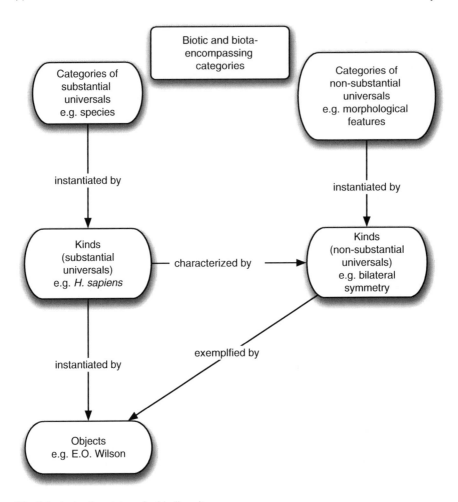

Fig. 3.1 A simple ontology for biodiversity

Until those discussions, I rely on the archetypal category of species as an emblematic stand-in. This aspect of my approach also resembles and respects common scientific parlance. It makes an attribution or characterization of biodiversity as arbitrary – or not – as the definition of the biological categories that underwrite our best scientific explanations of things biological.

Bryan Norton (2006, 53–55) attempts to distinguish between two types of definition for biodiversity – "inventory definitions" and "difference definitions"; and one might worry about whether my definition suffers for leaning towards the "wrong" type. Norton's characterization of inventory definitions is rather vague because constructed out of sample quotes (extracted from Takacs 1996) from interviews of scientists. But he asserts that it is exemplified by the "standard definition" of biodiversity as something like species diversity. On the other hand, he says, "Difference definitions… emphasize differences among entities rather than inventorying entities that exemplify differences."

Curiously, some familiar with my definition of biodiversity have identified it as an inventory definition, while others have called it a difference definition. I believe that both are correct. The weight that Norton himself places on this distinction[5] evaporates when one observes how tightly the two types of definition are joined at the hip. Inventory definitions – when the inventories are of reasonably well-defined kinds (such as species) – are squarely based on the "essential" differences that must exist between the kinds in order to qualify each as a distinctly countable entity. More precisely, an inventory is the simplest and therefore most intuitively appealing *measure* of differentness. Counting presupposes the prior individuation of that which is counted. And individuation, in turn, requires that sufficient, qualifying differences exist between the inventoried items to justify counting them as distinct. In other words, differences are built into the characteristics that distinguish the various kinds in an inventory.

The importance that Norton places on distinguishing these two types of definition also rides on his mistaking biodiversity for its measure, though it is also possible that he additionally or instead mistakes biodiversity for a measure of its *value*. Putting the latter possibility aside, if the existence of two or more distinguishable kinds of species constitutes a species-diverse state of affairs, the choice of a measure for this is a separate matter. On the one hand, one might choose to measure it by a species inventory count[6]; on the other hand, one might choose to represent each species by the relative abundance of conspecific individuals. Conflating a condition or state of affairs with its measure is one common category mistake that I take up in Sect. 4.1.2 (Measures and indexes).

A second category mistake is perhaps the most widespread, most persistent, and most responsible for derailing discussions about biodiversity – by transforming them into discussions about something else.[7] While arising from two or more kinds (instances) of a biotic or biota-encompassing category, "biodiversity" refers to the state of affairs arising from the *differences* whereby the kinds are distinguished. It does not refer to the kinds themselves, to the particular identities of these kinds, or to the particular objects that ultimately underlie the kinds.

Consider the species-diverse state of affairs that arises from differences between some set of individual organisms sufficient to warrant classifying them into two species rather than just one. For example, a boggy alpine meadow north of Mather Pass in California's Sierra Nevada is hopping with individuals of both *Bufo canoris* (Yosemite toad) and *Rana sierrae* (Sierra Nevada yellow-legged frog).

---

[5] According to Norton (2006, 54), inventory definitions are inferior to difference definitions for failing to capture the "complexities and interrelationships among biological entities" that are part of the true meaning of "biodiversity". On the other hand, he regards inventory definitions to have a persuasive power that difference definitions lack, and for this reason recognizes their utility and advocates their use in public rhetoric.

[6] Ecologists use the term "species richness" for "inventory count of species".

[7] The frequency with which Chap. 6 (Theories of biodiversity value) encounters this mistake is convincing evidence for this.

The amphibian diversity of this meadow does not require mentioning which species are involved. And indeed, south of Mather Pass, a similar meadow would more likely be home for *R. muscosa* (southern mountain yellow-legged frog) rather than *R. sierrae*. But so far as its amphibian diversity is concerned, that would be irrelevant. In short, propositions about species diversity (as one type of biodiversity) are not about the identities of particular species or even about species – except insofar as they contribute to the diversity.

This distinction is all the more evident – and important – when it comes to claims about value. To claim that the amphibian diversity of an alpine meadow is valuable is quite different from claiming that *B. canoris* or *R. sierrae* is valuable. It is entirely possible to value the diversity represented by these two easily distinguishable species without thinking much of either one as a species – except insofar as their joint existence is responsible for a more biodiverse state of affairs than would exist in the absence of either one. Valuing amphibian diversity, but neither *R. sierrae* nor *R. muscosa*, one would not hesitate to allow *R. sierrae* to trade places with *R. muscosa* across Mather Pass.[8] I am not trying to suggest that either species of frog is unimportant or valueless as a species. Rather, I am saying that the question of whether either one of these species or any other is valuable as the species should not be confused with the separate question of whether a species is important for its contribution to diversity – that is, whether (or not) species diversity is valuable.[9]

A third category mistake also merits attention here – both because it involves a mash-up of the first two category mistakes and because, like the second category mistake, it could be a basis for the mistaken impression that this book deals with an arbitrarily curtailed conception of biodiversity. The third category error arises from thinking that a difference definition of biodiversity entails a view of biodiversity as having to do with diversity; while an inventory definition entails a view of biodiversity as having to do with particular species. Insofar as inventory definitions do have legitimacy (at least as a measure of biodiversity), this account might, after all, appear to provide a legitimate basis for valuing *R. sierrae* (for example) as a species understood in terms of an inventory definition of biodiversity. The problem with this account is that "a species understood in terms of an inventory definition of biodiversity" is still a species understood as having characteristics that suffice to differentiate it from any other species. It is on that account that we are entitled to add it to our inventory without double counting. In other words, the two types of definitions more or less collapse into each other, reducing to irrelevance the claim that this book arbitrarily truncates its conception of biodiversity because it leans too heavily on differences rather than on inventories.

I have described a way to understand the "diversity" in "biodiversity" – as structured by distinguishable instances (the kinds) of biotic and biota-encompassing

---

[8] I am talking about the amphibian diversity of the meadows considered in isolation – not about the amphibian diversity of the entire mountain range or the world at large.

[9] This fundamental distinction is further discussed in Sects. 4.1.3 on Particular species and 4.1.4 on Particular ecosystems, as preparation for its Whac-A-Mole™-style tendency to pop back up no matter how many times previously whacked down.

categories. A biodiverse state of affairs obtains by virtue of some multiplicity of those kinds being instantiated, with the chain of instantiated universals ultimately grounded in actually existing objects – for example individual organisms, genes, or alleles.[10] Some forms of biodiversity – most notably "functional diversity" – somewhat strain this last requirement. But I believe that in the end, they can be satisfactorily accommodated.

Very many categories might enter in biodiversity; they could relate to each other in various interesting ways; and these relationships might be bases for defining yet more categories whose variety of kinds contributes to the diversity. Certainly, different scientists have different predilections for which categories to include and which to exclude under the biodiversity umbrella. These variations in choice of categories are not unimportant.[11] But they are less important than the invariance in scientific understanding of biodiversity precisely as a state of affairs with the sort of structure I have described. Biodiversity is a state of affairs that admits of more and less. Frequent recent agonizing about the possibility that a Sixth Great Extinction is underway evidences a widespread worry about a trend towards a less biodiverse state of the world. After all is said about exactly which categories legitimately contribute to biodiversity, which ones do not, and about how the "in" categories relate to each other in their contribution to biodiversity, most everyone will agree that the world's biodiversity is a matter of how many different biotic entities – of the relevant kinds – there are; and that this is ultimately rooted in the concurrent existence of biotic objects such as individual organisms. This is what I call "the biodiverse state of the world" – that is, the state of the world characterized in terms of which biodiverse state of affairs obtains in it.

Question: What could biodiversity be *other* than a state of affairs? One answer that has appealed to some: A process. Whatever a state of affairs is (once again, sidestepping ontological mischief), and with apologies to fans of Heraclitus and Alfred North Whitehead, I distinguish it from a process. Taking biodiversity to be a process is, I think, yet another and increasingly popular category mistake, which I take up in Sect. 4.1.5 (Biodiversity as process).

Even as it stands, without further elaboration, my disarmingly simple definition of "biodiversity" has some important, though seldom-discussed, implications. I explore a few these in the next few subsections.[12] Some more general and more deeply penetrating analysis must await Sect. 8.1 (The disvalue of the biodiversity project).

---

[10] States of affairs constitute another ontological category – one that has received a great deal of attention since the likes of Bertrand Russell and Ludwig Wittgenstein wrestled with facts, which are states of affairs that obtain. I ignore the philosophical niceties that bear on their ontological status because I believe that, however important they might be for a complete and coherent metaphysics, they matter little for what I have to say about biodiverse states of affairs.

[11] I later take up some details about which categories might be included, which might be excluded, and the structure of various categorical relationships.

[12] Wood (2000, Chapter 2) elaborates a conception of biodiversity not unlike mine. I disagree with Wood on a number of points, most notably his claims about why biodiversity is valuable. Nevertheless by a wide margin, his book is the most consistently clear and well-argued treatise on biodiversity and its value that I know. As a substantial added bonus, it provides a marvelous introduction to Canadian constitutional law.

3.1.1 Egalitarianism

The conception of biodiversity that I propose entails egalitarianism with respect to a category's kinds – at least so far as the contribution of each kind to that category of diversity. In the case of species, for example, my conception does not tolerate the sentiment that some species are more equal than others – so far as their contribution to species diversity is concerned.[13] Parasitic nematodes, whose parasitic way of life might be the most common on our planet, contribute as much as, say *Ursus maritimus* (polar bear).[14] Exotic species are entirely on a par with natives. So are hybrids, and so are hybrids that emerge from the genetic intermixing of exotics and natives. The contribution to species diversity of a new, genetically engineered species is no less a species than one that arises through, say, allopatric speciation. Nor, in that latter case, is the new species somehow less than a new species if a new highway were the blunt instrument of geographical separation that led to its speciation. No less species diversity-contributing respect is due to a new disease organism because it arises from selection for resistance to drugs.

Whether desirable or undesirable for other reasons, all kinds of organisms clearly contribute to species diversity. Insofar as species diversity has *value*, the contribution to diversity of a species *non grata* is indistinguishable from that of any other species. The perception that an organism is *non grata*, of course, indicates the presence of other, conflicting considerations in this assessment. But these considerations don't erase whatever status is appropriately awarded to any and every kind of species on grounds of biodiversity or at least, species diversity.

Resistance to the precept of egalitarian valuing of kinds within a category typically arises specifically with regard to the category of species. But "species" is just one category of kinds that might be said to contribute to biodiversity: Others that I will get to include features, functions, genes, and habitats. Insofar as diversity is truly what is valued, I know of no legitimate grounds – indeed, I know of no argument whatever – for resisting the implication of "one kind, one vote" within *any* category.

If there were good reasons for valuing the diversity in some category other than species, then these reasons might be good grounds for giving less consideration to the diversity of species and therefore to any particular species' contribution to biodiversity via species diversity. But these reasons would not be reasons for denying that an egalitarian principle applies to the kinds in every category – including the category that defines the gold standard. I will need to revisit the implications of this point below.

---

[13] Some biologists rebel against species egalitarianism. Warwick and Somerfield (2008), for example, put it right in the title of their paper – "All animals are equal, but some animals are more equal than others". The denial of species as equal contributors to diversity is often (as with Warwick and Somerfield) tied to a proposal that species diversity is subordinate in importance to the diversity of kinds in other categories, such as "phylogenetic diversity", which I discuss in Sect. 3.3.2.1 (Features).

[14] See Note 44 in Chap. 2 for the rationale for my practice of using scientific names of organisms.

3.1.2 Fungibility

Egalitarianism with respect to how each kind contributes to the diversity of its category makes kinds fungible with respect to their contribution. I use the term "fungible" in the sense that economics and law assign to it – namely, "admitting of substitution or replacement with no loss in what is at stake". Here, "what is at stake" is the diversity of kinds and the contribution of each kind to that diversity.

This means that biodiversity in any category that contributes to it is not bound to the identity or independent value of any of that category's particular kinds. For example, if one component of biodiversity is the diversity of species, then the species-diversity contribution of any particular species lies exclusively in its being some species or other that is distinguishable from every other species, but that is *in*distinguishable from any of them in its contribution to diversity. Therefore, species diversity on any spatial scale is conserved as one species emerges in the face of another's demise. In the extreme, both of two entirely *disjoint* sets of species could constitute an equally species-diverse state of affairs. A similar proposition holds for biological categories other than species.

Marine biologist John Spicer (2006, 73) conveys this notion of maintaining diversity within a category despite the flux of kinds – in the context of remarking on the stability of the diversity of Ordovician families (two levels up the taxonomic hierarchy from species) of organisms for a quarter of a billion years after the end of that period some 444 mya:

> That is not to say there was an absence of activity. The total number of people walking down the street may stay constant for hours, but it is not the same people all of the time. In the same way, families present at the end of the Ordovician were a completely different set from those a quarter of a billion years later. It's just that new families were being added at the same time, and in the same proportions, as old ones disappeared.[15]

That is, family diversity remained stable throughout that considerable duration, though the identities of the various families present from time to time was not at all stable.

Returning to kinds of species and narrowing attention to a shorter, non-geological time scale, this property (of constant diversity for a changing set of things) is also illustrated by the "dynamic equilibrium" in species, which renowned biologists Robert MacArthur and E.O. Wilson (1963) famously posited as part of their original theory of island biogeography.[16] According to that theory, the equilibrium on an isolated island is a matter of the stability of species diversity despite flux in

---

[15] With the exception of the rather confused Chap. 6 on the value of biodiversity, this delightful little book is highly recommended as a personal, engaging, and non-intimidating survey of much of the science related to biodiversity. It regularly brings out in relief important points that are lost in the more technical and detailed discussions that appear in the primary literature.

[16] The authors' understanding of species turnover in terms of dynamic equilibrium is also the least verified and most contentious part of their theory. However, it is the concept, not its truth value, that concerns me here.

the particular species present. Following colonization of an island, some species will go extinct while other species will immigrate to replace them. So while the numbers of species remain more or less constant over time, their identities change.

While dynamic equilibrium might not be a general biological law of nature, something like it has been observed in more isolated studies. Avian diversity, for example, has received relatively careful attention. On oceanic islands, within the continental United States, and in the state of Michigan, there have been enormous changes in composition while the diversity of bird species in these places has remained largely unchanged (Sax and Gaines 2003, 563).

In sum, the notion of biodiversity by itself gives no basis for investment in any one particular way in which a biological category might manifest diversity through a plurality of kinds. So far as the aspect of diversity anchored in that category is concerned, the identity of any instance-kind is irrelevant, making it (in this respect) completely interchangeable with any other kind within the category. Moreover, this fungibility of kinds within a category of biodiversity itself seems to carry over to biodiversity's value, too. For example, any given species has the same value insofar as that value has to do with its incremental contribution to species diversity.

I return in Chap. 5 (The calculus of biodiversity value) to some implications of species (and other kinds of) egalitarianism and fungibility for a calculus of biodiversity value. I am particularly interested in considering attempts to resist or nullify these implications when they are perceived to violate intuitive evaluations, according to which, for example, some species *are* more equal than others.

Discomfiture with the implications of principled egalitarianism and fungibility with respect to kinds in one category of diversity leads some to try to avoid these implications by positing the precedence of some other category. For example, those who cannot abide that all species were created equal argue for their unequal status on the grounds that phylogenetic or functional or phenotypical diversity (discussed in Sect. 3.3.2, Which categories and kinds qualify) takes precedence over species diversity in matters of natural value. But to arbitrarily posit a precedence ordering – phylogeny over species, or functions over species, and so on – merely to avoid the consequences of credible principles is unacceptable. In other words, this approach has a great burden, which consists in finding justifying grounds for the precedence ordering. Moreover, this move still does not escape the constraints of egalitarianism. The presumption continues to be that within any category – including the one at the top of the precedence list – there is no reason to prefer one kind to another.

A second maneuver to avoid egalitarian implications and the implications of fungibility attempts to dilute their force by bringing in criteria other than those that distinguish the instances of kinds, thus expanding the notion of biodiversity. This move is discussed in Sect. 4.2 (Accretive conceptions).

I am aware that my comments on egalitarianism and fungibility might not be happily received by those who are heavily invested in projects that depend on identifying differences that are supposed to justify differential treatment and triage in the natural world. Projects such as these might well constitute the preponderance of conservation and restoration efforts undertaken today. If you are pro-"native" and anti-"alien"; or if you are inclined to take sides when two natives have at it – for example, shoot the fox to save the shrike, as was done on San Clemente Island (discussed in Sect. 8.2.7.2) – then you must demand the moral space to make the case for "your guy". I am also aware that some might see my position as vulnerable to a *reductio*, which holds it to entail that if any species is to be saved then all must be saved; or claims it to entail that none should be saved because there are no credible criteria for selecting the proper proper subset. But neither of these propositions is credible.

My point is really a rather modest one: If biological diversity in any category is to be used to help assess what is valuable or most valuable in the natural world, then the principles used for that assessment must be part and parcel of a defensible evaluative framework. My purpose is to identify those principles that seem legitimate and those that are questionable. After that, I prefer to let the chips fall where they may.

3.1.3   Questionable Factors

Some factors that commonly creep into discussions of biodiversity and species diversity in particular have little to do with diversity. What follows is a brief and selective survey of them.

3.1.3.1   Abundances

Getting back to our simple ontology of biodiversity: The different kinds in a category are themselves universals and, as in the case of species, substantial universals. Each universal has its own identifiable instances, which is equivalent to saying that each instance can be individuated – that is, distinguished from every other. However, the multiple instantiations of a kind – that is, how many instances the kind has – has no direct bearing on the diversity of kinds in the given category.

In the case of the category species, the individuatable instances are individual species. A species, in turn, is instantiated by multiple individual organisms of that species – that species' abundance. The abstract principle stated above means that the abundance of individual organisms of any particular species has no direct bearing on species diversity because this is strictly a matter of there being distinguishable kinds of species.[17] If the abundance of a particular organism (kind) or its

---

[17]This does not preclude contingent connections between abundances of individuals that might figure in the diversity of their kinds. For example, an extremely low number of individuals of a particular species might entail a high likelihood of that species vanishing. This sort of connection is discussed at several points, including in Sect. 4.2.2 (Rarity).

abundance relative to that of other organisms factor into biodiversity, then this contribution must be justified independently of species diversity.[18]

This conclusion is a simple matter of defending a logically consistent conception of biodiversity. Yet it flies in the face of one of the most popular measures of biodiversity among ecologists. According to it, biodiversity is primarily a matter of species diversity. This, in turn, is measured in terms of a combination of species richness *and* the "evenness" of species counts – that is, how close in size are each species' population. It is possible that the evenness property is intended to be a manifestation of diversity in some category other than species. But I am not aware that anyone has ever identified this category. I myself cannot think of a reasonable candidate.

Oddly, widely *varying* abundances, considered as an extrinsic property of species, rather than evenness, would seem to constitute the more diverse state of affairs than one in which just one abundance level obtains. I return to this conundrum in Sect. 3.3.2.2 (Abundances (again)).

3.1.3.2 Abiotic Conditions

The "bio" in "biodiversity" says that it first and primarily has to do with multiple instantiations of *biotic* or *biota*-encompassing categories. In the first instance, it does not have to do with the abiotic (non-living) part of the world.

That is not to say that the abiotic world does not influence the biotic world and vice versa. They do influence each other, the influence is enormous in both directions, and ecologists devote careers to understanding these interactions. As just one example among many, those who study biogeochemical processes and cycles trace the flow of "stuff" – for example, energy, inorganic and organic compounds – through, about, and around the world. This stuff does not respect boundaries between the living and the non-living. Undeniably, the relative abundance of, say, reactive nitrogen in an organism's environment can profoundly effect which traits it expresses (phenotype), how it interacts with other organisms in that environment, and how they, in turn, act on and affect that environment. But the nitrogen cycle itself, or the relative abundance of reactive nitrogen in an environment, is not in itself part of biodiversity.

The abundance of reactive nitrogen is an interesting case in point because the anthropogenic production of reactive nitrogen – mostly in the form of nitrates in chemical fertilizers but also in the form of NO_x gases from combustion – has increased its presence in most biological systems; and it has reached places unreachable by ordinary or even extraordinary human transportation as the result of its atmospheric deposition. But the amount of reactive nitrogen in any particular

---

[18]Of course, a multiplicity of individual organisms of one species *is* biological diversity of another kind. Insofar as each individual has a unique genotype, it contributes genomic diversity. Or, looking at an even finer-grained level of organization, each individual organism may contribute to allelic diversity. However, this is not what is meant by authors who say that more nearly equal abundances of the various kinds of organism contributes to species diversity.

environment and even in every environment worldwide does not *logically* affect the assessment, in those environments, of the diversity of kinds in biotic or biota-encompassing categories. For example, it is possible to conceive of not just two equally biodiverse states of affairs, but two identically biodiverse states of affairs that obtain under conditions of nitrogen abundance and nitrogen scarcity.[19]

Moreover, the disconnect between abiotic factors and diversity is more than logical. It is far from clear that abiotic factors in general or the abundance of reactive nitrogen in particular has even a *causal* effect on the *diversity* of such central biological kinds as species or even their phenotypes. This contrasts with the effects it undoubtedly does have on *which* species might find a place suitable and *which* characteristics of a species emerge in that place.

I have used the phrase "biota-encompassing kinds" to allow biodiversity to include the diversity of kinds that are systematic or "structural" arrangements that include non-living parts, while also integrally including living parts. Ecosystems are the paradigmatic example of this. The word "encompassing" is an admittedly crude attempt to exclude categories of kinds that are logically detached, if not causally detached, from living organisms. Biodiversity is neither geodiversity nor chemodiversity, however much all three are causally interrelated.

3.1.3.3 Interactions

Within my conception of biodiversity as "diversity of kinds in biotic or biota-encompassing categories", it is possible to include such relationships as interactions between biotic elements as well as interactions between biotic elements and abiotic ones. There are two possible ways to understand how these relationships contribute to biodiversity. The most obvious one is in terms of kinds of relationships. A kind of *relationship*, in turn, is partly determined by the kinds of *entities* that are related – that is, the domains over which the relation is defined. I first consider relationships abstracted from the things related. Then I bring the related things back into the picture.

To focus the discussion, I propose to examine species interactions, and more specifically, interspecific "lifestyle" or trophic interactions – parasitism, mutualism (of various sorts), herbivory, and predation. These relationships are the stuff of trophic structures. The relationships are ordinarily defined over the domain (category) species. That is, the "whos" and "whoms" in "who parasitizes whom, who devours whom, and who munches on whom" are typically species. More formally, to a first-order approximation, this handful of distinguishable kinds of interaction comprises a handful of relations, each of which maps the domain of species onto itself.

---

[19]Riverine fluxes notoriously augment atmospheric deposition on a regional basis to produce eutrophic "dead zones". These areas are, in fact, quite alive – albeit with an altered specific constituency. For a discussion of this in the context of the value of biodiversity, see Sect. 6.3 (Biodiversity as service provider).

The diversity of these most basic kinds of trophic interaction has increased over time. There was a geological time before predation and herbivory – understood as activities in which the predator or the herbivore is a multicellular animal eukaryote (metazoan). The basic equipment required to successfully play either the role of muncher or munchee – for example, teeth or other seizing-and-masticating mouthparts (the better to eat you), hard exoskeletons (the better not to be eaten), and mineralized structures to support muscles (the better to catch or flee from you) – did not fully emerge until the beginning of the Cambrian period, some 545 mya, although the lineages of these creatures might extend back to 700–1,000 mya or perhaps even earlier. Of course, evolution did not stop in the Cambrian. The last 500 million years witnessed substantial evolutionary elaboration of the basic structures that "support" the basic kinds of trophic interaction. But none of this elaboration produced a new kind of basic trophic interaction. In other words, lifestyle-interactions have indeed increased in diversity since 1 bya. But not anytime recently.

Here one might object that the various elaborations have, in fact, produced new kinds of trophic interactions. In the Cambrian, they weren't yet "doing it" on land. So specifically terrestrial versions of these interactions – involving vascular plants, insects, birds, and mammals – were yet to come. And back in the drink, jawed fishes and baleen-equipped sifters were yet to evolve. But this objection entirely misses the point that, for example, jaws and baleens have not contributed to the diversity of trophic kinds.

At this point, the objector might reformulate her objection. "Lifestyle" or trophic interaction might be amenable to categorical slicing and dicing in other ways that produces a far greater diversity of relationships than the basic trophic interactions that I have so far focused on. For example, one might propose that kinds of trophic relationship should be individuated on the basis of the identities of the species that enter into them.[20] According to this proposal, each instance of the kind is bound to the particular "whos" and "whoms" – who eats whom, who lives inside whom, and who helps whom. Thus, the herbivory of *Diorhabda elongata* (the tamarisk leaf beetle) on *Tamarix* spp. qualifies as a different lifestyle from the herbivory of *Diatraea grandiosella* (southwestern corn borer) on *Zea mays* (corn). On this account, there is a far greater diversity in kinds of relationship. Also, this diversity is now more like species diversity in its enormous flux. As species emerge and go extinct, and even as species assemble and disassemble themselves into and out of communities, these kinds of relationship come and go.

However, the cost of a multitude of kinds is an extreme restrictiveness of each kind. There is, for example, the tamarisk-leaf-beetle-munches-on-tamarisk kind of

---

[20] Including the terms of the relation in the relationship kind is, of course, just one strategy for generating a diversity of relationship kinds from a collection that previously seemed meager. For example, one could avoid going down to the level of species by defining such kinds as "uses jaws to predate" or "uses baleen to ingest". But these alternative strategies fail in much the same way as the strategy involving related species. This might raise the suspicion that the rubric of "interactions" is really used to talk about the morphologies and phenotypes of the organisms that interact.

lifestyle. This, unsurprisingly, is a lifestyle that neither corn, nor the southwestern corn borer, nor any other species enter into. It might not be entirely irrational to describe the world in terms of kinds with such restricted scope. But it seems unhelpful. That might be why, to my knowledge, no one, and least of all biologists, utilize this descriptive framework.

It is tempting to say that the suggestion to consider tamarisk-leaf-beetle-munches-on-tamarisk interactions for the purposes of interaction diversity is not even logically consistent with the definition of biodiversity that I have proposed. One might come to this conclusion by way of analogy with the kinds of species. In Sect. 3.1 (The core concept), I suggested that it is an error to think that the diversity of species in a collection of species hinges on which particular species are collected together. The move to specialize species interactions – in terms of who is related to whom – might well be guilty of a similar error. However, it is harder to see the error in this current case, for we lack handy names – of the sort that we use to identify particular species – for the identities of the various (very specific kinds of) lifestyles. That is why I had to resort to such a grotesque moniker as "tamarisk-leaf-beetle-munches-on-tamarisk" interaction.

So it seems ill-advised to try to solve the problem of too broadly defined interaction kinds by embracing the opposite extreme. One way to ameliorate the problem of over-*narrow* kinds is to remove one or the other (but not both) of the related terms from the relation – thereby folding the relationship into a non-relational attribute of the other term. Thus, "munches on *Tamarix*" is an attribute of *Diorhabda elongata*. But it is still not an attribute of *Diatraea grandiosella*. But this kind still seems too narrow to be useful. And again, to my knowledge, no scientist uses it.

This reluctance on the part of scientists to adopt such a "species-attached" classification of interactions also might be due to the empirical evidence, which makes it hazardous to view interactions as an attribute, or at least as an unconditionally present attribute, of one of the interacting species. There is no denying the possibility that *Diorhabda elongata* might evolve a taste and even strong preference for some plant other than *Tamarix*. In general, such an interaction follows not solely from the evolving genetic makeup of a species, but from a variety of environmental conditions that might either facilitate or suppress its expression within a relatively fixed genotype.

Most emphatically, this move does not somehow increase the diversity of *species*, though this appears to be the subtext of many discussions about species *interactions*. The case for the twoness of those species is already sealed on species-distinguishing grounds, which omit all consideration of whether or not these particular species interact.

This turns us back to regarding kinds of interspecific relationships as divorced from the particular kinds of species that engage in them. In the context of my promiscuous ontology, this is a perfectly acceptable suggestion. In fact, it is one that dovetails into the increasingly popular ecological account of a "function" which itself might qualify as a diversity-rich category. On this account, a species interaction such as herbivory is a "function" and so adds to the diversity of functions. But there are other possible ways of looking at it. One alternative is to conceive of a species interaction as a "feature" or "trait" of one of each interacting species considered apart from all others, which adds to "phylogenetic diversity".[21] Both types of biodiversity are discussed in Sect. 3.3.2 (Which categories and kinds qualify). These various categories or components of biodiversity might be considered separately, or alongside each other, as well as alongside species diversity. But once again, it is not helpful to confuse them. It is certainly not helpful to conflate them all with species diversity.

3.1.3.4 Place

The conception of biodiversity on which I have settled says nothing specific about *where* the diversity is realized. There are many possibilities. It might be realized *in situ* (in "nature") or *ex situ* – usually thought to mean "in some human-created environment". A spectrum of possibilities spans these extremes, including parks or other places managed according to some plan, human-engineered landscapes (for example, designed to imitate forests or wetlands), zoos, gardens, farms, and gene banks. The notion of biological diversity by itself provides no obvious or *a priori* basis for discriminating among these various settings. This implies that biodiversity value is indifferent among them, too – other things being equal.

This is not to deny the legitimacy of attributing biodiversity to places. It *does* deny a different claim, expressed by Sarkar, that biodiversity is inextricably tied to the particular place where it is encountered. While a *kind* of place can be part of what makes up a "biota-encompassing" kind in some category (as in the discussion of anthropogenic biomes in Sect. 5.3, The moral force of biodiversity), the diversity of such kinds is (as always) independent of the identities of the instances of any of that category's kinds. Particular places, in their particular geographical location, with their particular history, do not legitimately enter into the diversity discussion. So far as kinds of places are concerned, it is hard to deny that biodiversity is enhanced rather than diminished by adding to the collection of kinds, ones that owe something (or a lot) to human intervention or design. I take another look at Sarkar's claim and some of its difficulties in Sect. 3.3.4.1 (Place (again)).

3.2 Characteristics

In this section, I glance at some writing on the ontological wall.

One might think that mere multiplicity of the kinds that instantiate a category does not exhaust the diversity to be found in that category or in its kinds. Different species, for example, might exhibit any number of diverse characteristics. In ontological

---

[21] As I observe in Sect. 3.3.2.1 (Features), the term "phylogenetic diversity" belies its meaning, which is something like "a measure of phenotypical diversity".

terms, they might exemplify non-substantial universals (i.e. properties), which do not have a role in making them instances of their kind according to the biological species definition. Surely, these non-species defining properties or characteristics – for example, that some species of animals are enormous and others miniscule, that some creatures have bilateral symmetry and others radial, that some organisms live in the sea and others on land, and so on – must contribute to the diversity of species.

The intuition underlying this point is entirely legitimate. But its expression – in the proposition that the varying characteristics of organisms contribute to *species* diversity – is not. For the species are different species according to criteria that do not directly take into account these characteristics.

But that does not preclude the possibility that there might be some category other than species whose kinds *are* distinguished on the basis of the characteristics whose diversity is at issue. My definition of biodiversity admits categories whose kinds are non-substantial universals (properties or characteristics) and I see no initial reason to exclude a category of characteristics or even several such categories whose diversity could be part of what "biodiversity" encompasses.

One might have a sense of great difficulty in the task of defining a category of characteristics and its kinds. Indeed, this might end up being something of a grab bag – body symmetry, body size, land-sea preference, etc. This has not deterred biologists from embracing this idea – typically under the umbrella of "phylogenetic diversity".[22] Not coincidentally, this notion already popped up in the discussion in Sect. 3.1.3.3 (Interactions); Sect. 3.3.2.1 (Features) considers it further. Difficulties arise in finding adequate qualifying conditions for kinds in this category. But this does not militate against its conceptual legitimacy.

3.3   Biological Categories and Kinds

3.3.1   Ta legomena in Biology

I have suggested grounding the concept of biodiversity in categories of biotic and biota-encompassing kinds that also ground our best scientific understanding of the biological world. It is reasonable to expect, or at least hope that the science that gives us these categories will also give us scientific criteria for defining the categories and (in doing this) distinguishing the kinds that are their instances. These criteria can then be utilized as the basis for defining diversity in the world as the diversity of kinds in the chosen categories. Diversity arises as the consequence of a multiplicity of such kinds. For species, species diversity arises as the consequence of a multiplicity of species.

Even the most established categories in biological science are problematic to some degree. Biologists chase their tails trying to accommodate ring species and

---

[22] See Note 21 for the required caveat about the use of this term.

species that reproduce asexually within the biologically defined category of species.[23] But despite its flaws, it mostly works well enough, and it has a heritage going back to Aristotle (though his concept was probably rather different).[24] For the purpose of defining the species diversity part of biodiversity, I take whatever flaws the biological species concept has to be irrelevant.

Speaking of that not-so-late, great, Greek biologist, categorizer, and philosopher: For Aristotle (350 B.C.E. a, I.4), there is a non-reducible set of *categories* that encompass *ta legomena* – literally, "what is said". Modern biological science does not quite match this fixed and clearly identifiable set of categories of things biological. Moreover, digital technology has made it apparent that fixed systems of classification are precarious constructions. David Weinberger, for one, presses the point that the digital storage of information in increasingly flexible schemes frees us to reclassify – in more flexible, interleaving, and non-hierarchical ways – "what is said" at the point it is "remembered" rather than when it is stored – according to our various and changing interests, and on the fly as our interests change.[25]

Weinberger presses his point too far. Even the most flexible object-oriented database requires good metadata and social agreements that combine to yield the technical agreements (conventions) for finding, accessing, and using these metadata as handles to retrieve the information that they describe.[26] In the biological sciences that underpin the concept of "biodiversity", the category-defining social and technical agreements have been relatively stable over a long period of time. These sciences – particularly the ones devoted to genetics, taxonomy, and the ecological and structural properties of places (which some authors group together in a "structural/ecological hierarchy") are still quite firmly rooted in several categorical hierarchies that for the most part have served well to describe biological phenomena and relationships. This is why one does well to build a model of biodiversity on these scientifically useful categories. This lends stability to the approach, does not require the invention of categories of unproven value, and does not exclude the possibility of later incorporating categories that are newly found to be useful.

I mentioned "categorical hierarchies" in the preceding paragraph. Kinds, as I define them, fall inside one category. But some categories do relate to each other, and the classical relationship is hierarchical. Species, that stalwart of biological categories, is one of several in a taxonomic hierarchy. Genus, the category above

---

[23] See Note 51 in Chap. 2 for a discussion of biological species and other species concepts.

[24] In light of the interest of modern biologists in "functional diversity", it is interesting to note that on one interpretation of his *Metaphysics*, Aristotle's notion of species focuses on justifying the classification of two individuals as members of the same kind in terms of their *functional* congruence.

[25] Although, as I say, Weinberger overstates his case, his discussion is valuable for pointing up changing categorizing behaviors in a world of media that are increasingly digital.

[26] Geoffrey Bowker (2006, Chapter 3, "Databasing the World: Biodiversity and the 2000s") provides an extremely perceptive account of "the problem of metadata".

species in this taxonomic hierarchy, is instantiated by kinds (genera), each of which is instantiated by one or more species in that genus (which makes it a category of substantial universals), each of which in turn is also instantiated by one or more individuals of that species (which are objects). We can say of an individual organism that it also instantiates the genus that the individual's species instantiates (making the genus also a substantial universal in my promiscuous ontology) – and (skipping up the taxonomic hierarchy) the family of that genus, the order of that family, the phylum of that order, the kingdom of that phylum, and the domain of that kingdom.

Within such a hierarchy of categories, it is clear that various categories of diversity are interrelated and interdependent. Most obviously, the differentiae that provide a diversity of kinds in a higher category such as phylum entail a diversity of kinds in a subcategory such as species. My purposes are not served by trying to disentangle these interdependencies of diversity. It suffices to say that hierarchies of this sort contribute various types of diversity to biodiversity.

Part of my reason for shunning this analysis is that some categories that might be relevant to biodiversity dwell outside the taxonomic hierarchy; they do not have any obvious logical, and certainly not a hierarchical, relationship with the taxonomic categories. This is true of such non-taxonomic categories as metapopulation, habitat, and community, among others that I list below. Also, their empirical relationships to the taxonomic domain are often neither obvious nor well known. This yields a picture of disparate categories that form (or might form) disparate hierarchies with a satisfying (or for those who covet reductive simplifications, *un*satisfying) Aristotelian irreducibility.

I alluded just above to possible hierarchies other than the standard taxonomic one that might feature in biodiversity. One candidate is a hierarchy of genetic "stuff" – including such categories as allele, gene, chromosome, and genome. Some biologists, such as fish biologists Paul Angermeier and James Karr (1994, 691), prefer to think of this hierarchy as independent of the taxonomic one. Others, such as Sahotra Sarkar, prefer to fold it into the taxonomic hierarchy – on the grounds that genotypes are related to intraspecific (allelic) diversity.

The genetic hierarchy interrelates with and so points to what some biologists refer to as a third, "structural/ecological" hierarchy.[27] Despite the "hierarchy" label and although some notion of spatial scaling is involved, this seems more a loose collection of empirically connected concepts than a hierarchy.[28] It includes such categories as subpopulation, metapopulation, population, assemblage, community, habitat, ecosystem, landscape, and biome. Unfortunately, though this might startle

---

[27] This compound term is borrowed from Sarkar (2005, 49, 64–66, 69–70); also Sarkar (2002, 136). He is one of the few writers who express some awareness of the incommensurability of various biological hierarchies.

[28] Angermeier and Karr (1994, 691), for example, join Sarkar in speaking of an "ecological" hierarchy.

nonscientists, these categories and their relationships lack consistent and even clear scientific characterizations. This frustrates attempts to use them in helping to frame biodiversity. I nevertheless follow those biologists who award this collection of categories "honorary hierarchy" status.

Hierarchies (both strict and honorary) do not exhaust the categories that biological science provides for framing biodiversity. A scan of the scientific literature reveals what might initially appear to be a hodgepodge of what I called "characteristics" (in the eponymous Sect. 3.2). These are categories of non-substantial universals that, as I observed, are sometimes confusingly tacked onto the category species and presented as increasing the diversity of species. The proper, or at least far more useful, way to regard them is as forming their own categories of non-substantial universals.

For example, the body plan (type of symmetry, appendage count, etc.), size, or some other *morphological* characteristic of one species might further distinguish it from another with which (by the biological definition of species) it cannot exchange genetic material. So this fact is not a matter of the diversity of species, but rather of the diversity of morphologies. In other words, morphology is a category that has a diversity of kinds. *Functional* differences (and samenesses) – for example, trophic function or role in nutrient cycling, which are now pervasively discussed in the ecological literature – bear similar treatment, as I remarked in Sect. 3.1.3.3 (Interactions).

I submit that we have a clearer-eyed picture of diversity by refraining from the unconvincing practice of folding these characteristics into the category of species, and instead giving them their own categories. The price of this conceptual clarity is a certain complexity. Morphology and functional role, for example, span taxonomic kinds such as species, families, and sometimes, even phyla. But I would say that this complexity accurately reflects the complexity in the diversity of biological things.

This simple and biology-friendly characterization of the differentness that goes into biodiversity has an important and immediate implication. Because biodiversity is invested in multiple categories – some logically related, others possibly empirically related, still others most likely completely independent – one should not expect to find any easily justifiable way to compare the contributions to biodiversity of the diversity in these various categories. Certainly, there is no reasonable expectation that differentness in one *hierarchy* of categories will be comparable to differentness in a category not in that hierarchy. What would the relative contributions to biodiversity be of, on the one hand, three as opposed to two species and, on the other hand, three as opposed to two body sizes? The question, I think, is barely comprehensible.

Thinking about different hierarchies of categories might obscure the fact that even for the relative contributions to biodiversity from two categories within a *single* hierarchy, there might be no reasonable accounting method. Within the taxonomic hierarchy, for example, how should one view the relative effects on biodiversity of the loss of multiple species in one species-rich genus with the loss of an entire genus containing but a few species in an otherwise family-laden order?[29] Any and all levels of taxa might enter into biodiversity. I find it hard to imagine any justified and satisfactory general accounting for how the contributions to diversity from various taxonomic scales relate to each other. From the example just tossed out, it seems clear that the difficulty is more than a matter of accounting for simple nesting and therefore overlapping redundancy within the hierarchy.

Finally, it is reasonable to think that the structure of biodiversity – as an amalgam whose relative contributions resist characterization – is reflected in biodiversity's value. Where there is an amalgamation of incommensurable values, there is the potential for conflicts. I explore the notion of a value calculus for biodiversity in Chap. 5 (The calculus of biodiversity value).

3.3.2   Which Categories and Kinds Qualify

What, then, qualifies as a "biotic or biota-encompassing category"? I have suggested some sort of informal criterion of scientific legitimacy. But I don't pretend to know how to precisely characterize this notion.

This book's primary interest in axiology might justify focusing on those categories whose diversity shows promise for harboring value. However, the helpfulness of this criterion is uncertain.

More certain is that the selection of categories for biodiversity should be guided by endorsable principles. In particular, any temptation must be resisted to focus on categories whose diversity of kinds is perceived to have *positive* value, while ignoring or withdrawing from consideration otherwise legitimate categories whose diversity of kinds might not appear so attractive. If, for example, species diversity turns out to dilute the pool of species that we most want to keep around with species *non grata*, that circumstance, in itself, provides insufficient reason to push the category of species aside. The danger in succumbing to this temptation is real because the scientifically defined categories are the inventions of biologists who bring to them their biases in how and where they find value in things biological.

At least as great a danger is the exclusion of a legitimate *kind* within a category – one that cannot legitimately be excluded from contributing to the diversity of that category's kinds, but which as a kind considered on its own terms or with respect to its relationship to people, might be negatively valued. In the case of species and functions, this is more than a danger. It happens and happens often, as when the contributions to species diversity of agricultural "pests" or disease organisms or "invasive" species are banished to their own marginalized categories.

Also, there is an important difference between figuring out what is valuable and what can be evaluated with relative ease. Caution is in order to avoid a selection

---

[29]This question and the difficulty of providing an adequate answer to it motivate some biologists to claim the precedence of phylogenetic diversity, which is a category of diversity that spans taxonomic categories. See Sect. 3.3.2.1 (Features).

of categories or kinds that is biased towards whatever is most easily assessed in the field.[30]

The question of which categories best capture our intuitive notions of biological diversity surely merits careful evaluation. But I am going to decline to engage in the analysis needed for it – except insofar as they fare well or not so well in supporting value claims with regard to their diversity. It is hard to know what the evaluating criteria should be or how the inclusion of any particular one might be justified. And in the absence of such criteria, it is hard to know whether the categories currently on offer from the fields of ecology and conservation biology are "right" or justified. But frankly, I don't know where else to turn but to these categories, which frame our best understanding of the biological world as well as our further pursuit of its understanding. Even in the absence of vetting criteria, it is possible to make some useful observations about limitations and confusions in the usual choices.

For one thing, the definition of each of the categories on offer from biological science is to some degree conventional. That is not to say that any category is arbitrary, for every biological term is constructed for its usefulness in describing, relating, and explaining biological phenomena. This is true even of the biological concept of species – that most venerable of diversity-carrying categories. As already pointed out, such phenomena as ring species and asexual reproduction (among other reasons) make the gene exchange-based biological definition ragged around the edges. But some concept of species, which embodies important distinctions between organisms, has dominated human thinking about biological variety for millennia. Species is a natural kind if there is any such thing.

There can be good and bad reasons for adopting a convention, and there are generally pretty good scientific ones for biology's taxonomic conventions. This is true despite ongoing friction between more conventional Linnaean views, which rely on morphological and phenotypical similarities, and more modern cladistic approaches,

---

[30] I make this obvious-sounding point – that an easy measure of a thing cannot be assumed to be its true measure – because operational definitions of "biodiversity" such as Sahotra Sarkar's can and do (Sarkar 2002, 148, among other places) stipulate "biodiversity" to mean (as I would paraphrase it) "that diversity which an algorithm computes on the basis of 'estimator surrogates', which are properties easily assessed in the field. His route to "estimator surrogates" is indirect. According to Sarkar, "true surrogates", not "estimator surrogates", are (as the term implies) stand-ins for the real (biodiversity) thing. But according to him, there really *is* no independent real biodiversity thing, only his operational definition that relies on *estimator* surrogates whose selection is justified mainly on the grounds that it is possible to assess them in the field. For a qualification of this critical analysis, see Note 2 in Chap. 2.

Of course, Sarkar might well have some idea of "what biodiversity really is" independent of its operationalization by means of his set cover problem heuristic – even though he does little to characterize any such idea. On possible clue is the answer to the question: What set does his heuristic aim to cover? One might suppose that it is the set of all extant species, though this supposition seems incongruent with Sarkar's dismissive attitude towards species richness as a central component of biodiversity. But perhaps any real inconsistency is avoided by viewing this dismissiveness as deriving from the valid observation that choosing each of a disjoint set of places for their independent species richness is unlikely to cover the entire set of species – due to the likelihood of a significant overlap in species between the places.

which look more to shared evolutionary ancestries. As I have already hinted, not a few biologists have recently balked at using species and their diversity as the dominating foundation for biodiversity. Their reasons are not always clear. Malaise with the conventional aspect might play a part. But greater discomfort seems tied to the egalitarian implications of adopting traditional taxonomic kinds as the gold standard for diversity value. In particular, some biologists feel that not every species contributes equally to diversity, or more precisely, that the differences that make two species two are not as value-worthy as differences that derive from other categories.

And indeed, why *should* species-distinguishing differences be the gold standard of biodiversity value? Even if the concept of the category species is not entirely "arbitrary", what justifies a single-minded focus on "bare" species diversity? In the next few subsections, I look at other categories of diversity, which even if they cannot reasonably be said to entirely displace the category species, might plausibly be said to modify the biodiversity picture offered by a "bare" species perspective.

3.3.2.1 Features

Biologists Andy Purvis and Andy Hector (2000, 214) are representative in their expression of malaise with banking biodiversity on species diversity. According to them, "[Linnaean] Taxonomic boundaries are not comparable among major groups". They first illustrate this point with juxtaposed cladograms. In increasing order of depth and complexity they are: 14 species of cichlid fish in 9 genera, 7 species in several families of anthropoid primates, and finally, 12 species within the single genus *Drosophila* (fruit flies). Next they present histograms showing the uneven distribution of species within the higher taxa of genus, family, and order. John Spicer sums this up nicely with his "nearest approximation" survey of "all creatures that on earth do dwell". For example:

> To the nearest approximation every [plant, animal, eukaryotic or prokaryotic] species … worldwide … is a jointed-legged animal with a hard outer skeleton, an arthropod (Spicer 2006, 16).
> To the nearest approximation, every plant is a flowering seed plant, an angiosperm (Spicer 2006, 27).
> The most common and species-rich backboned animal (vertebrates) worldwide … are, arguably, the fish… To the nearest approximation, every bony fish is a perch and so too is every fish (Spicer 2006, 28).

And hitting closer to home:

> To the nearest approximation, every mammal is a rodent (Spicer 2006, 29).

Marine biologists Richard Warwick and Paul Somerfield (2008, 184) escalate malaise to titular (and Orwellian) status: "All animals are equal, but some animals are more equal than others". They, too, express the intuition that whatever differentness is involved in the diversity of species should not dominate the notion of biodiversity. For them and for others of like mind, this intuition appears to flow from several assumptions. These include: that there is a positive and proportional correspondence between biodiversity and its value; that there is a total ordering (and perhaps even a metric) defined on the diversity and its value (by virtue of the proportional relationship); and that increments (or decrements) to the category species violate the proportional relationship property because they result in *dis*proportionate changes to biodiversity value.[31] I explore some of these assumptions later in Chap. 5 (The calculus of biodiversity value). For now, I just note that they are at the very least quite interesting assertions and that their truth value is far from obvious.

The intuitions that make scientists uneasy with species diversity as the gold standard of biodiversity head off along at least two divergent paths. The first path leads first down and then up the taxonomic hierarchy. Going down the hierarchy (assuming that the genetic hierarchy is an extension of the taxonomic at its lower extremity): Suppose that the value of species diversity is essentially derivative – an artifact of genetic (or allelic) variation or diversity, which is taken to be the truly valuable bedrock of biodiversity. Then one might justifiably view the value of species and their diversity as primarily a means of bundling and propagating this genetic (allelic) diversity. Thus, the first of two species might be more different from a third because genetically further removed – that is, contributing more different genes to the total collection embodied by these organisms. Then, going back up the hierarchy: This same proposition might be expressed taxonomically by showcasing the first species in its own genus. On the view from this path, affording the first species a compartment of its own at the higher, genus level is a way of expressing its special status as "more different". Another interesting implication of the gene-emphasizing view is that species with more (intraspecific) genetic variation must be considered to have greater value, other things being equal. This consideration is independent of and can conflict with the relative genetic isolation of one species relative to others. A case in point is *H. sapiens*, which has minimal intraspecific genetic diversity and yet is sufficiently genetically distinct with respect to other organisms to claim sole ownership of the genus *Homo*.

A greater number of biologists who eschew the species diversity gold standard take a second path. They follow the implications of their non-egalitarian intuitions with respect to species first up then around and finally back down the taxonomic hierarchy. This aptly describes the path taken by conservation biologist Daniel Faith, who joins Warwick and Somerfield in championing "phylogenetic diversity" (Faith's term, which I will use) or "taxonomic distinctness" (the term favored by Warwick and Somerfield).

Faith's use of the term "phylogenetic diversity" might make it sound as though he is concerned with something different from, say, Warwick and Somerfield. But this is a false impression, which results from two elements of confusion in Faith's presentation.

---

[31] A total ordering allows comparisons between all members of a set – here, the amount and value of diversity in all possible states with respect to the comparison operator "more" or "less". A metric would additionally provide the means of ascertaining how much more or less diverse, and therefore, how much more or less valuable, is one state with respect to another.

The first and less serious confusion is that what Faith calls "phylogenetic diversity" is in truth a *measure* of a certain type of diversity.[32] But it is the second confusion that is the key to the puzzle. Careful scrutiny of Faith's characterization of that "certain type of diversity" makes it clear that, despite his use of "phylogenetic diversity" in the title of his original paper (Faith 1992), he actually is concerned with *phenotypical* variety, which he prefers to call "feature diversity". Needless to say, the phylogenetic heritage of an organism has a big say in the phenotype that it expresses. But this evolutionary connection does not change the fact that time and again, Faith says that he is interested in measuring feature diversity.

Faith's account of his motivation confirms this focus on feature diversity. He (Faith 1992, 2) argues that by embracing feature diversity instead of species diversity as the gold standard, one is no longer bound by an egalitarian principle for species, according to which "each... species is of equal status". As another biologist (and obviously not a fan of rayfinned fish) put it to me, "it's an argument for targeting the 'living fossil' coelacanth for conservation over the umpteenth rayfinned fish." In other words, those organisms that happen to be the umpteenth anything are less equal than others. The underlying rationale, according to Faith (1992, 2), is that "Taxonomically distinct species... contribute more to the diversity of a given subset because they contribute different 'features'." What he (1992, 2) finds interesting about phylogeny and computations of phylogenetic distance is his belief in them as "an effective indicator of underlying feature diversity". In other words, while phenotypical diversity is the real concern, according to Faith, phylogenetic distance is a good measure of this diversity.

Faith's more recent writing on this topic (Faith 2007) is equally clear (or confused) in how it slides between phylogeny and phenotype – as measure of thing measured. He reiterates his view that computations of phylogenetic distance are valuable purely in their role of helping to determine the diversity of features. The diversity of features is what counts, but is difficult to measure directly; hence his interest in phylogenetic distance, which he claims to a measure of phenotypical difference.[33] Other like-minded scientists make remarkably similar statements. So, in focusing my discussion on features rather than on phylogeny, I am not making an elementary error. Rather, I am faithfully following the scientific discussion for which Faith and others use the rubric "phylogenetic diversity".

While Faith's formulation of phylogenetic distance has evolved over the years, the adaptive modifications are not relevant to this discussion. In his earliest, and still cited formulation, he proposes (Faith 1992, 4) to compute phylogenetic distance as "the sum

---

[32] More about this syndrome of confusion is in the Sect. 4.1.2 (Measures and indexes) in Chap. 4 (What biodiversity is not).

[33] In his later piece, Faith (2007, §3.2) states:

> Feature diversity can provide a basis for valuation, but it raises measurement challenges. Not only do we not know, in general, the future value of different features, but also we cannot even list the features for most species. Phylogenetic pattern provides one way to estimate and quantify variation at the feature level. A species complements others in representing additional evolutionary history (Faith 1994), as depicted in the branches of an estimated phylogeny.

of the lengths of the all the branches that are members of the corresponding minimum spanning path" in a cladogram. This graph-theoretic formulation essentially expresses a notion of "distance" in the phylogenetic "tree of life". Pick the species that is your starting point; then pick your species destination. A phylogenetic version of Google Maps then gives you the shortest travel distance (and route) between the two. The key point is that (according to its proponents) this distance is a *measure* of how distinct the two organisms are in "*features*" aka "traits"[34] – a category of kinds that, while bound up with (cladistic) taxonomy, is not entirely taxonomic. For Faith and like-minded biologists, species are not so much anonymous bundles and propagators of genes, as the gene-centrists would have it, as they are anonymous bundles and propagators of features or traits. While the feature-centrists obviously differ from the gene-centrists on which category should carry the most biodiversity weight, they join hands in relieving the category of species from primary responsibility in this.

Unfortunately, Faith and others who espouse the category of features as the gold standard of biodiversity typically neglect to supply any convincing justification for promoting it to preeminence while demoting species to a secondary role. I believe that the mere expression of contempt for "the umpteenth rayfinned fish" or the umpteenth arthropod, or umpteenth angiosperm, or umpteenth rodent, or umpteenth anything is entirely inadequate as a reason. The felt need to resort to such a hollow defense and the more general failure to supply any respectable reason, I have to believe, are bound up with the second confusion mentioned above: In a rush to describe various graph-theoretic assessments of phylogenetic distance, there is also a rush past any sober consideration of the merits of "feature diversity" as biodiversity's gold standard.

A third, especially confounding confusion plagues the discussion because obscuring what, in the end, it is really about. Whether phylogenetic distance is a good measure or "indicator" or "surrogate" or "proxy" for differences in features or for feature diversity seems to be an empirical question. The answer to that question depends on the definition of phylogenetic distance. But it also depends on some reasonable characterization of "feature" – a difficulty that Faith (2007, §3.2) acknowledges[35] but does nothing to address. If the question is an empirical one, then a "reasonable" characterization must (among other requirements) be independent of phylogenetic distance.

On top of all this, the correlation between phylogenetic distance and distinctiveness in phenotype, which is supposed to justify using the former to measure the latter, is an empirical, not a logical, relationship. A significant body of counterexamples undercuts confidence in its reliability. Perhaps most obvious and not uncommon are cases of convergent evolution – whereby different lineages converge on similar traits[36].

---

[34] The term "trait" is favored by Hooper et al. (2005). See, for example Box 1, p. 6 in their paper.

[35] See Note 33.

[36] It seems that somebody plans work so well that, not infrequently, phylogenetically distant organisms arrive at the same "idea" to deal with their environmental pressures. The most famous example of this among extant creatures is the wing. An example of both morphological and functional homoplasy, this flight-enabling structure of both birds and bats, whose last common ancestor did not itself have wings, spans the entire clade of amniotic animals.

The only visible light that might conceivably help Faith fill in the black hole of what features feature in the "list [of] the features for most species" comes in the form of hints from other biologists – mostly in passing and not necessarily when explicitly discussing biodiversity. These include the observation that features worth scientific attention include the morphological characteristics of a species[37] and "life history traits" such as "reproduction at earlier ages/smaller sizes, increased reproductive investment" (Darimont et al. 2009, 952). Some biologists (for example, Helmus et al. (2007, E68)) suggest that features are the basis of interspecific as well as abiotic interactions. Others (for example, Forest et al. (2007)) see features as any and all characteristics that underlie evolutionary adaptability.[38] This characterization can make it seem as though the set of features is or includes the entire set of relational characteristics of organisms packed (as described in Sect. 3.1.3.3, Interactions) into non-relational attributes.

Overall, "features" is a frustratingly ill-defined category with ill-defined kinds, which, as Faith (2007, §3.2) says, "we cannot even list". This situation falls far short of the presumptive requirement – for assessing feature diversity – of listing and characterizing all features at least for all extant species collectively. Sometimes, the set of features appears to boil down to "whatever characteristics one can observe in whatever species one can observe" – a distressingly casual approach to what biologists call "phenotype". Yet, while the most basic questions – about what phenotypical diversity (aka "phylogenetic diversity") could possibly mean – remain unanswered, the phylogenetic distance measures have acquired a life of their own. They take on the aura of legitimacy as they are picked up by proposals for summary biodiversity statistics such as the "quadratic entropy" that Carlo Ricotta (2005, 31–34; 2007, 29–30) builds from them.[39] What it is that they measure is anyone's guess.

At this point, it is useful to recall previous remarks on egalitarianism in Sect. 3.1.1 (Egalitarianism). Moving to a category other than species does not void the egalitarian and fungible principles that apply to the diversity-yielding kinds in the new

---

[37]Some might argue that the diversity of morphology can be captured taxonomically by the diversity of phyla. Phylum is the highest taxonomic category below kingdom, and for which the basic body plan and "life history" plan of a creature jointly tend to be defining characteristics. But convergent evolution (again) makes it possible for organisms even in different phyla to share "morphologically characteristic" features.

[38]Faith (2007, §3.2) joins Forest et al. (2007) in supposing that features undergird biodiversity's option value. Sometimes these authors appear to use "option value" as it is defined in neoclassical economics, which would tie features to economic value via the expected consumer surplus of the feature set of all organisms. Section 6.9 (Biodiversity options) treats the topic at length.

At other times, these authors use "option value" as a (confusing) way of characterizing an organism's "evolutionary potential". And sometimes they convey the impression that they believe that this second meaning coincides with the first. But the proposition that features, which enable an organism to adapt are also features for which one should expect high consumer demand is an empirical one. While isolated cases might seem to exemplify this connection, it is clearly untrue as a generalization – as the adaptability of pest and pestilence amply testify.

[39]This topic resurfaces in Sect. 4.1.2 (Measures and indexes).

category. A precedence of the diversity of *features* over the diversity of *species* might show that, *as carriers or bearers of features*, not all species are equal or interchangeable. However, no feature can be considered more equal than another in its contribution to *feature* diversity, except for arbitrary and therefore unacceptable reasons. And in the absence of any compelling countervailing reason, all *features* must be presumed to be interchangeable, so far as feature diversity is concerned. In fact, the egalitarian and fungible qualities of features are arguably stronger than those qualities for species. That is because, unlike the parade of species, which so far never has marched back a species that already passed by, the parade of features is entirely capable of performing this trick: Features, unlike species, can be resurrected. Features once exemplified, but now gone with extinction of the taxa that exemplified them, can be, and have been born again. This is precisely what makes Elvis taxa possible.

Still more conundrums lurk behind the notions of "features" and "traits" as these are related to "phenotype". Every high school biology student learns that an organism's phenotype is the product of the interaction of its genotype with its environment. The same organism (therefore, the same genotype) can express very different phenotypes in sufficiently different environments. One need not go beyond the most casual observation of a plant such as *Pinus albicaulis* (whitebark pine). It is startling to realize that it is indeed the same species that one sees on one hand in stunted, twisted, tenuously perched, rock-bound, lightning-split, snow-loaded, half-dead, pygmy, or even shrub form at tree line (or above) in the high mountains; and on the other hand in the lush, symmetrical, and grand form it assumes in botanical gardens at sea level. In fact, with a quick glance, it could be mistaken for *P. monticola* (western white pine) – another five-needle pine whose "normal" height is within the range of the whitebark's botanical garden phenotype.[40]

This basic biology has two implications for species and species diversity, which to my knowledge have not previously been noticed. First, it implies that diversity is increased by ensuring the presence of any one species in as many different environments as possible in order to encourage as many different phenotypical expressions as it is capable of. If phenotypical diversity is a good, then the introduction of *P. albicaulis* into botanical gardens at sea level should be viewed as just a first step in a direction that should look favorably upon, and even aid and abet species "invasions", which constitute a similar kind of "transplanting". Second, given that (as discussed in Sect. 3.1.3.1) abundances of kind instances have no direct bearing on the diversity of their kinds, it implies (if taken along with the assumption that species diversity is legitimately regarded as subservient) that the loss of a species has little effect on diversity when the kinds of observable characteristics that species can

---

[40] This example derives from a personal experience. Having become familiar with the isolated scraggly, beat-up whitebarks that barely cling to life at alpine tree line in California's Sierra Nevada, I visited Strybing Arboretum and Botanical Gardens in Golden Gate Park, San Francisco. There, before me at sea level, was an exotically lush and giant pine that I could not identify. Glancing down at the identifying plaque left me incredulous.

exhibit (in any environment) are also exhibited – either separately or collectively – by other species. *P. monticola* might "cover" the whitebark's botanical garden self, while *P. balfouriana* (foxtail pine, yet another five-needle pine) might do the same for its montane phenotype, making *P. albicaulis*'s contribution to feature diversity close to nil.

Like a number of other discussions of biodiversity in the scientific literature, the one regarding features and their diversity shows three common tendencies. The first is that it tries to account for a non-egalitarian attitude towards the kinds of one category of diversity (in this case, species) by claiming (with no discernible supporting argument) that another category (in this case, features) dominates and supersedes consideration of the first. This inclination might well express reluctance to accept a multi-dimensional conception of biodiversity that makes room for both categories without prejudice for either.[41] The second tendency is to rush past careful characterization of a biodiversity-contributing category and into the details of its measurement, computation, or both. The proper algorithm for computing the phylogenetic "driving distance" between two species has been a topic of intense investigation. In contrast, there has been little investigation into what the numerical output of any phylogenetic distance algorithm measures. This second tendency facilitates the third – which is to end up conflating the diversity in the category with its measurement.

3.3.2.2 Abundances (Again)

I noted in an earlier subsection (Sect. 3.1.3.1, Abundances) that the evenness of species abundances is commonly held, together with species richness, as a principal pillar of biodiversity.[42] I argued there that it is a logical error to think that abundances directly affect species diversity. How, then, can abundances, and in particular, the evenness of abundances, be justifiably thought to be integral to biodiversity? The case for this proposition seems precarious, at best.

Ecologist Stuart Hurlbert (1971, 579) notes that the more evenly distributed the numbers of individuals of the various kinds of species, the greater the likelihood of an interspecific encounter between randomly traveling individual (and mobile) creatures. In other words, an even distribution of instantiations of kinds would provide one individual organism with the likelihood of a greater number of more diverse encounters. Critically, the individual organism in question might happen to be a field biologist, who would consequently have the experience and the impression of a greater variety of organisms.

---

[41] This reluctance comes back up for discussion in Sect. 3.3.3 (Multiple dimensions).

[42] See, for example, Purvis and Hector (2000, 212–213). This discussion is one among many that start with the presumption that species richness and evenness are the uncontestable twin pillars of biodiversity.

This might be an adequate account of "diverse experiences for a field biologist working transects". But it is entirely inadequate as an account of "the diversity of organisms in some particular place or the world generally". I leave aside the obvious difficulty of determining what the account means for an oyster or groundsel, which do not get around like a biologist and are therefore most always on the passive end of encounters with other organisms.

If abundances are not part of species diversity, then one might suggest that they constitute or are part of a separate category that is the basis for another component of biodiversity. Unfortunately, this move leads to the antithesis of the standard ecological party line on abundances. If abundances are made into their own biological category – an aggregative attribute of the kind species – then the *diversity* of abundances would become a legitimate source of additional (abundance-related) biodiversity. To have some wide range of abundances among different species would be (other things being equal) a more biodiverse state of affairs than to have equal representation for each species.

The customary view on abundances also seems strangely at odds with how community ecology is thought to work. Insofar as most (perhaps all) species live in assemblages, and most (perhaps all) assemblages have trophic structure, it is thought that species abundances tend to diminish going up the trophic ladder.[43] Some species – for example detritus-processing fungi and bacteria – occur and function in enormous numbers. Others – for example, *U. maritimus* – function in very small numbers. Therefore, whatever value most communities have (independent of their diversity in the "structural/ecological" hierarchy) inevitably conflicts with whatever value there is in equal representation of each of the community's species. Of course, value conflicts do occur. But this particular conflict – between the good of species diversity (understood, in part, as approaching the ideal of equal abundance) and the good functioning of most every species as the kind of organism that it is – points to a fundamental incoherence.

Another riddle concerning the virtue of even abundances also seems to have gone unremarked. Again, according to the "standard formula", species evenness or the degree to which evenness is approximated is, alongside species richness, what counts for species diversity, biodiversity, and therefore, presumably their value. This means that if one wields Thor's thunderbolt to smite down individuals of all species – but in proportion to their current abundances – then one is left with the same set of relative abundances, the same species richness, the same biodiversity, and presumably, the same value for biodiversity.[44]

---

[43] This might not be true of species *biomass*: Substituting biomass for numbers of individual organisms might diminish these differences without necessarily eliminating them. Biomass is an aggregative "characteristic" of species that biologists are inclined to marshal (for legitimate scientific descriptions) when abundances of species do not yield the relationship they seek.

[44] It is also uncertain what one should think about the fact that, according to the same "standard formula", a world with twice the species and half the individuals is more biodiverse than a world with half the species and twice the individuals.

3.3.2.3   Functions

Along with the category "features", the category "functions" – shorthand for "functional traits" – provides a path of retreat from the species-egalitarian implications of biodiversity conceived as pure taxonomic (and particularly, species) variety. Also, like "features", it seems to be a vehicle for driving interactions and other relational aspects of organisms' behavior back into the diversity barn as *non*-relational "characteristics". It is also possible that this category is a means or viewed as one of straddling the taxonomic and "structural/ecological" hierarchies with the hope of merging them, or at least avoiding conflicts between their respective values.

The notion of a functional kind arises from the observation that the disappearance of some organisms from an environment can have little or no effect on which properties are attributable to it as an ecosystem (which is where the "structural/ecological" hierarchy comes in) or the degree to which any ecosystem property is manifested. At the other extreme, the arrival of some other organisms can be transformative. Though entrances and exits instigate changes that make the influence of organisms more apparent, that influence does not entirely depend on the time of an organism's arrival.[45] Traditionally, most attention has been devoted to species that are dominant in one of a couple of senses. This typically means either a species, which constitutes the greatest biomass in an ecosystem at some trophic level, or a "keystone species". The latter can be observed to exert such dominating trophic leverage – typically as top-level predators – that their influence can be determined without the need to monitor the effects of entrances and exits.

In the last decade or so, the idea of a species' function in an ecosystem has ballooned into the grander idea of an ecosystem service. I follow Hooper and his colleagues (2005) in taking an ecosystem service to be an ecosystem property of benefit to people.[46] On the scientific evidence, some relatively small contingent of all the species present in an ecosystem – those with the requisite functional traits – does the heavy functional lifting in rendering the ecosystem's services. Concern for functional diversity then becomes concern for the diversity of those organisms whose functional traits are responsible for performing the desired functions.

A broad understanding of an organism's "function" encompasses its trophic function or (simplifying somewhat) its "lifestyle" as (among others) a predator, herbivore, mutualist, or parasite. It also includes the organism's influence on such abiotic conditions as the availability of limited and organism-limiting resources, the

---

[45] I say "entirely" because the order of arrival is not irrelevant. A script for the eventual success of an organism can require an entrance that is fortuitously timed to coincide with conditions that enable it to gain a solid foothold.

[46] This is how Hooper et al. (2005, 7) put it:

> *Ecosystem services* are those properties of ecosystems that either directly or indirectly benefit human endeavors, such as maintaining hydrologic cycles, regulating climate, cleansing air and water, maintaining atmospheric composition, pollination, soil genesis, and storing and cycling of nutrients. [italics in the original]

See also Sect. 6.3 (Biodiversity as service provider).

disturbance regime (the patterns of change exerted by such forces as fire, wind, floods, disease epidemics, insect outbreaks, landslides, and avalanches), and even microclimate. Examples of resource effects include plants that fix nitrogen (for example, *Myrica faya* in Hawaii) and those that effect surface soil salinity and water availability (for example, *Tamarix* spp. in the North American southwest). Examples of disturbance influence include fire-inducing species (for example, *Bromus tecto-rum* – cheat grass in addition to *Tamarix* in western North America). An example of species-induced microclimate modulation is the influence of moss in boreal forests: The moss reduces heat flux into the soil, thereby stabilizing permafrost and sup-pressing rates of nutrient cycling.[47]

A major problem with this conception of function is that, while scientific discus-sions often present the locus of a function to be a certain kind of organism (species), a function does not just protrude from a creature or plant like some one of its append-ages. More often than not, a function is the result of multiple interactions of multiple kinds of organism with each other and with many factors in a shared abiotic environ-ment, in the absence of any one of which the function would not have appeared.

A neat study by Daniel Soluck and John Richardson (1997) nicely illustrates this detachment of function from any specific organism. These ecologists describe stream predatory invertebrates – stoneflies of the genera *Doroneuria* and *Calineuria* – whose predatory behavior encourages evasive behavior on the part of their prey. Flushed from hiding, the same prey becomes easy pickings for *Oncorynchus clarki* (cutthroat trout). Consequently, the trout gain weight.[48] In this case, the predatory interaction of one species aids and abets a predatory interaction of another (and changes its phenotype, to boot) that shares its prey. The end result is that the system functions in a way that increases the biomass of the trout. This, no doubt, is an ecosystem service for its increased production of tasty fish dinners. However, reinterpreting this phenomenon in terms of a trout-fattening functional trait possessed by stoneflies seems notably unhelpful.

The phrase "functional diversity" does not, by itself, appear to make any refer-ence to organisms. But the way in which Hooper et al. (2005) and others use the locution at times gives the impression that they intend it as shorthand for "the diver-sity of a complement of *species*, which collectively embody functional traits that manifest as one or another function in an ecosystem". This "meaning", however,

---

[47] See Chapin et al. (2000) for more examples of "functions" performed by organisms. Chapin et al. do not discuss functional diversity, as such. Rather, their discussion focuses on how species dif-ferentially affect the "functioning" of their environments. The projection of the functioning of an ecosystem onto the functioning of its inhabitants is the core of what Hooper et al. (2005) and others understand to be functional diversity.

It is not a coincidence that the cited examples involve "exotic" species. This is not because their function is necessarily either good or bad. Rather, it is a matter of an exotic's function being easier to discern because biologists have had the opportunity to observe the effects of their recent arrival on the scene. Sax et al. (2007) offer a fascinating discussion of how invaders are a fecund source of ecological and evolutionary insights.

[48] This work is also cited in Chapin et al. (2002, 274).

builds in an empirically testable hypothesis, which posits a certain direct kind of relationship – between the diversity of species and the diversity of functions, literally understood – that is mediated by the various "functional traits" inhering in different kinds of organism.

Biologists implicitly recognize the empirical nature of this thesis, while simultaneously signaling an uneasy awareness that they might be trying to "prove" it by definition. The relationship between the diversity of some complement of species and the manifestation of ecosystem functions is the subject of a large, long-standing, and ongoing body of research in numerous variations (for examples of which, see Hooper et al. 2005, 9). But even the scientists doing this research seem to be aware that something could be awry. They express their malaise by agonizing over the possibility that their results might merely reflect statistical "sampling effects" (Hooper et al. 2005, 9–10): Suppose that you just randomly start to throw various species at an ecosystem. If you keep on tossing in species, then the definition of the functional trait-mediated relationship dictates that, sooner or later, for any ecosystem function that you care to produce, you will toss in some species with the functional trait that gets that particular ecosystem job done. It will be sooner if you happen to toss in the right species sooner; later, otherwise. It is entirely a matter of the random draw of species. As a consequence, the biologists worry, the ordinal number of the eventually successful species toss will not reflect anything interesting about how species abundance and therefore species diversity determines functional diversity. What these scientists fail to see is that the specter of worrisome sampling effects is an artifact of the empirically untested assumption that throwing one kind of organism after another into the ecosystem mix is equivalent to throwing in the functional traits that are attached to these organisms. This equivalence is the crux of the "functional trait as appendage" model and it legitimizes the characterization of the experiments as "sampling the functional traits of organisms".

Yet, when scientists step outside the long shadow of biodiversity – where that normatively loaded topic does not figure in their stated hypothesis – they manage to free themselves from that model. In fact, a plethora of research concerned with how functions emerge from complex interactions shows just how untenable the model is. Arthropod ecologists William Snyder and Edward Evans (2006), for example, provide a fascinating survey of intricate trophic interactions centering on introduced arthropod generalist predators (AGP's). The functional focus is on "pest control". The dining habits of these insects and crustaceans tend to interconnect on multiple trophic levels. The negative or positive effects on the various organisms involved are complex, unpredictable, and often surprising. Often, there are competing positive and negative effects on a single organism. Snyder and Evans (2006, 97, Fig. 1) cite this case:

> In alfalfa fields in Utah (United States), invasive seven-spot ladybird beetles both suppress alfalfa weevil through direct predation and weaken weevil suppression by eating the aphids that provide food (honeydew) for weevil parasitoids, simultaneously enhancing and weakening weevil control.

There would seem to be little point in trying to understand this in terms of a weevil-controlling trait in ladybird beetles.

In another study involving alfalfa, Snyder and Anthony Ives (2001) detail how phenological considerations bearing on reproduction and parasitism lead to a long-term result that reverses the short-term one. In the beginning, when the alfalfa plants are short, climbing-challenged carabid beetles gobble up pea aphids. However, the beetles' fine dining experience concomitantly reduces the parasitism of already-present parasitoid wasps. The end result is that aphid populations rebound when the alfalfa plants grow to a height that put the now relatively parasite-free aphids out of the hungry beetles' reach. What explanatory purpose would be served by characterizing this nexus in terms of the alfalfa plant productivity-enhancing trait of the beetles?

It should be clear that, just as for purely taxonomic conceptions of diversity, functional conceptions (however defined) have to do with the *diversity* of kinds. Conceptual clarity demands straying into the just-described conceptual thicket of the functional traits of species. And it requires insisting that functional diversity focus on the diversity of kinds within functional categories or one grand, all-encompassing category of functions, rather than the diversity of kinds within, say, taxonomic categories. Just as for the diversity of species, the diversity of functions is not logically tied to the existence of particular, identified kinds (or some selected subset) of functions. So far as diversity is concerned, no kind of function can be justifiably excluded from consideration on the grounds that it is "bad" for some independent reason.

Left out of most discussions of functional diversity is the disconcerting reality that from the human point of view, for every beneficial function, there is one that is harmful. The productivity of food crops is beneficial. Not so, the productivity of weeds. Nor is it so for the crop-harming functions of various herbivores known as "pests". Also routinely omitted is consideration of the fact that the benefits of a function typically hinge on the degree to which a function is manifested. Too much is often as harmful as not at all. The nitrogen-fixing function of some plants can be a benefit. But not – according to some ecologists – when the supply of nitrogen is so abundant as to ensure the demise of plants whose prior success was the result of their ability to live in a nitrogen-impoverished environment.

Eutrophic coastal zones – areas that have large inputs of nitrogen and phosphorus that typically derive from agricultural runoff – offer an illuminating case in point. Eutrophication just might be the poster child for phenomena that are supposed to demonstrate how human activities negatively affect biodiversity in *both* taxonomic and functional ways. The gist of the standard narrative is that eutrophication causes a plunge in species diversity with a concomitant plunge in functional diversity. The unappealing side-effects – for example, the ugly algal blooms, toxin production, and fish kills – are consequences of this dual reduction in diversity.

Unfortunately, the premise – that there is a reduction in species and functional diversity – is entirely open to question. The plausibility of the narrative probably hinges on an equivocation, which conflates "reduction of functional diversity" with "reduction in the intensity of certain desired functions". For undoubtedly, eutrophication does change the mix of species and the set of functions that are most intensely manifested.

According to oceanographer Jeremy Jackson (2006, 34), these offshore regions are typically

> … inhabited by flagellates and microbes which are key components of the microbial loop and make better meals for jellyfish than for copepods and fish [which] also results in the deposit of vast quantities of unconsumed organic matter for burial on the sea floor. All of this carbon sinking may be good for global carbon credits against greenhouse gas emissions…

In other words, on the taxonomic side, eutrophic zones host a complement of species that differs from the copepod and fish-dominated collection that might previously have existed in these aquatic regions. But while there clearly is a collective shift in the species represented, there does not appear to be any clear or well-founded evidence for a decline in species *diversity*.

What about functions? Again, I am unaware of any before versus after accounting of functions that would evidence a reduction in functional *diversity*. Clearly, some functions are *added* – even some good functions. As Jackson suggests, the newly formed eutrophic regions add the functional kind "carbon sink" to their complement of functions. This has recently become viewed as a good function – an ecosystem service. As a result, at least one ecosystem service is added, too. Other functions are enhanced. The blooming of algae is a dramatic example of an intensification of the productivity function.

In this case, there is no doubt that the functioning of the system has dramatically changed along with its complement of organisms. But this is an entirely different assessment from the one according to which biodiversity has diminished.[49]

As another example, consider the riparian systems in the American southwest that recently have been transformed by the increasing presence of *Tamarix* (species of saltcedar). This phenomenon might be causing some changes in species composition and certainly in species abundances. Just as certainly, the functions exemplified by these riparian systems have changed. According to biologist Stuart Chapin III and his colleagues (2000, 235–236):

> Introduction of the deep-rooted salt cedar (*Tamarix* sp.) to the Mojave and Sonoran Deserts of North America increased the water and soil solutes accessed by vegetation, enhanced productivity, and increased surface litter and salts. This inhibited the regeneration of many native species, leading to a general reduction in biodiversity.

---

[49]One distinguished scientist dismisses my account on the grounds that, according to him, the new system is obviously "not as vibrant" as the old; it is sick and diseased – no longer in good health. But this is not so much of an objection to my analysis as an attempt to change the subject. In the absence of any account of what vibrancy is, one might think that it is a good state of ecosystem health. But ecosystem health is not biodiversity or a particular state of biodiversity; nor are these things logically equivalent. Moreover, a host of problems plagues attempts to characterize ecosystem health, especially insofar as its characterization leans on an analogy to human health. The biggest problem of all is in finding a good reason for why ecosystem health matters. Human health concerns people because people are concerned about their health. But ecosystems are definitely not concerned about their health. And so, if there are grounds for concern about ecosystem health, they cannot arise for the same reasons that justify concern about the health of other persons. This topic is an important, recurring theme. Section 7.2.3 (Home is where you make it – in situ, ex situ, and ecosystem-engineered), and more abstractly analyzed in Sects. 8.1.3 (The ends of the biodiversity project) and 8.1.5 (A metaethical gloss on biogeoengineering).

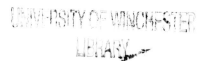

Scrutiny of the sources cited by these scientists makes it clear that none supply evidence for their claim of "a general reduction in biodiversity". The claim is a speculative leap from a 1970 *laboratory* study of salts exuded from the leaves of *T. aphylla* (Athel pine, a species of *Tamarix* that has established monoculture stands in some areas of southern Nevada). Chapin et al. cite no field study to support their claim. What is more, it appears to be false for both species and functional *diversity*. On the taxonomic (species) side, no species is known to have been extinguished (either locally or globally) as the result of the newly attained foothold of *Tamarix* in these habitats.[50] Furthermore, Chapin et al. do not take into account the fact that several *Tamarix* spp. (obviously, not present before their recent introduction and naturalization) now contribute to local species diversity. This is not to mention *Diorhabda elongata* (tamarisk leaf beetle), which conservation biologists intentionally imported in the hope that its herbivorous predilection for *Tamarix* will reduce or reverse the *Tamarix*-induced changes. So if anything, species richness has increased. On the other hand (as Dudley points out) there undoubtedly have been shifts in species abundances and consequently, in the compositional mix. But it is anything but clear whether this constitutes a reduction of biodiversity.[51] Undoubtedly, some species suffer in the *Tamarix*-altered environment. But others thrive. All persist. The famous (or notorious, from the viewpoint of biologists trying to eradicate *Tamarix*), case of thriving is that of the endangered *Empidonax traillii extimus* (southwestern willow flycatcher).

Something similar can be said about the functional side of the naturalization of *Tamarix* in the American southwest: Things have changed. One change is the increased productivity noted by Chapin et al. As with eutrophication, there is increased carbon sequestration. There are others, including functions connected with increased fire frequencies and the channelization of rivers. But to say that these riparian systems function differently (for better or for worse) is not the same as saying that they are functionally less diverse.

Section 6.3 (Biodiversity as service provider) further explores the question of how the value of biodiversity is affected by changes in ecosystem functions, which are regarded as "services".

---

[50]Tom Dudley, a leading researcher and expert in this area of conservation biology, sums up what is known about species extirpations in *Tamarix*-dominated systems in this way:

> As to local extirpations, again it is considered one of the factors causing declines in many riparian-dependent species, but I do not know of any studies or reports that have singled *Tamarix* as a sole causative factor. Particularly when an important factor contributing to *Tamarix* invasion is human-caused ecosystem modification (water diversion and regulation, agricultural development, etc.), this is especially difficult to tease out. In fact, there have been few riparian extinctions anyway since these species tend to be ruderal with fairly wide distributions, and only local endemicity in cases where there are extremely limited wetland sites with highly specialized conditions. [Personal communication]

See the further discussion of this in Sect. 3.3.2.3 (Functions).

[51]Part of the unclarity has to do with the unclear role of abundances in defining diversity, as discussed in Sects. 3.1.3.1 (Abundances) and 3.3.2.2 (Abundances (again)).

3.3.3 Multiple Dimensions

I do not claim to have described all categories or hierarchies of categories that might be said to contribute to biodiversity. In particular, I have refrained from delving into the diversity of kinds that Sahotra Sarkar and others characterize as belonging to categories in a "structural/ecological" hierarchy.[52]

This is partly for lack of space. But it is more for an even greater lack of conceptual clarity in defining these categories and their kinds. While it seems that there are some reasonable, if fuzzy criteria for distinguishing, say, populations of organisms, the same cannot be said for the kinds of such categories as community, habitat, ecosystem, landscape, and biome.[53] The lack of consensus about individuating criteria is not restricted to the kinds within any one of these categories. There is not even general agreement about how or even whether to distinguish the categories themselves. For example, some scientists (such as Angermeier and Karr 1994, 691) take community and ecosystem to be the same category. Many others do not.

The category "ecosystems" is the standard-bearer in this hierarchy. But this book primarily concerns itself with questions of value, and there seems to be little hope in making the diversity of ecosystems a bearer of value. Most if not all the positive value claims connected with ecosystems are caught up in the particular details of the particular workings of particular systems, which provide particular human-recognized benefits. This evaluation has little to do with the *diversity* of kinds in the category "ecosystems". If anything, it has something do with a noticeably complex arrangement of a wide variety of discernibly different kinds of parts (biotic and abiotic), which by definition requires a certain level of diversity in those parts. But it is not the diversity of parts that is doing the job. Rather, it is that particular collection of diverse parts interacting in just the way that those parts interact to produce the benefits that people enjoy.

One might try to take this into account by reinventing "ecosystem diversity" as "ecosystem complexity", taking the latter phrase to refer in some general way to diversity in other categories. But if the aim is to close the gap to human-recognized benefits, then this move accomplishes little. One need only consider such a basic ecosystem property as its productivity. Systems both simple and complex are productive. Some of the simplest, least diverse, and most productive are salt marshes. But not only does complexity not correlate with productivity, nor does productivity correlate with human benefits. The high productivity of *Tamarix*-altered habitats in the American southwest and that of algae-dominated eutrophic zones are generally not perceived as beneficial nor valued in any other way.

Despite this conceptual morass, I accept that, in principle, the diversity of kinds drawn from the categories of a "structural/ecological" hierarchy might contribute

---

[52]This "biological/structural" hierarchy is to be distinguished from the concepts of "place" and "scale", which I discuss in Sect. 3.3.4 (Place and Scale).

[53]This unfortunate circumstance is, I think, as significant a problem for ecological research as it is for those of us who are concerned about values that attach to ecological entities. The variable usage regularly confounds attempts by scientists to generalize across research efforts.

"components" or "dimensions" to biodiversity. More generally, I see biodiversity as analogous to a non-linear, multi-dimensional space. Each biotic or biota-encompassing category defines a dimension of diversity that is a matter of the diversity of its kinds. The dimensions are not all orthogonal (independently varying). Categories in one hierarchy such as the taxonomic categories are clearly interrelated and form closely aligned dimensions. But even categories from outside this hierarchy are not unrelated to it. Kinds of feature and kinds of function, for example, appear not to be completely independent of taxonomic kinds.

I do not wish to claim that there is an irreducible complexity in the concept of biodiversity, one that makes it logically impossible to reduce the concept to some single category of biotic-ness. But I feel secure in making the claim that the complexity is irreducible to simple or even complex-but-tractable formulas in practice. I also feel secure about the corollary that this kind of intractability is likely to persist for well over the lifetime of the average biologist or philosopher. If reductionist conceptions of biodiversity are not logical nonsense, right now and for the foreseeable future, they are practical nonsense.

At the beginning of this chapter, I promised to deliver a definition of "biodiversity" that is not so distended that it fails to carve out from the natural world something particularly value-worthy about it. I believe that my definition of biodiversity fulfills that promise to squeeze in the conceptual bellows. It certainly is a significant retreat from (recalling Bryan Norton's words, quoted in this book's Preliminaries) "all that we mean by, and value in, nature". Even so, there is no denying that my definition's multi-dimensional character makes it both far-reaching and complex. And my definition has some immediate implications. First, insofar as measurement of the diversity of any category is possible, it is unlikely that diversity in one category is commensurable with the diversity in another. As previously suggested, even categories within one hierarchy – for example, the diversity of species and the diversity of families two levels up the Linnaean ladder – quite likely elude rational comparative reckoning.

Second, I have required of my conception that it make sense of "more" and "less" in biodiversity. This is still possible insofar as it makes sense to restrict attention to the diversity of a particular category in isolation from other kinds. There can be a greater or a lesser variety of species; *ergo* more or less species diversity. Most discussion of biodiversity likely focuses on species for just this reason. Talk about "more" and "less" can be understood on these terms. Also, a partial ordering of states with respect to general biodiversity is conceivable and might even be possible.[54] For example, there might be situations in which a "Pareto improvement"[55] in diversity occurs – wherein some one or more categories clearly increase in the

---

[54] A set – here states of general biodiversity – is partially ordered when it is not always possible to say, of two distinct states, that one is more or less biodiverse, or that their biodiversity is equal. While comparisons of less and more might always be possible within a particular (totally ordered) category of diversity, one category might say "less" while another says "more".

[55] A "Pareto improvement" in neoclassical economics is a change wherein at least one individual's economic welfare increases but no one's decreases.

diversity, but the diversity of no category measurably or perceptibly decreases. However, a total ordering of states with respect to biodiversity is almost certainly not in the cards.

Third, the multidimensional nature of general biodiversity calls into question the legitimacy of any scalar rating or even (less stringently) ordinal ranking of biodiversity considered in all its messy complexity. One might compare the situation to that of consumer ratings services. To rank products, these services must take into account the relative value of ease of use, noisiness, repair record, and a raft of other properties. Ratings of biodiversity would have to indulge in similar judgments of the relative value of the diversity in each included category. As I think Chap. 6 (Theories of biodiversity value) demonstrates, the business of attributing value to diversity in any category is a tricky one. The prospect of comparing values across categories in a formulaic way is, I think, naïve and generally unhelpful.

Which has not kept a number of biologists from trying to do exactly that. In fact, there is a cottage industry that emits enormous volumes of papers entirely devoted to scalar grand summaries of biodiversity. In Sect. 4.1.2 (Measures and indexes) in the Chapter concerning What biodiversity is not, I further pursue the case for regarding these attempts as unhelpful verging on nonsensical.

3.3.4   *Place and Scale*

Limitations of space preclude a full treatment of how place and spatial scale fit in with the concept of biodiversity. But a few words on these topics are in order because these twin notions intrude prominently into many discussions of biodiversity. I am speaking of place and scale in a way that puts these outside the "structural/ecological" hierarchy of categories that works its way up from populations through communities to biomes. Place is here simply conceived as a bleeding chunk of land or water, not necessarily defined in terms of biological/scientific niceties – though a biodiverse place would contain many things that are so characterizable. Similarly, scale is here viewed as a hierarchy of the sizes of bleeding place chunks – ranging from "local" or even "very local" (less than 1 m^2), to "regional", on up to global.

3.3.4.1   **Place (Again)**

In the judgment of Sahotra Sarkar, biodiversity *content* and therefore biodiversity value[56] are both inextricably attached to some place where they (one must assume)

---

[56] According to Sarkar, biodiversity *content* is somehow weighted by "viability" (his term) to arrive at biodiversity *value*. Section 6.13.1 (Viability and endangerment) takes up the supposition that viability contributes to the value of biodiversity.

properly belong. Couching "value" in terms of "what matters for conservation", he (Sarkar 2002, 134) says:

> ...places – precise biogeographic allocations with their specific components, including all biological and cultural features – are what matter for conservation: they alone retain the heterogeneity that provides the intuition for bio*diversity*. [italics in the original]

This statement tightly couples biodiversity to the places where it exists. For Sarkar, this coupling is a consequence of two propositions. First is his definition of "biodiversity". Sarkar takes this term to mean: "that which should be conserved by applying an algorithm that embodies his definition of 'conserving biodiversity'". Second is the singular aim of such an algorithm – which is to select specific places for their biodiversity value. But Sarkar is confused about the implications of these precepts because, in fact, the net effect of his algorithm is to entirely detach biodiversity from place.

To show this, I must sketch Sarkar's algorithm.[57] In doing so, I am not concerned with its theoretical aim, but rather with its practical one – which appears to be to sufficiently well identify the differences between places (an evolution of Robert Whittaker's venerable (1974) notion of β-diversity) in a collection so as to effectively increase the *overall* diversity of a finally chosen subset of the collection. The subset chosen by the algorithm is its output.

The algorithm attempts to achieve its objective in an iterative way. In each iteration, it recruits for inclusion in the accumulating collection of places the next, still omitted place that is most different with respect to the existing collection taken as a unit. The newly selected place is added to the collective unit in preparation for the algorithm's next iteration. Thus, while each iteration requires that place-relative differences be assessed with respect to an already established collection of such places, this is merely a means to the end of as high as possible diversity in the final subset considered *globally*. Thus, it appears that Sarkar, contrary to his word, detaches biodiversity from place, after all – unless by "place" he means "the patchwork of places in the algorithm's output" or "the earth".

Other difficult questions are raised by Sarkar's insistence on tightly tying biodiversity to place; his answers seem vulnerable to serious objections. First, what qualifies as a "place"? Can it be a zoo, a garden, a gene bank, or an artificial wetland? A wetland in Keoladeo, India, whose existence was created by, and whose continued existence depends on, cattle grazing is one of Sarkar's favorite examples of a conservation-worthy biodiverse place.[58]

Second, how big or small can "a place" be? Sarkar acknowledges that this is an open and difficult question. But it is particularly problematic for his operationalized conception of biodiversity. This conception, he insists, must encompass migratory and other spatially dispersed "biological phenomena" (Sarkar's term).

---

[57]The algorithm's computation-theoretical significance – as a questionably effective heuristic to solve the NP-complete set cover problem – is characterized in Note 1 in Chap. 2.

[58]This interesting example is described more fully in Sect. 4.1.1 (Wilderness).

Some migrations span continents. So we are talking "very dispersed". The implication that, in the end, biodiversity is tied to place on a continental or global scale seems incongruent with Sarkar's intuition. And in any event, it makes the proposition of a place-diversity nexus considerably less interesting.

Third, Sarkar's attempt to push "valuable biodiversity" into the output of his algorithm raises a disturbing question about what his views entail for the practice of conservation. This is something else that he acknowledges, but brushes aside without substantive comment: How can we make sense of the biodiversity value of places that are not selected by the algorithm? The lack of a satisfactory answer to this question is made much worse by the fact that his heuristic[59] might actually arrive at a worst possible solution. A place that his heuristic omits might well be part of a minimal covering set, which is a correct solution to the problem.

Perhaps the most constructive interpretation of Sarkar's tight coupling of biodiversity and place is one that looks at place as a summing up of all the factors that influence the good functioning of organisms and that in turn, are influenced by their good functioning – whatever those factors might be. Unfortunately, this account is quite unhelpful – either because it leaves completely unspecified what the relevant factors are; or because it makes a tautology out of the proposition that couples together place and biodiversity: Place is coupled to biodiversity because place comprises whatever is coupled to (by way of influencing and being influenced by) biodiversity.

3.3.4.2 Scale

The discussion in the preceding subsection at several points trips across a consideration of the scale of a "place". That is because the two notions are closely linked.

The "place" in which some organisms, such as intertidal invertebrates, spend their life is measured in fractions of square meters. Scales of that order of magnitude are usually considered "local". The place for other organisms is measured in thousands of hectares. This is true, not just for migrating species, but also for wide-ranging ones such as *Ursus arctos horribilis* (grizzly bear). Anything more than a few dozen hectares would be considered "regional".[60] All scales are smaller than "global", which is the entire earth. Much biological heterogeneity arises because of some heterogeneity on every scale from the smallest local scale to the global. And many important "laws of biology" are highly qualified by scale.

---

[59] Recall that this heuristic is Sarkar's operational *definition* of biodiversity.

[60] Generally, the scale of "ecosystem functioning" – for example, productivity and nutrient cycling and retention – is "local" – often taken to be "less than 100 hectares". In fact, functions often present themselves in considerably less than this amount of space – as when scientists remark on an ecosystem comprising the bowels of a single carnivorous (pitcher) plant (Butler et al. 2008) or your bellybutton (http://www.wildlifeofyourbody.org/). However, regional and even global factors inevitably impinge on locales, as the global phenomena of CO_2 fertilization and N deposition (among others) make clear.

The relevance of scale to biodiversity is significant and this complicating factor transfers to the value picture. The phenomenon of species immigration (aka "exotic" species[61]) provides one salient example. There is accumulating evidence that localized and even regional diversity more often than not *increases* as the result of the immigration of exotics.[62] On the other hand, many scientists believe that specific diasporas are responsible for global biodiversity decline, though the force of this causal factor relative to others is uncertain.[63] Insofar as one places a value on biodiversity at each of these scales, conflicts are certainly possible and probably should be expected.

---

[61]Throughout this book, I consciously and conscientiously avoid the commonly used terms "invasive" and "alien" with regard to species, except when explicitly remarking on, or calling attention to how their built-in negative evaluative baggage illegitimately prejudices judgments of value. I prefer the aptly descriptive and less emotionally charged terms "immigrant", "recent immigrant", "recently naturalized", and "exotic".

[62]I think that caution is in order for evaluating the effect on global biodiversity of immigrating species. The vast majority of ecologists assert that "invading aliens" are responsible for global biodiversity declines as if it were well-established fact. But the evidence they cite for this tends to be highly anecdotal and few if any mention, let alone try to account for the extent to which species immigrations might engender new species. I revisit the neglected topic of speciation later in this book – primarily in Sects. 6.7 (Biodiversity as value generator) and 7.3 (Biodiversity value in human timeframes).

[63]For a summary of work on the consequences of species invasions for biodiversity, see Fridley et al. (2007). See also Gurevitch and Padilla (2004), who call for sober (re)evaluation of this question.

Chapter 4
What Biodiversity Is *Not*

This chapter is a compendium of common errors in conceptualizing biodiversity.[1] Its purpose, in part, is to clarify the positive space of what biodiversity is by way of contrast with the negative space of what it is *not*.

Each mistaken conception considered here can be classified in at least one of three categories. "Category mistakes" mistake biodiversity for something else entirely, such as wilderness. This is the territory of *The Man Who Mistook His Wife for a Hat*[2] except that no clinical diagnosis is involved. I include within this category the subcategory of "contractionist" conceptions that confuse biodiversity for a particular instantiation of one or more of its abstract parts – such as the identities of the particular species that constitute the current state of species diversity. Finally, "accretive conceptions" are ones in which biodiversity accretes properties, such as rarity and viability, which have little or nothing to do with variety in the world. While these properties might merit independent consideration, they are irrelevant for biodiversity and so it is illegitimate to allow them to influence biodiversity's assessment.

4.1 Category Mistakes

4.1.1 Wilderness

Biodiversity is not wilderness. Conserving biodiversity is not equivalent to conserving wilderness. And conserving biodiversity, under that description, is not even a means of conserving wilderness.

---

[1] I touch on some of these errors in the preceding chapter, but in this chapter elaborate on them in the company of others.

[2] This is the titular story in Oliver Sacks' well known 1985 recounting of various neurological disorders.

D.S. Maier, *What's So Good About Biodiversity?: A Call for Better Reasoning About Nature's Value*, DOI 10.1007/978-94-007-3991-8_4, © Springer Science+Business Media B.V. 2012

As stated in this book's Preliminaries, environmentalism (at least in North America) for a long while rallied around the banner of wilderness. Vexations in defining the concept of wilderness and in arguing for its value largely led to supplanting that environmental banner by the banner of biodiversity. The analysis in this book raises the question of whether that trade represents a bargain. But whether or not it does, the concepts *are* quite distinct.[3]

The common concept of wilderness leans on a distinction between the natural and the human cultural/artifactual that is embedded in the state of a place, with some possible adjustment for the history of how it arrived in that state. Most of all, a wilderness is a place where human design is absent, or at least largely insensible, or insensible to humans, or insensible to scientifically untrained humans, or insensible to those who are ignorant of how it came to be in its current state, or insensible to the scientifically untrained. Or – reflecting the vagueness of the concept – something like that. In contrast, biodiversity, properly considered, cuts across this signal distinction between two possible states of a place. Diversity in a biotic or biota-encompassing category might be wherever you find it – including in genetics labs, inner cities,[4] zoos, the halls of office buildings, artificially constructed "landscapes" – places that no one would mistake for wilderness. Therefore, as Sarkar correctly points out, the targets for the conservation of biodiversity are generally quite different from those of wilderness.[5]

Certainly, human design and intervention of the sort that destroys wilderness does not necessarily have a biodiversity-diminishing effect. Nor does the *lack* of discernible human design and intervention now and in the past have a biodiversity-enhancing one. In fact, human influence might just as well increase biodiversity, while human restraint might just as well decrease it.

Sarkar (for example, 2005, 42) is fond of citing the previously mentioned case of Keoladeo National Park in India, which is worth some attention in this context. According to Sarkar, part of Keoladeo became an "artificial" wetland – and therefore, not a wilderness – as the result of its dual use as a hunting preserve and as a grazing area for cattle. Subsequent exclusion of cattle led to the re-establishment of plants that choked off the bodies of water, which in turn led to declines in fish, and then birds. Subsequent to that, the re-introduction of cattle led to the reopening of the wetland and the return of the fish and birds.

There are many gaps in Sarkar's chronicle of Keoladeo. One is its failure to account for the plant diversity of the "natural" or "wild" state of this place, which

---

[3] Sahotra Sarkar (1999, 2005, 37–44) says most of what there is to say about wilderness versus biodiversity. Though Sarkar's discussion is polemically charged (he derides the wilderness as a "myth") and his analysis of the value of wilderness as resting solely on aesthetic grounds is one that I would challenge, Sarkar hits on the main points of distinction.

[4] While penning the first draft of this book, I monitored via live webcam (http://www2.ucsc.edu/scpbrg/nestcameras.htm) the mating and nesting activities of *Falco peregrinus* (peregrine falcon) atop high-rise buildings in San Francisco and San Jose, California.

[5] See, for example, Table 1 in Sarkar (1999, 406) or a later version of this, Table 2.3.1 in Sarkar (2005, 41).

might well offset the lesser variety of fish and birds. Nor is there an accounting for the variety of insects that almost certainly would accompany the "wild" plants and which might well account for the major part of that place's species diversity. Also, the cited declines are in fish and bird *populations*, not declines in the *diversity* of species. So, consistent with his account, there might well have been no decline at all in species diversity, only a mistaken belief in this as a result of a less-than-comprehensive accounting for all species combined with confusing species abundances for species diversity.

Despite the fact that Sarkar's analysis of Keoladeo is flawed in all these ways, it still illustrates (though perhaps without truly exemplifying) the valid principle that distinguishes biodiversity from wilderness.

4.1.2 Measures and Indexes

A thing is not the same as its measure. Biodiversity is not the same as its measure. Nor is biodiversity one and the same as an index that purports to summarize, represent, quantify, or rank it. Even for the simplest of all components of biodiversity – species diversity – there is a distinction between it and its measure in terms of species richness – a simple count or census. This particular measure certainly has strong intuitive appeal in capturing the diversity that arises from some number of distinct species. But by making a hair-splitting point, it is easy to appreciate the non-hair-splitting principle that even a species count is nothing more than a measure – and not the best measure – of species diversity: A better measure of the *diversity* would be, not the total number of species, but the total number *less one* because two distinct kinds (of species) define just one difference.[6]

To bring home the point that species richness is just one possible measure of species diversity, consider some other proposals for measuring it. One is the number of pairings between n distinct species, which is $(n * (n-1))$. It is hard to find grounds for preferring species counts to species pairings or vice versa, other than ease of computation. It is notable that while these two different measures produce different numbers, they yield identical rankings. But the two measures again yield different results when used to determine which of two pairs of species diverse states is *more* different. Suppose that <m, n> represents two different states of biodiversity – one with m species, the other with n species. According to the obvious difference arithmetic, the linear (simple count) measure says that the difference in diversity in <2, 3> is identical to the difference in <4, 5>: In each pair of states, there is a difference of precisely one species. However, (using the pairing formula above) the quadratic measure says that the states in the pair <4, 5> are more different than those in <2, 3>, since the diversity measures that correspond to <4, 5> are represented by <20, 12> (with a difference of 8), while the diversity measures for <2, 3> map to

[6] Wood (2000, 44) makes the same point.

<2, 6> (with a difference of just 4). Do the two states <4, 5> exhibit a greater difference in biodiversity than <2, 3>; or not? Which is correct? What does this disagreement even mean?

It is probably misleading to focus on a statistic for species richness because the rationale for such an easily understood object of measure is likewise most likely easy to understand. In contrast, the rationales for other measures and indexes occupy a different part of the spectrum, which ranges from elusive through obscure to totally opaque. Consider the Shannon-Wiener entropy or index – one of the most venerable of biodiversity statistics. This index, inspired by information theory, treats the individuals of a collection of species in some environment as transmitting bits of species "information". Each species is viewed as a species-message. The likelihood of "receiving" a species-message is the proportion of individuals of that species relative to individuals of all species. The Shannon-Wiener index therefore represents the species-information entropy of the system.

What entropy has to do with biodiversity in general or species diversity in particular is obscure. This would not matter if scientists were only concerned with correlating this strange information-theoretic statistic with some other ecosystem property. But that is not how they interpret their computations of it. Rather, they insist on saying that the entropy *is* biodiversity, that when the entropy is lower, this means that biodiversity is diminished (and by the way, this is bad), and that when it is higher, this means that biodiversity is greater (and by the way, this is good). The Shannon-Weaver entropy nonetheless remains in common scientific use to mean "the biodiversity of a system". Despite its opacity, it is wielded to justify real conservation decisions "for the good of biodiversity".

My description of the Shannon-Wiener entropy to this point has glossed over an important feature of it. It is one of many indexes that combine two or more measures of diversity. This combining of different measures inevitably incorporates judgments about the relative contribution of what is combined. The bases for these separate judgments, which are tacit judgments in value, are also obscure, rarely acknowledged, and even more rarely discussed. Shannon-Wiener entropy, for one, combines species richness with the evenness of species abundances. The index is maximized with maximum evenness – that is, when each species present is represented by an identical abundance of individuals. In fact, it is possible to *decrease* the value of the index by *adding* individuals of one of several previously equally abundant species. That is how entropy works. It is entirely unclear whether or not biodiversity or biodiversity value work that way, too.

Indexes, considered individually, do not just yield results that have a questionable relationship to the biodiversity that they claim to represent. As Stuart Hurlbert (1971, 578–579) points out, there also can be a substantial lack of concordance between indexes. Applied to two different communities, two indexes can disagree about which one is more biodiverse. This should be cause for substantial concern to scientists – even when "just" logical consistency, not value-laden conservation choices, is at stake. One community that is more biodiverse than a second community according to one index, and is less biodiverse than the second community according to another index, could actually be one and the same community at different times. In that case, one and the same community with one and the same history will have both increased and decreased in biodiversity.

Hurlbert's example of how one index can reverse the ranking of another compares the results of applying the Shannon-Wiener index to the results of applying a variant of another popular biodiversity index – namely, Simpson's index. The latter is a measure of the likelihood that two individuals in a hypothetical random drawing from a species-mixed collection will belong to different species. It differs in a fundamental way from the Shannon-Wiener entropy by emphasizing the dominance of the most abundant species and by discounting the evenness of species abundances. This is true to the point that, unlike Shannon-Wiener entropy, Simpson's index can *increase* if a relatively rare species goes extinct or is extirpated. Simpson's index can also *decrease* with the introduction of a relatively rare species.[7] This bears repeating. If Simpson's index is the measure of biodiversity, then it tells us that, in some situations, one method of increasing biodiversity is to ensure that a species goes extinct.

Indexes with increasingly arcane combinations of measures, increasing degrees of complexity, and concomitantly increasing degrees of remove from a comprehensible connection to biodiversity continue to be built. For example, as mentioned in Sect. 3.3.2.1 (Features), Carlo Ricotta (2005, 33–34, 2007) espouses an index in a class of "quadratic entropy" indexes. Ricotta proposes two overlaying interpretations of this sort of index. In the first instance it is supposed to be a Simpson-style characterization of how much community components (such as two kinds of organism) differ from each other – the "distance" of one to the other – considered as random pairs. The same formalism is also supposed to represent species richness.

In a single scalar value, a quadratic index expresses some functional combination of the number of species and their pairwise differentness. A quadratic index is, in equal measure, clever and formally elegant. But the price of cleverness and elegance is opaqueness. What does this scalar value mean? While he provides no satisfying answer to this question, Ricotta espouses using his quadratic entropy index to measure diversity in morphology and function. This makes it look as though part of the attraction of a quadratic index is that its pairwise differentness component facilely accommodates such a well-liked measure of pairwise differentness as phylogenetic distance. As discussed in Sect. 3.3.2.1, while some consider phylogenetic distance to be a measure of the diversity of features, this view is fraught with its own difficulties. And even Ricotta expresses misgivings about anomalies – some inherited from its Simpson index heritage – that make its coherent interpretation elusive.

I have tried to convey some sense of the nature of biodiversity measures and indexes by way of sketching a few of the simplest ones. There are many others and many arcane details about their definition, application, and comparison that I might have included. Fortunately, essentially none of these details are relevant to my discussion. That is because, so far as measures are concerned, when the measure is

---

[7] This violates the formal property of "monotonicity". It is dismayingly easy to generate this and other startling results with simple examples that one can feed into easily available tools for calculating Shannon-Wiener entropy and Simpson's index. One such tool is available at http://www.changbioscience.com/genetics/shannon.html.

of the diversity of a well-defined category such as "species", there is no need to resort to talk about the measure in order to talk intelligibly about the diversity. Even notions of "less" and "greater" are within easy conceptual reach without reference to measures. We have no problem understanding that the event of a species going extinct decreases global species diversity and that a speciation event increases it – without reference to an index.

On the other hand, one might contend that for the uncomfortably fuzzy categories of "trait", "feature", or "function", establishing a measure might help to sharpen things up. But that, of course, would be a mistake. If one cannot define general criteria for what qualifies as (say) a "feature", if one cannot identify what features there are in the biotic world and their relative contribution to the diversity of features, then pasting a number on the diversity of features is more than just unhelpful. It a distraction from realizing the need to remedy the absence of any adequate theoretical framework that would give that number some meaning.[8] A precise measure or index or ordinal ranking of who-knows-what or who-knows-why cannot ground a rational discussion about the value of biodiversity.[9]

Measure-combining indexes have even less relevance than "bare" measures to the concerns of this book. There are two salient reasons for this. First, as mentioned above with regard to Shannon-Wiener entropy, to the extent that an index formulaically weights various components of diversity in combining them, it builds in judgments about their relative importance. When the discussion moves to the "good" of biodiversity, those weights are cashed out as their relative contribution to the good. If those judgments of relative value had prior justification, then the summary index might serve as a kind of axiological shorthand. Unfortunately, justifications beyond brief expressions of intuition are, to my knowledge, entirely lacking. Moreover, the intuitions – such as those about species evenness – can be utterly bizarre, at least according to *my* intuition. As a consequence, the use of an

---

[8] For an interesting example of an elaborate formal evaluative framework for "attributes", see Nehring and Puppe (2002). Unfortunately, enthusiasm for their elegant "multi-attribute model of diversity" quickly subsides with the realization that they omit the same difficult part of the discussion that is also missing with regard to "features". While these authors propose a formalism for an evaluative diversity function, they provide no criteria determining what qualifies as an attribute that can be an argument for their function. They also do not offer criteria for why some arguments are evaluatively more dominant than others. And finally, they offer no good reason to think that it even makes sense to combine attribute values as arguments in a grand, total-diversity-valued function.

[9] There is a deceptive allure in numbers and in the notion that a thing represented by a number is, just by virtue of that, more real and better understood. In Chap. 4 of *The Little Prince* (de Saint-Exupéry 2000), after recounting the astronomical history of the asteroid that is the little prince's home, the narrator apologizes with the words:

> If I've told you these details about Asteroid B-612 and if I've given you its number, it is on account of grown-ups. Grown-ups like numbers. When you tell them about a new friend, they never ask questions about what really matters. They never ask: "What does this voice sound like? What games does he like best? Does he collect butterflies?" They ask you "How old is he? How many brothers does he have? How much does he weigh? How much money does his father make?" Only then do they think they know him.

index to justify a decision of what and how biodiversity *ought* to be conserved becomes a way of sidestepping the discussion about what justifies weighting the "components" of the biotic world, in the way that the index does, in making a conservation decision. It is a mistake to think that the mere invention, use, or publishing an index to represent some investigative result in a peer-reviewed scientific journal ensures that the value judgments it embeds are justified.

There is a second reason for why the details of measure-combining indexes are irrelevant. If biodiversity is the diversity of kinds in each of multiple biotic and biota-encompassing categories; if these categories interrelate in complex ways that make them irreducible in principle or practice; and if, *a fortiori*, they are irreducible to simple or even complex-but-tractable formulas (as suggested in Sect. 3.3.3, Multiple dimensions), then an index doesn't just hide unjustified value judgments about biodiversity. It is not just a distraction. Rather, it is practical nonsense. It is literally a game of numbers, which have no more descriptive power, meaning, or interest as a description of something in the world, let alone something of potentially great value, than the number that is the sum of a quark's charge and its mass. To the objection that this is an invalid analogy because charge and mass have different units, my response is "yes".

In short, the prevalent tendency to couch the value of biodiversity in terms of scalar-valued indexes, and for this to prop up the conclusions of "scientific" assessments of biodiversity, is quite troubling. One respect in which it is troubling has a parallel with neoclassical economics' use of scalar welfare indexes, which I described in Sect. 2.1.2.1 (Consequentialism). In both domains, the index seems to miss much that is both axiologically and morally relevant by purporting to be a uniformly applicable measure of all things, for all persons, in all circumstances – no matter how disparate. As a result, its axiological and moral relevance is obscured.

In another respect, the situation for measure-combining biodiversity indexes is worse than it is for economics. Economics, as I have observed, can at least lay claim to a "natural unit" of economic welfare when assessing various forms of capital, including "natural capital". This unit is grounded in the preferences that individuals express in real or imagined market transactions. This is the magic atomic glue that, however implausibly, binds such diverse things as Yosemite National Park, a coal-burning power plant, your safety, your favorite hand lotion, and your autonomy to a unified and uniform universe of values. In contrast, there appears to be no similarly plausible atom of value for biodiversity. This circumstance makes biodiversity indexes relatively inscrutable as comprehensive assessments of biodiversity and its value – compared to welfare indexes as comprehensive assessments of all that people value.

This comparison does not earn welfare indexes an endorsement or anything close to that. They are no more plausible than the premise that the enormously disparate entities whose value they summarize permit uniform and uniformly comparable valuations across a broad spectrum of circumstances. The issue of commensurability, too, is what most fundamentally makes the biodiversity index approach implausible – in light of such disparate, value-carrying categories as the diversity of species, phyla, habitats, alleles, functions, and features.

Juxtaposing the common use of indexes in modern economic theory and modern biodiversity theory points up one more troubling parallel between the two. I have just observed that the construction of a measure-combining biodiversity index requires a questionable commitment – to the position that the categories whose diversity enters into the index are commensurable. The use of the index as the representative of all biodiversity value carries with it another commitment – to the position that these various categories of diversity admit formulaic tradeoffs of the fungible "goods" (where the rules for computing the index or its algebraic expression define the formula). The consequences of this for biodiversity can be as unpalatable as they can be for economic welfare.

I think that this single, startling example suffices to make this point: Suppose that Shannon-Wiener entropy is taken to represent biodiversity and suppose, too, that different biodiversity values are ordered by the ordering of different Shannon-Wiener entropy values. Then one is committed to allowing that, by increasing the abundance of one species (for example, by somehow making it more fecund), it is possible to justify letting another go extinct.[10]

4.1.3   Particular Species

Identifiable species, represented by individuals of those species, are responsible for the species diversity of an ecosystem. A salt marsh, for example, has its plants, such as *Spartina* spp. (cord grass); mollusks, such as *Geukenzia demissa* (ribbed mussel); and birds, such as *Nycticorax nycticorax* (black-crowned night heron). But its state of species *diversity* does not depend on the presence of that particular complement of organisms. *A fortiori*, the diversity is not defined by or dependent on the particular identities of these particular organisms as being the "components"[11] of a biodiverse salt marsh. Were *Spartina* replaced by *Salicornia* (pickleweed), *G. demissa* replaced by *Melampus bidentatus* (pulmonate snail), and *N. nycticorax* replaced by *N. violaceus* (yellow-crowned night heron), then the species present in the marsh would all have changed, but it would be hard to deny that the species diversity would not have been affected at all.

Nevertheless, the error of conflating biodiversity with the identities of its "components" – species diversity melding into specific species and functional diversity sliding into specific functions – is ubiquitous in popular environmentalist and (perhaps more surprisingly) scientific discourse. The slide is often quite explicit.

---

[10] A simple example suffices to show this. The bracketed numbers are species abundances:

State 1: <100, 100, 50, 1>

State 2: <100, 100, 100>

State 2, thought lacking a species in State 1, has a higher Shannon-Wiener value than State 1. You can verify this by plugging these numbers into the calculator cited in Note 7.

[11] Here, "component" is used in a sense that is common in the scientific literature on biodiversity, where it refers to particular kinds or individuals. This should not be confused with the sense in which I have used the term – as "a category admitting a diversity of kinds".

As one of many examples, Chapin and his colleagues (2000, 234–235) define biodiversity to include "the particular species present (species composition)", as well as "interactions among species (non-additive effects)". Their definition gives them license to talk about how altering the species composition of an ecosystem sometimes changes how an ecosystem works – and to talk about this as a matter of changing biodiversity. But in fact, what they say has little to do with the *diversity* of species present or with the *diversity* of functions performed. They cite the example of the naturalization of *Tamarix* spp. in the American southwest, which has undoubtedly changed riparian habitats there. But as noted in Sect. 3.3.2.3 (Functions), the best scientific evidence undercuts their position on even its own terms: In this case, it shows that there has been little or perhaps no change in species diversity (as distinct from species abundances – that is, the size of their populations): The populations of some species have diminished. Others representing new species "on the block" (Sax and Gaines 2003, 563) have arrived or increased in number. In this much-discussed case, at least, there is so far no evidence of either an extirpation (local extinction), let alone a global extinction.[12]

For global diversity, the total loss of any particular species – its extinction – obviously decreases species diversity. But painful as this might be to acknowledge, this is a marginal decrease – by whatever amount that particular species contributes to species diversity and the diversity in any other biotic or biota-encompassing categories connected with it. This is important to keep in mind when discussing biodiversity's value. A particular species might have some value, even some great value that is distinct from its contribution to diversity. But that other value is a topic for a completely different philosophical debate – one that is not about biodiversity.

4.1.4 Particular Ecosystems

As we saw in the preceding subsection, talk about preserving species diversity often slides into talk about preserving some small set of particular organisms, or worse, into talk about changes in species abundances (that is, population sizes). Even more frequently, talk about ecosystem diversity often devolves into talk about preserving the current state of some particular ecosystem or other. The grounds for indulging in this biodiversity Newspeak might be a view of an ecosystem as something like a substrate for various other categories of diversity – including, perhaps, species diversity. But this defense merely provides a different proof of my general point – that despite appearances to the contrary, the diversity of the category ecosystem is not what is under consideration in most discussions of "ecosystem diversity".

In fact, my survey of the environmentalist and scientific literature leaves me in doubt about whether ecosystem diversity – in the sense of the diversity of kinds of ecosystem – is ever really discussed. The fuzziness of the concept of ecosystem as

---

[12] See Note 50 in Chap. 3.

a category might partly explain this. But in truth, under the heading "ecosystem diversity", one mostly finds discussion of more or less productivity, more or less stability, or a greater number or fewer resident and visiting species. As I have previously stated, I embrace in principle the idea that ecosystem diversity, properly understood as the diversity of kinds of ecosystems, could contribute to biodiversity. However, I have yet to encounter any serious consideration of this proposition.

4.1.5   Biodiversity as Process

Commentators sometimes transform biodiversity into a process. Bryan Norton (2001, 90, 91), for example, suggests that evaluating biodiversity by "the ranking of 'development paths'" avoids an "entity bias". That bias, he says, is responsible for various "destructive implications" – saliently, that it bogs biodiversity evaluation down in census taking: According to him, thinkers have "for too long emphasized 'being' at the expense of 'becoming'".

This might leave the reader scratching her head. And the head-scratching might not stop when, descending from these lofty Whiteheadean heights, Norton (2001, 90) explains that an inventory of kinds does not yield as "[deep a] source of value in nature" as what "might be called nature's 'creativity'". At times, he (Norton 2001, 91) seems to equate this creativity with productivity, which ecologists also routinely discuss, and which seems to have no straightforward relationship to the diversity of anything.[13] But productivity does connect with his central notion, straight out of neoclassical economics, of biodiversity as a "development path".

At other times, Norton's notion of the creative force of nature seems to be a matter of the unfettered working of natural laws (as he says, "the evolving processes of life"). But natural law needs no human assistance in continuing to operate as it always has and always will. At other junctures, Norton seems on the very brink of saying something like "biodiversity is the set of processes that engenders <blank> ". One senses that he really wants to fill in the <blank> with "those previously created species that have showed up in species censuses" (my words, not his). But perhaps sensing the obvious circularity and genetic fallacy (because the mere fact that those species came into being says nothing about their goodness) in this move he (Norton 2001, 91) instead substitutes "productivity" and "creativity" for <blank> .

Perhaps a more sympathetic reading of Norton is that he is not really interested in biodiversity at all. The vagueness of his discussion certainly leaves his real subject in doubt. He might be urging his readers to trade the concept of biodiversity in for another "keystone" concept (in the tradition of trading in "wilderness" for

---

[13] Ecologists find that the productivity of ecosystems has no consistent correlation with various kinds of diversity, including species diversity. For example, some of the most productive ecosystems (salt marshes among them) are also among the most species-lean.

"biodiversity") to which we can glue all or most natural or environmental value.[14] But even on this reading, Norton's exhortation (Norton 2001, 90) to pay primary attention to "change, process, and becoming" quickly hits a brick wall. He makes it clear that only *some* changes, processes and becomings are worthy of being preserved[15] – oddly, those that (Norton 2001, 91) "maintain, support, and repair damage to their parts". But what "parts" need repair? From what Norton says, one could think that the answer is: the set of species that, according to some unexplicated ideal, should show up in those much maligned censuses – that is, some particular biodiverse state of affairs (most likely some short time ago) before it was "damaged". Or one might think that the processes themselves need repair. But what constitutes their state of disrepair versus a state of proper functioning? Norton does not say. But his expression of concern about the capacity to "repair damage" and "heal wounds" puts him on a direct path to such (normative) notions as "ecosystem health" and "ecosystem integrity".

With "ecosystem health" and "ecosystem integrity", I believe that one leaves behind anything that can be recognized as "biodiversity", even in some extended, honorary sense. The move to these concepts is curiously reminiscent of the prior move from wilderness to biodiversity: In both cases, the successor concept differs utterly from, and cuts across, the critical defining categories of its predecessor. Also, in both cases, the successor's characterization and evaluation are no less formidable than those tasks were for its predecessor.

The notion of ecosystem health is not the most central concern of this book, but it has a way of intruding into arguments about natural value that begin with biodiversity, but then spiral away from all the problems connected with that subject. This notion already has popped up in Sect. 2.1.2.2 (Deontology) and it interjects itself later in the book. For now, this forward-looking comment suffices: The notions of "health" and "integrity" obviously build in their own evaluations or norms. Health and integrity are clearly the good and desirable norms. Their opposites – disease and disintegration – are bad. But the obvious analogy to the health of individual creatures or particularly to human health does not carry far, both because it is difficult to justify some set of properties as the norm for an ecosystem (the way it ought to be); and because even if such a norm existed, the ecosystem would have no morally compelling interest in achieving it.

Angermeier and Karr (1994, 692) offer a representative conception of ecosystem integrity in the form of "wholeness" and "appropriateness":

> Biological integrity refers to a system's wholeness, including elements and occurrence of all processes at appropriate rates.

---

[14] It really is not clear that Norton is entirely focused on either nature or the environment as most people would understand those terms. Norton (2001, 94) talks about "the forms of creativity that support a wide range [i.e. a diversity] of human opportunities". Here, it seems that he is thinking of some preferred processes whose value rests in their ability to produce diversity as exemplified by "a range of tree species" (species diversity) and a "variety of landscapes and settings available [to a house builder or developer]". This, it seems, is a concept of diversity that includes something like the diversity of economic development.

[15] One might wish to reflect on the oddness of "preserving change".

Unfortunately, they cannot offer an empirical test for "wholeness" or "appropriate-ness". In fact, in the absence of historical knowledge, there seems to be no viable scientific test for a state of "nature in balance", which one might presume to be the "healthy" state or state of "integrity".[16]

One suggestion for a test – from Norton (2001, 93) but not only from him – relies on a system's resistance to perturbation (stability) and/or a tendency for it to return to an initial, "healthy" state (resilience). But this proposal is inadequate for at least two reasons. First, most ecosystems would fail it – because, as a matter of fact, stasis or equilibrium appears to be the exception rather than the rule. That is as true now as it has been throughout geological history. Second, the test offers no non-arbitrary, scientific grounds for anointing any particular initial state as "healthy" and therefore the norm – in preference to some other state that "nature's creativity" could produce by acting on that state as a starting point. To sum this up from Norton's process view-point, the creativity of "nature as source" does not depend on its possession of any particular testable quality – including stability and resilience.[17]

However it is interpreted, in the end, Norton's suggestion for an ontological shift to processes for the most part seems to be a very confusing way to avoid talking about biodiversity and its value. It nevertheless re-emerges in Sect. 6.7 on Biodiversity as value generator.

4.2   Accretive Conceptions

The emblem of accretive definitions of biodiversity is Bryan Norton's "all that we mean by, and value in, nature". Norton is not alone in the broad sweep of his con-ception. In his fascinating sociological exploration of what prominent biologists think about biodiversity, David Takacs first asks these scientists to define the con-cept; then asks them what, in nature, is *not* covered by it. More often than not, the answers (from Takacs 1996, 75–81) in essence are, "not much, if anything".

In the Preliminaries, I suggested that a grossly distended conception of biodiver-sity-as-all-of-nature is both too broad (demanding a far more general and far-reaching discussion than "mere" diversity) and too narrow (for drawing a boundary between the natural and the artifactual – a boundary that biodiversity arguably crosses).

I am concerned in this section with some other, more modest conceptual accre-tions that sometimes make their way into conceptions of biodiversity. These tend to be properties or "characteristics" in the sense discussed in Sect. 3.2 on that topic. However, a characteristic in this sense can have little or nothing to do with diversity because, while a characteristic (unsurprisingly) characterizes a kind and might be

---

[16] I would claim (here without argument) that taking history into account still does not help to establish a viable test of ecosystem health.

[17] See the discussion of Mark Sagoff (2003, 99–103, especially) of this in the context of alien versus native species.

exemplified by the individuals that instantiate that kind, this in itself does not logically entail anything for the diversity of these kinds. Thus, rarity characterizes some species. But this characteristic of some species has no logical connection to the diversity of species, as such.

On the other hand, as Sect. 3.2 (Characteristics) suggested, a characteristic or property might be a kind (in my ontologically promiscuous sense) within another category – of non-substantial universals. On this account, rarity might be on one end of a spectrum of kinds in a category that extends to commonness at the other extreme. The diversity of this category would come from instantiations that span this range. That leaves rarity as just one characteristic among others, including commonness, in a diverse set. However, this interpretation is not at all in the spirit of what is commonly proposed.

According to another interpretation, what is proposed might be about the *value* of biodiversity rather than its definition. There are at least two ways to understand this. The first claims that (say) the rarity of a species affects the value of that particular species. That is, it is so much more precious for its rarity. Whether or not that claim has merit, it is clearly outside the domain of species *diversity* and therefore biodiversity insofar as this consists in species diversity. On the second understanding, the suggestion is that rarity does, in fact, bear on the value of, not one or another particular species, but on the value of species diversity. In the case of rarity, I am at a loss to think of how this is possible. But it is not entirely implausible for some properties such as endangerment/viability and causal history, which I take up in Sect. 6.13 on Viability and endangerment.

Rarity and the other characteristics briefly surveyed in this section are commonly represented as integral to the concept of biodiversity, and specifically, species diversity. My discussion takes this premise seriously.

4.2.1   Charisma and Cultural Symbolism

Charisma and cultural symbolism occupy close quarters. I take charisma to be a special case of cultural symbolism in which the potency of the symbol derives from popular perception (independent of justification) within a particular society.

Charisma and cultural symbolism can be props for subverting species egalitarianism. The troglobitic (therefore essentially never seen) *Batrisodes texanus*, unappetizingly but aptly dubbed the "coffin cave mold beetle", is unlikely to garner the cult following of *Panthera tigris tigris* (Bengal tiger) – which (subspecies) is the national animal of Bangladesh and (at the species level) of India, and is a symbol of power and grace in western cultures.[18] But the fact that the latter creature is charismatic and that the former is not has nothing to do with their (equal) contribution to species diversity.

---

[18] While diverse in charisma, both creatures share a vulnerability to extinction.

One might try to salvage some connection of charisma to biodiversity by broadening the perspective from *species* diversity to a more inclusive kind of biodiversity in which cultural meaning is itself a biota-encompassing category. This suggestion is open to more than one interpretation, but it seems unappealing no matter how interpreted. According to the first interpretation, cultural diversity would derive from the representation of different *kinds* of biological cultural symbols. But cultural symbols are both positive and negative; diversity requires both and as many as possible in between. On the negative side are, for example, symbols of pestilence (the bacterium *Yersinia pestis* and its flea hosts for bubonic plague) and symbols of filth (cockroaches). Perhaps just as unfortunate is the contribution to cultural diversity of the symbolic significance of whales to Japanese food culture and the symbolic significance of tigers to Chinese for the medicinal power of their genitalia. With these cases in mind, the idea of hanging value on cultural diversity loses much of its appeal.

According to a second interpretation, diversity derives from spanning different *degrees* of cultural symbolic significance – from the culturally meaningless to the most meaningful. This second interpretation says that any just-discovered species (which will have had no previous contact with human culture) would join *B. texanus* at the opposite end of the spectrum from *P. t. tigris*. But *all* these species would be contributing and *valued* as contributing to cultural diversity because they occupy different parts of the spectrum of cultural meaningfulness. Unfortunately, this interpretation quashes the likely motivation of featuring culturally heralded creatures over the unheralded.

In sum, it is not surprising that, even if these two interpretations of "cultural diversity" are coherent, neither, to my knowledge, has any serious constituency.

Even if there were a legitimate way to rig the cultural diversity scale in favor of the "more meaningful" or in favor of the "more positively meaningful", that notion would be difficult to define in a way that is both clear and justifiable. On account of the excitement connected with its novelty, the just-discovered organism might – at least by the natural philosophers who read the *Nature* paper that assigns the organism's unpronounceable pseudo-Latin scientific name – be considered as charismatic as *P. tigris*. On the other hand, it would resemble the anything-but-charismatic coffin cave mold beetle in its more or less complete disconnection from general cultural awareness. What justifiable principles of evaluation would legitimately adjudicate these conflicting considerations?

Perhaps I am belaboring a point that is abundantly obvious. Charisma and cultural meaningfulness largely hinge on the vagaries of human preferences, which are influenced by culture, marketing, individual tastes, and training, among other factors. No one would think that the biological diversity of the world really depends on the preferences of this or that group of persons. No one would think that an organism's contribution to diversity depends on its preferential standing in this or that human fan club. This is not to deny that these human constituencies exist. It is to deny that they are somehow part of biodiversity, enhance biodiversity, or enhance the value of biodiversity.

4.2.2 Rarity

Several distinct concepts of rarity are at times glued onto the conception of biodiversity. I touch on two prevalent ones.

4.2.2.1 Geographical Rarity

The concept of *geographical* rarity, also known as "endemism", plays a major role in the famous and influential definition of "biodiversity hotspot" proposed by environmentalist Norman Myers and his associates (2000).[19] According to Myers et al., the first qualifying condition for a hotspot is its inclusion of a minimum of 1,500 endemic plant species (0.6% of the estimated 250,000 species described at the time of their paper).[20]

As a general conservation strategy, which extends beyond the diversity of species, the suggestion to conserve hotspots is vulnerable to many criticisms, such as a number that conservation biologists Peter Kareiva and Michelle Marvier (2003) express. However, even as a strategy for more narrowly conserving species diversity, the attraction of focusing on endemism is problematic. It is possible that Myers and his colleagues offer their suggestion as a set of specific guidelines that are supposed to realize special efficiencies in running a Sarkar-class heuristic to solve the species set cover problem. In terms of that problem, any place containing endemics, by the definition of that term, covers some part of the universe of species that cannot otherwise be covered. But of course, the occurrence of one or more endemics in some place does not by itself entail that these and other species will be packed into that space with any remarkably great density. In fact, the definition of "hotspot" tacitly acknowledges this observation with its requirement, which is logically independent of that for endemism, for a great many different kinds of plants. Therefore,

---

[19] The term "biodiversity hotspot" traces back to Myers' previous work from 1988 (not cited). The notion of biodiversity hotspot described and discussed here is the later, revised version. Some version of this notion now appears in virtually every textbook on ecology and conservation biology.

[20] Vertebrates are *not* included in Myers' hotspot definition. However, his 25 selected hotspot areas show a considerable congruence between the degree of endemism of (the kingdom of) plants and (the subphylum of) vertebrates collectively (excluding fish). However, such a congruence does *not* obtain between the various non-fish classes of Vertebrata (Amphibia, Aves, Mammalia, and Reptilia) separately. On the other hand, the correlation with endemic species in other taxa – fish, insects, fungi, etc. is a matter of speculation – although again, there is some strong evidence for correlations between plant diversity and insect diversity. These details convey the flavor of the scientific scaffolding behind Myers' suggestion.

The second qualifying condition for Myers' biodiversity hotspots (not mentioned in the main text) is that no more than 30% of its original vegetation remains. This condition seems to be an attempt to capture some historical trend that points to endangerment, which I discuss separately in Sect. 6.13 (Other value influencing factors).

one might cover the same size or even a much larger set of species by just looking to those places that simply jam together lot of species, without any constraint regarding their endemism.

What then, really justifies Myers' concern with endemics? Why should a diversity of species that includes endemics be of greater value than a diversity of species that does not? The answer might be that endemism enhances biodiversity – somehow making biodiversity itself greater, or its value, or both. But how can one make sense of this proposition?

Taking a tack that by now will be familiar to the reader, one might propose that endemism incorporates into biodiversity the concept of differences in the geographical isolation of species. Diversity in this category would derive from distinguishable gradations of geographical rarity. On one end of the spectrum we place such species as *B. texanus*, which lives in just two karst caves in one Texas county. On the other end are ubiquitous species such as *Tyro alba* (barn owl), whose subspecies happily occupy every continent except Antarctica, and the similarly ubiquitous botanical *Senecio vulgaris* (common groundsel), which manages to flourish despite the most powerful herbicidal assaults on it that human ingenuity can muster.

But this new suggestion is for the diversity of a category that one might call "degrees of ubiquity", rather than the one particular extreme kind within this category – namely, "found in just one place" – that is attractive to Myers. This seems a relatively unappealing appendage to the collection of categories that contribute to biodiversity. In any event, it is clearly not what Myers and his followers have in mind.

It is hard to know what Myers et al. really do have in mind. Their proposal, like many other conservation proposals, leaves suppositions about value unexplained, despite the fact that it is quite obviously value-laden. My best guess is that they do have Sarkar's set cover model of biodiversity value in mind. This they overlay with a model of triage, which derives from these considerations: Global diversity in the biological world is unevenly packed. Therefore in biodiversity terms, conservationists would do best to let widespread organisms take their chances. At least some of these organisms' geographically dispersed populations are likely to beat the survival odds without human assistance. That would allow conservationists to focus their efforts on resisting forces that are making life difficult for the geographically rare. However, this reasoning reduces the significance of endemism to its role in assessing a species' survival risk. In that case, two more mysteries present themselves: First, what privileges endemism, among other factors that enter into the risk assessment, such as the requirement of a species for a very particular and fragile, though not uncommonly occurring, type of microenvironment? And second, at what point do the principles of triage dictate that conservation resources be dedicated to a surer bet? One possible answer to this second question is that endemics are too much at risk to justify efforts to save them.

In fact, regarding the first mystery, I know of no good reason to privilege endemism. Regarding the second, the claim that more species might persist if conservationists focus on saving the geographically rare ones is an empirical claim; and it is not at all obvious that it is true.

One other possible gloss on privileging geographically rare species remains: One might claim that an endemic species is more equal than one that is ubiquitous. Such a species simply and categorically counts for more when it comes to its contribution to biodiversity itself and (perhaps because of that) to biodiversity's value: If "you" have the good fortune to be a member of an endemic kind, then you are privileged by virtue of the privileged status of your kind; so much the worse for you if "your" kind is relatively widespread. But I cannot think of any principled way to defend this non-egalitarian position with regard to species.

4.2.2.2 Abundance Rarity

Sarkar (2005, 57, 162–163, 2006, 135) and others talk about, not geographical rarity, but *abundance* rarity, which is the condition of relatively low abundances of a species in a place or set of places.[21] Sarkar repeatedly expresses his preference for this attribute of a species as the single most important criterion for identifying an initial set of places with high biodiversity value for his biodiversity-defining conservation algorithm. That set is the starting point for the iterative selection of places to conserve with the aim of achieving a biodiverse cover set (Sarkar 2005, 162–164, 166, 211).[22]

This focus on low-abundance species is odd for several reasons. First, yet again, it has little apparent connection with biodiversity. There might be an empirical relationship between the presence of relatively less abundant species and diversity. But Sarkar offers us no evidence for this and I am not aware of any. Second, it is at odds with Sarkar's own consistent use of the equation

$$\text{biodiversity } value = \text{biodiversity } content \times \text{viability}$$

It is well known that low population numbers tend to make a population less viable. And so rarity in this sense of low abundance should *decrease* rather than increase Sarkar's estimation of that species' value, so far as its contribution to biodiversity is concerned. Third, extremes of abundance cut against the traditional view that, other dimensions of biodiversity held equal, a state of species equi-abundance is the most biodiverse state of affairs. The belief that rare species contribute disproportionately to biodiversity flatly contradicts this tradition. Of course, as I expressed in Sect. 3.3.2.2 (Abundances (again)), I myself am highly critical of the use of abundance statistics in defining biodiversity and particularly critical of the notion that equal abundances mean greater biodiversity. I note Sarkar's implicit deviation from this tradition because I think it worth observing that different scientists adopt contradictory positions with regard to biodiversity, with seeming unawareness of this.

---

[21] Sarkar also mentions *habitat* rarity. I skip over that form of rarity, though it is closely related to the important ecological notion of generalist versus specialist species. A specialist can be widely distributed geographically if its special habitat requirements are met in widely distributed places.

[22] The computational capabilities of Sarkar's algorithm are discussed in Note 1 in Chap. 2. Section 3.3.4.1 (Place (again)) provides a brief description of how the algorithm works.

4.2.3 *Uniqueness*

Some notion of uniqueness or distinctiveness often emerges in discussions of biodiversity and sometimes in discussions of biodiversity measures. Sometimes uniqueness is said to have something to do with each species' unique evolutionary trajectory.[23] But this conception does not seem to offer a criterion by which we might find some differentness, since each organism's (i.e. species') evolutionary history is, well, unique. Unfortunately, the notion is never well explained.

More plausibly, "unique" and "distinct" are the terms by which those who focus on a feature-based conception of biodiversity differentiate features. As explained in Sect. 3.3.2.1 (Features), whatever features are, phylogenetic "distance", computed by some formula such as the one offered by Faith (and described in that subsection), is said to measure the degree of the differences between features. Having covered that topic previously, I will not say more about it here.

According to another interpretation, the notion of uniqueness falls into the same conundrum that ensnares the other "accretive" suggestions considered earlier in this subsection. Uniqueness can be promoted to a category that admits different kinds or degrees. These would range from (say) the most phylogenetically isolated creatures, such as the splendid *Sphenodon punctatus* (New Zealand's tuatara) which has an entire taxonomic order all to itself and so has no kissing reptilian cousins,[24] to the reclusive *B. texanus*, which, despite its extreme geographical rarity as a species, has many dozens of cousins in its genus. But this interpretation seems as unattractive as the similar one for geographical rarity, and for the same reasons.

---

[23] Purvis and Hector (2000) seem to have something like this in mind.

[24] The status of various tuatara populations as a single species is the result of the latest genetic assessment, by Hay et al. (2010). Previously, a population confined to North Brother Island in the Cook Strait was accorded status as the separate species *S. guntheri*. As a result of its unusual status, the tuatara makes it into most current textbooks on ecology and conservation biology.

My personal favorite example of taxonomic uniqueness comes courtesy of the three known (as of the report by Obst et al. 2005) species of the genus *Symbion*. Each of these different creatures lives on the mouthparts of a different species of lobster. Together, they constitute the entire phylum (not mere order) Cycliophora, newly minted in 1995.

Chapter 5
The Calculus of Biodiversity Value

Enough said about the concept of biodiversity. Let's start talking value. Some value talk has already crept into the book, but from this point it is the main focus.

I begin this discussion in an unconventional way – by abstracting as much as possible from specific theories of the genesis of biodiversity's value. "As much as possible" is not "entirely". From time to time, I allow myself a forward reference to one or another theory – described in the next chapter – about how or why biodiversity might be valuable. But by avoiding most details of how different theories attach value to various diversity-engendering categories, I hope to intensify focus on some *formal properties* of biodiversity value, independent of how the value might arise. Under the umbrella of "formal properties" I include:

1. How changes in biodiversity relate to changes in its value.
2. How values arising from the diversity of kinds within different categories interrelate, and whether any conditions, other than biodiversity itself, affect biodiversity's value.
3. The moral force that biodiversity exerts on moral agents, given that:

   (a) The value of biodiversity can change – at least as the result of certain changes in biodiversity (according to some relationship or relationships described by (1))
   (b) The size and direction of change in value is influenced by the behavior of persons who might affect the size and direction of changes in biodiversity.

In the background of this discussion is the core concept – proposed in Sect. 3.1 (The core concept) – of biodiversity as a state of affairs in which there is a diversity of kinds within various biotic and biota-encompassing categories. I honor and quite centrally feature the common supposition that biodiversity admits of more or less. The relation more (or less) might hold of certain pairs of biodiverse states of affairs even if other pairs cannot be so compared. In that case, a partial ordering is still possible. In some circumstances, a diversity "Pareto improvement" might occur, as remarked in Sect. 3.3.3 (Multiple dimensions). And quite centrally to the majority

D.S. Maier, *What's So Good About Biodiversity?: A Call for Better Reasoning About Nature's Value*, DOI 10.1007/978-94-007-3991-8_5, © Springer Science+Business Media B.V. 2012

of scientific treatments of biodiversity, when attention is confined to a single category – say species – then comparisons of more and less seem to make a lot of sense. The question that largely occupies this chapter is: If biodiversity increases, does its value also increase; and if biodiversity decreases, then is there a corresponding diminishment of biodiversity's value?

5.1   How Biodiversity Relates to Its Value

This section presents several models for how biodiversity value might relate to biodiversity. The models are taken from the scientific and environmentalist literature, where their presentation is informal, at best. Most often, nothing on this topic is said explicitly. Rather, some model is entirely implicit in the forms of reasoning that are presented; and so the descriptions that I offer are my own reconstructions, which I believe, are required to make sense of the value reasoning that is nonetheless going on. It is useful to keep in mind that these models are not necessarily mutually exclusive. Some of them can be combined and in fact, appear to be combined in various ways in the literature.

I am not concerned to adjudicate among the various models. Rather, my concern here is to make the models explicit, and in doing so, to tease out and critically examine assumptions and implications that might not be obvious even to those whose reasoning these models guide.

5.1.1   The Incremental Model

According to the incremental model of biodiversity value, an incremental increase or decrease in biodiversity produces a directly related and similarly incremental increase or decrease in the value of biodiversity. Consider, for example, a speciation event that is the result of disruptive selection (which causes a divergence in traits within a single population) or allopatric speciation (in which traits diverge in spatially isolated populations) – in contrast to one that results from the emergence of a new chronospecies. A species-increasing speciation event incrementally increases species diversity. On the other hand, a specific extinction event (always) incrementally decreases species diversity. The incremental model says that the biodiversity *value* incrementally varies directly as the biodiversity.

This model makes perfect sense as an abstract proposition – whether or not it truly describes biodiversity's relationship with value. One should beware of granting it credence on the grounds that (as one might think) certain particular species are valuable for the species that they are. In considering the value of species diversity, what counts is that a species is valuable for its contribution to species diversity. Section 4.1.3 (Particular species) forewarns against making the category mistake of confusing these two different things.

The incremental model of biodiversity value (perhaps along with the quantum jump model, discussed in the next subsection), is the one that most easily lends itself to consequentialist treatment and, most easily of all, to economic reckoning. Market price is economics' representation of a commodity's value. And what determines the market price is its incremental economic value at the margins – that is, the value of the next "unit" available to the market. The unit of species diversity is a species. So its marginal value would be the gain in value that accompanied "buying", "consuming", or choosing to retain in the inventory of (natural) capital assets one more species. Or alternatively, the marginal value of species diversity would be the loss in value accompanied by "selling" one species or choosing to retain one less species as a (natural) capital asset.[1] Whatever else might be said about the merits of an economic assessment of biodiversity (or of anything else), it wholeheartedly (and I would say, correctly) embraces the fungibility of kinds within a biodiverse category such as species.

Of course, as soon as the incremental model tries to account for more than one category of diversity, it appears to be headed for trouble. As I have remarked before, there is likely no reasonable way to sum or compare the diversity of kinds in different categories. The standard economic solution – a cornerstone for an economic calculus of biodiversity (and every other) good – is to reduce the value of the diversity of kinds in each category to market-based human preferences. In a single stroke, this reductionist move appears to dissolve all conflicts into the seamless summing of positive and negative preferences involved in trading this for that. This removes all conceptual barriers to considering, for example, a trade in decreased species richness for an increase in abundance evenness. And it makes sense of striking a bargain that sacrifices some "function" for not just another function but a new species.

This facile economic maneuver skirts another major complication, which concerns the interdependence – in the real world – of these sorts of "natural commodities". The economic calculus treats them as if they are so many distinguishable items sitting on the shelves in a warehouse and that an independent choice for each one can be made to remove it or not. Putting aside complicating laws of nature, the maneuver of putting species in conceptual warehouses paves the way for application of the familiar tools of economic analysis whereby biodiversity value – or, more precisely, the value of each economically recognized category of biodiversity – enters into a "development path". The "development of biodiversity" thereby becomes one aspect of economic development. Economic analysis then proceeds to determine which development paths maximize biodiversity value, or satisfy any of several other (sometimes non-maximizing) criteria that an economist might conjure up.

Perhaps the ease with the machinery of economic analysis works with the incremental model of biodiversity value, combined with the deep penetration of

---

[1] A complete economic accounting would include the costs of "acquiring" species (including foregone opportunities for development), as well as the benefits of "selling" them (including actual development). Most people would identify a fair number of species as beneficial to "sell" for the perceived harm that these organisms inflict on people.

economic thinking into all aspects of modern western thought, creates a certain feeling of comfortable familiarity with that model. But that is far from a sound argument for its legitimacy. Even its basic assumption – that the value of diversity varies directly as the diversity – has no consistent substantive support. And that is not for lack of trying to find it. The multitude of scientific investigations that have attempted to find a consistent correlation (if not a causal connection) between biodiversity and a property that (for the purposes of argument) might be considered a "good" – for example, productivity, stability, resistance to invasion (though invasions more often than not *increase* local or regional species diversity), and the provision of "services" such as water filtering – consistently fail to find such a correlation. Worse yet, a property such as productivity is, in fact, a highly dubious candidate for a good, which would make any correlation of biodiversity to it – were there one – normatively irrelevant.

But ultimately, rationality obliges us to think that attempts to find the elusive correlation fail because none of the properties in question are consequent upon biodiversity or incremental changes in it. Rather, these particular properties are the result of very particular assemblages of biota in the company of very particular biota-encompassing conditions. In fact, this realization leads directly to the "just-so" model of biodiversity, which is the topic of Sect. 5.1.4.

5.1.2   The Quantum Jump Model

The quantum jump model can be viewed as a variant of the incremental model wherein changes in value tend to occur in quantum jumps that typically require more than incremental or decremental changes in biodiversity to trigger. The incremental loss of a single species might not affect the value of that incrementally diminished species diversity. That single loss might combine with multiple others – still leaving the value of the remaining biodiversity unaffected. At some point, however, the combined loss of species finally causes biodiversity's value to take a quantum jump down.

This model associates strongly with conceptions of biodiversity that focus on functional diversity through the lens of species diversity. Many scientific writings express the view that in some given ecosystem, more than one species performs "the same" valuable function. This leads to the thought that the function's continued operation depends, not on a single species, but on some indeterminate number greater than one. A quantum of functional value is then lost only when the function is suddenly reduced or ceases entirely. Of course, if the category of species is left out of the description of this phenomenon and it is viewed entirely in terms of functions and their diversity, then what is left is merely a decrement of functional diversity with a corresponding decrement in the value of its value. In other words, a characterization in terms of the quantum model devolves into an incremental one.

5.1.3   The Threshold Model

Just as the quantum jump model can be viewed as a special case of the incremental model, the threshold model can be viewed as a special case of the quantum jump model in which there is essentially a single quantum jump and it is very big. According to the threshold model, the value of biodiversity is a simple step function of biodiversity. Nonzero value is awarded only above a certain level of biodiversity. Below some minimum level of biodiversity, its value plummets to essentially nil.

The most widely cited presentation of the threshold model is due to ecologists Paul and Ann Ehrlich (1981, xi–xiv), who famously compared incremental reductions of biodiversity *qua* species diversity to popping the rivets of an airplane.[2] The plane keeps on flying for quite awhile. And then it doesn't. The concept has had wide play. Bryan Norton (1987, 67), for example, dresses it up as the "zero-infinity dilemma".

It is thinking in terms of the threshold model that most powerfully motivates turning to Safe Minimum Standards (SMS) or some other form of Precautionary Principle: If ongoing decreases in biodiversity result in increasing proximity to a value precipice, and if there is no climbing out after toppling over the brink (in other words, the consequences are "irreversible"), then despite no evident declines in value on approach, and despite uncertainty about the proximity, size, and even existence of the precipice, proponents claim that it makes sense to avoid getting too close. This case for precaution gains cogency insofar as there is convincing evidence for the existence of negative, biodiversity-diminishing feedback loops whose intensity increase with proximity to the precipice, thereby making a pullback increasingly difficult, even on relatively remote approach. The justification of a Precautionary Principle is a tricky and interesting topic of its own. I return to it in Sect. 6.3 (Biodiversity as service provider), where it is the ammunition for a last-ditch defense of the proposition that biodiversity is the foundation for ecosystem services. And it shows up again in considering the possibility of biodiversity-diminishing feedback in Sect. 6.4 (Biodiversity as (human) life sustainer).

5.1.4   The Just-So Model

According to the just-so model, biodiversity has value only when it is "just so". To start, this requires that biodiversity occur in the "appropriate" amount. But while this is a necessary condition, being "the right amount" does not suffice to qualify some biodiverse state of affairs as "just so". Whatever amount might be "the right amount", biodiversity cannot be "just so" unless it also comprises a certain set of species and other biotic kinds that constitute or at least approximate "the appropriate biodiverse state" of the world. And there is one more

---

[2] The Ehrlichs' book comes from the dark age before the neologism "biodiversity".

requirement: The appropriate biotic kinds in the appropriate abundances must also make their contribution to biodiversity in the appropriate places. Curiously, those who embrace the just-so model of biodiversity value tend to think that all these conditions are satisfied, and only satisfied by some relatively recent biodiverse state from with the world is now diverging.[3]

The just-so model is closely aligned with, and in significant respects merges into a "balance of nature" view of the natural world. The balance – so delicately poised on the razor edge of just the right organisms in just the right numbers in just the right places – is easily upset. There is no better illustration how easy it is to tip the fragile balance than the phenomenon of organisms recently arrived and recently naturalized in a place previously unvisited by them. According to the just-so model, exotics are not "appropriate" because the biotic constitution of their new home *before* their arrival defines what is appropriate for its biodiversity. No matter that some or many of those longer-time residents were themselves immigrants some longer time ago.

The just-so model is not creationist doctrine, but shares with it a central conception: Both doctrines make value attributions by reference to a particular standard, which is characterized by a particular set of ingredients in particular proportions in particular places. They differ chiefly only in their account of how this delicately poised ideal state of affairs was realized in the first place and in their conception of what justifies the choice of standard.

It is therefore surprising that the just-so model is at least implicit, and parts of it quite explicit, in a great deal of scientific literature. I have already cited Chapin et al. (2000, 234–235) as one example among many in which particular "components" of biodiversity – meaning the particular species found in ecosystems today – are held to be essential to its value. Otherwise, these ecosystems would not be "just so", though these authors stop short of using this locution.

For Chapin et al. and like-minded scientists and environmentalists, what seems to be at the root of just-so thinking is the category mistake of confusing diversity for the particular set of kinds that are diverse right now or at some recent time in the past. In the context of species diversity, this is fundamentally another case of confusing the identities of particular species now existing in a particular place with the diversity of species, which could derive from an entirely different ensemble. For Chapin and his colleagues, however, such an ensemble would not make a proper contribution to biodiversity, for it would not be "appropriate" in composition or place.

Insofar as the just-so model is based on this category mistake, it appears to be a non-starter as a candidate to represent biodiversity's value. But that is the start rather than the end of its problems. The central problem has to do with finding credible, generally endorsable principles that justify the gold standard according to which "appropriate" is judged. In connection with this, one set of difficulties has to do with precisely the sorts of biological and ecological truths that are also inconvenient for

---

[3] Though, there is considerable disagreement about which point in time defines the ideal model and what constitutes a better or worse approximation of it.

its companion doctrine of creationism. According to the best paleoecological science, not just particular species, but particular assemblages of organisms, are a matter of epiphenomenal happenstance. These compositional accidents are in constant and fairly rapid flux – not just over geological time but also over much smaller timeframes. The biotic world has been rearranged and remixed, and the terms of survival in it have been redefined multiple times – just within the short course of human history.

The more or less constant rearranging and remixing is as inconvenient for just-so value models of biodiversity value as it is for creationism because like creationism, the just-so conception seems to be invested not so much in the value of diversity as in the value of one particular state of affairs – namely, the "appropriate" one. Such a conception places a burden on both doctrines to offer legitimate reasons for why the current biodiverse state of affairs, some recent one, or one conjured up from whole cloth has a special status. Creationists have some handle on meeting this challenge because their theory of how the current state of affairs came about – as the bequest of a divine designer-creator – also serves as its justification. Just-so biodiversity theorists, on the other hand, cannot so easily appeal to their "genesis" story as a justificatory one. Why is the biodiverse state of the world at any selected point in its multi-billion-year history – including some point in the recent past – so special as to view any deviation from it as a deviation from the biodiverse state that ought to obtain?

In sum, although the origins of the just-so view are not theological, it shares with theology a great difficulty in finding credible reasons, as opposed to making a leap of faith. This situation is reminiscent of the situation with the "process" conception of biodiversity (Sect. 4.1.5, Biodiversity as process), which according to one interpretation, regards "parts" of the biotic world or biotic-part engendering processes to be damaged to the extent that they do not conform to some ideal that, unfortunately, is never characterized.

What *is* appropriate, so far as biodiversity is concerned? How much is enough, and why is such special status afforded to some particular amount (no more nor less) and some particular agglomeration and arrangement of biotic kinds in the places where they happen to have landed at some point in a multi-billion-year trajectory? On what basis can some kinds of species be privileged over others? For example, what makes "natives", which are somewhat less recently arrived immigrants, appropriate, while other kinds that are more recent are therefore deemed inappropriate "aliens"? And what justification is there for valuing – *on biodiversity grounds* – some essentially static, frozen state of affairs? What privileges some particular set of components – comprising the particular species, ecosystems, etc. which happen to have coalesced at some past time, perhaps recently – over other mixes of species or other kinds of ecosystems into which the recently coalesced state of affairs will inevitably metamorphose?

It is hard to imagine an account of just-so diversity that supplies non-arbitrary and broadly convincing answers to these questions. One might imagine claiming that the notions of "appropriate" amount, kinds, and components are time-relative. What is appropriate now might not have been appropriate yesterday and might not be appropriate tomorrow. So this view need not entail a creationist-style stasis after all. But while this response might avoid the objectionably static quality of the just-so model as I initially characterized it, it does so at the expense of adding yet one more necessary condition – a condition of temporal appropriateness – to the original set of conditions. Moreover this adjustment does not really escape the need to posit an inherent rightness of some starting point. For what is right for now inevitably depends on what was right for yesterday, what is right for tomorrow depends on what is right for today, and so on. So the question remains: Why should some peculiar biodiverse state of affairs be privileged in value because derived from the current or some recent one?[4]

One response on offer to this last question is that the flourishing of humankind arguably has depended and continues to depend on some particular state of biodiversity. There was, according a number of paleontologists (for example, Ambrose 1998; 2003), a pretty close call for *H. sapiens* when that species squeezed through a bottleneck some 70,000 years ago.[5] But things have been going spectacularly well for the species (just considered as such) since then. This, one might argue, is a basis for privileging some particular biodiverse state of the world that, along with those humans, has evolved into the current one, and for privileging whatever state of biodiversity evolves thereon after.

But by proposing to privilege "what is" on account of its origin, this response comes close to committing the genetic fallacy. It is also disappointing on two other independent counts. First, it seems to entail endorsing the spectacular rearrangement, and if the Sixth Great Extinction is indeed occurring, the significant diminishment of biodiversity that has accompanied humankind's ascendance. Worse, it seems to endorse yet more such rearrangement (and possibly diminishment) that seems likely to accompany the continued development that has been the adaptive means for that ascendancy. Second, this response makes "just-so" all but unrecognizable, having transmuted its meaning into "whatever amount of biodiversity serves human ascendance on the planet". The last stand on this might be a precautionary one – to the effect that human ascendance in particular and human purposes generally could be severely disrupted by a disruption of biodiversity beyond a certain threshold. But this, in turn, brings the discussion around to the threshold model of value, which must surmount its own considerable hurdles.

---

[4] This general failure to examine or justify the starting condition for a biodiversity norm is disconcertingly reminiscent of a similar deficiency whereby a Pareto improvement is regarded as a good without reference to a starting condition that might be both abhorrent in itself and an obstacle to any reasonable means of remedying the source of abhorrence.

[5] The anthropologist Stanley Ambrose and others theorize – not without controversy – that a super-eruption of Toba (at what is now Lake Toba in Sumatra, Indonesia) a bit over 73,000 years ago induced a volcanic winter that reduced human populations, presumed still concentrated in Africa, to a few as 15,000 individuals. The major and eventually world-encompassing human diaspora out of Africa began after *H. sapiens* emerged from this population bottleneck.

5.2   Value Interrelationships

In this section, I wish to make two simple, perhaps obvious, but oft-overlooked observations about how values that attach to different categories of diversity interrelate. My observations include and expand on some points already made (in Sects. 3.3.1 (Ta legomena in biology) and 3.3.3 on (Multiple dimensions)), that: (i) biodiversity is best understood as the diversity of kinds in multiple categories, (ii) there is no reasonable way to compare categories of kinds belonging to different categories, and (iii) there might well be contingent relationships between categories (and between kinds in different categories) that are complex and largely unknown. These latter (relationships) are subjects of a great deal of ecological research – for example on the relationships between species diversity on the one hand and on the other hand functions such as productivity, resistance to species immigration, and even functional *diversity*.

On the basis of the just-mentioned factors, it is hard to imagine anything more than a partial ordering (with respect to the relation more or less) of biodiverse states of the world, though some tighter ordering might occur within a single category and at some one chosen level of spatial granularity. This, in turn, makes it hard to imagine more than a partial ordering of biodiversity values.[6] Even a simple example entirely confined to the taxonomic hierarchy confirms how implausible is the projection of all of diversity onto a single scale and the similar distillation of its value: What credible basis could there be for assigning the relative positions of a new phylum with a complement of just three species in one genera[7] with respect to one species in a species-rich genus or family?[8]

As this hypothetical example suggests, it seems certain that the value contributions of diversity derived from different categories will sometimes conflict, though conflict is not strictly entailed by the incommensurability of the values. I am not speaking of the possibility or probability of conflicts between, for example, preserving wilderness and conserving biodiversity. Rather, I am saying that value in the diversity of kinds in one category might sometimes be retained only with the loss of value in the diversity of kinds in another category. This is obvious for the incremental model of value. The addition of one species might suppress another species' contribution of a function. Thus, the introduction of cattle in Sarkar's Keoladeo example (see Sect. 4.1.1, Wilderness) suppressed the functional contribution of grasses that choked out the waterways. Conversely, the elimination of a species

---

[6] It is hard to imagine, that is, except within the framework of neoclassical economics, which (as discussed in Sect. 4.1.2, Measures and indexes) seeks to reduce a plethora of different types of diversity value to a single gold standard.

[7] I have in mind the cycliophorans mentioned in Note 24 in Chap. 4.

[8] One might think of the snail darter, a species that occupies a crowded genus in a crowded family of the crowded (Perciformes) order of perch-like fish and which comes up for discussion again in the context of Biodiversity options (Sect. 6.9).

might add a function that it was previously suppressing – as evidenced by the result of (for a time) removing cattle from Keoladeo.

Taking species evenness to be integral to species diversity[9] only increases opportunities for conflict. Clearly, eliminating a rare species will tend to increase species evenness, though it will decrease species richness.[10]

Conflicts also arise at different spatial scales. An area that is broken into multiple and diverse fragments (by development of roads, for example) might engender greater genetic diversity in populations that consequently stop exchanging genes as the result of experiencing different adaptive pressures from their differing environments (the start of allopatric speciation). But this increased genetic diversity might come at the expense of the elimination of species whose individuals' survival requires space exceeding that of any of the remaining fragments.

As discussed in Sect. 4.1.2 (Measures and indexes), there is a substantial scientific literature on biodiversity measures that attempt a quantitative summary of all kinds of diversity within a single scalar value that produces a complete ordering (and actually more – a metric of differences). Such a summary would certainly seem to vastly simplify the problem of value attribution. With a complete ordering of "degrees" of biodiversity, one is tempted to see a concordant ordering of biodiversity value as a straightforward, irrefutable fact about the world. But as observed in that previous subsection, the analysis only appears to be simplified in this way. In actuality, the evaluation is still done and it is still complex. The only real difference is that it is covertly tucked into the formulaic prescription that dictates how disparate factors weigh into the final, scalar result.

Another methodology for dealing with value interrelationships attempts to circumnavigate the impossibility of comparing values found in multiple, incommensurable categories. I have in mind approaches generically known as "outranking methods" or methods for multicriterion decision-making. They are perhaps most famously discussed by Nobel laureate Kenneth Arrow and Hervé Raynaud (1986), though these economists were neither the first nor the last to explore this approach. Sarkar (2005, §7.2, 196–203) describes a recent, conservation-biological reincarnation, which he dubs "multiple criterion synchronization" (MCS).

Whatever the limitations of outranking, one would have hoped that MCS would attempt to place into a single framework the multiple, incommensurable value criteria associated with various intractably disparate categories of biodiversity. Unfortunately, MCS does not stick with this already formidable axiological task. Instead, it attempts to integrate this with what is essentially a theory of right action in which the values of cost and efficiency are themselves considered paramount.

---

[9] As discussed in Sects. 3.1.3.1 (Abundances) and 3.3.2.2 on (Abundances (again)), a conception of species diversity that combines species counts with the evenness of populations is well entrenched in scientific circles.

[10] See the example in Note 10 in Chap. 4 of how the commonly used Shannon-Wiener index can combine these conflicting factors to characterize the elimination of a species as an increase in biodiversity.

In this form, Sarkar and others take MCS to offer theoretical grounds for justifying triage decisions regarding which place to next "conserve for biodiversity".

I do not have the space for anything more than a brief description and an equally brief and incomplete appraisal of MCS. The basic idea is simple enough. One starts with a set of alternatives in a pool of choices. Alongside this set is another, non-singleton set of ranking criteria, each of which individually imposes a total order – presumed to be a value ordering – on the original set of alternatives. Of course, any pair of criteria might impose very different orders, up to and including one criterion completely reversing the ranking of the other. Out of this seeming chaos, the goal is to extract a criterion-independent total ordering – that is, a single, global, complete ranking – over all the alternatives.

At first glance, it might seem as though the gist of this idea is that one starts with multiple, incommensurable and very likely conflicting ways to value some alternatives; then magic happens; and then *voilà*, the alternatives are totally ordered. In fact, that is exactly correct. Unfortunately, there is no magic, and Arrow and Raynaud acknowledge this. They state that they wish to steer between the Scylla of open and footloose magic – such as "the majority method", whereby an agreement of an expert majority on a pairwise ranking removes all dissenting criteria – and the Charybdis of inflexible axioms that would exclude "the play of intuition and creativity." But it is hard to see how these comments provide a usable, let alone justifiable test of an outranking method's legitimacy. *Their* creative intuition leads to a "prudent order". In some sense intuitive to Arrow and Raynaud but not to me, this order is supposed to minimize pairwise strongest conflicts.

Sarkar, on the other hand, expresses discomfort with the idea of allowing this play of intuition. His proposal for MCS (his version of outranking) is to toss out, not ranking criteria (which is the suggestion of Arrow and Raynaud), but any *alternative* that gets a conflicting ranking from distinct criteria. That is, alternatives x and y are retained and alternative x is globally ranked no lower (or higher) than alternative y just in case no criterion ranks y ahead of (or behind) x. If any two criteria reverse x and y's rank, both x and y are said to be "non-dominated" and because of that, removed from consideration. In this case, there is no magic because none is required: Conflict is quite simply banished, not even by edict, but by definition.[11]

---

[11] See Sarkar (2005, 196–203). Sarkar (2005, 202) worries that there will still be so many elements in the non-dominated set that arbitrary (his word for "unjustifiable") criteria will be required for choice. I am skeptical of this and have the opposite worry – that no two alternatives will satisfy Sarkar's "non-domination" requirement. In the latter case, there will be as many non-dominated sets as there are alternatives and all will be singletons.

I believe that the my worries differ from Sarkar's because he takes what I think is a hard-to-justify, restricted view of biodiversity that excludes many conflicts that a more open view would allow. In fact, at times he appears to reduce his set of basic biodiversity-establishing criteria to precisely one – one "surrogate", in his terminology – and that criterion is rarity of species. At that point, he is talking about something that I cannot recognize as being about biodiversity. (See Sect. 4.2.2, Rarity.)

5.3   The Moral Force of Biodiversity

The single most salient and consistently emphasized part of the public discussion of
biodiversity's value is an expressed imperative to conserve it. Those with just-so
leanings are likely to emphasize their goal of optimizing biodiversity by bringing it
into alignment with some ideal. Sometimes this is supposed to involve tossing additional species into a particular ecosystem mix. But from a more sweeping perspective, the just-so goal is usually thought to be best served by conserving something
close what biodiversity currently exists. Not increasing it. Not maximizing it –
unless that is taken to mean "maximize conservation of biodiversity".

There is something axiologically odd about ignoring the potential for increases
in biodiversity that aim to enhance its value. That is especially true according to the
incremental model of biodiversity value, whose basic principle holds an incremental increase in biodiversity to increase value as much as an incremental decrease
decreases it. This indifference of the incremental way of thinking to the direction
of change in "biodiversity supply" is most saliently expressed in economic analyses, which routinely consider the effect on price and economic welfare of both
increases and decreases in the supply of a commodity. The threshold model reinforces this tendency to ignore biodiversity increases. According to it, any increase
"north" of the threshold does not contribute value. But from the viewpoint of the
incremental model, the situation is reminiscent of the time reversal symmetry of
most laws of physics: Substituting $-t$ for t yields a valid solution, but one that is
ordinarily disregarded.

One might think that this tendency to focus on the conservation of biodiversity
rather than on bolstering or otherwise enhancing it is due to a large differential
between the ease of the acting so as to decrease it – often without thinking or trying,
as the collateral damage of development projects – compared to the difficulty or
even near impossibility of adding to it. But the relative difficulty of beefing up biodiversity as compared to whacking it down is both exaggerated and obscuring of
several truths.

First, additions to biodiversity are occurring all the time and possibly at high
rates. Many additions come "without thinking or trying". For example, a nascent
body of research on adaptive changes in various organisms is uncovering speciations that are occurring behind our backs and under our noses.[12] Moreover, the
means are within fairly easy reach to consciously design and implement projects
that substantially step up biodiversity "production". I will list some possible and
plausible means shortly.

An incremental model of biodiversity value supposes that more biodiversity is
good. The minimal implication for right action is that one ought to act so as to
increase biodiversity – at least when there is little cost of doing so. If the good that

---

[12] See Sect. 6.7 (Biodiversity as value generator) for remarks on how rapidly adaptations can occur
and Sect. 7.3 (Biodiversity value in human timeframes) for more reflections on, and implications
of rapid evolution.

can be achieved is judged to be sufficiently great, then some large effort might be justified. Furthermore, there is no reason to interpret this mandate for increasing biodiversity as requiring that currently existing kinds be conserved. Even the mere conservation of current biodiversity does not require the conservation of any particular extant kind. The incremental calculus – especially as embodied in economic cost benefit analysis – argues for spending a lesser effort to "produce" more biodiversity rather than a greater effort to prevent a lesser amount of biodiversity from vanishing. On the threshold model, these means to increased biodiversity would also be justified for furthering the end of putting the threshold at the greater distance ensured by the greater biodiversity.

Attitudes within the scientific and environmental communities are oddly conflicted with regard to the suggestion, implicit in some value models, that it would be a good thing to promote the production or introduction of species to offset losses or even to swing the balance in the positive direction. Species introductions, in particular, are often espoused for a variety of reasons – for example, in the name of better recreation, to re-create the past, for better human health (Sect. 6.5.2, Biodiversity as safeguard against infection), or to control some other, unwanted organism.

But there is a sometime lack of enthusiasm for increasing biodiversity, and I can think of several reasons for this. One is a reluctance to fully accept the fungibility and commensurability of kinds on which the suggestion depends. The phrase "offset losses" is thinly veiled code for familiar economic marketplace accounting. While this approach is enormously attractive for its power to in principle reduce complex judgments to simple arithmetic, some will resist its implication, once it is realized, that new species can and under some circumstances should be allowed to compensate for the loss of others. Yet this resistance regularly folds back on itself when "conservationists" define their goals quite selectively in terms of specific species, let the chips fall where they may for any others.

But as I have argued (in Sects. 3.1.1 (Egalitarianism) and 3.1.2 (Fungibility)) for better or for worse, commensurability and fungibility are built into the notion of diversity – at least as restricted to kinds in a single category such as species. If one species replaces another in some environment, then it is hard to reasonably deny that the end result is an equally species-diverse state of affairs. This appears to leave no room for any change in value, so far as species diversity goes: The only grounds for saying that one species-biodiverse state of affairs is better than another, equally species-biodiverse state of affairs must be on some grounds other than species diversity.

One can look for such grounds outside the domain of biodiversity entirely. Maybe there are legitimate reasons for preferring the continued existence of any extant organism to any new organism or set of organisms that might emerge to replace it in the biodiverse mix. I shall return to that thought momentarily. But also (as suggested earlier), another category of biodiversity can be brought into play. Perhaps, so far as overall biodiversity value is concerned, the diversity of kinds in other categories – for example, a higher taxon, such as order – carries more weight (value) than the diversity of species. This combines with the circumstance that, as Purvis and Hector (2000, 214) point out, "subtaxa within taxa are often distributed very unevenly".

This, according to them, entails that "Taxonomic boundaries are not comparable between major groups." Therefore, some species *are* more equal than others, after all. And while this appears to be non-egalitarianism with respect to species, it is not non-egalitarian in some higher-order taxonomic sense, which still offers some principle of just species distribution among the higher taxa.

Behind this reasoning is some presumption about the evaluative precedence of taxa. The higher the taxon, the higher its precedence. For example, there are far fewer kinds of (the taxon) order than kinds of species. Therefore, if a species such as *Sphenodon punctatus*, which is the only one in its order, vanishes, there is a significant loss of order diversity. On the other hand, as the result of that same species' extinction, the (let's say) 10 million species diminish to just 9,999,999 – a barely detectable difference. The loss of a species in a species-rich order would not have this effect. In this case, one could still say that the tuatara contributes to species diversity just as much as any other species. But its contribution to order diversity is more important. The principle of species egalitarianism still holds. But the emergence of another species in another, species-rich order would not compensate for the loss of the tuatara so far as order diversity is concerned.

Similar reasoning can be (and has been) applied to the diversity of functions when one particular species is primarily responsible for some kind of function and a functionally non-equivalent substitute (quite literally) just will not do. Or again, if focus shifts to the diversity of morphological features, then the extinction of the last radially symmetric organism will not be compensated by the emergence of another bilaterally symmetric one.

Of course, one can also appeal to the value of a particular species as the species that it is, and not its value as contributing to the diversity of species, functions, features, or some other category of diversity. It is entirely legitimate to bring this up. But in doing so, one wanders outside the bounds that circumscribe consideration of the value of species diversity or biodiversity more generally. The move rejoins the debate of the moral value of a species, which was well underway before the more recent one about biodiversity. That might be a very good thing; but this book is not about that more venerable discussion. Rather, it is concerned to see how far natural or environmental value can be viewed as a matter of the world's biodiversity, because this latter concern has received little careful attention.

What else can account for less than universal enthusiasm for the idea of promoting the proliferation of kinds to achieve higher levels of biodiversity? I have suggested that this is a presumptive obligation on incrementalist principles. One possible explanation is that acts of increasing biodiversity are permissible or supererogatory, but not obligatory. But if the rationale for this relegation to permissible or supererogatory status is the supposed insuperable difficulty of acting in ways that actually do promote biodiversity, then as I have already suggested and elaborate below, this is mistaken.

A third reason for not fully embracing the implications of an incremental model of value could hinge on some notion of which methods are "appropriate" and which ones are not appropriate for increasing biodiversity. In the context of the moral category of obligation, what is appropriate must relate to methods involving actions and behaviors that are under the control of moral agents. The idea is that if a state of affairs with greater biodiversity is obtained through "inappropriate" or "illegitimate" means, then its value cannot have increased. On the other hand, a like increase through legitimate means would indeed be value-increasing. This is a variation on the ethical theme that the value of ill-gotten gains is tainted.

Of course, this line of thought cannot be understood without some definition of an "appropriate" or "legitimate" method. It must also supply some account of why certain causal histories of biodiversity increases – the ones that use appropriate methods – effect an increase in biodiversity value, whereas other biodiversity-increasing causal histories – ones that use inappropriate methods as a means – either leave the value of biodiversity unchanged, or even diminish it.

To assess this notion of "appropriate increase", as well as my previous and related claims concerning the practical possibility of increasing biodiversity, it is useful to consider a list of plausible and possible ways for people to act so as to increase biodiversity:

1. Resurrect extinct species (suggested, for example, by Judson 2008a). As I recount below, this is likely not beyond even current capabilities for the significant set of cases in which the extinction occurred within a few tens of thousands of years ago. There might be a greater reliability in this approach compared to projects to design never-before-existing transgenic species, because the old designs are known to have worked – at least in one place at one time. A newly invented organism does not come with any such assurance. Furthermore, resurrection science would remove a principal onus of extinction. No species need be lost forever, so long as we act to resurrect it soon enough.
2. Ramp up the technology-based production of species – most promisingly through genetic manipulation. Investments in the science and technology involved would undoubtedly enhance success rates in creating transgenic species. Agricultural research is already hard at work doing this. But why restrict this work to species that have agricultural utility? The seeming ease with which scientists created GFP (green fluorescent protein) rabbits (and now also mice, frogs, nematodes, flies, bacteria, and plants[13]) demonstrates an established capability in this regard. There is the potential to literally flood the species market with new products.
3. Encourage polyploid speciation, which can be virtually "instantaneous" – that is, a single-generation event.[14] Recent estimates are that between 47% and 100% of all angiosperms (flowering plants) arose in this way. Moreover, "neopolyploid" speciation proceeds at startlingly high rates. Without human assistance, these range from 15% of all new species of angiosperms to 31% of pteridophytes (ferns and their allies) (Wood et al. 2009, 13875). Human

---

[13] Though none of these see-in-the-dark creatures strictly constitute new species, they are surely new phenotypes. And it is not hard to imagine some cases in which sexual selection – for examples, females rejecting the advances of phosphorescing males – does lead to speciation.

[14] The issue of polyploid speciation is somewhat complicated by the fact that gene flows across "ploidy levels" might continue after the creation of hybrids. See Slotte et al. (2008).

assistance might well increase polyploid speciation rates substantially – witness Georgii Karpechenko's (1928) production of a new species born of a cabbage and a radish (from the two distinct genera, *Brassica* and *Raphanus*) in 1928.

4. Invest in gene banks. I have omitted all but scant mention of the genetic category of diversity. But it *is* a category of biodiversity on many accounts, and it is not uncommonly mentioned in the biodiversity literature. Of course, genetic diversity might be preserved in banks quite as well as in a viable species – or even better and with less cost than required to maintain a viable population of individuals. Moreover, genes in a bank cannot evolve, whereas those carried by living individuals of a species can and do. In this way, gene banks achieve the ultimate of conservation. This should appeal to those for whom a just-so model of biodiversity value is also appealing.

5. Introduce existing species to new places that might suit them and where they are *un*likely to displace "natives". The idea would be to increase local, regional, or even global diversity – with speciation making global gains possible. Sober scientific assessment of previous, unplanned, and inadvertent human-assisted migrations and immigrations shows that immigrations often, and perhaps more often than not, increase species diversity – at least locally and regionally.[15] Studies of modern species immigrations as well as of immigrations from the geological past are beginning to give scientists a working idea of which species are good candidates for effecting local diversity increases and under which conditions of the target environment this is likely.[16]

6. Increase the phenotypical diversity of existing organisms by strategic introduction of some of their populations to environments that are identified as different in some significant respect or even extreme. This suggestion is, like the previous one, but an extension of what people have been doing with increasing intensity over the last 70,000 years or so, but without putting much if any thought into it. By putting more thought into it, yet more diversity-increasing changes can be realized. The efficacy of this method requires some understanding of the conditions in which different phenotypes present themselves. But as Dov Sax and his colleagues (2007, 467) and Stephen Jay Gould (1998) make clear, rather than being "optimally" adapted for the environment in which they evolved, organisms have been just good enough – at least until now. In an environment more amenable for an organism, previously unattainable phenotypes can (and sometimes do) emerge.[17] Of course, an increase in phenotypical diversity could also result from less optimal or even severe environments. New phenotypes can reflect the increased stress of a new environment on an organism, or at least a

---

[15] For example, Sax and Gaines (2008) provide an account of how immigrations have generally increased species richness on oceanic islands. This is consistent with the disappearance of some "native" species, which, in the case of these islands, were just earlier immigrants.

[16] See Vellend et al. (2007) for some of the various ways in which speciation can result from species immigrations.

[17] See the example of *P. albicaulis* (whitebark pine) in Sect. 3.3.2.1 (Features). This is typical of many examples of exotics thriving more in their new home than in their "homeland".

shift in which adaptive capabilities count the most for the organism's procreative success in the new environment.

7. Encourage speciation by creating novel *habitats* that are known to favor or stimulate it. It is well known that the emergence of new habitats is principal among the factors responsible for the emergence of new species. As John Spicer (2006, 78) remarks,

> In terms of the origin of species, it would appear that the easier the access to new and/or different environments, the greater was the likelihood of new species arising – and that has happened throughout the history of life on earth.

Again, people have been busy creating new habitats for at least 70,000 years, though without much thought about the effects on biodiversity. Most recently, these projects have been euphemistically dubbed "land conversion". Of course, these efforts pursue the purpose of accommodating the desires and needs of people rather than that of stimulating evolutionary change. But evolutionary change is nonetheless stimulated.

Most saliently, we humans have our cities. These glass-canyoned environments, for example, select for birds with big brains (Maklakov et al. 2011). And there is reason to believe that overall size is also an urban advantage both because smaller birds and their nests are more vulnerable to larger predators and because the scarcity of their favored foods throws them into more direct competition for resources with larger birds. There is no reason not to expect that this will ultimately create species of city birds, which are incompatible with their country bumpkin cousins. A city version of *Peromyscus leucopus* (white-footed mouse) is also turning the city mouse/country mouse fable into reality (Munshi-South and Kharchenko 2010). The pollution associated with cities is also well known to exert adaptive pressures on organisms. The best-known and perhaps best-studied case is the industrial melanism of *Biston betularia* (peppered moth). But among a myriad other such cases, grasses have adapted themselves to grow amidst mine tailings (Antonovics et al. 1971) and mosquitoes adapted themselves to DDT (Gladwell 2001). Moreover, there is evidence that stripping away a diversity-encouraging characteristic of a habitat – including the removal of "environmental toxins" – can concomitantly strip away diversity (which Levinton et al. 2003 describe in the benthic worm *Limnodrilus hoffmeisteri*).

Of course, I don't recommend dumping toxins to promote biodiversity. Yet the study of the adaptive reaction of various organisms to "new resources and other materials in a species' environment" (Myers 1996a, 40) has joined a large body of evidence that suggests that biodiversity would be well served by a project to create "start from scratch" reserves with novel habitat features. A significant contribution could also come from isolating extant habitats – carefully managed, of course, so as to isolate only self-sustaining populations. From Spicer (2006, 79), again:

> The erection of slightly smaller-scale barriers [than those that isolated Australia] such as mountains, hills, valleys, streams, lakes and channels has also acted in the past as a successful contraceptive, preventing different populations of the same species reproducing with each other.

This is well within human means. People routinely construct all sorts of environmental barriers – for example, highways and building developments. In fact, as Norman Myers (1996a, 40) discusses, most likely more than two-thirds of all speciation events are allopatric speciations – occurring as the result of geographical separation.

8. I have mostly avoided looking for biodiversity in the "structural/ecological" hierarchy in general, and ecosystem diversity in particular. But this category undoubtedly offers fertile ground for sprouting biological diversity. Consider the notion of anthropogenic biomes, which is beginning to achieve some purchase in terrestrial ecological science. The invention of this category is intended to replace (mere) biome as a way to acknowledge the fact that there are essentially no places on the planet that are devoid of past and ongoing human influence. Therefore (as noted by Erle Ellis and Navin Ramankutty 2008, 440, 443), there are few if any remaining places whose ecology can be understood in isolation from "new forms of human-ecosystem interaction" and "humans [creating] landscape heterogeneity". So as these authors suggest, humans are in some significant sense the creators – intentionally or not – of most of the planet's biomes.[18] The current kinds of biomes are largely unintentional creations – the inadvertent side effects of other human activities. But there is no reason why human creativity could not seize on a nascent understanding of this genesis to generate many more kinds intentionally. The more different ways that humans design cities, agricultural landscapes, etc., the greater the diversity in this category will be.

9. Design bioengineering and breeding programs to genetically adjust the tolerances and capacities of organisms whose native adaptive capacity does not suffice to adjust sufficiently fast to ongoing human-generated incursions that are unlikely to subside. One might think that a program to bioengineer tolerance to higher temperatures might help *Ochotona* spp. (pikas) – creatures that are running out of time and room (to run uphill) to adapt to warming alpine temperatures. Insofar as amphibians are declining from exposure to toxins such as atrazine (the world's most popular herbicide and not likely to vanish soon), we might consider bioengineering toxin resistant variants of the sort that sometimes develop on their own. In this case, it appears the scientists already have a bead on a specific genetic target.[19]

10. Accelerate evolutionary adaptations by establishing conditions that are known to increase rates of mutation. This is obviously more of a shotgun approach. And at least one obvious means – opening up instead of closing the stratospheric ozone hole[20] – does not shield humans from the scattershot. But more

---

[18] Ellis, in particular, emphasizes the human creative role as a source of human responsibility for nature. This is an important topic that I take up later in Sect. 8.1.2 on Responsibility for nature.

[19] The research of Hayes et al. 2003 suggests the approach of developing a strain of amphibians that resist the expression of the aromatase gene, which demasculinizes and feminizes males, and is responsible for declines in reproduction.

[20] This creative suggestion is due to Darren Domsky.

highly localized high radiation zones exist at such places as the Hanford Site and Chernobyl.[21] The radiation in these places was not introduced with biodiversity in mind. They nonetheless are causing elevated rates of mutation. And they demonstrate the availability of the basic principle, even if some greater degree of control is undoubtedly desirable in applying the principle for the sake of increasing biodiversity.

11. Increase human predation. Evolutionary ecologist Chris Darimont and his colleagues (2009) argue that human predation (short of extinction) is one of the most forceful drivers of phenotypical change. Much of that change is genetically underwritten. That is, it is the result of species adapting to persistently large predatory pressure. Applied differentially to different populations, this could well be an efficient stimulus for the divergent adaptation of isolated populations, and eventually, speciation.

12. Increase nitrogen fertilization in N-limited habitats to reduce the competitive exclusion that inhibits diversity in many places. This is a more benign variation on Myers' observation that "new resources and other materials in a species' environment" are an effective means of stimulating adaptive change.

13. Replace relatively simple ecosystems – such as wetlands and salt marshes – that are known to harbor relatively small amounts of biodiversity with artificially constructed habitats that build in greater complexity with a correspondingly greater potential for diversity.

14. Log or burn forests on a staggered schedule to produce areas that are concurrently in various successional stages. Each stage of succession typically harbors a different assemblage of organisms. To have each successional stage exist concurrently with every other ensures an overall increase in species diversity at any one time.

I will get back to assessing what these sorts of projects mean with regard to "appropriate increase" later in this section. First, I wish to assess the practical possibility of actually engaging in them. But even before that, let me note that even if every one of the suggestions were frivolous figments of my imagination, they would still serve to make an important point: Someone who is uncomfortable with them and who yet views biodiversity's value on an incremental model – and most especially someone one who fully embraces the price of a species to reflect the marginal value of species diversity – is obligated to supply reasons for this discomfort. And those reasons would necessarily have to come from considerations that do not have to do with biodiversity.

But in fact, none of these imagined projects are frivolous. None of them appears to be beyond human capabilities. Most of them are within reach now. As a matter of

---

[21] The Hanford Site was established as part of the Manhattan project to produce plutonium. It contains an immense amount of radioactive nuclear waste from its years as a production facility. It is now largely a preserve comprising the Saddle Mountain Wildlife Refuge and Hanford Reach National Monument. Chernobyl, of course, was the site of history's first horrific nuclear disaster in 1986.

fact, many of them have actually received serious consideration. And in some cases, that consideration has been serious enough to engender either a funded scientific project or an artistic one that merits thoughtful evaluation. It is worth recounting some of these.

I start with perhaps the only suggestion that might initially seem to be fantasy – resurrecting extinct species (item (1)). But in fact, several resurrection projects are well underway. These variously propose to use cloning (sometimes combined with genetic manipulation), hybridization, and "back-breeding".

The first step in all these resurrection projects is to sequence the genome of the extinct species. Like most biological macromolecules, DNA tends to disintegrate without a living organism's constant repair. Still, while estimates of its longevity vary, scientists have found that DNA remains largely intact for several tens of thousands of years under commonly encountered conditions. This timeframe encompasses many, if not most of the creatures that went extinct during the period of most intense human activity. *Mammuthus primigenius* (woolly mammoth) did not disappear in most of its range until 10,000 years ago, with a vestigial population persisting until just 3,700 years ago (according to Vartanyan et al. 1993, 340). *Bos primigenius* (auroch, several subspecies) went extinct only in the early seventeenth century. And *Equus quagga quagga*, a subspecies of *E. quagga*, the plains zebra, vanished only 100 years ago. These relatively recent dates of disappearance mean that viable genetic material might well be available and that modern, related lineages might not greatly diverge from the extinct ones. These two considerations combine to encourage efforts to resurrect these creatures.

As of 2008, sequencing of the mammoth genome was well under way at The Mammoth Genome Project at the Pennsylvania State University (Miller et al. 2008). Once the genome is in hand, there are various ways to proceed. In the absence of viable mammoth sperm (which would be hard to find but could be used to produce a mammoth-elephant hybrid) or a fully intact cell nucleus (which could be used for cloning), scientists propose a genetic modification scheme. That would involve modifying the approximately 400,000 sites on the genome for *Loxodonta africana* (African elephant) to produce a mammoth nucleus usable for cloning. The Polish Foundation for Recreating the Aurochs (PFOT) in Poland also proposes (Czechowicz 2007) to use DNA – directly from bones of aurochs in museums – to re-create aurochs via cloning.

Cloning in any of its variants has the distinct disadvantage that a viable reproducing population will require quite a few (perhaps one hundred or more) successfully cloned individuals. This puts aside the other difficulties of recovering the requisite intact cell nuclear material. The cloning technique for mammoths would also rely on implantation and *in vitro* development in an elephant mother. There are obvious difficulties – including actually recovering suitable genetic material and worries about whether the mother of a surrogate species might produce a viable offspring.

With the current explosion of developments in genetic research, it is not implausible that these difficulties might soon be swept aside. Even so, this is not the only route to resurrection. For example, if mammoth sperm could be recovered (a very big "if"), then insemination of an elephant with an X chromosome-carrying sperm might produce a female mammoth-elephant hybrid. The female hybrid might then be used to produce another, more "mammothy" hybrid. The second hybrid could be used in a breeding program to select out the remaining genetic variations that distinguish modern African elephants from woolly mammoths.

The relative ease of breeding partly motivates projects – a separate, non-cloning one for aurochs and the one for quaggas – that only use "back-breeding". This seems most attractive when the target genome (of the extinct species or subspecies) is sufficiently conserved in an extant close relative – *Bos primigenius taurus* (domesticated cattle) for the auroch and *E. quagga* (plains zebra) for the quagga. That relative can then be bred to weed out the genetic variations that developed after the extant lineage departed from the extinct one or from their common ancestor. If the extant species contains enough of the extinct species' genome, then over the course of some number of generations, breeding might hope to approach the extinct species' genotype and achieve phenotypical resemblance, too. The Quagga Project (Harvey 2008) in the Western Cape of South Africa was launched in 1987.[22] As of 2005, it had produced a third generation foal with uncanny resemblance to *E. quagga quagga*.[23] As of 2010, The Consortium for Experimentation, Dissemination and Application of Innovative Biotechniques (ConSDABI) in Benevento, Italy had completed a first round of breeding selected cattle for auroch characteristics (Sandy 2010 and Shanks 2010).

Serious skepticism has been expressed about the means employed in all these projects. Criticism is not restricted to the ones that involve cloning and hybridization. Some scientists doubt that back-breeding can obtain sufficient genetic similarity to justify claims of recreation. However, given rapidly developing exploits in genetic manipulation, the means of any of these projects are likely to be substantially enhanced and relatively soon. In sum, it seems foolish to dismiss any of them as "impossible" or fantasy.[24]

Along similar lines, the Brazilian-born American "bio artist" Eduardo Kac (2000) says that he regards the creation of such a creature as Alba, his GFP bunny, as an "extension of the concepts of biodiversity and evolution" (item (2)). His interest, as embodied in Alba, is in creating novel kinds of creatures, not in recreating old, lost ones.

There are several very serious and seriously funded gene and DNA banks (item (4)). Perhaps the best known and best funded is the Svalbard Global Seed Vault in Norway. But this effort is just one among many like-minded ones. These include plant DNA blanks at the Kew Royal Botanical Gardens in the UK, the New York

---

[22] See also http://www.quaggaproject.org/.

[23] Of course, "uncanny resemblance", which might involve morphology and phenotype more generally, might not suffice to justify "resurrected species". That depends on how one defines "species".

[24] This is not to say that resurrection projects are not very hard. And it is safe to say that the obstacles for resurrecting long-extinct creatures are probably very formidable, given the unlikelihood of finding sufficiently intact chromosomal material older than a few tens of thousands of years.

Botanical Garden, the Missouri Botanical Garden, the Graduate School of Biotechnology at Korea University in Seoul, the Jardim Botânico do Rio de Janeiro, the South African National Biodiversity Institute at Kirstenbosch, the National Institute of Agrobiological Sciences in Japan, and the Nationaal Herbarium Nederland. For zoology, there are the Frozen Ark project for endangered fauna, the Ambrose Monell Cryo Collection in the American Museum of Natural History's Sackler Institute for Comparative Genomics, the Frozen Zoo® at the San Diego Zoo, the Australian Frozen Zoo of the Animal Gene Storage Resource Centre of Australia at Monash University in Melbourne, the Conservation Genome Resource Bank for Korean Wildlife at Seoul National University, the Austrian Farm Animal DNA Bank, and the National Plant, Fungi and Animal DNA Bank in Poland.

The "Pleistocene Rewilding" project (Donlan et al. 2006, 664), though mostly focused on North American megafauna, acknowledge the potential for global biodiversity gains along the lines of items (5), (6), and (7). Specifically, its proponents reluctantly recognize the diversity-fostering effect of roads and other infrastructure (which they would prefer to see removed):

> The most straightforward conservation advantage of Pleistocene rewilding would be to enhance the persistence of endangered large vertebrates with a multicontinent system of reserves inspired by evolutionary and ecological history. This has been a positive approach to the conservation of rare species, as illustrated by the reintroduction of Przewalski horses (*Equus caballus przewalski*) from North American and European zoos to a semiwild state in their native habitats in central Asia. Additional viable populations could also enlarge the possibilities for adaptation by target species to global change as well as provide the selective regimes that have fostered existing genotypes. *Range fragmentation arguably might provide opportunities for speciation*, but that potential "positive" effect on biodiversity is surely countered by the threat of small population size, failure to adapt, and stochastic extinction. [italics added]

The caution expressed at the end of the last-quoted sentence is short-lived. Immediately afterwards (Donlan et al. 2006, 665), the authors cannot contain their "excitement" for "actively restoring evolutionary processes" as a salient goal of rewilding. They contrast this vision with "merely slowing down the rate of biodiversity loss". All told, these rewilding advocates convey the distinct impression that a major component in what makes "restoring evolutionary processes"… "an exciting new platform for conservation biology" is, in large part, its potential for promoting the adaptation and evolution of creatures in new habitats, which are likely to yield increases in biodiversity.[25]

The rewilding proposal splits the world of conservation biology – or at least it initially appears to do that. I don't have the space to comment on the details of the controversy here – except to remark on a certain incongruence in the complaints of its critics. This is one of those areas where those oddly conflicted attitudes – regarding the rightness of arranging for biodiversity to increase – are on display. For it is striking

---

[25] The inspiration for this project goes back at least to paleoecologists Paul Martin and David Burney (1999) whose exhortation to "Bring back the elephants!" channeled the "reintroduction" strategies espoused prior to that by conservation biologists Michael Soulé and Reed Noss.

for how much critics agree with proponents on such basic principles as that gestures at re-creating the past are good things, while disagreeing mainly on which past is the best one to try to re-create and on how much of the past is good to bring into the present.[26] However, one vital point seems to be lost on such critics of rewilding as Luiz G. R. Oliveira-Santos and Fernando A. S. Fernandez (2010), who hold up the specter of "Frankenstein" ecosystems resulting from novel interactions involving the "reintroduced" creatures: Their Frankenstein is the rewilders' welcome fruit. Proponents of rewilding justify their proposed means of setting transplanted creatures on a new, evolutionary path as one of its salient goods. As Sect. 6.7 (Biodiversity as value generator) brings out, the rewilders are not alone in playing on the theme of engendering biodiversity by setting evolution on a new course.

The rewilders are not alone in promoting projects to introduce novel species into new habitats. Though not all projects have biodiversity "enhancement" as a primary goal, many are still likely to have this effect. These include the routine introduction of game fish in various places for the benefit of sports fishers. Also included are "biological control" projects such as the biological control of *Tamarix* with *Diorhabda elongata* (the tamarisk leaf beetle) in the American southwest. These lauded intentional introductions, along with reviled unintentional ones, as well as the reviled intentional ones, have the potential for producing phenotypical and genetic novelty (hence diversification) in new populations of the introduced species. The possibility that new species might evolve from these introduced populations cannot be dismissed.

Finally, a variation on item (6) involves arranging for the cohabitation of populations of two closely related kinds of organisms that do not normally associate. Biologists have long been aware of the phenomenon of "character displacement". This involves phenotypical changes in the sympatric populations of two species with respect to allopatric conspecifics. These changes can be both significant and rapid.[27]

Eduardo Kac once again wields the barbed prod that pushes us towards an understanding of the creative potential of created habitats, which he calls "biotopes" (item (7)).

---

[26] The biologist Tim Caro (2007, 281), for one, senses this underlying basic agreement. He identifies the "target date" – before or after 1492 – as the crux of the controversy, while pointing out that "restoring past ecological and evolutionary processes, [is] something that conservation biologists [generally] acknowledge to be important." The date of re-creation plays a critical role in the minds of conservationists, who regard any creature who did not live at a certain geographical coordinate after the "correct" date as an "alien" and therefore *creatura non grata*. The debate about the status of *Gopherus flavomarginatus* (the Bolson tortoise) with respect to Big Bend National Park, where the Pleistocene rewilders wish to re-introduce it, largely hinges on just this question. Something like that particular terrestrial member of Testudinidae (but larger) evidently lived at that location, but not since the late Pleistocene/early Holocene.

[27] See, for example, Diamond et al. (1989) who recount the fascinating evolutionary history of two species of myzomelid honeyeaters with initially non-overlapping ranges. A volcanic eruption-induced "chance meeting" of populations of both species best explains a dramatic increase in the size difference between the birds in sympatric populations. In each other's presence, birds of the larger species got larger while birds of the smaller species got smaller – over the course of just three centuries.

Kac (2006) requisitions the term – meaning "a region from a habitat associated with a particular kind of biological community" – from ecology. Describing one of his major exhibitions, he says:

"Specimen of Secrecy about Marvelous Discoveries" is a series of works comprised of what I call "biotopes", i.e., living pieces that change during the exhibition in response to internal metabolism and environmental conditions, including temperature, relative humidity, airflow, and light levels in the exhibition space. Each of my biotopes is literally a self-sustaining ecology comprised of thousands of very small living beings in a medium of earth, water, and other materials. I orchestrate the metabolism of this diverse microbial life in order to produce the constantly evolving living works.

For Kac, biotopes are ways to "orchestrate" a kind of aleatoric performance by organic and living materials placed inside a glass case. But the case is an artifact of Kac's need to accommodate the constraints of a museum display. There is no such barrier to a more expansive expression of Kac's vision at landscape scales.

Kac's work also reminds us that, insofar as biodiversity is taken to include the diversity of habitats, one must not disregard how creatively potent people can be (item (8)). Perhaps this kind of biodiversity enhancement cannot be strictly separated from item (7). But it seems that the diversity of biomes can be considered on its own and without central regard to the diversity of organisms that inhabit them. Kac's biotopes also remind us that this is another way in which "habitat conversion" is a friend, not the enemy, of biodiversity. All that need be added is a quantum of Kac's creative energy to conceive and create novel biomes that are interestingly diverse.

So far as I am aware, no serious consideration has been given to a project to bioengineer *wild* species to better live with human incursions (item (9)). However, the basic idea behind this suggestion is far from original. In its last two reports on "Impacts, Adaptation and Vulnerability" (2001 and 2007), Working Group 2 (WG2) of the Intergovernmental Panel on Climate Change (IPCC) (Parry et al. 2007) suggested bioengineering as a means of adapting crops to hotter or drier climes. This sort of engineering for stress tolerance is already routine in the world of agriculture. The most famous example is Monsanto's "Roundup Ready" crops, which are designed to tolerate glyphosate, the vegetation-destroying ingredient in Monsanto's signature herbicide. The trivial extension to engineering for other forms of stress tolerance – ones that might help in adapting to climate change – has not been lost on Monsanto (Cheikh et al. 2000).[28] And it is straightforward to imagine applying this idea to, say, pika spp. for better tolerance of their increasingly hotter living conditions.

As I already mentioned in connection with items (5), (6), and (7), some suggestions for projects to increase biodiversity derive their inspiration from human projects that happen to have this side-effect, but that are pursued for a different purpose and without this effect in mind. Several other suggestions on the list are similar in this regard. People have inadvertently increased mutation rates by depleting stratospheric ozone (item (10)). They certainly have increased predatory pressures

---

[28] All contributing authors were Monsanto employees at the time of publication.

on many species by dint of hunting and fishing practices in various oceanic as well as freshwater fisheries (item (11)). For those creatures that survive their human predators, scientists are beginning to notice phenotypical changes, and it would be unsurprising if some of these changes eventually were to engender new species. That this is happening is relatively uncontroversial. Only the rate at which it is likely to occur is still a matter of scientific debate (Darimont et al. 2009). Human activities have vastly increased the amount of reactive nitrogen virtually everywhere on the planet (item (12)) – though, for the purpose of increasing biodiversity, one would want to be selective about the environments where it would increase rather than decrease biodiversity.[29] Finally, developers who see in salt marshes multimillion-dollar development projects – for building structures or as inevitable casualties of resource extraction – are only too happy to greenwash their projects with habitat reconstruction in places with less development potential (item (13)). It is not hard to believe that an ecosystem engineer with Eduardo Kac's imagination could invent a "reconstruction" whose complexity and diversity well exceeds the original.

It seems that the plausibility, possibility, even viability of pursuing most if not all of these projects cannot reasonably be denied (which is not to say that they would not encounter difficulties, and in some cases end in failure). If biodiversity is good, and if projects such as these would increase it, and if increasing biodiversity increases its good, then what possible objection could there be to pursuing them? I've already suggested one answer: In this domain of conflicted attitudes, scientists and environmentalists generally do in fact embrace many elements in many of these proposals, though they do not express them so boldly as I do. The other answer returns to the question of "appropriateness". This is the last expression of resistance from those whose value model for biodiversity would seem to encourage their pursuit. There are several variations on this theme.

According to the first variation, "not appropriate" means "not natural". "Not natural", in turn, is an historical property of states of affairs that are brought about by means that are "not natural". Thus, starting from the means: The proposed methods are "not natural"; therefore any biodiversity produced by their application is also "not natural".

The merits of this objection and the question of whether or not it commits the "genetic fallacy" are less of a concern for me than the fact that the objection effectively changes the subject. One moment, we thought we were discussing biodiversity and its value. The next moment we find ourselves in the middle of the chaotic *mêlée* over the relative merits of the natural versus the artifactual, which pre-dated all the discussion about biodiversity.[30] It is conceivable that in the final analysis, the discussion of biodiversity and its value cashes out as one way to engage in this debate.

---

[29] This is not to deny that figuring out how to selectively fertilize to achieve greater biodiversity would require some effort beyond deploying an already-proven ability to fertilize everything, everywhere.

[30] I do something very much like this in Chap. 8. But this is part of a move to push biodiversity off the table, not salvage it, as the principal emblem of nature's value.

But then biodiversity loses its interest as a new and independent locus of environmental or natural value.

A survey of the literature on biodiversity by scientists and environmentalists who hold biodiversity to be emblematic of nature's value reveals a great amount of natural/artifactual fence straddling. As noted in this book's Preliminaries, Bryan Norton (2006) talks about protecting biodiversity as something equivalent to protecting all *natural* value. Many of the biologists surveyed by David Takacs (1996) (see Sect. 4.2, Accretive conceptions) are prone to agree. And yet, many biologists who form the opposing camp – including Sahotra Sarkar and others who try (however unsuccessfully) to divorce biodiversity from "what is natural" – emphasize that biodiversity is completely compatible with and even an integral design element in a wide range of human projects. In fact, this is part and parcel of this latter group's rallying cry to abandon wilderness as the central locus of nature's value and to embrace biodiversity in its stead. I return to the tension of straddling this divide in Sect. 7.2 (From the natural/artifactual trap to nature as science project).

Several other variations on the theme of "appropriateness" can be seen to flow from a just-so model of biodiversity value. The first and most obvious of these is that my biodiversity-increasing methods cannot produce a biodiverse state of affairs that is "just so". Unfortunately, "just so" rests on some notion of what is "appropriate" for biodiversity value. As a result, the just-so theory of value does not so much offer any reason for objecting to biodiversity-increasing projects as merely restate the objection.

Another just-so variation on "appropriateness" draws on uncertainties in the outcome of any of the suggested interventions. Our state of knowledge is such that we cannot know whether any of them will result in just the right kinds and right levels of diversity. Of course, this reservation is still firmly based on the questionable notion that only the right (appropriate) kinds and levels are valuable.

Yet another just-so variation falls into the trap, endemic to this model of biodiversity value, of focusing on specific kinds and thereby losing sight of the central issue of their diversity. The objection goes: While increases in biodiversity that result from obvious human intervention might indeed increase the diversity of kinds in various biological categories, the particular kinds that are newly created are inappropriate; and this blemishes their worth. This objection founders for at least two reasons. First and most obviously, it once again merely shifts the problem of defining "appropriate". The shift is from attributing the attribute "appropriate" from biodiversity to the particular kinds that fall within the biodiversity-contributing categories. There is no evident progress for understanding "appropriate" in that. Second, it once again commits a category mistake that will be familiar from several previous encounters in this book. Diversity, and therefore any value that attaches to diversity, has no investment in the identities of the entities (kinds) that are diverse. To focus on the identities of these kinds is to take the focus away from diversity.

There is another odd feature of the notion of "appropriate" kinds and "appropriate" levels of diversity. The notion seems to be a shield designed quite specifically to deflect suggestions to actively engage in behavior that increases biodiversity. But it is unclear why the shield would not work equally well to deflect many actions aimed at conserving or preventing decrease in biodiversity. Sarkar's example of letting the cows back into Keoladeo to "conserve" the wetland is precisely an example of this. There appears to be no significant difference between that maneuver, and the suggestion of item (7) in the list above to create new habitats with the aim of encouraging either new species or new phenotypes to emerge.

This, of course, brings us back around to the conflicts in attitude that permeate this discussion. Other examples come from sometimes-heroic efforts to save species by, for example, sequestering individuals (in "bio-secure" lockdown) and raising individuals in zoos.[31] Could it be that these species have reached the "natural" and "appropriate" winding-down that is the fate of every species, perhaps some 5–10 million years after its start? If "appropriate" biodiversity (diversity worthy of value) only arises by "natural" means, then it would seem that biodiversity that is retained by "unnatural" means is as tainted in value as "unnaturally" induced increases.

Section 6.13.3 (Causal factors and history) returns to the proposition that, although biodiversity is a state of affairs, causal factors still affect its value. And Chap. 8 finally digs in to sort out why, in fact, it might not be appropriate to generate biodiversity – whether via the sorts of methods that Sarkar advocates, or via one or another of my proposed methods.

---

[31] One striking example comes from recent attempts to save *Leptodactylus fallax* (Montserrat island "mountain chicken" frogs), as described in Aldred (2009).

Chapter 6
Theories of Biodiversity Value

It's time to survey theories of the source of biodiversity's value. My survey is not exhaustive. But it covers the theories most frequently offered by the scientific and environmentalist communities and the ones that are most robust, in my judgment. Also, I would say that these constitute a sufficiently diverse sample to give a good general sense of how suitable a peg biodiversity might be for hanging environmental or natural value.[1]

As prelude to this survey, I bound the discussion by recalling where biodiversity fits with respect to some other candidates for environmental "goods": Biodiversity does not have to do with individual organisms except insofar as individuals contribute to genomic diversity. But the usual arguments for the good of individual (nonhuman) organisms that hinge on an individual's rationality, consciousness, sentience, desires, needs, or the individual's "good of its own" offer no obvious support for the good of the *diversity* of rational, conscious, sentient, needy, or telic organisms. Any moral consideration afforded to individual organisms by dint of one or another of the aforementioned capabilities does not obviously extend to the *diversity* of individuals that possess these capabilities. One might suppose that a diversity of creatures[2] is required to make possible a diversity of *kinds* of rationality, consciousness, sentience, etc. For example, human sentience, crotaline sentience, formacid sentience,

---

[1] Among other theories of biodiversity value that I omit are ones that build on religious beliefs. Three reasons, not out-of-hand dismissiveness, guided this choice. First, insofar as various western religions find grounds for a just-so model of biodiversity, I do, in fact, deal with a salient element of their position. On the other hand, I think that other, mostly eastern religions do not view biodiversity in anything like the way that western-trained scientists and environmentalists discuss it and which is the main topic of this book. Finally, I feel that a fair treatment of any religious approach would require a substantial opening out of this already substantial book, which puts such a treatment beyond its scope.

[2] With the possible exception of the possession of a *telos*, these considerations are predominantly geared towards promoting the moral status of a small proper subset of organisms – a select group of multicellular *animals* – while demoting the status of the vast majority of (all other) organisms.

D.S. Maier, *What's So Good About Biodiversity?: A Call for Better Reasoning About Nature's Value*, DOI 10.1007/978-94-007-3991-8_6,
© Springer Science+Business Media B.V. 2012

and so on are possible only with some diverse complement of creatures (people, pit vipers, and ants) that possess these various forms of sentience. But then the diversity of modes of sentience must be shown to be a good. It is hard to imagine a convincing argument for this proposition; and none has been forthcoming.

As I suggested in Sect. 3.1 (The core concept), species diversity is as central to a conception of biodiversity as it is to modern biology – whatever the flaws of the species concept. The good of a species is notoriously difficult to rationalize, since with few exceptions, there is no obvious sense in which a species leads a purposive existence in the sense that is commonly taken as requisite for grounding the value of a species' individuals.[3] But that difficulty is also irrelevant to the question of how biodiversity gets its value because the good of any particular species as the species that it is, like the good of individual organisms, lies outside the domain of concern for biodiversity insofar as it consists of the diversity of species. Of course, the loss of any particular species *does* constitute a loss – an incremental one – in species diversity. It might also constitute (more than the loss of a single individual of that species) a loss in allelic diversity, in feature diversity, or in functional diversity. But the value of that species for species diversity or biodiversity generally is restricted to that species' contribution to just that – diversity.

As an orienting device and as an introduction to problematic syndromes that affect the project to locate value in biodiversity, it is useful to reflect briefly on the upper and lower bounds for biodiversity. At the lower bound, zero biodiversity means no people – not a recommended option. Just considering bare human survival (and leaving aside other needs and desires), it seems safe to say that some sufficient, nonzero amount of biodiversity is infinitely valuable for what seems to be its essential instrumental role in ensuring the continuation of at least one particular species – *Homo sapiens*.[4] Yet even that "safe" assumption comes with caveats. It is difficult to know what that sufficient amount might be – a topic that Sect. 6.4 (Biodiversity as (human) life sustainer) explores. In fact, the amount might be impossible to determine, except by fatal experiment – especially if the boundary between adequate and fatal is thin and almost impossible to perceive on approach. This kind of precautionary consideration comes up for discussion in Sect. 6.3 (Biodiversity as service provider).

Determining a lower bound is problematic for others reasons, saliently including the fact that species diversity in itself says nothing about which species are included. It seems entirely possible that some larger number of species will, with higher likelihood, include some combination that places some terrible burden on humanity. It could even be the death, or at least the infirmity, of us all – a possibility that is

---

[3] Populations of colonial animals such as ants and other social insects of the family Formicidae – already mentioned in Sect. 2.3.1 (Abstraction) – might constitute an isolated and (because not the entire species) qualified exception.

[4] It should be clear that even if this *end* – the persistence and flourishing of one peculiar primate – is supposed to be served by biodiversity, biodiversity does not serve to *justify* it. No matter. I assume along with most others that human existence can be justified on other grounds, and not just on the basis of the diversity of the human genome.

taken up Sect. 6.5.2 (Biodiversity as safeguard against infection).[5] Considerations such as these suggest that the trickier problem might be setting an upper bound – a maximum recommended therapeutic dose of biodiversity. But at some point, difficulties with lower and upper bounds and everything in between should raise suspicions that biodiversity is simply not the sort of thing that admits of credible norms along these or any other lines.

This last difficulty stems from the fact that a specification of biodiversity is also a curtain of ignorance, which conceals the exact identities of the biodiverse kinds. The difficulty is a reflection of a considerable temptation to offer reasons for why biodiversity might be valuable – including its role in mankind's survival – which toe or pass over the line between the diversity of kinds and the particular kinds in the biotic world that happen to benefit people. That temptation is in evidence throughout this chapter.

This chapter mainly concerns itself with theories according to which biodiversity is either a constituent good (as with Sect. 6.8, Biodiversity as font of knowledge) or (as with human survival) an instrumental one. A good of the former type competes with other, similar complementary goods, whose realization is sometimes mutually exclusive. A good of the latter type has a provisional quality because, as with any good that is good only as a means to an end, its value is contingent on the absence of other means or their inferiority.[6]

Finally (for this chapter's prefatory remarks), so far as I can tell, there is no *a priori* reason for why every category of diversity that figures into biodiversity should be valuable; and this supposition could well turn out to be false. The diversity of species might, in fact, be of little consequence while the diversity of orders is really important according to some norm. Or Daniel Faith might be right and what really only counts is phylogenetic diversity. In short, it is possible that the diversity of kinds in some categories is valuable while diversity in others is not. Also, as I remarked in Sect. 3.3.3 (Multiple dimensions), the situation is complicated insofar as different categories of biodiversity are likely not to be independent, but interrelated in various complex ways; and so this is likely to be true of their values, too.

Perhaps there is a presumption that the diversity of kinds in any category that is interesting to scientists – something that correlates with other, interesting biological properties and phenomena – will be valuable, if for no other reason than that studying such diversity can increase scientific knowledge. This supposition seems to lie behind much of the "scientific" discussion of biodiversity's value. I grant it provisionally and only for the purpose of engaging with many of the theories that appear in this chapter.

---

[5] This point owes to some comments made by Jeffrey Lockwood.

[6] Perhaps one can even imagine humans becoming autotrophic – by acquiring the capability of manufacturing, from inorganic materials, all the organic compounds that are critical for humans to consume for their health. This is but a few steps beyond the nascent capability – in the lab at least – of growing disembodied meat (Britten 2009).

6.1   Unspecified "Moral Reasons"

In their paper on "Biodiversity Studies", Paul Ehrlich and E.O. Wilson (1991, 760) express their belief that

> Because *Homo sapiens* is the dominant species on Earth, we and many others think that people have an absolute moral responsibility to protect what are our only known living companions in the universe.

Wilson has echoed this sentiment many times over and with great eloquence.

Certainly, a "moral responsibility" is precisely what advocates of biodiversity wish to find and justify. Unfortunately, merely making the claim for such a responsibility does not constitute an argument for it. Nor does the fact that people have the capacity to destroy biodiversity by itself entail a moral obligation to refrain from doing this. It is easily within my power to smash the coffee mug containing the black, aromatic elixir that fuels my writing of this book. But I am unaware of any moral obligation to refrain from doing so – though it might be extremely imprudent, so long as my writing task is incomplete.

There is a contrary tradition of thought – stemming from the Stoics, promulgated though Augustine, and some might say, on through welfare economists – that human dominion is evidence that all of earth's goods exist for man's pleasure, to use as he sees fit. There is no reason to believe that biodiversity is somehow exempt from this rather different brand of "stewardship". The bare matter of fact that *H. sapiens* is a dominant species on the planet lends as little (or as much) support to this Stoic interpretation of its moral implications as that fact lends support to the moral duties that Ehrlich and Wilson seek to squeeze from it.

Perhaps Ehrlich, Wilson, and others who say similar things have in mind duties to nonhuman individuals. Or perhaps they are thinking of duties to each and every particular species that happens to exist at the moment. If so, they are mistaking the value of those individuals or the value of those extant species for the value of biodiversity.

It is also possible that Ehrlich and Wilson presume that valuing each and every species entails valuing biodiversity. Unfortunately, the logic behind this presumption is faulty. While saving each and every species would, as a matter of fact, save biodiversity in the sense of preserving the current biodiverse state of affairs, the parallel inference for value commits the fallacy of composition. There is no guarantee that species diversity has independent value as a collection of species, just because each one of the species is valuable.

Or again, Ehrlich and Wilson might believe that some (possibly great) amount of biodiversity is a necessary condition for any individual organism of any species to thrive. But this appears to be a doubtful proposition, which I revisit in the special form of Biodiversity as (human) life sustainer (Sect. 6.4). And in any event, there is no evidence that this is what these authors have in mind.

6.2   Biodiversity as Resource

One of the most common claims on behalf of biodiversity is for its enormous value as a resource, and one that is critically important. Ehrlich and Wilson (1991, 760) represent this position when they note

> … that humanity has already obtained enormous direct economic benefits from biodiversity in the form of foods, medicines, and industrial products, and has the potential for gaining many more…

There are really two reasons here on offer – first, biodiversity as past and current resource; and second, biodiversity as potential future resource. Glossing over this distinction does not compromise my discussion in this section, though it can become critically important in the context of sophisticated economic analysis. The economic reckoning of future goods can involve such niceties as their discounting or (as discussed in Sect. 6.9, Biodiversity options) wrapping them in the complexities of option value.

Confusion between particular species and the diversity of species is immediately evident in the representation of biodiversity as resource. Some particular species are good for people to eat. Because people need food in order to survive, those species might qualify as critically important. Other species have been found to have value for their production of chemicals of pharmacological value. But particular species have yielded the benefits of providing sustenance and the means to restore health – not biodiversity, nor specifically, species diversity.

For the sake of trying to explore what Ehrlich and Wilson might be driving at, I overlook this confusion and presume that their position involves something more like the claim that: a great diversity of organisms increases the odds that at least some few of them are or will be around that are good to eat, that some few others of them do or will provide good medicines, and that some few others do or will provide good building materials. There remains an apparent assumption that the resource-providing organisms are a random sample of all organisms. This is almost certainly untrue and I return to this matter of fact shortly. But putting this objection aside (and alongside the previously noted category confusion), this is still a singularly unconvincing defense of the value of species diversity.

The fact is that an extraordinarily tiny minority of organisms has benefited humanity as resource, now and previously. The majority of this minority – especially when it comes to food (which I discuss just below) and medicine (which I discuss in Sect. 6.5.1) – are highly likely to persist even in the face of a general decline in biodiversity. Furthermore, there is little reason to believe that this circumstance will change in the future. These facts combine with the other that any economic resource competes with other economic demands. As a consequence, from an economic point of view (which includes both resource and "service" value, the topic of Sect. 6.3, Biodiversity as service provider), there is scarcely ever justification for not letting a species go extinct – even if the effort or cost required to save it is minimal. Certainly, many if not most of the symbolic creatures – such as *Ursus maritimus* (polar bear) and *Eubalaena* spp. (right whales) – fall into this category. When, as in the case of both these creatures, there is, in fact, a significant economic cost to saving them – for polar bears, reversing climate warming,[7] for right whales, slowing down or rerouting the ships that traverse their thoroughfares – then the mere possibility of a future benefit from their incremental contribution to species diversity is an essentially nil "expected net present value" (to use the standard economic jargon) by comparison. Daniel Faith joins other conservation biologists in supposing that biodiversity as a resource has (positive) *option value*. But demonstrating that requires showing that biodiversity commands a *premium* over its expected value to consumers, which standard it already fails to meet. Where Faith leaves off with option value, James Maclaurin and Kim Sterelny pick up. I take up their treatment of this theory of biodiversity value in Sect. 6.9 (Biodiversity options).[8]

There is yet another objection to the resource rationale. Insofar as conserving biodiversity preserves the likelihood of conserving one or more valuable resources in the future, it also preserves the likelihood of conserving creatures that are destructive of resources or otherwise harmful. Disease organisms, "pests", and destructive parasites contribute to biodiversity (or at least species diversity) at least as much as (and possibly much more than), for example, the trees that provide good building materials. In fact, because parasitism might well be the predominant "lifestyle" on the planet – by some estimates, outnumbering free-living species by a factor of four (Stiling 2002, 193) – conserving biodiversity is far more likely to ensure that parasitic creatures continue to be in good supply. Parasites even come with a diversity bonus – namely, the species on which they are parasitic (their hosts). Polyphagous parasites deliver multiple bonuses.[9]

In addition, and contrary to the random sample assumption, food for people – the most essential of resources for humans – is actually supplied by organisms in a set that is vanishingly small in the total (species) diversity picture, and that predominantly are carefully maintained and managed by humans on farms. Reliable recent estimates (Khoshbakht and Hammer 2008) are that there are around 7,000 cultivated

---

[7] In May 2009, the Obama administration in the United States reaffirmed the preceding (George W. Bush) administration's position that the Endangered Species Act, as applied to polar bear, is not a basis for curbing the greenhouse gas emissions that are the root cause of that species' demise. See Revkin (2009). The rationale, supplied by Interior Secretary Ken Salazar, was largely economic (in terms of the impacts on cement manufacturing and regulatory difficulties) – though it is not at all clear that the ESA admits this sort of test as legitimate.

[8] Faith (2007) seems to regard mere mention of "option value" as sufficient to counter current uncertainty regarding what value biodiversity might have, either as resource or service. Unfortunately, he ignores the wider economic picture: Uncertainty in biodiversity supply, uncertainty in biodiversity demand, and expected consumer surplus heavily influenced by the relative certainty of savings realized by avoiding the cost of conservation can easily push (garden-variety) option value into negative territory.

[9] Of course, people like some parasites – such as the wasps that pollinate commercially valuable *Ficus* trees.

crop species of plants. That is only about 2% of the estimated 320,000 kinds of plants on earth (according to Spicer 2006, 27).[10] Yet that percentage is enormous in comparison to that represented by the number of livestock species. There are an estimated 7,600 breeds (in the 2006 Global Databank for Farm Animal Genetic Resources of the FAO – the Food and Agricultural Organization of the United Nations) of perhaps 40 species.[11] The contribution of these creatures to species diversity borders on infinitesimal in the context of over 9 million other animal species.[12]

A similar consideration applies to biodiversity as medicinal resource, which Sect. 6.5.1 (Biodiversity as pharmacopoeia) takes up in detail.

One final consideration might be the most devastating for the argument for biodiversity as food resource: There is perhaps no better understood and no greater conflict involving biodiversity than the one between it and the production of food. Food production involving crops or tended livestock takes land away from the many other creatures that might otherwise live on it. Chopping or burning down a forest to put the soil under the plow is the most common source of "habitat conversion". That is the euphemistic term for the phenomenon that scientists routinely cite as, by far, the single most decisive force in the global reduction of biodiversity.[13] And agriculture dominates the three leading causes of habitat destruction or "conversion" – which also (Dirzo and Raven 2003, 159) include "extraction activities (mining, fishing, logging, and harvesting), and the development of infrastructure (such as human settlements, industry, roads, dams, and power lines)."[14]

Empirical support for the conflict between biodiversity and food for humans is reinforced at a macro level by the "species-area effect", or as Robert MacArthur and E.O. Wilson (1963) put it in their seminal paper, "the fauna-area curve". MacArthur

---

[10] Dirzo and Raven (2003, 142, 159) note that estimates of the number of flowering plants (the angiosperms which dominate the plant world) range from 250,000 to more than 500,000 species.

[11] This figure is due to Drucker et al. (2005, 11), who give a lower estimate than the FAO's for the number of breeds, though agree with the FAO's figure for the number of livestock species.

[12] Agroecologists have lately stressed the prudence of maintaining some greater rather than lesser variety of strains of agriculturally important plants and breeds of agriculturally important animals. But they are talking varieties of plants and animals in the hundreds or at most a few thousands. These are created and maintained as domesticated organisms, and unlikely to be much affected even by a great extinction event. Insofar as wild species are thought to be needed for future crop plant varieties, seeds (along with other *ex situ* methods of conservation) serve quite well. Kew Garden's Millenial Seed Bank alone projects having one-fourth of all wild plant species represented in their seed bank by 2020 (http://www.kew.org/ucm/groups/public/documents/document/ppcont_016021.pdf).

[13] See, for example, Sala et al. (2000, 1771), who summarize their findings on "drivers" of biodiversity loss in their Tables 6.1 and 6.2.

[14] The numbers cited by Dirzo and Raven come from the year 2000 Red List of Threatened Species due to the World Conservation Union, Gland Switzerland (IUCN). See also the discussion of "land transformation" in Vitousek et al. (1997, 494–495).

and Wilson originally proposed this principle, which posits a positive relationship between species richness and land area (proportional to a fractional exponent of the area), in the context of island biogeography. But the correlation has been found to hold in a wide range of different ecosystems, making it one of the more generally applicable and most highly verified principles in ecology. These days, great effort is devoted to finding ways to "de-intensify" the use of land given over for the production of food. There is much talk about ecologically less destructive agricultural practices and systems.[15] But "less destructive" means "still very destructive, indeed". They might somewhat ameliorate the effect on biodiversity. But they cannot undercut the general and overwhelmingly dominant principle. Food for people decreases biodiversity. Or conversely, biodiversity is the enemy of human food resources.

One might counter that sometimes, predators and parasites are introduced into agricultural systems to suppress pests; and this has a biodiversity-enhancing effect. Also, there is evidence that sometimes, cultivation of a greater diversity of crops enhances food production overall. But even if these propositions were always and not just sometimes true, no one could honestly suggest that more hectares should be appropriated for cultivation in order to increase biodiversity.[16]

Taken together, these considerations should relieve worries that the loss of biodiversity will inevitably mean the loss of valuable resources. On the evidence, great diversity – of species, at least – is not of any great benefit, considered as either actual or potential resource. Quite to the contrary, attempts to maintain biodiversity are fundamentally at odds with efforts to produce food – the most valuable of all resources.

6.3   Biodiversity as Service Provider

In their discussion of "Biodiversity Studies", Ehrlich and Wilson (1991, 760–761) prominently feature what has lately become the most popularly cited value attached to biodiversity. That is

> … the array of essential services provided by natural ecosystems, of which diverse species are the key working parts. Ecosystem services include maintenance of the gaseous composition of the atmosphere, preventing changes in the mix of gases from being too rapid for the biota to adjust… The generation and maintenance of soils is another crucial service… Soil ecosystems… are… providers of two more services: disposal of wastes and cycling of nutrients… Another… is the control of… species that can attack crops or domestic animals…

---

[15] Less destructive agricultural practices attempt to retain local pollinators and enemies of crop pests. But this barely registers in the balance of biodiversity after the existing flora of a large swath of land is ripped out (which also rips out dependent fauna) and replaced by a set of food crops that is considerably less diverse (even when those crops are highly diverse by agroecological standards). Moreover, the opportunities for accommodating local organisms are limited by the fact that almost all agricultural crops everywhere are exotics whose interaction (beneficial or otherwise) with local organisms is uncertain. And many crops do not require insect pollinators.

[16] This clarification owes to a challenge made by Jeffrey Lockwood.

My discussion of ecosystem services observes the consensus usage of ecologist David Hooper and his colleagues (2005, 7) in their survey on the "Effects of Biodiversity on Ecosystem Functioning":

> *Ecosystem services* are those properties of ecosystems that either directly or indirectly benefit human endeavors, such as maintaining hydrologic cycles, regulating climate, cleansing air and water, maintaining atmospheric composition, pollination, soil genesis, and storing and cycling of nutrients. ...*Ecosystem properties* include both sizes of compartments (e.g. pools of materials such as carbon or organic matter) and rates of processes (e.g. fluxes of materials and energy among compartments). [italics in the original]

This setup merits two observations. First, the very definition of ecosystem services prejudices a discussion, which focuses on them to the exclusion of ecosystem *dis*services. There is no non-arbitrary reason not to balance the discussion of the good that ecosystems do (their services) with the bad (their disservices).

The second observation is that the thesis about the value of *biodiversity*-provided ecosystem services proceeds in two steps. The first step is the biasing one, which simply identifies and selectively focuses attention on particular ecosystem services, whose positive value for humanity is guaranteed by the definition of "ecosystem services". But even aside from the obvious bias of failing to bring ecosystem disservices into the discussion on an equal footing with ecosystem services, the attribution of positive value to identified ecosystem services does not automatically clinch the proposition that these services should be maintained in their current form. Making the case for that requires that an ecosystem service value be not just positive, but greater than the value of what might replace the ecosystem if it were "developed" for some other purpose despite the sacrifice of the service. But this broader evaluative context does not even need to grant that the service is sacrificed. Nothing precludes considering how that service might otherwise be provided in another location or by some other method; or still provided by the original ecosystem despite drastic changes to it.

The second step in the value proposition for biodiversity-provided services is of greater concern for the central theme of this book. It requires that a connection from ecosystem services to biodiversity be established. That is, it requires support for the thesis that a service performed by an ecosystem – one whose value is arguably great enough to merit continuation – generally and critically depends on that ecosystem's biodiversity, as it currently exists. Doubts about this thesis can arise from doubts about either of the two steps that lead to the second observation. The thesis can fail either by failing to make a non-biased case for the existence of an ecosystem property whose benefit is not outweighed by concomitant disservices; or it can fail by failing to make the case that a service critically depends on some significant biodiversity. I discuss both sources of failure, but focus the second, which tries to link biodiversity to the (ecosystem) service sector of the human economy.

Recent years have seen increasing promotion of the thesis that there *is* a link between biodiversity (mainly, but not exclusively considered as species diversity) and ecosystem services. For some time, this view has been the gold standard for at least two of the three world's largest transnational so-called "conservation" mega-organizations – The Nature Conservancy and the World Wildlife Fund. In 2009, the third of the three "conservation" mega-organizations – Conservation

International – joined the ecosystem services fold. Many large corporations express their enthusiasm for this approach by sponsoring these dominant organizations.[17]

It is both easy and revealing to understand why corporations welcome the proposition that what really counts in ecosystems is their services, and that biodiversity counts precisely insofar as it contributes to these services. This guiding principle provides an apparently environmental justification for the most commercially profitable environment- and ecosystem-altering activities. The imprimatur and public presence of TNC, WWF, and CI serve to deflect questions of the principle's environmental credibility.[18] Unfortunately, as I will show, the implications for the environment in general and biodiversity in particular are, by most any standard, entirely unattractive, if not devastating, to nature and its value.[19]

The ecosystem-services view of natural value is built atop two exculpating principles that grant permission to activities that impinge on the environment. According to the first and most basic principle, if the activity does not compromise some one or more identified services, then the activity is environmentally permissible. According to the second principle, if the first principle does not apply because an ecosystem service is compromised or removed, then the activity is still permitted – provided that there is an alternative way to provide that service. The alternative service provider is allowed to be another ecosystem constructed specifically for that purpose. But the logic of ecosystem services does not require that a surrogate be a re-creation in any way other than in a capability to render the service.[20] In both principle and practice, the engineering details are left open.

---

[17] I sometimes use scare quotes for the word "conservation" (and sometimes utilize the term "neo-conservation") in reference to these organizations as well as to allied practices that are increasingly mainstream in conservation biology. The scare in these quotes has to do with the fact that conservation, as these parties understand it, has much to do with managing and developing the planet for the production of stuff, services, and a few creature that excite the public, biologists, hunters, or fishers. It has little to do with what one might have thought to be the core meaning of "conservation" as "preserving the natural world".

[18] See Note 25 for more on the relationship between TNC and the international mining giant Rio Tinto.

[19] Those who bet nature on its provision of ecosystem services nevertheless try to weasel out of the terrible implications of a consistent application of their own principles by suggesting the ecosystem-service argument is merely a trump card, which can be pulled out when the need arises. I point out in this section that this is an arbitrary and therefore illegitimate move. Section 8.2.3 ("Living from" nature, uniqueness, and modal robustness) provides a more complete perspective, from which one can see that this card is pulled from a house of cards.

[20] The position set out by the three major "conservation" organizations (and others) is actually more radical than my description in the main text suggests. The added radicalization derives from the principles of "mitigation banking" and "habitat banking", which justify any action that compromises an environmental service on the mere *promise* that the identified, affected service will be restored *somehow or other* and this will be done *someplace or other*. There is a fairly uniformly bad track record in actually making "withdrawals" on banked mitigation efforts and banked habitats. But whether or not promises tend to be kept in actual practice, there remains the serious normative question of whether or legitimate norms concerning the value of nature justify the destruction of a natural habitat in exchange for the promise to reinstate a service or two.

So far as species diversity is concerned, the first exculpating principle essentially says that an organism's value depends on its ability to pay its own way as an indispensable contributor to a valuable ecosystem service that is not otherwise more efficiently provided. It is important to notice that this is an extraordinarily high standard for a species to meet in order to justify itself. It requires that some valued service would not be robustly maintained in the species' absence. For perspective: few species listed as "endangered" through the Endangered Species Act would meet it.

Viewing this standard from a broader economic viewpoint makes it higher yet. That is because an unbiased assessment of how biodiversity affects services cannot justifiably restrict its attention to *ecosystem* services. There is no good reason not to broaden the purview to economically valuable services generally. When that is done, the assessment must countenance the fact that biodiversity often is an *obstacle* to valuable (non-ecosystem) services. Which brings back into view ecosystem disservices of at least one variety – namely, those that obstruct other services.

But this is still not the whole story. Biodiversity is often the enemy of ecosystem services, too. Water purification would go a lot better were it not for the microbes and flora that make people sick. Of course, such an observation does not clinch the case for biodiversity being a disservice, for a greater amount of biodiversity might, in fact, not include those nasty organisms. But this is hardly a vindication of ecosystem service principles of valuation. For it is often fairly clear that, even when biodiversity is not an obstacle to an ecosystem service, it fails to pitch in; and that, according to those evaluative principles, makes its removal permissible.[21]

Still, most ecosystem *dis*services have to do with how ecosystem properties negatively impinge on other kinds of valuable services. Vociferous opposition to listing many if not most candidates for "endangered" under the Endangered Species Act arises precisely on the grounds of obstructing economically valuable services. This was behind Alaska Governor Sarah Palin's announcement (Bryson 2009) that the state of Alaska would sue to have the endangered Cook Inlet population of belugas delisted.[22] Cook Inlet is an area where exploding gas and oil development has spurred planning to expand the port of Anchorage and possibly build a new bridge, the Knik Arm version of the "Bridge to Nowhere" across the Inlet. The beluga swims – at a typically unhurried 3–9 kph – squarely in the way of the services that promise substantial increases in economic welfare. Whatever small economic benefit the little white whale contributes derives mainly from the amusement it affords people in marine "parks". So far as Palin and most Alaskans are concerned, the whales should stay there and clear of the development of vastly greater economic goods in the Cook Inlet. There it is unequivocally an economic liability.[23]

After Palin quit her job as governor later in 2009, Anchorage mayor Dan Sullivan joined with several other local mayors to take up the cause of litigating the delisting

---

[21] I am indebted to Jeffrey Lockwood for a remark that forced me to clarify this point.

[22] The suit was eventually filed, as reported by Joling (2010).

[23] This commentary is derived from an online version, available at http://environmentalvalues. blogspot.com/2009/01/beluga-isnt-that-caviar.html.

of the Cook Inlet belugas. Like Palin, they made their case squarely on the grounds that the large mammals impede (non-ecosystem) services such as the transportation of oil by ship and efficient vehicular transportation of people across Cook Inlet. The belugas' disservice consists in the fact (Hunter 2009) that they "could have big negative effects on government projects and activities in and near the Inlet, as well as fisheries and oil and gas development." These anti-beluga partisans do not challenge the service framework for assessing values. Quite to the contrary, they embrace the framework as the preeminent arbiter of value and deploy their arguments within it.

There is no principled way to arbitrarily disallow consideration of the negative impacts of biodiversity on *non*-ecosystem services that might outweigh biodiversity's positive effects on ecosystem services. But even if one condones this unprincipled exclusion, the evidence is at best shaky for the proposition that biodiversity has much if any beneficial effect. The science simply does not strongly support this sanguine assessment. Even if a species is not performing *dis*services (in the restricted sense of "ecosystem disservices"), if it is rare and isolated, it is unlikely to be performing any real positive service.[24] Many ecosystem services, such as the oft-cited provision of potable water to New York City (discussed in Sect. 2.2.1, The bare assertion fallacy), make minimal demands on biodiversity. While this conclusion seems difficult to avoid, this unavoidable conclusion is routinely avoided: A high degree of biodiversity is often not critical for delivering ecosystem services; and furthermore, many species in many ecosystems are the bane of human existence. Therefore, even when the service-contribution standard of evaluation is arbitrarily confined to ecosystem services, what this standard shields is *service-providing* diversity, while blithely leaving biodiversity-at-large to suffer or, more likely, be decimated. When non-ecosystem services and disservices are brought back into the equation (as in the preceding paragraph), the status of biodiversity-at-large is even more dramatically reduced. I further explore the evidence for how biodiversity relates to ecosystem services shortly.

So far, I have dwelt on the first and more fundamental exculpating principle of ecosystem service evaluation, whose essence might be expressed as "no service for organisms that give no service". When the first principle does not apply, the second one often does. It legitimizes the development of a landscape or habitat and its concomitant eviction of resident organisms when a surrogate provider of that habitat's services can be engineered elsewhere, "elsehow", or both. In itself, this principle of substitution does not *logically* entail anything one way or the other about the net affect on biodiversity. But the surrogate – even when it is some other habitat – might be, and as a matter of fact often is, less biodiverse than the original habitat. Even if the surrogate is a thing constructed from soil and some number of living

---

[24] For example, Díaz et al. (2006, 1300–1301) say that "… rarer species are likely to have small effects at any point in time." These authors confront the fact that rare species don't count for much in the way of services by falling back on an unexplicated concept of "biotic integrity". It is entirely unclear what "biotic integrity" might mean. (On this, see Sect. 4.1.5, Biodiversity as process.) But whatever it is, these authors suggest that it should be the goal of conservation rather than "simply maximizing the number of species present".

bio-parts – for example, when a natural wetland in one place is sacrificed in the name of condominium development and an artifactual wetland created someplace else – and even if the "same" services, such as those of filtering water are reinstated by the human-made edition, there seems to be little evidence that the living complement of resident creatures is also routinely re-created *en masse*.

Furthermore, some *service* surrogates are quite obviously not also *biodiversity* surrogates. One might naturally assume that the surrogate is a cleverly constructed re-creation of the original, complete with the original's complement of denizens, just sited some place that is economically more "efficient". But as I noted in my original mention of the second exculpating principle, nothing in the ecosystem services paradigm requires this. It is the service that counts. And so the surrogate could just as well be constructed from steel, concrete, pumps, and bulldozed mounds of gravel. It need not bear any particular resemblance – in physical appearance or in its biotic or abiotic component parts – to its predecessor service provider, provided that it "does the job", or it is thought to do so.

This problem – that biodiversity can and does suffer in the name of more economically efficient maintenance of service levels – is little acknowledged, let alone addressed in the ecosystem services literature. But two recent addenda to the ecosystem service framework can be reimagined as responses. The first response is to sever the previously supposed direct link between biodiversity and of ecosystem services and to treat these two matters separately. This leads to the recently popularized notion that "biodiversity banking", "biodiversity offsets", or "biodiversity development"[25] can ensure "no net loss" of biodiversity; and this is considered quite separately from the question of maintaining services. Nevertheless, "no net loss" tends to be rationalized via the fictional ability of people to create a duplicate habitat from whole cloth. It makes Creators out of ecosystem engineers, who populate their Creation with the appropriate complement of bio-parts. Though the Creators of this Creation relieve the bio-parts of service-rendering expectations, they breathe into them another expectation. These bio-parts are supposed to coalesce with the abiotic bits into something recognizable as "the same" as the developed-out-of-existence home of their sacrificed brethren.[26]

The second and more recent response takes off in the opposite direction, to presume a tight *bi*directional link between ecosystem services and biodiversity.

---

[25] The term "biodiversity development" and its unblinking usage as one type among others of economic development might sound like fiction to someone not inoculated with the serum of Natural Capitalism. Sadly, it is not my invention, but rather a quite real touchstone of current-day conservation. See The Nature Conservancy Leadership Council (2008) (which features a presentation by Rio Tinto's manager of biodiversity offsets) and Richards (2005). Section 8.1.8.1 (Biogeoengineering as right action) further explores this concept.

[26] It is hard to exaggerate the enthusiasm with which this fiction is embraced and used for self-praise and self-aggrandizement by the most ravenous developers, such as the international mining company Rio Tinto. This is made possible by such organizations as The Nature Conservancy, which, while receiving donations from Rio Tinto, eagerly supplies and endorses the "conservation" rationale for that company's extraction practices. Because this topic crosses the boundary that circumscribes "biodiversity as service provider", I will not pursue it here.

This is supposed to make it essentially unnecessary to distinguish the two for purposes of conservation. Not only is biodiversity said to have value as a means of ensuring ecosystem services, but also it is said that "conserving" ecosystem services is among the surest means of conserving biodiversity. In fact, it is said that focusing on ecosystem services better ensures that biodiversity is conserved than trying to conserve biodiversity directly – that is, with biodiversity specifically in mind as the goal.[27] In effect, this response simply denies, without accompanying reasons for denying, that biodiversity fails to track ecosystem service levels. Unfortunately, as I have already observed, there is clear and abundant evidence that ecosystem services can and do continue in substantially transformed landscapes with substantially reduced biodiversity.

Of these two responses, the second is emblematic of the consensus supposition that biodiversity derives its value from ecosystem services because the biodiversity of ecosystems and their services are joined at the hip. This supposition suffers from much the same *logical* lacunae as the similar proposal, examined in the preceding section, for Biodiversity as resource. But before examining these lapses in logic, let's take a look at the *empirical* basis of the claim and various lapses in accounting for the facts. In their paper on the "Effects of Biodiversity on Ecosystem Functioning", David Hooper and his colleagues (2005, 4) express a scientific disciplinary consensus in saying that they

> ... *are certain* [that]... More species are needed to insure a stable supply of ecosystem goods and services as spatial and temporal variability increases, which typically occurs as longer time periods and larger areas are considered. [italics added]

The proposition is presented as if it were a general law of ecology. But while some evidence instantiates the generalization, much other evidence does not. In other words, the generalization is false: It crumbles under the weight of the full complement of scientific evidence and in the absence of any credible way to qualify it to account for exceptions.

It is undoubtedly true that some (though, as I shall shortly observe, not all) properties regarded as services depend on certain species – sometimes acting single-handedly, sometimes in certain combinations ("assemblages"). But Hooper et al.'s claim that "more species are needed" is undermined by an abundance of counterexamples. One counterexample is perhaps *the* classic textbook example of the salt mash – a relatively species-impoverished type of ecosystem that is often highlighted as a service provider. These habitats, dominated by naturally occurring (not human-designed) monocultures of *Spartina* and *Juncus* and a few other salient

---

[27] This is the entire point of the paper by Goldman et al. (2008), for example. Originating from the epicenter of "The Natural Capital Project", which is closely linked to TNC, it contends that greater biodiversity ensues from "protecting" ecosystem services. In this context, "protecting" must be interpreted in the very odd way that can be understood only in terms of the economic calculus, which permits new service providers to be created (or the mere promise of creating them) in exchange for a license to destroy existing ones.

species (mollusks and others) provide (for example) water-filtering services.[28] It does not appear that more species must be recruited to get this job done. A second counterexample or class of counterexamples is provided by exotic species, which, more often than not, are "experiments" in increased biodiversity at local and regional scales. The hard evidence shows that much of the time, the arrival of new species in an ecosystem does not cause "natives" to disappear; and so the new arrivals must be supposed to increase biodiversity or at least species diversity.

But what about the ecosystem functions? Sometimes they do not change upon the arrival of a new visitor. But sometimes they do. In the neat tripartite classification scheme due to Peter Vitousek (summarized in Vitousek and Walker 1989, 262), "invaders can cause changes in overall resource availability, in the trophic structure of an area, or in disturbance frequency or intensity".[29] Vitousek and Walker (1989) chronicle the dramatic changes that *Myrica faya*, an actinorhizal nitrogen fixing tree, brought to Hawai'i's young volcanic soils by vastly increasing the previously meager supply of the nutrient nitrogen to coresident organisms. In this dramatic case and others not so dramatic, the increase in biodiversity is accompanied by a change in ecosystem properties. But that proposition is entirely different from the one that says that "more species are needed" to provide desired services.

One might object that I have misinterpreted the "more species are needed" dictum – that it was not meant to apply to the effects of adding species, but to the effects of removing them. But this supposition is not congruent with what Hooper et al. say. Another "certainty", according to them (Hooper et al. 2005, 4) is that:

> Some ecosystem properties are initially insensitive to species loss because (a) ecosystems may have multiple species that carry out similar functional roles, (b) some species may contribute relatively little to ecosystem properties, or (c) properties may be primarily controlled by abiotic environmental conditions.

This passage indicates that species are often dispensable, so far as their role in providing services is concerned. This hardly supports the case for biodiversity or the need for more species as the key to providing services. One might further object that the key word in this "certain" proposition is "initially". Emphasizing the importance of the word, one might interpret Hooper et al. to be saying that a further diminishment of biodiversity *might* seriously compromise an initially unaffected service. This line of thought leads to the complex and perplexing realm of precautionary argument.

I return to the important topic of framing principles of precautionary prudence at the end of this section. But the last-quoted statement by Hooper and colleagues provides an opportunity to first show that the application of any precautionary principle in the domain of ecosystem services is extraordinarily restricted or even marginalized. To that end, I characterize two classes of ecosystem services for which precaution with respect to biodiversity is quite difficult to justify because they are

---

[28] See, for example, Stiling (2002, 292), Kareiva and Marvier (2003, 347), and some related remarks in Sect. 3.3.3 (Multiple dimensions).

[29] See also Note 34 for more on the work described by Vitousek and Walker (1989).

largely indifferent to biodiversity. Yet the services in the union of these two classes constitute the vast majority.

The services in the first class are characterized by radical indifference to which species, the number of species, and the diversity of species that provide them. Any species in a large pool of candidates will do, without compromising the quality and level of the service. Moreover, that pool of candidates is so large so as to be practically inexhaustible. Many could vanish and species diversity could plummet – again, without detriment to the service. Services in this class are also indifferent to which particular *ecosystems* render them. Thus, for example, pretty much any vegetation in pretty much any kind of ecosystem suffices when it comes to sequestering carbon, filtering water, cycling water, mitigating floods, and controlling erosion. A place can be wiped fairly clean of vegetative and other diversity – say as the result of strip mining or a volcanic eruption. But if it still can support vegetation, then any single species of recolonizing plant – even one that never before grew in that place (that is, an exotic) – is likely to do well in rendering any one or more of these services. Unless all vegetation in all places capable of growing it vanish, services in this class are highly likely to persist.

The scientific literature contains occasional glimmers of awareness about this class of services and its significance. David Ehrenfeld (1988, 214–215), for one, generates a large shaft of light:

> The sad fact that few conservationists care to face is that many species, perhaps most, do not seem to have any conventional value at all, even hidden conventional value. True, we cannot be sure which particular species fall into this category, but it is hard to deny that there must be a great many of them. And unfortunately the species whose members are the fewest in number, the rarest, the most narrowly distributed – in short, the ones most likely to become extinct – are obviously the ones least likely to be missed by the biosphere. Many of these species were never common or ecologically influential; by no stretch of the imagination can we make them out to be vital cogs in the ecological machine.

Another glint of awareness comes from Norman Myers (1996b, 2764), who correctly diagnoses one of the maladies that underlies Ehrenfeld's disheartening assessment: "While biodiversity often plays a key role, the services can also derive from biomass and other attributes of biotas." In other words, biodiversity counts for little in providing a service if what really counts is a diversity-independent attribute such as biomass. Unfortunately, there has been no full reckoning, let alone a clear one, of what this implies for the thesis that "ecosystem services depend on biodiversity".

In the second class of biodiversity-indifferent services are ones that are unusually robust because they depend on unusually robust interrelationships or "functional groups". While these functional groups might change in composition from time to time or from place to place, they tend to require species that reproduce easily, are abundant, and are unusually adaptable. They are the species whose ubiquity spans even dramatic ecosystem changes in the face of which many more vulnerable species succumb. David Wardle (1999) and his colleagues provide one example – of nutrient recycling service by soil detritivores. They find that, in a grassland setting, rates of decomposition, measured as microbial respiration, depend on the presence or absence of plants, but not on their diversity. The implication is that the persistence of the recycling service depends on the grassland continuing to have grasses *simpliciter*, no matter how great or lacking in their diversity.

Also focusing on detritus recycling, E.O. Wilson (1987 and reprised elsewhere) claims that if all the insects vanished, all humanity would likewise vanish. But this is hardly an argument for the *diversity* of species.[30] Arthropods most likely represent the vast majority of (non-microbe) species on the planet. If you toss in all the other invertebrate animals that perform the detritus-processing function that Wilson has in mind in making his claim, and beef up this collection with the fungal and bacterial organisms that also do a substantial share of it, then but a miniscule number of organisms remain. A more sober interpretation of Wilson's remarks is that they argue for preserving some very small portion of an overwhelming abundance of detritivores – a selection that would likely include sufficiently many well-adapted arthropods and other organisms that recycle adequate amounts of detritus.

What sort of hard evidence supports the claim that maintaining biodiversity is essential to maintaining essential services? One of the most cited experiments designed to answer the question of "how many is enough" is that of David Tilman and his colleagues (2001). Tilman's work examines the consolidated productivity of various combinations of grasses. In it, the combinations are chosen at random from a pool. The pool, in turn, is apparently chosen for its diversity (within the realm of grasses), but according to no other identified or identifiable rationale. Tilman and his colleagues find that, under these conditions, the values for this one property (productivity measured as biomass) increase with increasing biodiversity. These increases are mostly realized with the assemblage of the first few species, but continue with diminishing returns up to the experiment's grand total of 16 species of grasses.

What does this say about the role of biodiversity in ecosystem services? Grass productivity might be considered a service if the grasses in question somehow fed the economy – for example, by feeding cattle. On the other hand, grass productivity would be considered a disservice were these grasses considered weeds or "aliens" (that is, exotics viewed negatively), which insinuated themselves, unwanted, into other agricultural efforts or into manicured lawns. I return below to this difficulty – the several respects in which classifying "properties" as services or disservices is highly problematic. In the current context, it is more important to reflect on what resemblance, if any, Tilman's "ecosystems" – the very simple creations of his experimental design – have to the "self-assembled" ecosystems that are the concern of most environmentalists. Tilman's work undoubtedly lends plausibility to the supposition that 16 species of grass in neatly planned experimental plots might, *in toto*, create more biomass than one or two species. But to claim that this says anything at all about natural ecosystems and how their diversity relates to the services they provide to humanity would be an inductive leap that no scientist could in good conscience defend.

---

[30] I am sympathetically supposing that Wilson takes himself to be presenting *some* argument for defending, as he does in his paper, the value of the diversity of invertebrates.

To be scrupulously clear about this, the phrase "ecosystem service" is not to be found in the paper by Tilman et al. However, the phrase is promoted to titular status in the paper by Hooper et al., who are among the myriads who cite Tilman's research in the context of trying to connect biodiversity to ecosystem services. They are representative of a literature that invokes Tilman's and other, similar work to urge that changes in biodiversity might well compromise, as Ehrlich and Wilson (1991, 760) say, "essential services provided by natural ecosystems, of which diverse species are the key working parts". When explicitly referring to Tilman's work, Hooper et al. use the word "properties", which in the case of Tilman's work refers to productivity as measured by biomass. They never once justify their leap from "biodiversity affects properties" to "reduced biodiversity compromises critical services". Yet like many others who cite Tilman's work, they regard it and similar work to be a springboard for launching them across this logical void. In this case, the void spans the chasm between "16 arbitrarily selected grasses produce more biomass than one" to "essential services require diverse species".

In the end, Hooper and his colleagues give some signs of understanding that there is no credible basis for this leap in logic. They do not make any clear statement to this effect. However, they review several logically possible candidates for characterizing how biodiversity might relate to the properties or functioning of ecosystems. The possibilities range from no influence, to a direct variation, to an asymptotic correlation, and end with "idiosyncratic patterns". For all but the last candidate, there is contravening evidence, not just a lack of it. That leaves "idiosyncratic patterns", which is a scientific gloss for: irrelevance – because the truly relevant factors might be species composition, trophic structure, nutrient availability, or some other factor that has little to do with biodiversity. In other words, the leap looks plausible only when "biodiversity" is misunderstood in terms of category mistakes or unless its meaning is distended beyond recognition to encompass any and all function-determining factors.[31]

In their final analysis of the "Effects of Biodiversity on Ecosystem Functioning", Hooper and colleagues (2005, 8–15) say that they cannot find any convincing evidence or reason to prefer any one of the above candidate characterizations to another. But this inconclusive summation does not adequately reflect the real gist of the stubborn science that they recount. The correct scientific inference, I believe, is that the last class of possible "relationships" – "idiosyncratic patterns" – wins the lottery, for that is the "relationship" that denies that any real relation exists. It says, in agreement with my reading of Hooper et al., that there is no known consistently characterizable connection between biodiversity and ecosystem functioning, despite persistent attempts to find one. A fortiori, there is no consistently characterizable relationship between biodiversity and ecosystem services. The inference that this warrants is that the biodiversity of an ecosystem is simply the wrong lens through which to view its functioning. This is not to deny that some properties, processes, and functions – including ones that are regarded to be services – depend on specific

---

[31] In the latter case, the claim – that functions of organisms influence the functions or processes observed where they live – becomes tautologous. See Díaz et al. (2006, 1300).

interactions of specific resident organisms with each other and with their specific abiotic environment. It *is* to deny that this has something to do with the *diversity* of those "working parts".

In the absence of credible scientific evidence, the case that services depend on biodiversity rests on major lacunae in logic, concept, and representations of fact. Some of these will be familiar from previous discussion of Biodiversity as resource (Sect. 6.2) and from Sect. 3.3.2.3 (regarding Functions). One pervasive form of reasoning goes from a "job" done by some one species or group of species to the conclusion that it is diversity that is responsible for this beneficial industry. The plausibility of this inferred relationship tacitly but heavily leans on the functional-trait-as-appendage view of organisms, which folds the observed functions and properties of ecosystems into their denizens. This picture of ecosystems depicts certain species as having service-providing attributes that enable them to pitch in and do their job, while others not well endowed with such attributes sit it out on the functional sidelines – slackers or free riders, as it were.

Unfortunately, this picture is oversimplified to the point of being misleading. There simply is no fixed "job" that a species inherently performs independent of the biotic and abiotic environments in which it finds itself. Examples of the disutility of this conception abound. One, related in Sect. 3.3.2.3, involves trophic interactions between invertebrates and trout, which this conception would describe as the trout-fattening function of stoneflies. Another (previously mentioned in this section) elegantly illustrates how easily this picture can mislead. A nitrogen-fixing plant might not provide any significant nitrogen-enriching or fertilization function in a place where some other source of reactive nitrogen (deposition, for example) already provides an ample supply (Vitousek and Walker 1989). When agricultural crops benefit from bee pollination, then bees can provide a pollination function. But replace the crop with one that is not bee-pollinated, and like an invisible appendage, that function is summarily amputated from those busy and unsuspecting insects.[32]

At least as pervasive is a class of problems characterized by an unwarranted leap from the beneficial activities of a specific organism to the alleged service-providing benefits of species diversity. Often the burden of providing the service – for example the "pollination service" proffered by *Apis mellifera* (European or western honey bee) in North America – falls largely upon the "shoulders" (or species-appropriate body part) of a single species.[33] But it is hard to connect this circumstance or the economic value that derives from it with the claimed role of the *diversity* of

---

[32] This example is from a much-heralded study of how native bees in Costa Rican forests adjoining a coffee plantation were observed to increase coffee yields. Subsequently, coffee prices plummeted and the fields were replanted with pineapple, which does not benefit from bee pollinators. The bees were thus put out of work. Their service value, as well as the service value of the peripheral forest that housed them, thereby plummeted to zero. See Ricketts et al. (2004) for the first part of this story. There has been considerably less fanfare surrounding the story's conclusion. See also Note 38 in Chap. 8 and Sect. 8.2.3 ("Living from" nature, uniqueness, and modal robustness) for the broader significance of this story.

[33] It is worth noting that this case is representative of ones in which hard work earns a reprieve from derogation as an "alien": *A. mellifera* is exotic in North America.

species or the diversity of anything else. Consider that an enormous portion of the world's productivity – including the production of perhaps one-fifth of the world's atmospheric oxygen – can be credited to species of the single genus *Prochlorococcus* – chlorophyll b-containing marine cyanobacteria. So this genus might be regarded as providing breathable air-producing services. But diversity is clearly not the key to this. Indeed, there appears to be little convincing evidence that most functions vital to or highly valued by people could not be performed by a tiny fraction of the world's various kinds of organisms.

All told, there seems to be a gross superfluidity of species – a true *embarras des richesses* – associated with the majority of ecosystem services. In terms of the two classes of biodiversity-indifferent service that I distinguished earlier in connection with the "more species are needed" refrain: Either the work (for a class one service, such as carbon sequestration) is handily performed by more or less any organism blindly drawn from an enormous pool; or (for a class two service such as nutrient recycling), the work is handled by a small, species-indifferent, but robust and functionally coherent set of species. Some scientists who press the case for biodiversity's value as the basis for ecosystem services seem to realize this. They therefore shift their focus from species diversity to functional diversity. While removing (or adding) any number of species might leave a function largely intact, removing the function itself (thereby decreasing functional diversity) obviously would not. The diversity of functions would intuitively seem to be far more closely related to the provision of services ("properties" with functional benefits) than to species or species diversity.

I would say that the relationship – between the diversity of functions and the provision of services – is so close that it borders on, or even crosses over into, tautology. Sandra Díaz, David Tilman, and others (2006, 1300) present this near tautology as if it were a scientifically discovered empirical fact about the world:

> The evidence available indicates that it is the functional composition – that is, the identity, abundance, and range of species traits – that appears to cause the effects of biodiversity on many ecosystem services.

In this context, it is evident that "species traits" means "functional traits", which in turn means "the traits that are evidenced by functions or processes observed in the species' habitat". So it seems that Díaz and colleagues end up saying that the traits of organisms that yield identifiable ecosystem functions yield ecosystem functions – and particularly, the preferred class of functions known as "ecosystem services".

Tautologies are uninteresting. But let me try to honor the literal meaning of Díaz et al. by reinterpreting what they say as an argument to conserve *functional* diversity. According to this interpretation, care must be taken not to deprive humanity of a function that provides a valuable service – such as (in addition to the production of oxygen) pollination and the recycling of nutrients. The species involved in these functions – *A. mellifera* in the case of pollination, and arthropods et al. in the case recycling detritus – are merely means to the more proximate means (of a diversity of functions) to the final end of providing people with these services.

But this line of reasoning nosedives back into the logical lacuna that confuses a kind for a diversity of kinds. Previously encountered in connection with claims about the need for a diversity of species, it reemerges here in connection with functions. It is this particular function – for example, the pollination of an extremely narrow range of cultivated plants – that is valued, not the *diversity* of functions. And the value placed on a function such as pollination does not seem to be merely, if at all, a matter of its incremental contribution to the diversity of all functions. But let's make the same sort of provisional assumption made for the *resource* value (as contrasted with service value) of species diversity. For the sake of argument, let's consider the possibility that the more functions provided and the greater their diversity, the greater the likelihood that the set of all functions will implement services prized by humans.

Unfortunately, this suggestion to salvage the case for functional diversity is also unconvincing, and for reasons similar to those that make unconvincing the like claim for the resource value of species diversity. The major problem is that functions that are *dis*valued by humans crash the party. Lots of them. On what principled (as opposed to arbitrary) grounds can one exclude, for example, the vegetation-reducing function (otherwise known as "herbivory") performed by *Diatraea grandiosella* (southwestern corn borer)? Or the human population reducing function performed by deadly microbes such as *Mycobacterium tuberculosis* and *Plasmodium falciparum*? Or along the same lines, the disease-transporting function performed by *Anopheles* and *Aedes* spp. vectors?

Despite the last example, "serving human health" features prominently in support of the thesis that biodiversity underwrites ecosystem services. There is undoubtedly much in the biological world that influences human health. But it is doubtful that any sort of diversity is unambiguously beneficial in this regard. Discussions that tout the human health services provided by biodiversity invariably fail to take into account what I hinted at above – namely, the diversity of disease organisms and their vectors. An explosion of claims for the health benefits of biodiversity warrants a separate discussion of this topic, which appears in Sect. 6.5 (Biodiversity as a cornerstone of human health).

The ecosystem service theory of biodiversity value also encounters difficulties because the sometimes very particular context required for some services to be services makes them considerably less than an unqualified good. Notably, certain disservices are necessary conditions for some services. The vector function of mosquitoes, a disservice for *H. sapiens*, is a good *resource* for species of the infraorder Anisoptera (dragonflies), *Progne subis* (Purple Martin), and various species of the suborder Microchiroptera (microbats). These creatures are thereby given a chance to provide their (good) disease vector-reducing function on behalf of *H. sapiens*. This situation should not be unfamiliar. It is of a piece with the economic reckoning of situations commonly encountered in the service economy. Economic welfare can be and has been bolstered by such services as those for cleaning up oil spills and other "environmental disasters". This uncomfortable contingency of (good) services on disservices or on other conditions that are quite undesirable is yet another central issue that is routinely ignored by those who try to connect biodiversity to services.

I now return to the several respects in which classifying "properties" as services or disservices is highly problematic, mentioned in connection with David Tilman's grassland experiments. One way to characterize the problem is as a certain porosity in the containers that ostensibly compartmentalize kinds of service, and even that separate services from disservices. A radical compartmental leakage occurs because the same kind of function can have (positive) value in one context and disvalue in another. Nutrient retention and nitrogen fixation can be services when agriculture is served by fertilization efficiency. But these are viewed as disservices when they impinge on the success of favored plants and other organisms whose adaptive advantage derives from their ability to thrive in nitrogen-impoverished soils.[34] Pollination services might be pollination disservices when what is pollinated is considered a noxious weed. Weed control services, provided by an arthropod herbivore on noxious plants, can be food production disservices when applied to farm crops. In fact, the "noxious weed" and "farm crop" can, in different places, be one and the same species. They can even be one and the same plant in the same place. Population ecologists Lesley Campbell and Allison Snow (2009, 498) provide one example:

> Volunteer populations occur when unharvested seeds from a previous crop germinate and grow in and around agricultural fields. For example, canola harvesting can drop more than a thousand seeds per acre, and volunteers often compete with subsequent crops…

Flood control services are desired when floods endanger human lives, structures, and property whose dry state is valued. But not when agriculture depends on regular flooding. The flood control that the Aswan High Dam offers is, according to many environmental precepts, a major disaster.[35]

One might object that the ecosystem that the Aswan High Dam regulates is an engineered ecosystem. This objection seems irrelevant. Nothing in the definition of "ecosystem service" disqualifies engineered ecosystems from supplying ecosystem services. However, it is possible that *Castor canadensis* (North American beaver) makes this point nearly irrelevant. Beavers historically have been systematically extirpated in North America for their dispensing a number of ecosystem disservices, including tree-destroying, field- and road-flooding, water-stealing, and water-rights-violating disservices. But lately, the same beaver activities are being viewed (for example, in Taylor 2009) as providing water storage services. In fact, enthusiasm for how efficiently these creatures supply these services has led to talk of a

---

[34] For example, Vitousek et al. (1997, 742) suggest that nitrogen deposition is reducing biodiversity – by (among other effects) transforming heathlands, with their characteristically nitrogen-poor soils, into more commonly encountered grasslands. Vitousek and Walker (1989) report that the relatively recent appearance of *Myrica faya* – a nitrogen-fixer – in Hawaii altered ecosystem processes in ways that most conservationists view as bad. Therefore, their contribution of nitrogen to the environment is viewed as a disservice.

[35] For a perceptive history of the Aswan dams, see Hughes (2009, 175–181). Hughes' book accomplishes the seemingly impossible task of surveying humankind's impact on the environment – globally and over essentially the entire course of human history.

different kind of engineering that involves introducing beavers into places where there's a beaver-friendly river and a desire for water storage (Groc 2010).

So it is not the diversity of properties or functions that is valued, but rather the particular benefits of a carefully selected, desired group of them, dubbed "services". And these are not always welcome; they are positively valued only in "appropriate" contexts. With this again firmly in mind, one might still insist that these services *depend* on some other kind of diversity – and the most likely kind is the diversity of species. But having circled back to supposing that species diversity anchors ecosystem services, one again runs into the brick wall of scientific evidence that simply does not support the proposition of a dependent relationship of ecosystem services on species diversity, let alone any kind of consistent correlation between them. The fact is that many functions in the human-coveted set – that is, many services – derive from relatively non-diverse landscapes. And many of these landscapes are highly human-engineered according to a specification that requires the weeding out of previous, naturally occurring resident organisms. Moreover, the addition of a species (variously tagged "exotic", "alien", "invasive", or "means of restoration") that increases species diversity in an ecosystem is just as likely to do any of: nothing (so far as recognizable functions are concerned); remove a desirable function; alter some characteristic of an existing, desirable or undesirable function (such as its intensity); or add an undesirable one – as it is to add a desirable function or (by some metric) strengthen or intensify an existing (desirable) one.

Productivity – the focus of Tilman's experiments – illustrates all this. This property or function and its relationship to species diversity, species composition, nutrient limits, and a host of other and possibly interacting ecological factors is probably one of the most highly studied because one of the most tractable to measure. However in and of itself, productivity is really of no value to humanity. This is easily seen by reference to the many plants that people classify as weeds. Some of them, such as kudzu in the southeastern United States, are enormously productive. Productivity is only a service when it involves the production of what people *like* – because it is good to eat, because you can build things from it, because it provides shade, because it anchors the soil, or because it performs some other, desired chore.

Even then, productivity is a service only if what is produced is in quantities that people also like. Producing either too much or too little can be a disservice. Juxtaposing production-as-service with expressions of alarm about enormously productive ecosystems illustrates this point. Eutrophication in coastal zones (briefly touched upon in Sect. 3.3.2.3 on Functions) is one of those cases for alarm. These offshore environments often arise from the anthropogenic, albeit unintentional, fertilization of algae by fertilizer that washes into their aquatic environment from upriver farms. Just like terrestrial crops and other plants whose growth is limited by the availability of nitrogen and phosphorus, the algae sometimes respond with an enormous increase in primary productivity – an "algal bloom".

However, unlike the terrestrial crops fertilized upriver, people do not eat these aquatic algae. It is ugly. Because it tends to reduce oxygen levels, it makes life hard for the fish and crustaceans that would otherwise thrive in these environs.

People like to eat those meaty creatures, not the algae. Worse, this fertilization often favors algal species that produce toxins, which harm, not just the fish and crustaceans that people like on their dinner plates, but people directly. Quite obviously, this productivity is *not* good and therefore it is *not* an ecosystem service. Rather, it is *bad* productivity and an ecosystem *dis*service.

The production of toxins provides yet another illustration of compartmental leakage. In the context of what E.O. Wilson (2002, 126) calls "nature's pharmacopoeia" (discussed separately in Sect. 6.5.1), Wilson (2002, 127) joins many others in lauding the production of toxins by assorted creatures as part of "biodiversity's bounty". But it is clear that toxin production, like primary production, is a function that is neither inherently good nor bad. Insofar as these functions are tied to biodiversity, only a logic of contextual convenience separates diametrically opposed conclusions – that biodiversity is good for people or that it is bad for people.[36]

It is worth dwelling on aquatic (including marine) eutrophication for another reason, which has to do entirely with matters of fact that bear on value claims relating to biodiversity. Human-induced aquatic eutrophication is frequently cited as a terrible bad for its deleterious effect on ecosystem services. And the bad of eutrophication is said to be the result of the erosion of biodiversity. But what is said in these contexts must be distinguished from actual evidence. In extreme cases, eutrophic zones become hypoxic (oxygen depleted); and these areas are labeled "dead zones". A layperson would be forgiven for conjuring up a zombie movie, or a moonscape of total and desolate lifelessness, or at least in the context of a discussion about biodiversity, for thinking that this term implies a serious loss of species. But this is not so obviously the case.

The real science is much more subtle and complex. In a report on eutrophication in the Baltic Sea, marine ecologists Jesper Andersen and Janet Pawlak (2006, 13) provide some hint of the real complexity:

> In terms of the biological response to hypoxia, the lethally low concentration of oxygen depends on the species. Fish and crustaceans have higher requirements for oxygen and they react very quickly to a lack of oxygen. Other species can tolerate low dissolved oxygen for longer periods. Under conditions of hypoxia, the benthic responses involve a change from communities of large, slow-growing, slowly reproducing species to communities of small organisms with a rapid turnover rate. Hypoxic and anoxic (total lack of oxygen) conditions also may result in the formation and release of hydrogen sulphide (H_2S) from the sediments, which is lethal to higher organisms.

Unstated in this passage (and in the report) is the likelihood that the hydrogen sulphide is produced by (diverse) sulfate-reducing microbes in the sediment. Certain other microbes – chemoautotrophs capable of oxidizing hydrogen sulfide or methane for energy – are known to thrive in environments like this. These organisms, in turn, might support protozoans and metazoans, such as sponges. Of course, all these

---

[36] This vexing characteristic of arguments for the value of biodiversity – that, when carefully considered, they cut both ways – is a recurrent theme. It recurs again quite prominently in the discussion of the health benefits of biodiversity in Sect. 6.5 on Biodiversity as a cornerstone of human health.

are "lower organisms". But this derogatory classification is irrelevant so far as the diversity of species or functions is concerned. In fact, the type of benthos just described routinely occurs in the deep ocean's seabed, which is "naturally" oxygen-depleted and lacks sunlight – signal characteristics that that environment shares with the seabed in "dead zones". However, I am not aware of any research that attempts to determine whether benthos-resembling communities exist in "dead zones", and if such communities do exist, what diversity of species they support.

Let's return to the more general case of eutrophication, which does not always lead to hypoxia. Stripping away a commentary that relies more on its use of normatively loaded words than on reasons leaves this story: There are changes in the species composition in several categories – including phytoplankton, zooplankton, fish, submerged vegetation, and benthic macrofauna. Even the story for algal species is mixed and complex, apparently depending on the characteristics of the herbivores that are present.[37] Generally, the biomass of phytoplankton (and their production), zooplankton, and benthic animals on shallow bottoms above the halocline *increases*. It is not at all clear that the diversity of creatures diminishes overall. A scientifically justified claim to this effect would require a careful census of all organisms – lower, higher, great, small, and miniscule – both before and after eutrophication. Again, I am not aware that any such survey has been undertaken.[38]

"Stability" joins "productivity" as another ecosystem property that purports to be a biodiversity-dependent service. Unfortunately, discussions of how stability relates to biodiversity have a strong tendency to forgo logical consistency in favor of normative commentary.[39] The first thing to note about the stability of ecosystems is that there is no known general positive or negative correlation of this property to any kind of diversity – including species diversity.[40] Moreover, it's hard to know what service is implied by stability, other than that of preserving the *status quo*.

---

[37] See, for example, Mazumder (2009, 143–157).

[38] Nor have marine biologists whom I've contacted been able or willing to point to evidence for the thesis that eutrophication reduces biodiversity overall. The pervasive claim is rather that "biodiversity is degraded". But "degraded" does not mean "reduced", and "degraded biodiversity" has less to do with biodiversity than with an expression of displeasure with the changed composition of a eutrophic environment. The scientist mentioned in Note 49 of Chap. 3, who changes the subject from biodiversity to vibrancy, also seems to be doing no more than expressing his displeasure with eutrophic ecosystems. I suggested that the basis of his displeasure cannot be reduced biodiversity. But nor can it legitimately be reduced vibrancy. That is because neither he nor anyone else has provided a respectable account of what a vibrant ecosystem is, why vibrant ecosystems are good, and what implications this has for the Sonoran Desert and Antarctica. Exploring these issues might be interesting and valuable, but this exploration would apparently have little to do with biodiversity.

[39] The concept of "stability" has many components or interpretations – including resistance to change and resilience after perturbation. I don't try to dissect the concept here because my discussion cuts across these multiple interpretations.

[40] The scientific literature that discusses ecosystem stability is enormous. My impression is that there is no general thesis about how stability and species diversity are related that either does not have substantial disconfirming evidence or is not so highly qualified that it essentially turns into a description of one special case study.

What really seems to be at issue with stability is that at least certain particular deviations from the *status quo* are perceived to be bad. Perhaps the most-discussed example of this is the form of stability that consists in the resistance of an ecosystem to the immigration of species new it. In what follows I largely have this particular form of stability in mind.

Two formidable problems confront those who propose that species diversity "enchances" an ecosystem's stability considered as a *status quo*-maintaining service.[41] The first, as noted just above, is an empirical matter. The preponderance of evidence simply does not support the empirically falsifiable generalization that greater species diversity diminishes the naturalization chances of a newcomer. As noted by Thomas Stohlgren and some colleagues (2003) and reiterated by ecologist Jason Fridley and his colleagues (2007, 7–10) in a survey on this topic, it is often the case that "the rich get richer".[42] Fridley et al. (2007, 10) propose a list of eight factors that might enhance an ecosystem's resistance to invasion. Only one of them has to do with its existing level of species diversity. Oftentimes, it appears that the conditions that were most likely responsible for engendering a rich diversity of "natives" are also quite welcoming to latecomers, too – and in a way that does not displace or entirely displace the early-comers.

The second problem has again to do with a failure to bridge the normative gap. The property of resistance to immigration might indeed be found to a lesser or greater degree in different ecosystems. But it is entirely unclear why this should be considered, without a great deal of contextual qualification, a "service" or (what is implied by "service") "a good". Most immigrants increase rather than decrease local and regional species diversity and relatively few result in global extinctions. So the anti-immigrant sentiment seems antithetical to a purely biodiversity-valuing point of view.[43] But aside from that, there are many contexts in which people value, not immigration resistance, but immigration acceptance. Fish and game managers routinely introduce species (particularly fish) to satisfy interests in (mostly) recreational predation.[44] Ironically, biologists intent on controlling previous, recent immigrants, introduce immigrants – known as "biological controls" – of their own.

---

[41] Those familiar with the history of ecology will recognize this as a restatement of the eminent English ecologist Charles Elton's (Elton 1958, 145) famous richness-stability hypothesis:

··· the balance of relatively simple communities of plants and animals is more easily upset than richer ones.

[42] The phrase "the rich get richer" is stolen from the title of the paper by Stohlgren et al. (2003), which is devoted to an exposition of this phenomenon.

[43] The issue of how species diversity might increase or decrease as the result of immigrations is a complex one, which I cannot adequately address in this book. An excellent short synopsis is due to Sax and Gaines (2003).

[44] Not infrequently, wildlife managers take advantage of the phenomenon that the introduction of exotics increases species diversity without compromising the fate of natives. Reporting on Florida exotics, Burkhard Bilger (2009), cites Paul Shafland, the director of Florida Fish and Wildlife's Non-Native Fish Laboratory, as stating that:

Florida's waterways are home to more than 30 species of exotic freshwater fish … and their total biomass nearly tripled between 1980 and 2007. Yet the number of native fish hasn't changed in that period; nor have any natives gone extinct.

In cases such as these, the ecosystem's immigrant-accepting property is valued: It enhances human endeavors and therefore meets the criterion for being an ecosystem service.

Of course, a newly arriving organism can and sometimes does upset the ecosystem *status quo* in some negative way other than by negatively affecting the ecosystem's species diversity. While a newcomer can instigate one or more new functions that are desirable, it can also precipitate a decline of one or more functions that might be regarded as human services. This possibility of declines in desirable functions might be why resistance to immigration is sometimes regarded as an ecosystem service.

This is not infrequently the basis of complaints about recently naturalized species, such as *Tamarix* species in the American southwest. As previously observed (in Sect. 3.3.2.3, Functions), according to the best experts, that plant's recent arrival and dominance in some riparian systems has not extirpated any other species. However, it does tend to change the way ecosystems function. For example, the presence of *Tamarix* promotes fires, and so the naturalization of *Tamarix* species can be said to compromise a fire-suppression function, which might be regarded as an "ecosystem service". But if it is the newly arrived Tamarix that is responsible for the disservice of suppressing this ecosystem service, then this is not the result of a decrease in species diversity nor, in any obvious way, of any other type of diversity. In other words, the changes that *Tamarix* spp. bring to the ecosystems where it settles have little to do with the biodiversity of those ecosystems.

All in all, the theory that ecosystem services are critically tied to biodiversity is singularly unpromising. The empirical evidence does not support it, for the complete body of facts includes a raft of disservices, which handicap human endeavors rather than benefit them. Also, the nature of many services places them in two broad classes of service that are extremely insensitive to diversity. The logic also does not support it, for the rationales rely on invalid inferences such as those that infer the need for biodiversity from the contribution of a single species. Even the foundational concept of "ecosystem service" seems ill-equipped to shoulder the weight of generalizations that allege the dependence of ecosystem services on biodiversity, for it does not account for the differing circumstances that sometimes make a property a service, and other times a disservice.

Finally, one plausible view of progress in ecology has clear implications that substantially deflate any remaining enthusiasm for the ecosystem service theory of biodiversity value. The value of some thing that provides a service falls off precipitously when another, more efficient (more effective for the cost) service-provider is found.[45] One need not speculate on the possibility of robotic but functionally equivalent surrogates for the current complement of service-providing organisms. For some time to come – but almost certainly not forever – humankind will rely on biological

---

Bilger goes on to say:

> In 1984, Shafland spearheaded the introduction of the South American butterfly peacock bass to Florida, arguing that it would both control invasive tilapia and make a superb sport fish…
> "Nothing has been displaced," Shafland said. "We're just changing the carrying capacity."

[45] As noted in Sect. 2.1.1 (Concepts and categories of value), all instrumental goods are fragile in this way.

organisms, perhaps with modest genetic tinkering, to play service-rendering roles in ecosystems that are also tinkered with. But continual advances in science will increasingly demonstrate ways to rely on fewer and fewer, less and less diverse, relatively unreliable and fickle working bio-parts. As this happens, from an ecosystem service point of view, there will be an increasing superfluidity of organisms that constitute inefficiencies in the "development" of these services. As such, they will be not just dispensable, but liabilities – obstacles or at least excess baggage that humankind will be happy and perhaps even obligated to jettison and get out of the way of development that promotes human well-being. If this speculation is true, then the organisms that now are still regarded as key service-providers will, with time, join the already much larger contingent that are already of little service consequence or are disservice-providers.

Viewed from this perspective, the argument for the value of biodiversity as the basis for ecosystem services is, at bottom, an argument from scientific ignorance. The value derives from the provision of services that *so far* are not otherwise cost-effectively rendered. This is distinctly reminiscent of the argument for God's existence on the grounds that God is required to explain that which science *so far* does not. This type of theological argument is slowly but inexorably eroded by progressively more and more satisfying scientific explanations. Similarly, as science advances beyond the relative ignorance presumed by Hooper and colleagues, it will explain in greater and greater detail what minimally diverse biological elements need be retained to provide coveted services. That knowledge will enable a more and more "efficient" use of biodiversity. This will progressively diminish the service value of any great amount of biodiversity, which will be seen as more and more superfluous and more and more a barrier to progress. This line of reasoning leaves aside the completely separate assault on this value of "nature's services" by the nonbiological surrogates that technology surely will eventually provide.

Of course, one can challenge this speculation about the progress and potential of ecology. One can maintain that the functions of ecological systems are so individual, so contingent on locally unique conditions, so confounded by the myriad, overlapping spatiotemporal scales in which the myriad interactions between organisms and never-duplicated environments play out, that one cannot expect to ever arrive at ecological laws that would ground a general program of streamlining ecosystems for their service value.[46] If one is willing to embrace this radical skepticism about the possibility or likelihood of finding general ecological laws, *and* if one has a like skepticism about the likelihood of attaining a sufficiently detailed knowledge of a sufficient number of ecosystems on a case-by-case basis, then (and I believe, only then) one last argument for the proposition that biodiversity underwrites vital ecosystem services remains available. That argument, mentioned at the outset of this section, is based on precautionary prudence.[47]

---

[46] An argument to this effect is made in Reiners and Lockwood (2010). See, in particular, Chaps. 2 and 10 of their book.

[47] A comment by Jeffrey Lockwood prompted me to realize the need for this discussion.

The ecologist Paul Ehrlich has vigorously pursued a precautionary argument for preserving biodiversity over the course of his distinguished career. In its original presentation with Anne Ehrlich (which predates the invention of the neologism "biodiversity"), the Ehrlichs do not base the argument on radical ecological ignorance. In fact, they (Ehrlich and Ehrlich 1981, 95) confidently posit that the link between rates of extinction (and hence, one might suppose, rates of species diversity loss) and ecosystem services is ironclad:

> … *all* [ecosystem services] will be threatened if the rate of extinctions continues to increase…[48] [italics in the original]

On the other hand, an assessment of technology as stuck in a state of more or less permanent incompetence is already part of the argument. According to the Ehrlichs (1981, 96), technology at best can supply a "partial substitution" for a lost ecosystem service, for the ecosystem "nearly always does it better".

A more recent version of the Ehrlichs' thesis retains the earlier one's dismal assessment of technology's capabilities. However, the previous confidence in the tight link between biodiversity and ecosystem services gives way to a confession of ecological ignorance – at least as the current state of the art. With another group of distinguished colleagues, Paul Ehrlich (1997, 101) states that

> No one knows precisely which, or approximately how many, species are required to sustain human life; but to say… that "there is no credible argument … that … all or even most of the species we are concerned to protect are essential to the functioning of the ecological systems on which we depend" is dangerously absurd. Until science can say *which* species are essential in the long term, we exterminate *any* at our peril.[49] [italics in the original]

Given its very strong claim about sustaining human life, this passage could just as well be discussed in the following Sect. 6.4 (Biodiversity as (human) life sustainer). I discuss it here where it can get a more favorable hearing as the much weaker claim that every extinction puts humanity at risk, not necessarily for absolute survival, but for losing a relatively low-cost and effective way to provide a service, considered apart from how critical it might be.

Let's recall two considerations that burden this argument from the outset. First is the empirical consideration that the majority of ecosystem services fall into one of two classes characterized by very weak sensitivity to biodiversity. According to Ehrlich et al. (1997, 101) the "critical life support services"

> …include the purification of air and water; the mitigation of droughts and floods; the generation and preservation of soils and renewal of their fertility; the detoxification and decomposition of wastes; the pollination of crops and natural vegetation; control of the vast majority of potential agricultural pests; and partial control of climate.

---

[48] While I believe that the Ehrlichs' precautionary argument is flawed, this should not be interpreted as a general criticism of their extraordinarily prescient book.

[49] My taking exception to this part of the argument by Ehrlich and his colleagues should not be misinterpreted as my taking exception to any other point in this response to Mark Sagoff's answer (Sagoff 1997) to the question, "Do We Consume Too Much?". However, it is odd that Norman Myers, a coauthor of this piece, elsewhere (Myers 1996b, 2764) expresses his awareness of one major reason why this precautionary logic does not hold up.

In this list, air and water purification, flood mitigation, and climate control (via carbon sequestration), as I said before, "are radically indifferent to which species, the number of species, or the diversity of species that provide them". Furthermore, soil renewal and waste decomposition depend on "functional groups [that] might change in composition from time to time or from place to place, [yet] tend to require species that reproduce easily, are abundant, and are unusually adaptable. They are the species whose ubiquity spans even dramatic ecosystem changes in the face of which many more vulnerable species succumb." This leaves only pollination and pest control as "critical" services that might, in fact, be vulnerable to species extinctions, because pollinators and insectivores (for controlling agricultural pests) can be specialized.

The second consideration observes the epistemological condition that some of the foremost practitioners of ecology see as defining the bleak prospects for their profession. While this view is not universally endorsed, those with a more sanguine view unburden the precautionary argument only to make it less relevant. If ecology will figure out how and to what extent the ranks of species can be thinned without any palpable effect on services, then it seems sufficiently prudent to simply follow the science.

The net effect of these considerations is to marginalize precautionary considerations: From the point of view of how biodiversity affects ecosystem services, little seems to hang on an argument from precautionary prudence for the simple reason that there is little space in which it can legitimately operate. Services such as pollination and pest control, which might be exempt from the empirical disqualification, are exempt precisely because they are performed by well-known, specialized "players". As a consequence, these services fail to avoid the epistemological disqualification. Science already knows, or there is good reason to suppose, that in the near term it will come to know "who" these specialized players are, along with their specialized requirements for survival. That knowledge, in turn, reveals which species are *dispensable* so far as the service is concerned. In such cases, what counts, so far as ecosystem services are concerned, is some modest and not particularly diverse set of particular species.

If matters of fact and epistemology relegate precautionary arguments to a nearly irrelevant corner, then a notable lack of any credible defense invites suspicion that there is none. Ehrlich et al. do nothing to elaborate, let alone justify, the precautionary logic behind their claim, "Until science can say *which* species are essential in the long term, we exterminate *any* at our peril." Declaiming (Ehrlich et al. 1997, 101) that it is "dangerously absurd" to question their precautionary stance does not constitute a credible defense of it.

Still, the defense of biodiversity's value so often devolves into precautionary exhortations that is worthwhile trying to understand in a general way why appealing to a Precautionary Principle is a hard row to hoe – even aside from its restricted "application space" in the domain of ecosystem services. The demands on precautionary logic and the difficulties in meeting them will make plain why it is not at all absurd to challenge Ehrlich et al.'s precautionary proclamation.

Many versions of Precautionary Principle are possible and many versions have been proffered. But it is relatively safe to say that no reasonable Precautionary Principle lacks certain basic ingredients. In the context of taking precautions to prevent species extinctions from disrupting ecosystem services, key ingredients are[50]:

1. **A threat of harm that is considered serious, great, or catastrophic.** It is unclear just how "serious" the harm must be to justify precautionary action. The envisaged outcome must fall outside some vague boundary that circumscribes the domain of "the acceptable". It is more clear that a principle that urges precautions against small (and "acceptable") harms would be difficult to defend because consistent adherence to such a principle would arguably result in greater harm (from paralysis) than the harms the precautions are supposed to avert. Most of us would say that we would cause ourselves greater harm by always refusing medications to avoid their side effects; or by refusing to drive to a job interview to avoid the additional risk of being in a traffic accident.

   A number of considerations bear on how serious the harm must be and how serious a threat it must constitute in order for precautionary logic to apply. They include:

   (a) The ability to avoid the harm altogether by severing the first link in an otherwise uncertain causal chain that leads from species extinction to disruption of service. In this regard, one must consider how one species might substitute for another in maintaining a particular, desired function. However questionable this might be as a general principle, many biologists regard this kind of substitution as a basic tool for conserving and restoring ecosystem functions. There is also the possibility of resurrecting species, though this possibility and its attendant controversies need not burden the current discussion. Proponents of The Natural Capital Project (for example, Goldman et al. 2008, 9446) even promote substituting entire ecosystems and habitats (under the rubrics "mitigation banking" and "habitat banking") as not merely acceptable but a positive selling point for their economically oriented doctrine.

   (b) The ability to ameliorate the harm, sometimes known as "adaptability". This is the context in which technological substitutes are relevant.[51]

   (c) The scale and pervasiveness of the harm.

   (d) How the harm is distributed, which leads to considerations of distributive justice.

2. **Some high degree of *uncertainty*.** Care must be taken to distinguish uncertainty from "mere" risk as an epistemic condition in which (European Environment Agency 2001, 170, Box 16.1) "… the adequate empirical or theoretical basis for assigning probabilities to outcomes does not exist."[52] When risks are known, a different, non-precautionary logic is appropriate for taking them into account.

---

[50] This explication is indebted to Stephen Gardiner's in Gardiner (2006).

[51] See also Sect. 2.1.2.1 (Consequentialism) regarding technological substitution.

[52] For a more detailed discussion of the distinction between risk and uncertainty (and ignorance), see Sect. 6.9.2 (Risk, uncertainty and ignorance).

On the other side of the spectrum, that a harm can be (merely) imagined to occur, or that its occurrence is a logical possibility, or that its occurring is consistent with known science cannot reasonably ground precautionary action. That would be as indefensible as stretching precaution to cover small harms and for the same reason described in item (1): It would excessively inhibit actions that have a high likelihood of rendering great benefit.

Several kinds of uncertainty can be involved in precautionary logic. First, there is uncertainty in whether or not the harm will occur at all. Item (1a) bears on this kind of uncertainty by suggesting a method, already a part of current conservation and restoration practice, for substantially reducing it – perhaps to the point of making precautionary logic inappropriate.

Second, if the harm (in the limited sense of an ecosystem no longer serving a particular human purpose) does occur, there is uncertainty about how seriously this occurrence should be taken, given the other considerations in item (1) and particularly the ability to adapt (item (1b)). Ehrlich et al. make a broad claim about the futility of technological adaptation. For support of this claim, these authors refer to the failure of the Biosphere 2 experiment. But that experiment was an attempt to construct a completely self-contained, human-friendly ecosystem from whole cloth rather than an experiment in extirpating or substituting for some components of one that is already functioning. And it is certainly at odds with the Natural Capital Project doctrines such as those that permit the substitution of habitats. So the relevance Biosphere 2 to their case is marginal, at best.

Third, in addition to uncertainty about whether the harm will occur and how seriously such an occurrence should be taken, precautionary logic presumes uncertainty about what links and what sequence of links might form a causal chain leading up to an occurrence. Most discussions of the Precautionary Principle focus on this particular kind of uncertainty.[53] The precautionary argument that warns against allowing extinctions must conceive of a plausible (though uncertain) causal chain that, starting from an extinction event, leads to a deleterious effect on an ecosystem service. As I have already said, this account must rest on something more than mere logical possibility or consistency with known science.

The "sweet spot" between blind speculation and known risk is extremely difficult to find. Ehrlich et al. are not alone in failing to supply such an account. In the narrow range of cases for which it is plausible to link a service disruption to an extinction, it seems that the causal chain is typically short and well known – that is, not at all uncertain. If a particular crop relies on a particular pollinator, then with relative certainty or at least with a scientifically ascertainable likelihood, the extirpation of the pollinator will leave the crop unpollinated. Because it is so central to the precautionary case against extinctions, I will revisit this dilemma – of attempting to cast an extinction as a Pascal's wager – immediately after this list of Precautionary Principle ingredients.

---

[53] Most saliently, the Wingspread Statement on the Precautionary Principle (http://www.sehn.org/wing.html) emphasizes the condition that "some cause and effect relationships are not fully established scientifically".

Finally, it is important to emphasize the epistemic contingency that attaches to precautionary logic's grounding in uncertainty. Uncertainty reflects a current state of knowledge. Should science ever find a sound basis for assigning probabilities for a causal chain leading to a harm in ecosystem services, then a discussion of risk analysis and risk management will properly supplant precautionary logic. As I have pointed out, science has already pushed beyond a state of uncertainty for some cases within the restricted, presumptive application space for the Precautionary Principle.

3. **An "appropriate" response to the threat and whether this response is merely permissible, advisable, or obligatory.** Ehrlich et al. can be interpreted to suggest that the appropriate response is to do what is possible to avoid letting any species go extinct and that this is *obligatory*, no matter what the cost. (Keep in mind that cost-benefit analysis applies in the domain of problems with understood risk, not in the Precautionary Principle's domain of uncertainty.) This exhortation is both dramatic and difficult to defend. Not the least reason for its vulnerability to objections is the fact that in some contexts such as pest control, ecosystem services are enhanced by species extirpations (local extinctions) rather than damaged by them, or so it is commonly thought. So in its unqualified form, the Precautionary Principle (on this interpretation) would seem to entail obligations that sometimes compromise ecosystem services by way of the precautionary obligation to not let any species go extinct.

The obvious response to this criticism is to insist that sometimes the precautionary obligation to prevent species extinctions applies; sometimes it does not. But far from being a defense of the Precautionary Principle, this response makes it clear how its plausibility as a general action-guiding principle hinges on an accompanying framework that provides a principled way to distinguish circumstances in which precautionary measures are obligatory, advisable, permissible, or proscribed. It is a nontrivial matter to construct such a framework; and it has yet to be done – by Ehrlich et al., or by anyone else.

These difficulties suggest that it might serve a Precautionary Principle well to retrench its domain yet further. At the cost of leaving out promising but difficult cases, one might try to define conditions that sharply qualify a candidate for precautionary prudence and that decisively rule out implausible and irrational applications of it. One obvious candidate condition is one (of the three) that John Rawls (1971, 154) proposes must qualify application of his "maximin" rule (for choosing principles of justice)[54]:

… the person choosing has a conception of the good such that he cares very little, if anything, for what he might gain above the minimum stipend that he can, in fact, be sure of by following the maximin rule. It is not worthwhile for him to take a chance for the sake of a further advantage, especially when it may turn out that he loses much that is important to him.

---

[54] Rawls' other two conditions are ones already described. One is the requirement for true uncertainty – that is, the absence of adequately known likelihoods for outcomes. The other is the requirement for serious outcomes – that is, (Rawls 1971, 154) "outcomes that one can hardly expect".

But this is anything but a panacea. Adopting this "nothing to lose" condition as a qualifying one for precaution would rule out most, if not most all situations where the extinction of a species hangs in the balance. For in these situations, it is precisely for significant economic gain that the possibility of extinctions is routinely ignored. Think again about Sarah Palin's beluga whales.

So it seems that even a much more limited program of precaution is likely to encounter serious problems.

All in all, I believe that whatever force there is behind Ehrlich et al.'s precautionary argument derives from their implicit supposition of a threshold that is the backdrop for a Pascal's wager on the removal of the next species.[55] The terms of the wager are that

(i)   there is radical uncertainty about the outcome of the next extinction, no matter what the circumstances; but that
(ii) one plausible outcome is the irremediable loss of a vital ecosystem service; and that
(iii) there is little to lose by preventing the next extinction, thereby avoiding such a dire outcome.

I have argued that there are fatal objections to all three terms of this wager: So far as terms (i) and (ii) are concerned, we are most often justified in thinking that the removal of a species will have absolutely no deleterious effect. In cases where we believe that a deleterious effect might result, the belief is oftentimes not at all uncertain; we have a good handle on the likelihood with which the harm might occur. As for term (iii), there is often much to lose by doing what is needed to prevent an extinction. And finally, term (ii)'s claim for irremediable loss presumes a radically broad and unqualified inability to adapt that has, at best, fragile support.

6.4   Biodiversity as (Human) Life Sustainer

In an interview that David Takacs presents in his sociological study of how scientists conceive of biodiversity, biologist and conservationist Reed Noss (Takacs 1996, 75) says that biodiversity "is life, and all that sustains life." Not a few of the other biologists interviewed by Takacs express a similar sentiment, which promotes the value of biodiversity for reasons that range from "providing services" to "the conditions required for all life". As mentioned at the outset of this chapter, zero biodiversity would logically and with certainty remove life, though perhaps not its possibility. After all, there was a time – over 3.5 bya – when life *was* just a possibility.

At face value, the claim that biodiversity is "all that sustains life" borders on hyperbolic nonsense. The conditions for life are set by many factors that have little

---

[55] Jeffrey Lockwood suggested this interpretation to me.

or nothing to do with anything biological, let alone biological diversity. To mention just a couple: The characteristics of the sun as a star and the distance of the earth from it; and plate tectonics, which drive the carbon cycle. So surely, biodiversity cannot be *all* that sustains life.[56]

One can rework Noss' statement to say that biodiversity constitutes "the *biological* or *biotic* factors that sustain life". In this form, the proposition might not be hyperbole, but it is next to useless in characterizing what biodiversity is supposed to be. The best geological and ecological evidence points to the conclusion that life will persist for a very long time under very many conditions – including the many different ones that have actually existed for the last 3.5 billion years of life on earth. This is true even in the face of extreme sorts of changes. Should all the methane currently trapped in clathrates suddenly be released, the earth would be bathed in a methane-rich atmosphere, as it was at the dawn of life. That would not offer a particularly friendly environment for most of the many respiring organisms that currently inhabit the planet. But very likely, methanogenic organisms, driven to obscurity some 2.5 bya by the photosynthesizing ones that transformed the methane-rich troposphere into an oxygen-rich one, will once again thrive.

One could rework Noss' statement in another way to restrict the biodiversity it mentions to "that which has accompanied the presence of *H. sapiens* on the planet". Its value then derives from all the biological factors that have sustained life that humanity has known from human prehistory onwards. Human presence is extremely recent, geologically speaking – less than 200,000 years – so this is a considerable restriction. Yet even during humankind's brief tenure, not just the composition of biodiversity (which species exist), but also species diversity in its proper sense, without regard to composition, have dramatically changed. These changes were largely induced by characteristically human activities and behavior. Focusing on just the direct effects on other species, conservation biologist Martin Jenkins (2003, 1177) remarks,

> There is growing consensus that from around 40,000 to 50,000 years ago onward, humans have been directly or indirectly responsible for the extinction in many parts of the world of all or most of the larger terrestrial animal species.

The component set of species has undergone transformation due to human influences. The component set of ecosystems has been concomitantly transformed. This is a matter of humanity's transformation of "the lay of the land" and of its biogeochemistry. It is the major point of the concept of anthropogenic biomes (mentioned in Sect. 5.3, The moral force of biodiversity), none of which existed 70,000 years ago. The biomes from that past time are now extinct, like many of the species that occupied them, and partly on account of the extinction of those species. In other words, whatever biological conditions have sustained life over the last 200,000 years have also sustained so many changes in life that the planet now is hard to recognize as a later biotic and environmental version of its former self.

---

[56] Peter Ward and Donald Brownlee (2000) offer a more complete and fascinating accounting.

This is a serious blow to the supposition that biodiversity, just as it was at some point arbitrarily selected within the interval of human tenure, was essential to sustaining life from that point onward.

One might insist that the concern for biodiversity should be restricted even further – to the sustaining of life (just) as we know it right now in the early twenty-first century. This additional restriction finally reaches a confluence with the just-so model of biodiversity value whose attendant problems are recounted in Sect. 5.1.4 (The just-so model). In the current context, it also reaches a tautologous and there-fore uninteresting conclusion: Biodiversity as we know it now is all that sustains biodiversity as we know it now.

Yet another reworking of Noss' statement would place a different restriction on biodiversity – as that which sustains not all life, but just *human life*. However, from an historical perspective, none of the transformational changes in the particular kinds that are diverse – nor the changes in their diversity – hindered humanity from emerging from a bottleneck population of perhaps 15,000 individuals 70,000 or so years ago, to grow to its current population size. Now nearing a population of 7 bil-lion, *H. sapiens* has become the world's apex species[57] to boot. That is not just "sustaining human life". It is a spectacular flourishing of a species by any purely biological standard.[58]

To give it some degree of plausibility, I can only interpret this last reworking of Noss' statement as a way to reinterpret the threshold model of biodiversity value. That is, at some point not yet in the experience of *H. sapiens*, sufficient change (most likely reduction) in biodiversity will pass the point beyond which human life will not be possible – even though it is nearly certain that other life forms will still flourish.

The question then becomes, where is the threshold? Although the most easy-to-notice organisms, and particularly megafauna, have undoubtedly suffered easy-to-notice declines in increasingly human-dominated landscapes, *H. sapiens* clearly has not suffered as a biological species on account of that. Moreover, the planet still stands at something near an all-time earth history high point in species diversity – indeed, at an all-time high for diversity considered at almost every taxonomic level.[59] Also, we know that some species – particularly the most adaptable generalists, such as *H. sapiens* – have often survived dramatic extinction events and squeezed through

---

[57] I mean "apex" in the sense of being the dominant, unmoved mover of ecosystems on a more widespread basis through a greater variety of means than any other organism.

[58] Of course, the flourishing of a species by a "biological standard" does not guarantee the flourish-ing of individuals of the species. The ability of human *individuals* to flourish is likely to decline if their numbers increase well beyond the current 7 billion. Also, my remark about the population-measured success of humans is not meant to imply (falsely) that *H. sapiens* outstrips all other species in this department. A comment by Jeffrey Lockwood prompted this clarification.

[59] See, for example, Dirzo and Raven (2003, 140), Purvis and Hector (2000, 214–215), and Spicer (2006, 71–78). Estimates of historical trends in diversity predominantly build on the pioneering work of Sepkoski (1982).

the narrowest of population bottlenecks. So the most straightforward inductions argue *against*, not *for*, any immediate danger to the continuation of human life.

The absence of convincing evidence that there is a threshold, let alone that one is near, suggests that the supposed threshold must be reached via an uncertain sequence of causal links. This plainly rehearses the "uncertainty" ingredient of a Safe Minimum Standard or other form of Precautionary Principle. And so, the current discussion reverts back to that one explored in the preceding section. At least one other ingredient of a Precautionary Principle – the identification of (though not a fully elaborated causal sequence for bringing about) significant potential harm – is surely present in this case. But a responsible deployment of the Precautionary Principle also requires that credible threat of potential harm be established. The evidence available today seems to fall substantially short of that standard.

Evidence for a negative feedback loop that might wildly accelerate processes leading up to the total demise of humanity would contribute to the case of legitimizing a Precautionary Principle. But with one possible exception, there is none. And speculation, based on "what if" conjectures, swings well wide of the "sweet spot" that a Precautionary Principle must find between blind speculation and known risk. Neither the mere logical possibility of a negative feedback loop nor the fact that such a thing is consistent with known science suffices in this regard.

The exception takes the form of credible evidence for the phenomenon of coextinction. This is (according to Koh et al. 2004, 1632)

> … the loss of a species (the affiliate) upon the loss of another [which can include] the process of the loss of parasitic insects with the loss of their hosts [as well as] the demise of a broader array of interacting species, including predators with their prey and specialist herbivores with their host plants.

Evolutionary biologist Lian Pin Koh and his colleagues (2004, 1633) claim to have identified some 200 such extinctions. Moreover, coextinction gains theoretical credence as a straightforward consequence of the existence of very specialized parasitic (including parasitoid) and mutualistic relationships between species. But I believe that this level of evidence does not yet rise close to the level of "credible threat to human existence".

Koh and his colleagues (2004, 1632) describe various well-known interdependences – of "pollinating *Ficus* wasps and *Ficus*, parasites and their hosts, butterflies and their larval host plants, and ant butterflies and their host ants." As Sir Robert May (1995, 16), the famous Australian physicist-turned-ecologist, points out,

> … it could reasonably be argued that for each species of metazoan or vascular plant there is at least one specialized species of parasitic nematode and protozoan, along with at least one species of bacterium and virus. Thus an estimate of plant and animal diversity can be multiplied by five, at a stroke.

How worrisome this is hinges critically on the degree of specialization, the degree to which the species are dependent on one another for their well-being (because some mutualisms are probably not critical to survival), whether the parasitic or mutualistic dependencies form cascading patterns, whether those cascades lead specifically to undermining resources or services that are vital to human existence, and whether or not there are other factors – such as the ability to express adaptive behavior that might sustain an affiliate species in the absence of its host – that might thereby inhibit a cascade or feedback loop. These critical factors are not addressed in Koh et al.'s work or elsewhere, to my knowledge.

In short, arguing for biodiversity value on the basis of a threshold for human survival does not achieve the threshold for sound practical reasoning. Abandoning the threshold model, one might argue that reductions of biodiversity might, more modestly, result in humankind's foregoing an economic bounty of services (or resources). This brings the discussion back around to Biodiversity as resource and Biodiversity as service provider – the notably unconvincing theories of value discussed in Sects. 6.2 and 6.3.

6.5 Biodiversity as a Cornerstone of Human Health

Among arguments that try to make a case for biodiversity's value, the ecosystem service one likely ranks first in frequency of occurrence, with biodiversity as resource taking second. In both categories, arguments that deal with issues of human health have recently come to dominate. Biodiversity is supposed to provide the service of safeguarding humans against infection. And it is supposed to be an indispensable pharmacological resource. I collect most of these arguments regarding health issues together in this separate section.

6.5.1 Biodiversity as Pharmacopoeia

Without biodiversity, your drug cabinet would be bare. To open the door to your drug cabinet is to open the door to biodiversity. That, in essence, is the claim made by those who urge (Newman et al. 2008, 117) that

> … biodiversity… provides us with medicines that relieve our physical suffering and treat, and in some cases even cure, our diseases.

Or (Cox 2009, 269):

> Over 50% of Western pharmaceuticals are derived from biodiversity.

Or (Cox 2009, 278):

> Historically, biodiversity has been the major source of pharmaceuticals, and today is relied on by 85% of the world's population for primary health care.

Again (Grifo et al. 1997, 131):

> Extinction of biological diversity risks the loss of the raw materials for existing and new weapons in the fight to alleviate human suffering and prevent death.

According to these authors and others, biodiversity is invaluable as an indispensable source of beneficial drugs.

I shall examine the details of arguments for this claim. But before turning to that, I wish to note an element common to all its formulations – an element that opens a window on all the maladies that plague the arguments made on its behalf. That element is a misstatement. Medicines are not derived from biodiversity. They are derived from specific plants, animals, and microbes. Furthermore, the "species richness" in the set of organisms from which medicines derive is tiny. Consequently, the claim amounts to a form of the category error discussed in Sect. 4.1.3 (Particular species), which mistakes biodiversity for some number (and in the case of providing medicines, some very small number) of particular species. This observation should end the discussion. But arguments from biodiversity's pharmaceutical value – both actual and potential – make a number of other interesting missteps, which are also worth examining.

There are two main argument threads for biodiversity's value as nature's pharmacopoeia:

1. The argument from the number of drugs that actually do owe their original discovery or current manufacture, in some part, to one or another organism. This includes the large number of folk remedies used around the world. A variant of this argument (which I take up first) starts with a premise, not about the number of drugs, but about the number of usage instances – typically in the form of number of prescriptions dispensed. The two numbers obviously convey different information.

2. The argument from the *potential* for deriving new pharmaceuticals from the huge set of organisms not yet evaluated. This includes the claim that these organisms would be indispensable for such derivation. It also tacitly includes the claim that the potential for finding medicines of sufficiently great benefit justifies forgoing other opportunities for realizing the more certain benefits of development that might impinge on the ability to find the medicine-yielding organisms.

I take up each thread in turn. First thread (1), which runs along actual use of medicines.

There is an obvious gap between a number representing how many organism-derived drugs there are or how many organism-derived drug instances are dispensed by pharmacies (on the one hand) to the supposition that the beneficial use of drugs would diminish and human suffering thereby increase as a consequence of a severely diminished diversity of species (on the other hand). To see why, I start by looking at the numbers in the "numbers premises".[60]

---

[60] It is worth noting that it takes serious historical research to classify a drug's origins – that is, to determine whether or not a critter or plant somehow led to its development. Drugs initially derived from organisms are often completely synthesized for commercial production. There are several other tricky problems in coming up with summary numbers, as Grifo et al. (1997, 135–140) discuss. These problems are dealt with in different ways (or not at all) by different authors. This makes it very difficult to compare numbers between surveys. An additional difficulty is that the identities of the top drugs change dramatically from one decade to the next. The work by Grifo and her coauthors is based on 1993 data. That is why I supplement their results with more recently available data.

What are the numbers? Let's first look at the numbers of prescriptions dispensed. Pharmacognosy expert David Newman and his colleagues (2008, 117) cite a 1997 study by biologist Francesca Grifo and her colleagues (1997), based on 1993 data from the commercial U.S. healthcare industry information vendor IMS America:

> In the United States… half or more of the most prescribed medicines come from natural sources, either directly, or indirectly when these natural compounds serve as models or as chemical templates for new drugs.

What Grifo et al. (1997, 136) actually state is that

> 57% of the top 150 brand names prescribed in [the 1993 time period] contained at least one major active compound now or once derived or patterned after compounds derived from biological diversity.

E.O. Wilson (2002, 118–119) recites a 40% ballpark number, without citing any reference. It might well be based on data more recent than Grifo's. The differences in these numbers, though not insubstantial, do not affect my consideration of them.[61]

To be clear, I am (first) talking about the number of dispensed prescriptions, not the number of drug types prescribed, nor the total sales or profits. Also, as suggested by Newman et al. and as reflected in Grifo et al. (1997, 137, Table 6.2, "Origins of Top 150 Prescription Drugs"), I interpret "comes from nature" to mean "natural product" or "semi-synthetic" drug – that is (Grifo et al. 1997, 136), a drug *not* "entirely synthesized without specific reference [via either discovery or current manufacture] to a compound found in nature".

This immediately raises the question: To what organisms are we indebted for this cornucopia of drugs? Wilson's answer (2002, 118, 119) – that it is due to "wild species" – is at once a red herring and quite misleading. It certainly does not address the overriding question: Is there any credible reason to believe that reductions in biodiversity of the sort that E.O. Wilson and others foretell will threaten human health and welfare because of the health-benefiting drugs that we might consequently forfeit?

That question is addressed by Grifo et al.'s analysis of 1993 drug data. The summary (Grifo et al. 1997, 138, "Table 6.3: Derivative Organism"; 144ff., "Appendix to Chapter 6: Origin of Pharmaceuticals Index") show that the "derivative organisms" from which the top five drugs (again, in the sense of "most frequently prescribed") partly derive are *Equus caballus*, *Ovis aires*, *Penicillium notatum*, and "various mammals", with number 4 on their "Top 5" list being completely synthetic. That is, the top featured organisms are the common domesticated horse – specifically, pregnant mares, whose urine is the main ingredient of the estrogen supplement premarin (PREgnant MAREs' urINe); the common domesticated sheep, for the anti-hypothyroid hormone drug synthroid (levothyroxine); and Sir Alexander Fleming's famous, ubiquitous, and easily cultured bread mold, which continues to be the inspiration and source of a variety of semi-synthetic, *Penicillium*

---

[61] Yet earlier studies report a 25% figure, but they are restricted to drugs whose derivative organisms are plants, not animals or microbes. But even the highest figure of 57% does not deflect the thrust of my argument.

species-based antibiotics, most notably amoxicillin. Grifo et al. give part credit to "various mammals" for ranitidine (Zantac), a drug in the class of H_2-receptor antagonists for treating gastrointestinal reflux. Ranitidine's connection to animals is somewhat tenuous. Zantac's development by the pharmaceutical firm Glaxo was by rational drug design, based on their model of the histamine H_2-receptor. Evidently, mammals such as rats and dogs played a role as laboratory subjects for initial screening for effects. The number 4 drug in Grifo et al.'s survey is nifedipine, an antianginal that is an all-synthetic affair. This "Top 5" list is unrepresentative only in its emphasis on animals and microbes, as opposed to plants, which tend to dominate the rest of the "derived from nature" list.

A survey using more recent (2008) data shows major changes in the drug lineup. But the collection of "derivative organisms" gives a similar impression. In this survey, the cholesterol-lowering statin atorvastatin (Lipitor) now dominates prescription dispensing numbers. *Penicillium* spp. (*P. citrinum* and *P. brevicompactum*) once again figure in the derivation list, this time as the organisms from which compactin, the first known statin, was isolated. Ethnobotanist Paul Cox (2009, 274) appears to suggest that red yeast from rice can also claim part credit for statins. The analgesic hydrocodone (Vicodin), in various formulations, is also now a dominant drug. It is a semi-synthetic opioid derived from naturally occurring opiates in *Papaver somniferum* (the opium poppy) – one of the most widely and easily cultivated plants in the world. Amoxicillin from *P. notatum* continues its dominance in the world of antibiotics. Premarin has fallen away with the discovery of life-threatening side effects. Zantac has yielded to omeprazole (Prolisec) and esomeprazole (Nexium, a mixture of D- and L-isomers of omeprazole), which like ranitidine (Zantac) has, at best, a tenuous connection to living organisms.

In short, the facts do not merely fail to support but strongly contradict Wilson's general claim, to the effect that some significant amount of wild biodiversity is a cornerstone of medicine. Some "wild species" do offer up interesting bioactive molecules. But domesticated ones often do, too. And in any event, that is not the issue. Simply put, an extinction far more massive than even the most dire predictions from such experts as Wilson would, with near certainty, leave us with domesticated horses, domesticated sheep, bread mold, rats, dogs, and poppies.

Of course, the proposition that 50% (or so) of the prescriptions dispensed are derived from organisms is not equivalent to the proposition that 50% of the pharmaceuticals are so derived, though sometimes (as with Cox 2009, 269) the two are conflated. The latter proposition is therefore worth an independent look to see if it might ground a more convincing case for the need for some non-trivial amount of biodiversity.

Unfortunately, this other way of accounting for drug usage is not up to the task, either. According to the World Resources Institute, of the 119 plant-derived drugs used worldwide in 1991, just 90 of the 270,000 described plant species and perhaps 320,000 estimated different plants (according to Spicer 2006, 27) can be credited with even a peripheral role. Grifo and her coauthors (1997, 136, 138, "Table 6.3: Derivitive Organism") find 86 drugs in the top 150 (for the United States) that derive from some living thing. But many organisms are counted multiple times. All told, just 20 different species appear on their list (excluding *H. sapiens*, which is listed as a "derivative organism") plus "various mammals", which I infer refers to lab animals used for testing. This is hardly an impressive representation of the 10–100 million organisms that dwell on the planet.

Those who nevertheless persist in trying to make the case for "biodiversity as pharmacopoeia" pursue two auxiliary lines of argument. The first one abandons the numbers to present a sequence of anecdotes about "biodiversity-derived" drugs or just bioactive molecules and their benefits or potential benefits. This is accompanied by tales of bioprospecting, and exciting drugs-in-the-works. Examples (of unproven drugs) that find a place in almost every discussion include a possible cure for leukemia (fucoidan from various brown algae), promise for those suffering with HIV-AIDS (calanolide A from *Calophyllum langigerum* and *C. teysmannii*, cousins of the rubber tree), and anti-malarials from *Artemesia annua* (sweet wormwood).[62]

Discussions such as these draw in a larger entourage of organisms. Unfortunately, this approach relies on the excitement that these anecdotes generate, rather than even modest, let alone convincing, inductive evidence for the case that our pharmacopoeia and consequent good health relies on some significant amount of biodiversity that is in danger of being lost. Some of the excitement comes from a small number of exceptional stories of biomedical sleuthing – involving the near-disappearance of a species with possible biomedical value. These stories are very popular and receive a large number of tellings. The story of *C. langigerum* and calanolide A is one such story.[63] But the more common stories are about species that are common and abundant and often easily cultivated – such as wormwood and a raft of the brown algaes.

Another significant quantum of excitement comes from pharmacologically active materials that are stimulating scientific imaginations, but in fact are not efficacious medicines whose benefits outweigh their harms. Mention of this also belongs with the discussion of the argument from potential medicines, below. I include it here because the organisms involved are already known and identified; and they are routinely misrepresented as evidence for the medicinal value of these organisms.

Pumiliotoxins, initially isolated from *Dendrobates pumilio* (Panamanian poison frog) and subsequently found in many other Dendrobatidae, are typical in both the amount of enthusiastic interest that they generate and their failure to actually become viable medications. Their myotonic and cardiotonic effects – that is, their ability to affect heart contractions – has resulted in ubiquitous citing of these toxins as showcase examples of potential medicines in the future's medicine cabinet. Underemphasized and often unmentioned is the toxicity of these alkaloids, which has precluded their

---

[62] To be clear, a tiny number of successful drugs do derive from unexpected organisms; and recitations of their stories are legion. One cannot avoid encountering mention of *Taxus brevifolia* (Pacific yew), which yielded the chemotherapeutic mitosis inhibitor paclitaxel (Taxol). Nor can one fail to read about *Catharanthus roseus* (rosy periwinkle), which yielded vinblastine and vincristine – also mitotic inhibitors and also effective against cancer (in their case, Hodgkin's lymphoma).

[63] It is difficult to avoid the story of calanolide A. It is featured in Newman et al. (2008, 131–132) and Wilson (2002, 123–124). Nor does Cox (2009, 271) neglect its mention.

entering even Phase I (safety) trials for actual medical use (Chivian and Bernstein 2008c, 214). Epibatidine has a similar story. Initially discovered in *Epipedobates tricolor* (Ecuardorian poison frog), it was found to be a powerful analgesic in mice. Unfortunately, epibatidine itself is far too toxic to give to humans. Derivatives of it have not fared much better. Abbott Labs got the farthest with their ABT-594 (Tebanicline) (Chivian and Bernstein 2008c, 215). But it had too many adverse gastrointestinal side effects to make it out of Phase 2 trials.

The pumiliotoxins and epibatidine are the rule, not the exception. The vast majority of bioactive substances never even make it to clinical trial.[64] A pharmacological profile is required (at least by the FDA in the United States), and more often than not, it is too unpromising to proceed. The substance must be tested for first acute toxicity and then short-term toxicity (2 weeks to 3 months) in (typically) at least two species of nonhuman animals. Further testing in animals can require up to several years, because not all adverse effects present quickly. These tests are often failed, too.

Those substances that do make it to trial almost never make it past Phase I (safety and dose) or Phase II (short-term effectiveness and side effects). And although it is surprisingly difficult to find consistent data on the success rates of clinical trials, a 50% figure seems to split the differences of the few available numbers. Since this percentage applies to each one of the three trial phases, one might expect something like a 12.5% rate of success for the tiny number of substances that pass pharmacological and toxicological tests to make it into initial clinical trials. This is merely to say that the excitement surrounding this research is not equivalent to, or even remotely supported by, evidence that human health depends on the organisms involved.

A second auxiliary line of argument for "biodiversity as pharmacopoeia" focuses on "traditional therapeutics". More than medicines in industrialized regions, these remedies rely on a variety of organisms, some of which are threatened. Moreover, people in the industrialized world increasingly embrace folk remedies. Here is how Peter Canter and his colleagues (2005, 180) (the source for similar comments by Meyerson et al. (2009)) set out the case for this:

> The World Health Organization has estimated that more than 80% of the world's population in developing countries depends primarily on herbal medicine for basic healthcare needs. The use of herbal medicines in developed countries is also growing and 25% of the UK population takes herbal medicines regularly. Approximately two thirds of the 50 000 different medicinal plant species in use are collected from the wild and, in Europe, only 10% of medicinal species used commercially are cultivated. There is growing concern about diminishing populations, loss of genetic diversity, local extinctions and habitat degradation. Well-known species threatened by wild harvesting include *Arcostaphylos uva-ursa* (bearberry), *Piper methysticum* (kava), and *Glycyrrhiza glabra* (liquorice). Between 4000 and 10 000 medicinal species might now be endangered.

---

[64] A summary of the United States drug approval protocol can be found in Chin and Lee (2008, 32–33).

The last claim about the number of "medicinal species [that] might now be endangered" is apparently derived by projecting another estimate – that 8% of all plants are threatened – onto the subgroup of plants that are regarded as medicinal (on the accounting of Schippmann et al. 2002, 4). Also worth mentioning is that Canter includes "aromatic" plants used in perfumes as contributing to "medicinal species". That might inflate the accounting for medicinal value, strictly construed. On the other hand, the number might be deflated because based on plants (no animals or microbes) alone.

Putting these admittedly relevant details aside, this account by Canter joins similar ones in skating past the single most important consideration in the matter. That is the question of the efficacy of these "medicines" – that is, whether or not they are actually promoting human health beyond the psychological benefit of the ministrations of traditional healers who typically dispense them. And if they do have some real salutary effect, there remains the question of whether their pharmacological benefit exceeds that of other, easily produced medicines. It is a sad circumstance that not a few have been found to be poor alternatives for other medications that are often ignored, even when they are available, in favor of the less efficacious ones. For example, ecologist Mark Tanaka and his colleagues (2009) express concern that, "… in Nigeria, witchcraft and traditional remedies of unknown efficacy are widely employed as treatments for malaria, instead of, or delaying access to, modern medicines of proven effectiveness." This is aside from the separate concern, expressed by these authors, for the outright toxic effect of other traditional medicines, which contain heavy metals, for example. In any case, the answers to these questions regarding efficacy are far from clear.

Nor is the increasing over-the-counter use of traditional herbal remedies in industrialized countries credible evidence for their efficacy. Efficacy is not proven by popular votes, which are often cast with unawareness of toxicity and side effects. For example, Schippmann et al. (2002, 214) raise the concern that

> …many herbal remedies like ginseng (*Panax quinquefolium*), gingko (Gingko biloba), valerian (*Valeriana officinalis*), kava (*Piper methysticum*), or St. John's wort (*Hypercum perforatum*), very popular in the West, are more toxic than previously believed, and present dangerous interactions with prescription drugs…

Of course, these points – regarding a pharmacuetical's efficacy, the absence of adequate or even more effective alternatives, and whether its benefits extend to people generally – are not restricted to "traditional" medicines. They apply quite generally, so that even if the gap between drug numbers and biodiversity were bridged, this would still not secure the argument that connects pharmacological benefits to substantial biodiversity.

But in the case of a traditional medicine especially, even if all these formidable obstacles for establishing medicinal value were surmounted, there might still be good reason to refrain from using it as a medicinal resource. Those reasons have to do with the consequent threat to the medicine-supplying organism or to its environment on account of its extraction for medicinal use. In fact, with alarming frequency some set of organisms, if not biodiversity, is put at risk by intense use of traditional medicines, independent of their merit. There is meager evidence for the efficacy of some number of them, such as the Chinese use of the horns of various rhinoceros spp. for fever, convulsions, and delirium; the bones of *Panthera tigris* (tiger) for joint ailments; and the gall bladders of *Ursus* spp. (Asiatic black bears) for liver ailments and headaches. Musk glands from *Moschus* spp. (musk deer), in enormous demand for various Western homeopathic medicines as well as for perfume, is of similarly questionable benefit.[65]

The use of a resource need not entail using it up. But it is a real danger, and an imminent one in cases such as the ones just cited. Attempts to ameliorate this risk by resorting to medicinal agriculture can exacerbate other risks associated with the "land conversion" that accompanies all commercial agriculture. The conversion typically banishes many or most of the organisms that formerly made a home in that piece of real estate. Additionally, the medicinal plants will often be exotics where they are planted. Some will likely escape and become "invasive weeds". *H. perforatum* (St. John's wort), for example, has acquired this prejudicial label in both Australia and Canada (Newman et al. 2008, 153) – though when examined without prejudice for their status of as "aliens", their effect on biodiversity is uncertain. In other cases, there is the risk of transforming natural areas into industrial gardens. Precisely this is apparently under consideration for *Pseudopterogorgia elisabethae*, a soft coral that produces pseudopterosins, which have use in topical anti-inflammatories (Newman et al. 2008, 146). These examples collectively point up the possibility of a direct conflict between the sometimes marginal human health benefits that derive from organisms-as-medicinal-resource on the one hand, and the welfare of the medicinal organisms themselves, other organisms that might suffer for their cultivation, and the extractive environment, on the other.

Furthermore, in some cases, the use of traditional medicines is in all likelihood directly responsible for *declines* in human health. This appears to have been the case for the outbreak of severe acute respiratory syndrome (SARS) in southern China in 2002 and 2003. It seems likely (Li et al. 2005) that horseshoe bats in the genus *Rhinolophus*, sold in Chinese markets for use in traditional medicines as well as food, constituted the original reservoir of the SARS virus.

I now finally turn to the second of the two main argument threads for biodiversity's value as nature's pharmacopoeia – the case for biodiversity as an indispensable source for *future* medicines (item (2) at the start of this section). This argument is typically posed as the specter of losses in biodiversity that will forever deprive humans of the means to assuage their pain and cure their ills and will therefore condemn the human race to eternal, disease-ridden desperation. Built on conjecture, this is an argument of last resort.

Many writers present this argument, but none better than E.O. Wilson. He (Wilson 2002, 125) asks his reader to consider a two-dimensional matrix.

---

[65] On top of this, the question of general, long-term, and indispensable benefit is brought into sharp focus by such dominant drugs as Lipitor and Prolisec (and Nexium) in the United States, where their availability might well encourage people to make diet-related decisions that contribute to the continuation and proliferation of the maladies and that make these drugs attractive.

The vertical dimension comprises a list of the millions (Wilson modestly says "thousands") of species of plants, animals, and microbes. The horizontal axis identifies all possible functions (though Wilson does not explicitly or immediately restrict these functions to medicinal ones), which he suggests are in the hundreds. Wilson then asks his reader to imagine filling in this matrix.

Thinking uncritically, one might imagine every cell filled in, producing Order(1 million species)×Order(1,000 functions)=Order(1 billion) gifts of nature. Thinking more critically, the matrix gets very sparse and the cornucopia suddenly gives way to leanness. That is because several considerations prune away several large chunks of the matrix.

First, while domesticated animals such as horses and sheep figure prominently in supplying a few widely used pharmaceuticals in medicine cabinets now, wild animals, or at least non-sessile wild animals, do not figure prominently in the search for new medicines. An impression to the contrary might come from the most problematic medicinal exploitation, recounted above. While megafauna tend to be the targets in the best known anecdotes, this is likely to be a matter of human observational and emotional bias. It is not unlikely that people tend to find exploitation of such animals more disturbing than exploitation of plants, with a concomitant increase in the degree to which animal exploitation registers in human awareness.[66]

In fact, the featuring of horses and sheep in a few superstar medications notwithstanding, few non-sessile animals figure in medicine overall. As Cox (2009, 269–270) notes, there might be good reason that

> … both scientists and indigenous peoples direct the majority of their attention to sessile organisms, particularly plants and marine invertebrates. While perhaps the immobility of such organisms facilitates ease of mapping and subsequent recollection, it appears that sessile organisms also produce the most potent bioactive molecules.

Cox (2009, 270) goes on to explain that

> Sessile organisms must mediate their interaction with the world – including parasites, predators, and competitors – primarily with chemicals. Evolutionary pressures have selected for toxins that fulfill this protective role.

In one telling stroke, this consideration lops out from the original matrix the preponderance of rows for non-sessile animals. Among the organisms lopped out are all arthropods – the vast majority of non-bacterial organisms on the planet. This points up a serious deficiency of the matrix representation. Its implication of "equal opportunity" for each cell's pharmacological potential is not realized.

Moreover, the equal opportunity supposition is confounded, not just by differences between kingdoms (such as Plantae) and phyla (such as Arthropoda), but also by the

[66] A similar principle might also diminish the visibility of the exploitation of fungi. The collection of *Ophiocordyceps sinensis*, the caterpillar fungus, on the Tibetan Plateau for medicinal purposes has exploded with a 5–30-fold increase in its commercial value (since the 1980s), with uncertain effects on that organism and its habitat.

similarities between species within genera and even between genera within families. It is not uncommon for different but closely related species to manufacture very similar or even identical bioactive molecules. Precisely these similarities are the pivot points of exciting and oft-repeated tales of losing an organism that initially appeared to be the sole source of some bioactive compound. This is true of calanolide A, which was found in *Calophyllum teysmannii* when its source in *C. lanigerum* seemed to vanish (Newman et al. 2008, 131).[67] It is also true, for example, of diazonamides, anticancer agents originally found in one species of sea squirt from the genus *Diazona*. That species could not be relocated after the initial assay. However, other species in the genus were eventually found to produce the same compounds.

Of course, the tellers of these "narrow escape" stories wish to persuade us that, with the disappearance of each organism, we risk losing the drug that might save our life. That possibility cannot be eliminated *a priori*. But time after time, the stories instead illustrate the unlikelihood that any organism has a monopoly on the manufacture of a coveted molecule. This is anecdotal evidence, but only by way of reinterpreting the original anecdotes with greater clarity. It is anecdotal evidence that even the decimation of biodiversity is unlikely to substantially diminish the very modest pharmacopoeia it offers.

Second, serious consideration is due to the magnitude of the time and effort that is required to fill in whatever portion of the matrix that remains. Newman et al. refer to the adaptive evolution of life over the past 3.5 billion years as a field version of combinatorial chemistry. As a consequence, they (Newman et al. 2008, 118) sanguinely proclaim "that in many cases clinical trials have already, in essence been done".

Nothing could be further from the truth. Medicines for humans require preclinical pharmacological and toxicological testing followed by clinical testing with human subjects. This takes huge resources in the forms of funding, the dedicated expertise of research scientists, and most importantly for this discussion, time. For the testing of drugs for human use, the first 3.5 billion years count for naught. Screening technologies might improve and become more efficient. But the best of these technologies cannot make the trials go faster. The time required for them is not a matter of technological limits at which new technology can chip away.[68]

The time required for careful testing is not the only time constraint that is largely unchangeable. For species that are actually "in the wild", there is the significant challenge of finding and collecting individual organisms. It is hard to imagine major new efficiencies in such efforts that do not have the side effect of destroying what is sought as the result of gross, habitat-altering incursions into the homes of the target organisms that this would require.

---

[67] As mentioned in Note 63, calanolide A's story is ubiquitous in the literature. See also Wilson (2002, 123–124), who recounts a somewhat different version of its tale.

[68] I hope that my description, earlier in this subsection, of the rigors of the pre-trial and trial protocols give some sense of how formidable these hurdles are and how seldom they are surmounted.

Third, there is good reason to suppose that expertise in molecule modeling and in molecule synthesis will improve. This is not to say that rational drug design – using such techniques as combinatorial chemistry and computer-aided design – and fabrication is a perfect substitute for fortuitous discoveries in the natural world. It *is* to say that whatever balance there is between the two approaches is likely to inexorably tilt more and more towards rational design as pharmacologically-inclined biomolecular engineers get better at it.

Finally, insofar as the argument for biodiversity as a potential medicinal resource is an economic one, it must soberly account for the vanishingly low probability that the medicinal benefits will actually be realized. That vanishingly low probability entails a vanishingly small expected net present value, which a cost-benefit analysis must weigh against the expected net present value of the benefits of economic development forgone in order to ensure the protection of the medicine-yielding natural resources. The economic analysis, when honestly done, does not appear to give the answer that environmentalists want. A big hint that this is so comes from "big pharma", whose sole *raison d'être* is economic gain. As Cox (2009, 270) complains,

> … there is no large pharmaceutical firm that currently bases a majority or even a significant component of its research program on searching for new molecules from rainforests.

Pharmaceutical firms ignore two salient considerations that contravene the pursuit of medicines "in nature": They have no interest in the possibly greater benefits of developing a promising locale in some way that might conflict with extracting drugs from it. Nor do they have an interest in avoiding the harm that might be done to the natural environment in pursuing a drug residing in some one of its residents. This builds into their assessment a significant bias towards developing medicines "from nature". Despite this, the pharmaceutical industry generally finds that it does not pay to include species-harboring areas in their asset portfolio. Of course, insofar as species-harboring areas have other, highly beneficial uses that do not accommodate the continued presence of these species, the economic rationale to retain them as a potential medicinal resource is only diminished.

Considerations previously presented in this subsection further weigh against an economic case for preserving biodiversity for its potential medicinal value. Much medicine "from nature" comes from domesticated species, ones that *can* be domesticated, ones that can be cultured or cultivated, or from new creations that are the product of selective breeding and genetic modification. Medicines initially found "in the wild" are subsequently synthesized. As for the discovery of new compounds, the apparent redundancy of their production by various different species appears to allow that a great extinction is more likely to increase the difficulty of finding them rather than to cause them to be lost entirely. And to extract their medicinal value, the species that remain need only be represented by a few individuals in zoos, aquaria, or seed banks (see Chivian et al. 2008, 201).

Taken together, these multiple considerations militate heavily against incurring the cost of preserving large populations of a large number of species in the wild for their quite marginal medicinal potential.

6.5.2   Biodiversity as Safeguard Against Infection

Not all human diseases are caused by infections, but a great many are, including a great many serious ones. According to the World Health Organization, more than one quarter of all human mortality is due to infection.

The notion that biodiversity helps to protect humans against infectious disease is, on the face of it, very odd. After all, infectious agents[69] – pathogens and parasites – are themselves organisms. For once, it would be accurate to say that they constitute a very large component of biodiversity – in the sense that these categories encompass a stunning diversity of organisms. In the most straightforward way, it seems that people would lead far healthier lives if this "component of biodiversity" were summarily extirpated. Of course, care must be taken *not* to say that biodiversity makes people sick. That is done by the huge set of organisms that make their enormous overall contribution to species diversity.

The pathogens and parasites that directly infect people represent only one aspect of the diversity of organisms that conspire to cause the human misery of disease. The direct agents of infection often, even typically, require the support of other species. Ecologists Ryan Hechinger and Kevin Lafferty (2005, 1059) performed studies that provide evidence for the sensible hypothesis "that rich communities and high abundance may foster parasitism."

This means that the pathogens and the parasites that infect people are not the only organisms whose absence might benefit human health. That is because the epidemiological situation is almost always more complicated than merely putting a "bug" together with a person. About 60% of all infectious agents reside and multiply in other animal hosts, known as "reservoirs", before being transmitted to people (Molyneux et al. 2008, 287). Infectious diseases involving transmission of the infectious agent from nonhuman vertebrates to human hosts are known as "zoonoses". Moreover, the transmission is often not direct, but via a vector – yet another animal – most commonly a cold-blooded arthropod or mollusk. In a particular place and for a particular vector-transmitted zoonotic disease system, a *variety* of host species, pathogen and parasite species, and vector species can be present to create a complex dynamic involving the interactions among the various populations of multiple species in these three functionally characterized groups. Different pathogens present in a disease system can even interact – as the result of their varying and sequential effects (morbidity and mortality) on the hosts that they share.

The complexity does not stop there. Yet other organisms that do not enter directly into the epidemiological equations affect populations of the species that are the direct players in the disease cycle. Predators might prey on some one or another of

---

[69] In this subsection, I use "infectious agent" to mean "agent that can infect individuals of the species *H. sapiens*". Many pathogens and parasites infect nonhuman creatures, but not humans. The group of infectious agents involved in diseases known as "zoonoses" infect humans as well as other creatures.

these species.[70] Plants might provide suitable breeding habitats or food for other, more directly involved players.[71] Changes in the mix of all these species or in their varying populations enter into the dynamic. As I will suggest, extirpating oak trees in the northeastern United States might go a long way towards reducing the incidence of Lyme disease there – even though oaks are not hosts, pathogens or parasites, or vectors for this disease.

Finally, changes in the habitat of a place – typically wrought by human activities – impinge on these populations and consequently on the dynamics of their interactions. This almost always produces some change in rates of human infection. So this is the mazy route that is implicitly traversed by the oft-expressed proposition that "ecosystem disturbances affect human health via changes in biodiversity".

An enormous gap separates "ecosystem disturbance" from "bad effect on human health". Ecosystem disturbances can affect the mix of organisms involved in human infections, with some consequent effect on human health. But this says nothing about whether the health changes are for the better or for the worse. Nor does it say anything about whether these changes are accompanied by increases or decreases in the variety of the organisms involved. Nor does it determine even whether these effects on other organisms are the cause of human ill health or are collateral damage.

All combinations of better/worse and increased/decreased biodiversity occur. Sometimes the "biodiversity changes" are simultaneously both up and down within the same disease system. This happens when decreases in predators result in increases in the prey species that constitute disease reservoirs. Even then, the effect on human health is uncertain. That can depend on such other factors[72] as the "competence" of the reservoir species – that is, its relative ability of this nonhuman host to infect a vector that subsequently transmits the pathogen or parasite to a person. Reduced predation that results in an increase in a prey species that is a particularly

---

[70] An example of how "outsiders" can dramatically affect the incidence of infection is the relative absence of predators of *Peromyscus leucopus* (white-footed mouse) in the fragmented and converted forestlands of the eastern and central United States. *P. leucopus* is an especially competent reservoir of *Borrelia burgdorferi*, the Lyme bacterium. Large populations of this mouse therefore increase the likelihood that local populations of the local tick *Ixodes scapularis* (black-legged tick) will be infected by biting them, and will subsequently bite and infect people. This zoonotic disease system and a few others involving Lyme disease are among the most thoroughly studied and most discussed. See Molyneux et al. (2008, 305–306), Rapport et al. (2009, 44–45), and Thomas et al. (2009, 232–233). I return to Lyme disease later in this subsection.

[71] An example is the intentional introduction of *Erythrina micropteryx* (immortelle tree) into Trinidad from Peru in order to shade cocoa. *E. micropteryx* also created a home for various bromeliads, which in turn, provided water reservoirs for the breeding of the malaria vector mosquito *Anopheles bellator*. See Molyneux et al. (2008, 297–298), Thomas et al. (2009, 232–233), and Dobson et al. (2006, 714–715). I return to this example, too.

[72] These other factors can make things extremely complex. Dobson et al. (2006, 716) discuss tick-borne encephalitis. In this particular case, the effective pool of pathogens to which humans are exposed is "diluted" only by the *combination* of two host reservoirs – one small, the other large, both at high density – which produces the desired protective shield for humans.

incompetent disease reservoir could be a boon to human health. Causal chains are also linked – or unlinked – by dint of such circumstances as the timing or sequence of events, which have little bearing on which creatures interact in a "disease system", their number, or their diversity.

This kind of complexity, which quickly transcends mere diversity of organisms and even biological diversity more broadly conceived, makes analyses that revolve around biodiversity appear frivolous – a distraction from the real epidemiological science of complex disease systems. Yet, not only does "biodiversity" figure centrally in the discussion of these systems, it is often promoted to titular status. Mere inclusion of a paper (Molyneux et al. 2008) on "Ecosystem Disturbance, Biodiversity Loss, and Human Infectious Disease" in a book with the title *Sustaining Life: How Human Health Depends on Biodiversity* will lead, or actually, mislead, many readers into thinking that losses in biodiversity jeopardize their health. But one need not rely on this inference when the authors introduce their paper by citing Rachel Carson's words (from an April 13, 1963 broadcast of *C.B.S. Reports* on *The Silent Spring of Rachel Carson*): "Man is a part of nature, and his war against nature is inevitably a war against himself." Insofar as the war is a war against biodiversity and the consequences have to do with rates of human infection, that proposition is, quite simply, false.

But the "dependence" of humans on "biodiversity" for their health and the consequent adverse health affect of biodiversity "loss" is not just a story that one must piece together from cryptic headlines. Experts often stake out their general claims – that greater biodiversity is protective, that reduced biodiversity is riskier, and that, notably, "habitat modification" precipitates such risk – quite explicitly, clearly, and directly. Ecologist/epidemiologist Andrew Dobson and his colleagues (2006, 718) state that there is

> ...a strong selfish motivation to conserve biological diversity – our health may depend upon it.

and (Dobson et al. 2006, 714):

> ... it may be sensible to conserve biological diversity for the purely selfish reasons of protecting human health.

and again (Dobson et al. 2006, 717):

> ... significant threats to human health may be buffered by the presence of a diversity of other species.

Ecologist/economist David Rapport and his colleagues (2009, 50) are eager to "underscore how biodiversity can buffer exposure to disease" and further "underscore how habitat modification can facilitate disease".

As Keesing et al. (2006, 489) explain, there is significant confusion surrounding the term "dilution effect", although these authors might well add to it. They propose to define "dilution" in terms of the "net effect" of "a decrease of disease risk due to an increase in diversity". Unfortunately, using this definition has the effect of skipping past the most critical empirical question – of whether or not the change in diversity causally affects infection rates (the alternative that the definition builds into itself), or is merely correlated with infection rate changes.

It is unfortunate that what these scientists choose to "underscore" is both terribly incomplete and terribly prejudicial. It is only by failing to also underscore a raft of inconveniently contradictory evidence that they make the thesis of an inverse relation between biodiversity and infection rates appear plausible.

The remainder of this subsection has two parts. It first samples some of the science to illustrate how the argument succumbs to fallacies of accident by ignoring essential conditions and facts that bear on the case. It then concludes with a brief characterization of the real complexities of disease dynamics, focusing on zoonotic systems. I think that a sense of the actual epidemiological science, stripped of unjustified inferences from it, helps to make clear why the strategy of arguing from biodiversity to simple disease-protecting conclusions is a nonstarter.

I start with Dobson et al.'s (2006, 714) relatively straightforward presentation of the case for "biodiversity reducing disease risk". These authors focus on zooprophylaxis, the introduction of animals to divert disease vectors from people to other animals. Zooprophylaxis is a fancily named version of the "bite him (the nonhuman animal), not me (the person)" trick.

On what grounds, exactly, do Dobson et al. suggest that biodiversity reduces the risk that persons will get infected? They argue by induction from precisely one case; and this showcase example of how biodiversity benefits human health is *Bos primigenius* – the domesticated cow. They urge their case along by suggesting that the status of cows as sacred in India might be due to the disease protection that they supposedly confer.

I am mainly concerned with the informal and inductive logic that these authors use to go from the proposition that people who keep cattle have less risk of malarial infection to the proposition that biodiversity protects against disease. But before examining this logic, I should note that not a few serious studies provide strong evidence against the premise. Among them, one study (Bøgh et al. 2002) found that the protective barrier apparently afforded by cattle to cattle owners was, in fact, a matter of their greater wealth (and consequently, better access to good health care) relative to non-cattle owners. Another study (Saul 2003) found that while the cattle diverted mosquito bites, they also afforded such copious blood meals that mosquito populations flourished along with rates of (mosquito) survival. Once again, there was no prophylactic effect due to the cattle. Many other factors and conditions bear on whether or not cows in the vicinity are protective shields for humans. One factor is that cattle harbor a large variety of diseases that are transmissible to humans, including (Pelzer and Currin 2009) cryptosporidiosis, *Escherichia coli* infections, giardiasis, leptospirosis, Q fever, ringworm, salmonellosis, and tuberculosis.[73] The many factors and conditions that bear on the proposition that cattle are protective shields make it hopelessly and misleadingly simpleminded. The discussion of some of the intricacies of host/pathogen/vector dynamics at the end of this subsection should make plain why this it so.

---

[73] In fact, Rothschild et al. (2001) present evidence for *M. tuberculosis* in bison dated 18 millennia ago. This suggests that the disease originated in cattle or their ancestors.

In short, the major premise's truth value is questionable, at best. Momentarily putting that aside, a couple of interrelated questions arise. Answering them requires a reasonably clear rendering of the hypothesis for which the allegedly protective benefit of introduced cattle is supposed to be evidence. The hypothesis seems to be: "The introduction of populations of nonhuman species that are alternative blood meals for a vector of a human disease decreases the vulnerability of people as blood meals, thereby decreasing the incidence of the disease in people." Given that, the first question is: What are the implications of this hypothesis for recommended human behavior and action? One obvious answer is one that authors on this topic never mention: Wherever vector born diseases adversely affect human health, one should introduce some one or more decoy species that will divert the bites of the vectors.[74] In other words, this is a call for species introductions that serve a human health purpose.

The second question, which comes in two versions, can be viewed as a corollary of the first: What, exactly, does the introduction of cattle (or other species) have to do with biodiversity? Or: how can the observed phenomena be understood in terms of biodiversity? The answer to the first version of the question is "not very much". To the second version, the answer is "not very well". It is not biodiversity that is the decoy for mosquito bites, but those large, domesticated, turf-compacting, flora-removing, water-fowling, methane-burping ruminants. To characterize the situation as a matter of increasing biodiversity via the introduction of that creature is to commit one of the category mistakes discussed in Sect. 4.1. This is another point at which the discussion really should end. But I will play along and past the category mistake to make a number of additional observations, which point up how the argument is built on fallacies of accident (discussed in Sect. 2.2.3).

Not all examples of zooprophylaxis involve domesticated animals. Nor do all of them involve species introductions; many involve species extirpations – for example, getting rid of creatures that are disease reservoirs. I shall expand my discussion to include extirpations and "wild" creatures, shortly; but for now, I confine my attention to the introduction of domesticated creatures and cultivated organisms. Considered in one way, moving a bunch of bovines into the neighborhood clearly has a rather marginal effect on biodiversity. It is marginal because there is no dearth of cows (or almost any other domesticated organism) and so certainly no increase in global biodiversity by herding them into a new location. So far as local diversity is concerned – without taking into account the effect that a beast such as a cow has on other creatures in its vicinity – there is at most an increase of precisely one otherwise extremely abundant and common species.

Of course, this incremental effect on local species diversity is far from the end of a story involving bringing in the cows. Taking into account how cattle affect the fate of other organisms and the most basic characteristics of their residence makes the biodiversity picture far more complicated. Recall Sahotra Sarkar's (unsubstantiated)

---

[74] This suggestion takes its cue from the World Health Organization, which for decades has urged a more modest version of this proposition. Bøgh et al. (2002, 593) remark that it "has recommended the use of cattle for zooprophylaxis as a protective measure against malaria since 1982".

claim that cattle increase biodiversity by maintaining the Keoladeo wetland (Sect. 4.1.1, Wilderness). This type of claim might gain credence from some studies, though not always ones free of conflicts of interest – such as a study (Marty 2005) claiming that grazing cattle can maintain native biodiversity in vernal pools in the western United States.[75] And sometimes, a single, sentimental favorite species can hang on, courtesy of four-legged mowing devices.[76] But the overwhelming body of solid evidence does not point at all towards such a clearly sanguine conclusion about bovine effects on biodiversity or the natural environment – a marquee case of disregard for countervailing evidence.

Some of the best documentation for bovine influence on biodiversity comes from the western United States. There, livestock grazing nearly matches the combined effect of mining and logging in contributing to the demise of over one-fifth of all species that are federally classified as threatened and endangered. That includes fully one-third of all endangered plants (Wilcove et al. 1998, 610). And although species diversity might not be affected by a change in the species mix, one study (Kimball and Schiffman 2003), showing that native plants tend to be quite vulnerable to cow herbivory as compared to non-native plants, makes clear that cows alter the mix.

The causal influence of cows on grazed turf are multiple and, in combination, often dominant: The direct effects on other species include competition for forage (with bighorn sheep and pronghorn, for example), blowing the cover for ground-nesting grassland birds (such as mountain plovers and at least two species of grouse) and small mammals (such as prairie dogs), and the systematic extermination of potential bovine predators such as wolves. This leads to indirect effects on species that have a predator, competitive, mutualist, or commensal relationship with one of those directly affected. Among animals, that includes black-footed ferret, swift fox, and Mexican spotted owl (Miller et al. 1994, 678–679; Salvo 2009). Among plants, it includes flora that cows prefer *not* to munch on.

Cattle are responsible for wholesale modifications of habitats, which are hard to view in the sanguine way that the creation or maintenance of a wetland might be. The effects on riparian habitats are dramatically transforming. These include deposition of pathogens in streams, as well as increases in nutrient levels, turbidity, and temperature – all of which affect the viability of a host of aquatic creatures ranging from invertebrates through amphibians to fish. Grazing alters the morphology of stream banks – downcutting them and reducing their stability, as well as the number and quality of pools that salmonids (among other fish) depend on. Grazing changes stream hydrology by increasing runoff and changing flow patterns. At the same time, it exposes bare ground, which is compacted and more easily eroded. Woody and herbaceous plants suffer (Belsky et al. 1999).[77] On the other hand, algal populations in the stream tend to flourish. So do populations of nonnative plants – typically

---

[75] This study was done on behalf of The Nature Conservancy, which has a large stake in accommodating cattle interests.

[76] See Nash (2009), which describes biologist Stuart Weiss lauding cows as "keystone herbivores" and the saviors of *Euphydryas editha bayensis*, the bay checkerspot butterfly.

[77] For the effects on upland habitats, see Belsky and Blumenthal (1997).

as a result of their relative resilience to trampling and their ability to take advantage of altered fire regimes. And the claim has been made that cows create pockets of standing water that provide breeding opportunities for some insects (including disease vectors) as well as amphibians who are pleased to dine on those insects.

All things considered, the jury is out on the net direct and indirect effect of cattle grazing on biodiversity. The issue is further confounded by the fact that different grazing disciplines (when, where, in the company of what other beasts, and in what numbers the cattle are grazed) might have different effects on the grazed habitat. But I have gone into these details to emphasize what the perceptive reader will already have realized: This discussion is entirely irrelevant to the proposition that biodiversity has a positive effect on human health. For even if cattle do provide a zoonotic decoy for some human disease vectors, and even if untold numbers of creatures of untold numbers of species sprang up in the footprints of every bovine, this would do absolutely nothing to show that biodiversity is zooprophylactic. Rather, it would show that the introduction of one species both moderates a human disease and fertilizes the biodiverse tree of life. The matter of human health and the matter of biodiversity would be correlates that happen to stem from the same cause. But the biodiversity would otherwise have absolutely no connection to the health benefits for humans.

Therefore, Dobson et al.'s argument for the health benefits of biodiversity is completely based on the fallacy of correlation – except insofar as it can be shown to rely on one common domesticated beast. So much the worse for their argument if it turns out that cows figuratively trample biodiversity as a consequence of their literal trampling of the turf on which they graze. When scientists such as Dobson join E.O. Wilson in agonizing over a possible Sixth Great Extinction, it is doubtful that they have in mind the urgency of saving domesticated cattle.

This brings me back around to the truth value of Dobson et al.'s major premise. Are there circumstances in which introduced cattle exacerbate instead of ameliorate disease? To answer this question, it helps to consider the disease-affecting properties of cattle introductions with dynamics that differ from those in the sort introduction that occupies Dobson and his colleagues. Molyneux et al. (2008, 306) relate that

> In Uganda... the expansion and movement of cattle populations into areas previously inhabited by native ungulates (a large group of mammals that have hoofs, e.g., antelopes and cows), combined with the invasion of abandoned cropland by the nonnative plant *Lantana camara*, is believed to have contributed to changes in tsetse fly (*Glossina*) distribution that initiated epidemics of African sleeping sickness (ASS) in the 1980's... ...the introduction of cattle... provided a highly competent reservoir host for a subspecies of the parasite that causes ASS, *Trypanosoma brucei rhodesiense*. *G. fuscipes* is a generalist vector that will feed on cattle, as it will on any available host. The movement of cattle in Uganda continues to this day to influence the spread of sleeping sickness in that country.

Unlike Dobson et al. and Molyneux et al. do not praise diversity-enhancing additions of *Bos primigenius* and *L. camara* (Spanish fig, an intentionally imported ornamental) to Uganda. It suits their (Molyneux et al. 2008, 306) particular purpose to point out that these are "nonnative alien and invasive species". However, this classification is entirely beside the biodiversity point. At least, it is beside the point for Dobson et al., who suggest that cows might be the sacred heroes (or heroines) of disease prevention.

Within a broader purview of domesticated beasts (beyond just cows) and cultivated crops, the biodiversity literature mostly portrays these organisms as health villains rather than as health heroes. Various species of the genus *Sus* (pigs) seem particularly adept at transmitting various diseases to people. They stand accused (and perhaps convicted) of this accomplice role in connection with irrigated rice fields, home to various *Culex* spp. mosquitoes that harbor the viral pathogen for Japanese encephalitis in various parts of Asia (Molyneux et al. 2008, 301–302); and in connection with fruit orchards in Malaysia, where Nipah virus-infected pteropid fruit bats find sustenance (Molyneux et al. 2008, 303–304).

In the first of two ironic twists, closer scrutiny of the scientific literature reveals its tendency to hold up as villains not just the pigs but also "human encroachment on biodiversity". This is "science-speak" for the observation that when people move into the vicinity of a diverse collection of creatures, there is a significant likelihood that the new neighbors will be carriers of pathogens and parasites capable of infecting humans. In other words, this is evidence that a large and diverse collection of creatures is generally bad for human health.

There are few completely reliable "rules" of ecology. Rapoport's rule is probably as reliable as any – at least for terrestrial (as opposed to marine) systems. It states that biodiversity increases as the distance to the equator decreases. As a corollary, this rule also applies to parasitic and infectious diseases (PID's): The diversity of PID's is greatest in low latitudes (Guernier et al. 2004). So it is not surprising that human infection rates are highest in the tropics. The straightforward conclusion is that, so far as infections are concerned and on a global scale, biodiversity is bad for human health. In fact, it is very bad.

In a second ironic twist, this is probably good news for biodiversity. That is, it is good news if it encourages people not to venture into and change the habitat of creatures that might make them sick. It could also be bad news for biodiversity if it encourages people to venture forth anyway, while trying to exterminate any and every living thing that might play a role in the causal chain that ends in human infection. As mentioned in Sect. 2.2.4 (The fallacy of correlation), this is the inclination of villagers in Cameroon who view the preemptive extirpation of both species of *Pan* and of *G. gorilla* – local primate neighbors who carry the Ebola virus – to be in their health interest.

I have pursued at length the example of cows and other domesticated creatures viewed as offering health protective services. A second example comprises a family of narratives. These narratives do not involve the intentional introduction of a species, which characterized as "greater biodiversity", is supposed to serve as a prophylactic shield against human infection. Instead, they start from the intentional modification of habitat. The most frequently encountered variant involves deforestation, and so that is what I take up. A typical story line threads its way through changes (not necessarily reductions) in local biodiversity that result from the changed habitat. The story concludes by noting an increased incidence of human infection. This (Molyneux et al. 2008, 297) is taken to be inductive evidence that "deforestation increases the risk of human infectious disease".

Of course, deforestation is a radical form of "habitat conversion", which inevitably leads to changes in the array of species that reside in a place and in their relative numbers. Equally true is that sometimes, species that are bad for human health move into deforested or fragmented areas. The standard arguments "underscore" these cases. Thus, in a number of cases in Southeast Asia and Amazonia, wholesale removal of trees has favored *Anopheles* spp. over the previous, more benign mosquito residents. The newcomers are more effective transmitters of the more virulent species of malarial *Plasmodium* (genus) parasites (Molyneux et al. 2008, 295–296). Similarly, the removal of trees in Cameroon has shifted the balance from one snail species, *Bulinus forskalii*, which hosts a relatively non-virulent schistosome (a trematode), to *B. truncatus*, which effectively hosts *Schistosoma haematobium*. This latter schistosome readily infects the human urinary tract (Molyneux et al. 2008, 297).

Is this inductive evidence that "deforestation increases the risk of human infectious disease"? The appearance that this is evidence is sustained only if one ignores the real causal factors that bear on these cases. In the case of the malaria vectors, one might think that much more depends on *how* the deforestation is done. If it is done in a way that also ensures continued good drainage and that reduces or eliminates the standing water that favors malaria-carrying *Anopheles* spp., then, apparently, the influx of those species could be avoided. Perhaps the lesson is that care must be taken in *how* deforestation is carried out, with particular attention given to installing proper drainage systems. Another possible lesson is that consideration should be given to planting forests where there are malarial outbreaks as a means of reducing their frequency or intensity. I shall say more about reforestation and revegetation shortly.

More to the point of claims for the health-preserving effects of biodiversity: is this inductive evidence for the proposition that a change in biodiversity increases the risk of human infectious disease? The evidence presented does not suffice to answer this question with assurance. But with high probability, the answer is, again, no. If there has been a mere shift in the relative size of the populations of different species of mosquitoes, and no species have been locally extirpated, then these circumstances say nothing about whether the *diversity* of species has changed. Even if, in particular places, the more benign species are entirely displaced by more aggressive *Anopheles* spp., then these latter species might more than make up for the local loss of the previous resident species. Again, the local diversity would be undiminished. And finally, there is no indication that, in any case, the mosquitoes that previously dominated locally went globally extinct. Under any of these conditions, even to say that a "component of biodiversity changed" is, at best, a very confused and confusing way to say that, although the effects on diversity are entirely uncertain, the particular combination of creatures in a specific local mix has changed.

What are we to make of the Cameroon snails? In this case, unlike the mosquito case, no additional *sine qua non* for the outbreak of disease is immediately evident. *B. truncatus*, the snail principally responsible for urinary tract schistosomiasis, likes sun-exposed water bodies; the relatively benign *B. forskalii* does not. So one might think that this is a case in point for the thesis that deforestation causes disease.

But not when one considers what has been left out of the story. There is no general law of nature or one special to ecology that says that vector species favored by deforestation are more likely to transmit disease.[78] In fact, just the opposite is sometimes true. *Re*forestation can, just as legitimately (or really, illegitimately) be said to *in*crease the risk of human disease. I have already noted (Note 71) how the introduction of *Erythrina micropteryx* (immortelle tree) into Trinidad did just that – by providing suitable habitats for bromeliads, which in turn provide suitable breeding habitat for *An. bellator*, a malaria vector. If this case is mistakenly said to not count, for the irrelevant reason that *E. micropteryx* is an "alien" in Trinidad, then other examples will serve.

One such example is the reforestation of New England – with "native" trees. It is fair to say that it is the reforestation in that region that has led to a serious risk of Lyme infection there. This synopsis of the plot – which starts from the reforestation – might come as a surprise to those familiar with the usual narrative, which skips over the reforestation prequel and begins with the fragmentation of the reforested landscape by roads and other human structures associated with towns. (This disease system is briefly described in Note 70).

Predators and, more generally, larger animals higher up in the trophic structure tend to have relatively small populations merely by virtue of the demands of their trophic position. They also tend to require relatively more contiguous territory to meet their dietary needs. As a result, these creatures are disproportionately affected by the fragmentation of their habitat. It is this relative reduction in predators and larger competitors of *Peromyscus leucopus* (white-footed mouse) that is said to have produced a surge in populations of that small rodent. As it happens, *P. leucopus* is also an especially competent reservoir of *Borrelia burgdorferi*, the Lyme spirochete bacterium. Therefore, the presence of large populations of this rodent increases the likelihood that local populations of *Ixodes scapularis* (black-legged tick), the Lyme vector in this area, will be infected and will in turn infect people. To repeat, according to the usual, truncated version of the story, forest fragmentation causes Lyme disease.

The Lyme disease system in New England and Lyme disease elsewhere are perhaps the most heavily researched zoonotic disease systems in the world. One need not plunge into gory details of the science to notice that the usual narrative contains prejudicially selective "underscoring". It begins with the arbitrary starting point: It is a story about the ills of biodiversity-reducing forest fragmentation, not the ills of regenerating a forest. More central to this discussion, there is no law of nature that says that the small species favored by fragmentation will be more competent reservoirs of the disease. In fact, another small species, *Sceloporus occidentalis* (Western fence lizard) predominates as the target of tick bites in the United States Pacific and southwest regions (Rapport et al. 2009, 45).[79] These creatures are not particularly susceptible to Lyme infection, and so probably

---

[78] There *is* speculation about this – to the effect that generalist vectors that have fewer strong biting preferences and that therefore are more likely to bite people, tend to be the pioneers in the modified landscape.

[79] In that part of the American West, the principal tick is *I. pacificus* (Gubler et al. 2001, 225).

reduce the frequency of Lyme disease in western U.S. ticks and therefore, in western U.S. humans, too.

Finally, it also should be noted that a much larger animal, *Odocoileus virginia* (white-tailed deer), not *P. leucopus*, is the primary host for ticks in the northeast, though not a particularly competent reservoir for the Lyme bacterium (Gubler et al. 2001, 225). This fact connects back to the prequel story, which reveals reforestation to be a prior cause of Lyme disease in the U.S. northeast. Research suggests that, at the root, the real culprits are the acorns of *Quercus* spp. (northeastern oaks). Ecologist Clive Jones and his colleagues found that populations of *I. scapularis* surged eightfold in acorn-rich plots, perhaps as the result of the deer spending more time enjoying the repast in acorn-rich environs. Densities of *P. leucopus* also surged with the abundance of mast, as did their Lyme infection rates in this tick-rich environment (Jones et al. 1998, 1024–1025). With this, the risk of human infection also increases. Applying the logic of convenience that infects the standard narrative to this more complete narrative might lead to the conclusion that *Quercus* spp. are bad for human health. This logic would target those grand trees for extirpation in areas where they nourish the nonhuman hosts that put humans at risk for Lyme disease. In fact, because they do not move and are slow growing, oak trees would be easier targets for extermination than deer, mice, or ticks. This consideration combines with the economic bonanza of valuable building material to make it likely that the removal of oaks is the most economically efficient means of reducing Lyme risk to humans in the U.S. northeast.

I cite one other example of the planting of native vegetation that has led to disease outbreaks: In several Mediterranean countries, including the southern Jordan Valley, cases of zoonotic cutaneous leishmaniasis surged as the result of planting native Chenopods (plants in the goosefoot family). The newly vegetated landscapes provided good homes for both rodent hosts – *Psammomys obesus* (sand rat) and *Meriones tristrami* (Tristram's Jird, on the *IUCN Red List of Threatened Species*), and the phlebotomine sandfly vectors for protozoan parasites in the genus *Leishmania*. This "biodiversity" was unwelcome and led to the uprooting of the recently reintroduced plants and the destruction of the rodents' burrows (Rapport et al. 2009, 50).[80] The habitat modification, in turn, led to the reduction of the cutaneous leishmaniasis. Evidently, habitat modification can cut both ways. In Jordan, the choice was to do the equivalent of uprooting oaks in New England.

As I said at the outset of this subsection, the notion that biodiversity forms a kind of infection-shielding cocoon for humanity is very odd on its face. Some initially odd-seeming hypotheses do turn out to be true. But what is known in disease science – the uncut, unexpurgated version – suggests that the cocoon hypothesis is not so lucky. In fact, it seems doomed to be a nonstarter. With an eye towards giving a sense of why this is so, I conclude this subsection with a glance at some of the science of zoonotic disease systems that involve vector transmission.

Some zoonotic disease systems involve a parasite with a complex life cycle that requires a diverse collection of host species. The life cycle of such a parasite

---

[80] See also Kamhawi et al. (1993).

proceeds in a sequence of stages. Each stage requires one particular species of mollusk or vertebrate host. This is common for trematode flatworms, which "flow" from one species of host to the next, quite often winding up in a vertebrate. The cycle starts with a free-swimming ciliated miracidium, which enters a mollusk, where the miracidium produces sac-like sporocysts and possibly rediae, the embryonic form. These latter mature into cercaria – the larval form with a swimming tail, which propels it into a second host – typically another mollusk, a copepod, or a vertebrate carnivore (amphibian, fish, bird, or mammal). It develops into an adult there, or possibly within yet a third host. A third host is typically a vertebrate carnivore, which receives its unwanted visitor by eating the second host. The trematode cannot survive the extirpation of any one of its specialized host species.[81] But clearly, the final vertebrate host, which could be *H. sapiens*, would be healthier for the absence of any of the trematode's upstream hosts. This is a case where a variety of hosts is not just conducive to vertebrate infection; it is essential.

The epidemiological equations for zoonotic disease systems show that multiple factors are critical for determining the incidence of human infection. Many factors have nothing to do with the number of species of host, pathogens or parasites, or vectors.[82] Entering into the equations are: the rates of encounter between each vector and healthy individuals of each of its human and nonhuman host species; the varying rates of transmission for each vector/host pair, given the probability of transmission on an encounter; densities (not just abundances) of the (one or more) vector species; properties of the multiple hosts that affect the efficiency of their transmission of an infection directly (not via a vector) from one host individual to another of the same species; the properties of the various hosts that affect the direct transmission of the infection from one host individual to an individual of *another* species; and whether the transmission in each of these various cases follows a frequency-dependent or a density-dependent paradigm. Each of these multiple factors must be added to the already non-trivial epidemiology that describes a simple one pathogen/one host system. For each host, the epidemiological equations must take into account its rate of recovery, mortality, and whether or not (or to what degree) recovery removes an individual from the pool of susceptible individuals. Changes in any one of these many factors, and even the precise sequence in which the changes occur can affect human infection rates, and whether they increase or decrease. Finally, causal factors for many of these changes can be changes in the populations of species that are not hosts, pathogens or parasites, or vectors. This includes, at the extreme edges, the introduction or extirpation of species.

---

[81] Hechinger and Lafferty (2005) focus on such a system in which birds are the ultimate vertebrate host.

[82] I mostly follow community ecologist Felicia Keesing and her colleagues (2006) in giving some sense of the various complexities of species interactions that, in the end, determine human infection rates.

Consider various ways in which the introduction of a species into a disease system can – and in some cases have been observed to – increase the incidence of disease:

1. A new predator induces populations of its prey to pack themselves more densely in areas that offer the best protection without necessarily changing the size of the prey's population. If the prey species is a pathogen host, this can increase rates of encounter between infected and susceptible host individuals, thereby providing a larger and more fertile breeding ground for the pathogen.
2. A new species is a food resource or a mutualist for a nonhuman disease host or vector, leading to a more robust population of a key infectious agent.
3. A new host species is a far more competent reservoir for a disease than any host previously present, leading to far higher incidence of the pathogen or parasite in vectors that transmit it to people. This is part (but only part) of the story of the Lyme bacterium hosted by *P. leucopus*. As another example, the rabies virus cannot be sustained in humans alone because humans rarely communicate the disease directly to each other. Introduce *Procyon lotor* (raccoon) and rabies becomes viable (Keesing et al. 2006, 491).
4. A new host species, even if not a particularly competent disease reservoir, helps sustain vector populations that still feed copiously from competent reservoirs. This is the role of *O. virginia* (while-tailed deer) in some Lyme systems such as the well-studied one in the northeast U.S. It is also the role of *Cervus elaphus* (red deer), the primary host of louping ill, whose transmission vector is *I. ricinus*, another tick. The number or density of deer must be just "right". Too low, and the tick populations decline. Too high, and the deer draw too many bites from viremic hosts, such as *Lagopus lagopus scotica* (red grouse) (Keesing et al. 2006, 494).
5. In a disease system in which interspecific transmission rates exceed intraspecific transmission rates, a new host species increases the prevalence of infections in all nonhuman hosts through interspecific transmission. There are several examples of this, including rabies once again: Populations of *Canis adustus* (side-striped jackal) in Zimbabwe could not support rabies, except via frequent re-inoculation by rabid domesticated dogs (Keesing et al. 2006, 492).
6. An additional vector increases disease risk. The presence of two tick vectors of Lyme disease in California – *I. spinipalpis* and *I. pacificus* (Western black-legged tick) – increases the risk of Lyme disease relative to areas where only one tick species resides. Similarly, two mosquito vectors of West Nile viral encephalitis – *Culex tarsalis*, which feeds on birds and maintains high rates of avian infection, and *C. pipiens*, which bites both birds and people – are jointly responsible for high human infection rates (Molyneux et al. 2008, 307).

Of course, adding a nonhuman host species to a zoonotic disease system does sometimes cause a "dilution effect". And sometimes a dilution effect reduces rates of human infection. When the newcomer is a relatively incompetent carrier or transmitter of the disease in question, it can supply enough of a vector's blood meals to reduce the disease's overall transmission to humans. But the effect is not a matter of the mere presence of the "decoy" species. I have given many examples to show that a salutary result hinges critically on multiple properties of all the organisms involved in the disease system. This includes whether or not the dilution effect is more significant than the increase in pathogen populations that a new host might foster (point (4) in the list above).

This range of possibilities provides some better perspective on the notion that the diversity of species is prophylactic medicine. Posed as an unqualified generalization, this proposition is quite categorically false. Mostly, the diversity of species is quite irrelevant to the question of how much disease spreads to humans.

Sometimes more species can reduce human infections. Sometimes more species can increase human infection rates. Whether or not people get infections is determined by conditions and causal chains that either wind up directing pathogens and parasites into human bodies; or not. The "right" conditions for infection can involve more or fewer species. But the number and diversity of species in the causal chain leading up to infection is entirely irrelevant as a causal factor in itself. This is another way of getting back to saying that, in the end, the proposition that biodiversity serves to protect human health is based on a category mistake.

6.6   Biodiversity as Progenitor of Biophilia[83]

In "biophilia" we have a neologism to pair with "biodiversity". In fact, pairing these two concepts is exactly what E.O. Wilson and Stephen Kellert – the two most distinguished and vocal proponents of "the biophilia hypothesis" – set out to accomplish.

The term "biophilia" might have originated with Erich Fromm's use of it as a "normal biological impulse" or state, which he contrasted with the "psychopathological phenomenon" of necrophilia. This usage is somewhat removed from the notion that Wilson (1984) later popularized in his eponymous book. But in retrospect, Fromm's explication (Fromm 1973, 406) can be seen to contain the seeds of Wilson's later extensions:

> Biophilia is the passionate love of life and of all that is alive; it is the wish to further growth, whether in a person, a plant, an idea, or a social group.

Fromm here allows biophilia to be a projection from a person's love of her own life and her love of other individual persons, to loving the life of other organisms (plants) and (metaphorically) the life of social groups.

Wilson (1996, 165) cultivates this germ of an idea into the definition of biophilia as the speculative hypothesis that there exists "the innately emotional affiliation of human beings to other living organisms". In the fertile mind of Kellert (2005, 49), it

---

[83] My thinking on this topic owes much to an unpublished paper that Dan Haybron presented in a 2008 conference and in subsequent verbal and email conversations with him.

develops and branches into "the inclination to value nature". Based on this speculation, Wilson and Kellert fear great psychological damage will accompany great damage to biodiversity. Wilson (1996, 170) exhorts

> psychologists... to consider biophilia on more urgent terms. What, they should ask, will happen to the human psyche when such a defining part of the human evolutionary experience is diminished or erased?

Several major obstacles stand in the way of connecting biophilia to biodiversity. A cursory glance at the two rather different definitions already cited reveals the first of those obstacles. The definition of biophilia, like that of biodiversity, is anything but clear. The various definitions of various proponents and even the same proponent at different times are not obviously equivalent. Second, biophilia is pure speculation. There is no direct or clear evidence for the existence of biophilia, however defined. Proponents urge that the principles of evolutionary psychology are amenable to its existence. But it is a long way from saying that something is consistent with natural law to saying that it, in fact, exists or must exist. The third difficulty has to do with connecting biophilia to any good connected with "the natural environment" for current-day humans. Let us grant for a moment that biophilia exists and that the principles of evolutionary psychology can account for its coming into existence. Even then, the mere "fact" of an evolved tendency – even one that for a long time conferred adaptive advantages – does not, by itself, make that tendency good or worth nurturing *now*. This point is obvious, for example, from human tendencies to harm or flee a person whose appearance is unfamiliar and solely on that account. One needs to beware of committing genetic fallacies.

The first three difficulties might already be fatal for the biophilia hypothesis even before any attempt is made to connect it to biodiversity. But the fourth and fifth difficulties might make them nearly irrelevant: Even if there were a clear definition of biophilia – for example, as a collection of conative or affective tendencies, or as a collection of functional capabilities; even if there were convincing evidence that these tendencies or capabilities actually exist and came to be built into human genes in the evolutionary course of things as adaptively advantageous characteristics; and even if there were a convincing argument to the effect that these tendencies constitute a good that ought to be nurtured in the lives of people now, then there remains a fourth challenge, which resembles the one faced by the ecosystem services paradigm of natural value: Insofar as biophilia is taken seriously, it appears to value, at best, a seriously fractured and truncated natural world, which thereby makes permissible behavior and actions that might lead to this result. Finally, even if *this* obstacle were surmounted, there remains the fifth and final challenge of leaping across the chasm that still separates *biodiversity* from biophilia.

Why would any significantly biodiverse state of the world be needed to satisfy biophilic inclinations? I will suggest that the biological diversity required for the purpose of nurturing biophilic tendencies appears to be vanishingly small. I will also suggest that, insofar as biophilic needs are thwarted by aversive reactions, biophilia points towards extirpating elements that would otherwise be threatening. And more generally, the demands on environments for nurturing biophilia (as specified by its advocates) are so minimal that they entail the superfluidity of anything that would pass for a truly natural environment.

My discussion focuses on the first (definitional) and the fourth and fifth (biophilia-natural good and biophilia-biodiversity spanning) problems. However, I touch on the second and third by way of getting from the first to the last two.

Let's reconsider the definition. Wilson (1996, 165) elaborates the brief one already cited by saying that

> From the scant evidence concerning its nature, biophilia is not a single instinct but a complex of *learning rules*... The *feelings* molded by the learning rules fall along several emotional spectra, from attraction to aversion, awe to indifference, and peacefulness to fear-driven anxiety... When human beings remove themselves from the *natural environment*, the learning rules are not replaced by modern versions... [italics added]

As Wilson uses the term "feelings", it seems to be an umbrella covering both conative tendencies (preferences, wants, desires, and urges) and affective ones (approval, pleasure, fulfillment, happiness, and the like). He makes clear that these "feelings" have, as either their object or source, "the natural environment".

The meaning of "learning rules" is more elusive. Wilson's explication of this phrase gives a sense that he believes that these are genetically encoded dispositions or tendencies whose presence is the result of the adaptive advantages they have conferred upon humans living in the cultures of their societies. They are rules in the sense that they might or might not be invoked – depending on the availability of a proper environment to stimulate or encourage their use. I believe that it is in this sense, too, that both Wilson and Kellert refer to them as "weak". This interpretation is reinforced by Kellert's apparently interchangeable use of "weak genetic tendencies" and "learning rules". Kellert (2005, 49–50) muddies these waters by also tossing into the stew "genetically encoded values" and "the inclination to value nature", which he appears to regard as additional equivalents of "learning rules". I make sense of these phrases as ill-chosen alternative ways to characterize affective "feelings" (again) of approval and disapproval.

One aspect of the definition of biophilia as learning rules quite directly subverts the case for biophilic value as a good for people and as a good for natural environments (difficulties three and four). While the speculative theory of the genesis of biophilic feelings posits their evolution as beneficial to the species *H. sapiens*, they are not necessarily feelings that are pleasant or that a person would desire to have. And they are not necessarily of benefit to nature. Wilson (1996, 167–169) writes at length about aversive reactions to snakes. Kellert (2009, 118) suggests that biophilic values have a "negativist perspective", which manifests when "snakes, spiders, large predators, swamps, steep precipices, lightning, and others" incite "apprehension and avoidance" or even "aversive reactions" that "provoke abusive behavior".

In sum, Wilson and Kellert seem to agree on a definition of biophilia as a set of dispositions towards "the natural environment". The dispositions include affective ones that incline humans towards both positive and negative feelings towards natural objects. Also involved are conative dispositions that incline humans to desire or seek out some things natural and to avoid others. Affective dispositions and conative ones can mix in any combination. Sometimes people are inclined to seek out – in order to destroy – natural objects that evoke negative feelings.[84]

A key phrase in Wilson's elaborated definition is "the natural environment". A lot rests on this phrase because it must be understood in a way that helps to fill the gap between biophilia and some coherent view of the natural world. It must also be understood so as to bridge the chasm between biophilia and biodiversity. So how is "the natural environment" to be understood? Kellert says some things about "nature" that shed light on this. For him "nature" is an enormous umbrella and many things camp out beneath it. Among them is "self-sustaining nature", which he takes to be more or less equivalent to "relatively undisturbed nature". This includes (Kellert 2009, 101–103) everything from what one some might call "wilderness" to (perhaps surprisingly) urban parks and gardens where human-made structures are not excessively intrusive. But for Kellert (2009, 104–111), "nature" also covers "domesticated nature", which includes your pet dog and the potted plant in your cubicle; also "neighborhood or community nature", which is the unbuilt, manicured lawn-covered space between your house and your neighbor's. In essence, nature is roughly anything nonhuman not made in a factory. Furthermore (Kellert 2009, 99),

> … the term *natural diversity*… encompasses any form of direct, indirect, or symbolic experience of the nonhuman world. [italics in the original]

Here, I presume that Kellert is not promulgating a tenet of Berkeleyan idealist metaphysics wherein real things are (literally) conceived as boiling down to collections of our ideas-as-symbols of them. Rather, I think it safe to assume that he intends to describe some kind of *experience* of natural diversity. He takes that experience to be equivalent to experiential contact with anything nonhuman and not made in a factory – although that description might be challenged were that aforementioned manicured lawn a "Roundup-ready" variety.

I now turn to the problem of how one might get from the biophilic starting point, so defined, to an affirmation of the value of biodiversity (the chasm of problem five) by way of finding a "good" in biophilia. But before doing that, a few words are in order about evidence for biophilia as a descriptive hypothesis about innate tendencies that are the product of evolution. This is the second difficulty mentioned above,

---

[84] In an unpublished paper, Dan Haybron perceptively distinguishes between a "weak" version versus a "strong" version of the biophilia hypothesis. The weak form supposes a desire – a mere liking. The strong, more tenuously conjectural version of the hypothesis supposes a need. My treatment presumes something in between – a desire whose fulfillment is actually beneficial, whether or not this constitutes the satisfaction of a full-fledged need. I believe that something like this intermediate form is what is most easily extracted from the writings of biophilia's proponents.

Haybron also distinguishes between mere "contact with nature" versus some more significant "active engagement with nature", wherein a person engages in a way that involves acute skill, knowledge, and awareness. Both of Haybron's distinctions (weak versus strong and mere contact versus engagement) are important for a full understanding of the scope and limits of the concept of biophilia. However, these distinctions are peripheral for my central and more limited purpose of determining whether there is any way to connect biophilia (in any form) to biodiversity.

These qualifications owe to personal correspondence with Haybron on his work.

stripped of any normative veneer. Wilson's definition of biophilia (cited above) concedes "scant evidence" for it. Indeed, he presents no independent evidence for it at all. In lieu of that, he (Wilson 1996, 166) states that biophilia is a logical implication of evolutionary theory; it is, as he says, "compelled by pure evolutionary logic". Unfortunately, the "logic" is a story about how people (as well as Old World monkeys and apes) might have come to be leery of snakes. While informed by Wilson's formidable grasp of evolutionary biology, this is really a "Just So" story – an unfalsifiable genesis narrative about how cercophithecids and hominines might have acquired such a trait as a consequence of the adaptive advantages that it might have conferred upon them.[85]

Even if Wilson's "Just So" story is accepted without question, it still does not answer a critical question. That question concerns whether or not the biophilic trait is still functional in modern humans. "Evolutionary logic" might lead one to speculate that, given the societies and environments in which people have now lived for thousands of years, the trait has, for some time, not conferred much advantage. Could it therefore have become largely vestigial? A credible answer to this question requires evidence.

Kellert gets past genesis stories and tries to address the need for evidence, presenting a number of studies to this end. But their cogency in support of biophilia is underwhelming. This is partly because, as he (Kellert 2009, 107) admits, "Few of these studies have been rigorously conducted." It is also partly because, rigor aside, their results do little to support the normative burden that is subsequently placed on them. As I shall show, they have little to do with "self-sustaining nature" and apparently nothing at all to do with biodiversity.

For the sake of further discussion, I suppose that Wilson's "Just So" story of biophilia's genesis solves the second difficulty. With regard to the third difficulty, let me also provisionally suppose that biophilia is not (yet) vestigial in humans. This lets me move on to normative part of the third difficulty, which has to do with extracting some kind of value from this allegedly non-vestigial tendency. Is the exercise of biophilic tendencies by people a good and furthermore a good that accrues to people in and only in a "natural environment"? The mostly anecdotal evidence that Kellert presents might at first make one think so. That "evidence" is largely a tale about the apparently pleasant effects that "nonhuman things not made in a factory" have on people – a walk in a park, greenery outside a hospital window, grass instead of concrete between adjacent houses.

But even if one receives Kellert's evidence with uncritical acceptance, chinks already begin to appear in the biophilia-based case for nature being a good for people. Kellert (2009, 106) inadvertently helps to identify one chink in the course of making his case on the grounds that "natural lighting, natural ventilation, [and] natural materials" in buildings "enhance worker comfort, satisfaction, [and] physical and mental well-being." This makes it apparent that, whatever benefits Kellert

---

[85] I am here using "Just So" in the sense of Rudyard Kipling's account of the genesis of the camel's hump and the leopard's spots. This is but a very distant relative to the "just so" model of the calculus of biodiversity value, described in Sect. 5.1.4 (The just-so model).

believes might accrue from the exercise of biophilic tendencies, the need for "something nonhuman not made in a factory" might well be supplanted by "something nonhuman that *can* be made in a factory if it adequately simulates a 'natural environment'". This raises the general question of whether or not "the real thing" is indispensable, so far as the satisfaction of biophilic tendencies is concerned. On Kellert's account, this seems quite doubtful. Kellert (2009, 102) also supposes that aesthetic benefits follow from biophilic tendencies. Insofar as these tendencies are satisfied by such fabrications as potted plants and Kellert's criteria qualify artificial ones, this is another case in point.

A second chink in Kellert's case – for biophilia as the basis for the good of nature – has to do with his (and Wilson's) acknowledgement that nature is not always so good for people; that it is not always pleasant or attractive; and that it threatens us and leads to very unpleasant, aversive reactions. It is difficult to understand how snakes, dark forests, and animals that threaten to eat us can be understood to evince the kind of positive affiliation on which biophilia advocates build their case. It seems that, in the name of biophilia, one is committed to say, "so much the worse for snakes, dark forests, and big toothy animals."

The last observation merges into the fourth difficulty, which is whether or not biophilia is a credible basis for valuing the natural world. The first part of biophilia's answer seems to be that the whole of nature cannot be regarded as valuable, since at least those elements that biophilic inclinations reject must likewise be rejected and therefore excluded. So in the absence of other considerations, biophilia seems to endorse a very "patchy", human-selected and architected view of the natural world, at best. At least, a more sophisticated argument is required to break the connection between natural value and an affirmative orientation whose focus is narrow and restricted to the relatively small portions of the natural world that people are likely to regard as human-friendly.[86]

I leave off an exploration of what such an argument might look like in favor of examining the fifth and final obstacle – the chasm between biophilia and biodiversity – that most directly relates to the central topic of this chapter. This chasm is, I think, unbridged; and most likely, unbridgeable.

E.O. Wilson doesn't even attempt a bridge. In a kind of Chewbacca defense, he (Wilson 1996, 170) leaps without explanation from a speculation on the possible psychological implications of limited opportunities to exercise biophilic inclinations directly to a lament about the global loss of biodiversity. Does loss of biodiversity figure importantly, or even in some limited way in the loss of such biophilic opportunities? Wilson offers no argument at all to persuade us that it does. Moreover, given the

---

[86] The most promising such argument is one that Dan Haybron develops in the unpublished paper mentioned in Note 83. It makes the conjecture that some active engagement with nature might be essential for the full development of certain human capacities, which in turn, are part of a fully realized human life. Unfortunately, Haybron explicitly excludes dispositions that do not have a pro-orientation towards nature; so he is not taking in the full compass of biophilia, as I understand it. However, I see a potential for this approach to bring the negative dispositions back into the biophilic fold, and even strengthen Haybron's case as a consequence.

sorts of "natural diversity" that, on his account, suffice to provide such opportunities, it would be extremely surprising if the loss of even the entire 30% of species that Wilson and others consider at risk would have any palpable biophilic effect.

If anything, the existence of a bridge between biophilia and biodiversity is made more dubious by Kellert's attempts (Kellert 2009, 106–109) to deal with the difficulties that Wilson ignores. Kellert's arguments are plagued by difficulties that fall into several categories. First, virtually all the arguments for the benefits of exercising biophilic tendencies are so easily satisfied that they make no meaningful requirement on even "self-sustaining nature", let alone biological diversity. That a window looking out on a tree is better for a surgery patient than a window looking out on a brick wall; that domesticated animals make nice companions; that potted palms are welcome additions to cubicles at work; that natural lighting is favored over fluorescents; that groomed parks are nice places to walk; that suburban developments featuring more "open space" are favored over ones that do not: none of these things have any, even remote connection to biodiversity in the sense that makes environmentalists such as Wilson lament the Sixth Great Extinction. It is difficult to imagine how even adding a *Seventh* Great Extinction would affect our ability to have far more than enough biological diversity to have potted plants and natural lighting.

Ethnobiologist Alain Froment (2009, 213) puts it this way:

> Psychologically, the contemplation of a "natural" landscape is recognized as excellent for mental health, but biodiversity is not a factor here. First, most of the landscape, such as a garden or the countryside, is not "natural", but humanized. Second, viewing an environment poor in biodiversity, such as bears on the Arctic Circle, may be more mentally beneficial than a rich environment like a jungle, which may cause anguish in some. For the "civilized" world, forests (from the Latin *foris*, "outside") are savage (from the Latin *silva*, "forest") and wild jungles (from the Hindi *jangal*, "uninhabited space") can generate anxiety. There is, then, no direct correlation between the relaxing role and comfort provided by nature, and wealth of biodiversity.[87]

Second, the exercise of some biophilic inclinations militates directly *against* biodiversity. People want clear ponds and fast-moving streams (Kellert 2009, 102). That is a strike against inviting beavers "back in"; for some, it is strike three after the first two strikes of chewing on trees and flooding farmers' fields (Taylor 2009).[88] The many creatures that elicit aversive responses would not do so if they were extinguished. Many people do indeed feel the beckoning call of "nature", which they satisfy with behavior that is detrimental to biodiversity. The inclinations of off-road vehicle users in the American West come immediately to mind. These people truly appreciate the majestic backdrops – in fact, regard them as essential – for the pleasures of their "habitat-converting" sport. So much the worse for the creatures that once lived there.

---

[87] Of course, beholding a Polar bear in close proximity might be substantial cause for anguish, too. This observation doesn't so much undermine Froment's point as it underscores the more damaging one for biophilia – namely, that it has no respect for biodiversity.

[88] For more on the various services and disservices that *Castor Canadensis* offers, see Sect. 6.3 (Biodiversity as service provider).

The third and last point connects back to Wilson's lament for the loss of biodiversity. Even if one grants each and every biophilic benefit and even if (contrary to fact and even Kellert's account) each benefit requires "relatively undisturbed nature" in Kellert's sense, then these benefits still are easily and readily available in a world undergoing the kind of mass extinction that E.O. Wilson and others decry. There is no solid case for the proposition that biophilic tendencies really exist. Nor is there a solid case for the proposition that, if they do exist, then their exercise constitutes an integral part of human flourishing. But even supposing the truth of these propositions, biodiversity – at least the sort of significant biological diversity that truly concerns biologists and environmentalists, as opposed to having natural interior lighting or a nice lawn to separate you from your neighbor – would be entirely dispensable for biophilic benefits.

The unbridgeable gap between biophilia and biodiversity is not held open by an imagined theoretical replacement of the experience of biodiversity by a functioning set of Delgado buttons. Nor does the gap persist on account of a supposed prospect for replacing the current biodiverse state of the world with some other one equally biodiverse. The problem is that real biodiversity is mostly and perhaps entirely irrelevant to biophilia.

6.7   Biodiversity as Value Generator

Several accounts of biodiversity value attempt to locate it as a kind of meta-value, which derives from its capability and performance as a value-*generating* engine. The forestry biologist and philosopher Paul Wood, for one, tries to make a case for biodiversity's value as primarily a matter of its being, more specifically, a *biodiversity*-engendering engine. He (Wood 2000, 51–57) tries to avoid the obvious apparent circularity in this proposition by saying that the biodiversity-engendering capabilities of biodiversity constitute a tertiary value which is a precondition for the secondary value of adaptive evolution, which in turn, is a precondition for maintaining a range of biological resources, which is biodiversity's proximate value. I address the question of circularity shortly.

Bryan Norton (2001, 90–94) also talks about creativity as the core value of biodiversity. He (Norton 2001, 90–91) has in mind

> … the processes that have created and sustained the species and elements that currently exist, rather than … the species and elements themselves.

His discussion is opaque about whether biodiversity is the cause or the effect of this creativity. But it is plausible to interpret his position as essentially that of Wood – namely, that biodiversity is the fuel for a process that engenders more biodiversity whose components are good stuff (resources) or have good properties (offer services).[89]

---

[89] As in Sect. 6.3 (Biodiversity as service provider), I utilize the consensus definition of "ecosystem services" offered by Hooper et al. (2005, 7).

Sahotra Sarkar (2005, 103) seems to bark up much the same value tree as Wood and Norton when he speaks of diversity as engendering a kind of valuable "novelty", by which he means something that "will contribute something new to science".[90]

This proposal for biodiversity's value is unconvincing in light of two observations. First, in its several variations, the value of biodiversity-as-value-generator devolves into the value of what is generated. Going "one level up" does not remove the burden of justifying the value of whatever is generated at the lower level. Of course, finding that what is generated is valuable lends legitimacy to the claim that whatever is capable of ensuring a steady supply of that value-laden product is itself valuable as a means to that end. But Wood, Norton, and Sarkar all suggest that what biodiversity generates is … biodiversity. That is, according to them, biodiversity is valuable as a means to the end of biodiversity.

All three advocates seem to realize that this point in their argument remains far from a successful conclusion. Wood pushes ahead by saying that ultimately, generated biodiversity constitutes resources. Norton says (among a dizzying assortment of other things) that the generation of biodiversity is a kind of productivity. For Sarkar, it is (again, among other things) the "stuff" of biological science. In each instance, the case must ultimately be anchored in these ends. Unfortunately, there is ample reason to be skeptical that any of these proposals is a reliably firm final anchor for value.

Norton's notion of productivity is a non-starter, for as discussed in Sect. 6.3 (Biodiversity as service provider), there is both bad and good productivity. Perhaps the most promising of these proposals is Wood's – that generated biodiversity is tantamount to generated resources. But it is hard to see how diversity in itself constitutes a resource, except by courtesy of the entities that constitute the diverse set; and Sect. 6.2 (Biodiversity as resource) shows how tenuous is the case for connecting these two different things. Often recited is the proposition that (for example) a new species is a kind of "raw material", which necessarily gives human-kind new options for constructing solutions to problems. After all, the more and varied kinds of raw materials, the more design choices we have.[91]

The very phrasing of this recitation prejudices the issue by suppressing a couple of questions that admit contravening answers. First, what good reason is there to believe that, say, an encounter with a new species is likely to present itself as a resource for humankind's use rather than as an impediment to humankind's general good – say, by eating resources, by carrying a previously unknown disease, or by just getting in the way of human development? A novel organism is not necessarily a benign hammer waiting to be picked up when humankind discovers a nail that it

---

[90] This statement is part of Sarkar's theory of "transformative value", discussed separately in Sect. 6.10 (Biodiversity as transformative). His statement also suggests the value proposition of biodiversity as contributing to human knowledge. That topic is also treated separately – in Sect. 6.8 (Biodiversity as font of knowledge).

[91] I am indebted to Jeffrey Lockwood for forcing me to clarify, tighten, and properly qualify my argument here and in the remainder of this section.

can drive. It could just as well be a malignant hammer that picks itself up and whacks away at human well-being.

The second question is twin to the first. To what extent does the recitation presume a strictly additive conception involving a continual adding to a stockpile, each element of which (and its quality as stockpiled resource) does not otherwise change? This picture is at war with itself. It presumes a creative dynamic that generates new resources; upon their creation, the dynamics fall away and the resources enter a static state in which their resource-providing qualities are thereafter frozen.

By asking these two questions, one can see that the plea for novelty rests on an unsupported and perhaps unsupportable assumption – namely that the new will serve humans at least as well as the old. The plausibility of this assumption requires a picture of an evolving world in which resources are continually replenished; services continue undiminished, uninterrupted, and perhaps even augmented. But from a naturalistic perspective (which I suppose those who promote the novelty thesis to have), no purpose, and particularly, no benign or human-benefiting purpose, can be justifiably presumed. And no resource can be assumed to be immune to some novel circumstance that transforms it into something entirely unhelpful, a burden, or worse.

One could legitimately ask why it is not just as likely that the creative forces of nature might create a nightmare world for humans – say a world with a methane-dominated atmosphere (again), or one that contains a crafty predator with a taste for human flesh – as that it create a paradise. Just over 70,000 years ago, one nightmare scenario (though probably not predominantly biotic in origin) did play out, and it brought the recently emerged species *H. sapiens* to the brink of extinction (Ambrose 1998, 2003).[92] But so far as I can tell, an unbiased review of geological history does not favor one "creative" possibility over the other. Moreover, the question of whether or not "the creative forces of nature" favor or disfavor the retention of biodiversity is open. The current biodiverse state of affairs (not the diversity of it, *per se*) notably features an apex species with a penchant for re-engineering everything around it for what it perceives to be its own considered, near-term benefit. Taking that into account points towards the likelihood that those creative forces will lead to a novel state of significantly diminished biodiversity.

The second observation of the two mentioned at this section's outset derives from the notion of trying to evade the preceding criticisms by taking the forces of "creativity" or "novelty" seriously as independent bearers of value that do not ultimately depend on the value of the novel resources or services that they create. It is possible that there is a case to be made for this. But the rationale is not obvious, I cannot construct it, and none of the above-cited authors even attempt to sketch such a case. Why should something have positive value just because it is new? Why should novelty be more valued than what we have now? No satisfactory answers are forthcoming.

---

[92] See Note 5 in Chap. 5 for a brief account of the paleontological background for this.

Sarkar might at first appear to offer an answer to this question by suggesting that novelty feeds scientific knowledge (which I discuss separately in Sect. 6.8, Biodiversity as font of knowledge). But further reflection reveals how precarious this position is. An opportunity to study the natural world undoubtedly serves a deep-seated human need to know about the world. But there seems to be no reason think that one kind of world – say, a more biodiverse one or one with a greater rate of novel creature creations – would serve that need better than one in which biodiversity wildly fluctuates and sometimes (perhaps around now) plummets, with an attendant decrease in the rate of novel creations. The point of inquiry is to understand how the world works, whatever its dynamics, whatever the engines of change, and whatever the current or future state of biodiversity.

Moreover, the notion of "novelty" is pliable in a way that just as legitimately permits one to argue that a dramatic reduction in diverse biotic and biota-encompassing kinds would provide the most novel of circumstances. Though successors to kinds that are now being swept aside might take some number of human generations to fully evolve (an issue that I address separately below), this does not detract from the novelty of actually being at the start of such a dramatic biotic event. Moreover, even the first incremental step in "recovery" would be more novel – and possibly far more novel – than maintaining the *status quo*. Realizing this, in turn, makes it easier to call into question the independent value of novelty, as I already have.

Finally, something should be said about the diversity that might arise out of current extinctions. If a great extinction event is indeed now underway, by most accounts, it differs in some significant respects from previous ones. For example, it probably involves a greater preponderance of plants; and many of the facilitating conditions can, in some measure, be traced to the behavior and activities of one species that did not exist during any prior great extinction. Therefore, it, it is risky to draw any inferences based on induction from previous such events. With that as preface, consider the suggestion, sometimes made, that current reductions in kinds of species constitute the single most powerful way to encourage the generation of entirely new kinds in every category of biodiversity. On this account, a great extinction should be welcome as a way to renew and refresh biodiversity – a way to attain a diversity of diversities, strung out along the planet's timeline.

The standard response is to dismiss this suggestion as a clever but insubstantial argument, which surreptitiously and illegitimately trades geological timeframes for shorter, more human-relevant ones measured in numbers of human generations. In the short term relevant for people, it is said, we are screwed out of biodiversity. But this is a misleadingly incomplete account of what is actually going on. It is true that full recovery – in the sense of re-attaining a similar level of species diversity in the largest organisms most palpable to humans – is likely to be measured in expansive timeframes. But adaptive changes in many organisms will occur (and are now occurring) quite rapidly; in some cases, these changes will engender (and are engendering) new species; and these near-term, species-engendering changes will be (and sometimes are) accelerated by some of the same forces that are simultaneously causing the extinction of other organisms. In other words, it is likely that, despite the extinctions, biodiversity overall will not suffer nearly so much in the relatively near future as one might think by focusing exclusively on the extinctions.

For example: There is increasing and increasingly compelling evidence (for example, Sax et al. 2007; Vellend et al. 2007) that species immigrations can and do trigger rapid selective adaptation and rapid changes in phenotypical expression. Possible mechanisms include (viable) hybridization, "disruptive selection" of natives (whereby traits at the extremes are favored significantly over intermediate ones), and the adaptive transformation of the immigrants in their new and geographically disconnected environment (which can lead to allopatric speciation).[93] Various of these mechanisms significantly push the evolution of plants, which are disproportionately vulnerable to current-day extinction pressures. For example, in studies of *Hypericum perforatum* (St. John's wort), ecologist John Maron and his colleagues (2004) find evidence that evolution can sometimes be quite rapid for a recent "transplant" that finds itself in a novel environment. Jason Sexton and his colleagues (2002) examined the (in)famous case of "invasive" tamarisk in the U.S. southwest and found "surprisingly high levels of genetic variation. This along with other factors, such as persistence long enough to experience adaptive evolution", they (Sexton et al. 2002, 1652) say, grounds a "potential for evolutionary increases in invasive traits and plasticity" that may "greatly influence their future invasiveness" (Sexton et al. 2002, 1658).

In the realm of animals, entomologist Anna Himler and colleagues (2011, 254) report that the invasion of an invader – *Bemisia tabaci* (sweet potato whitefly) by a *Rickettsia* bacterium – induced a dramatic shift in the whitefly phenotype, whereby it "produced more offspring, had higher survival to adulthood, developed faster, and produced a higher proportion of daughters". Biologist Olivia Judson (2008a) relates a variety of other cases of rapid evolution, some of which do not involve recent immigrants. But one that does is the assisted immigration of *Podarcis sicula* (wall lizard) from the Croatian island of Pod Kopište to nearby Pod Mrčaru. Geographically isolated in their new island getaway and perhaps on their way to allopatric speciation, these creatures quickly evolved cecal valves. This "suggests [to Judson] that arrival in a new environment can result in dramatic changes to an organism within fewer than 40 lifetimes."[94] She (Judson 2008a) concludes that:

> At least one other lesson can be drawn from all these studies. Natural selection has its most dramatic effects when an organism's environment is perturbed in some sustained way – prolonged droughts, the arrival of species that compete for food, warmer winters, the use of pesticides. If we humans continue to increase our impact on the globe, we're likely to see lots more evolution. And soon.

This kind of evidence suggests that some novelty emerges quite quickly in rapidly reproducing and adapting organisms. Consider that the preponderance of the

---

[93] See Sax and Gaines (2003) for a description of hybridization processes.

[94] This figure overshoots the number reported by Herrel et al. (2008, 4793), the source for Judson's account of *P. sicula*. Those researchers document 30 generations of the lizard, which emerged over the course of a 36-year study.

planet's organisms – lizards are one speciose group, but insects account for something like half the planet's non-bacterial species – meet this description. This means that one can plausibly expect some considerable amount of novelty to emerge within the lifetime of currently respiring humans, some within a few generations, and much more within human historical frameworks. Something like "full recovery" of diversity within the most species-diverse groups might well occur, and well within the "lifetime" of the human species. And very likely much of this recovery will occur well before current events become ancient history.

These considerations tend to be lost in frequent recitations that, according to the geological record, it takes from 5 to 10 million years to "recover" from a mass extinction event (Kirchner and Weil 2000). More recent work (Brayard et al. 2009) on the Permian extinction event – which finds that ammonoid cephalopods recovered in something more like 1–2 million years – might call for a reevaluation of the larger numbers.[95] While still long, the shorter timeframe is significantly shorter by virtue of falling more comfortably within the likely lifetime of the human species.[96] But it is still two orders of magnitude longer than human historical timeframes, which one might suppose to be on the order of 10,000 years.

I revisit the topic of timeframes for recovery from a great extinction event in Sect. 7.3 (Biodiversity value in human timeframes). For now, let's suppose that something that one might be willing to call "full recovery" of the diversity of life forms might span the entire lifetime of our species. How much weight should this carry? The answer to that question, I believe, must find the relative weight of two other considerations. The first consideration is one that I suggested above: Much evolutionary working out is already well under way, likely at accelerated rates; and much recovery of diversity is likely to occur well within human historical timeframes – though perhaps among organisms (such as reptiles and insects) that some might (unjustifiably) tend to disregard or discount. The second consideration reconnects with the love of novelty expressed by Wood, Norton, and Sarkar. It involves a sober and unbiased assessment of the extent to which this love can be justified, given a sober and unbiased assessment of whether novel modes of being bode ill or well. This assessment is strongly reminiscent of the basic conundrum of evaluating diversity itself: As Cowper's poem (Sect. 2.3.2, The value of diversity in general) brings home, there is good variety; but there is also bad variety. Much the same can be said about novelty.

---

[95] But once again, inductive caution is called for.

[96] In the background of my discussion is a picture of species lifetimes that looks something like this: According to some standard estimates, the average lifetime of invertebrate species (or at least marine invertebrates) is 5–10 million years, plant species around 3.5 million years (Niklas et al. 1983), small mammals perhaps 2.5 million years (van Dam et al. 2006, 687), and megafaunal mammal species such as *H. sapiens* around perhaps 1 million (May 1995, 14). For mammals, longevity honors of 16 million years go to such smaller members of the class as mole and dormouse (Liow et al. 2008, 6099). Within this temporal framework, *H. sapiens* is still at the beginning of its run on the planet – begun around 200,000 years ago. Combined with the uncertainty in "recovery" times (in the narrow sense that prejudicially excludes the vast majority of rapidly evolving organisms), which might be shorter than previously supposed, these numbers make it plausible to suppose that people might be around to witness a "full recovery" of a major extinction event.

All things considered, it is hard to avoid an ironic suspicion that Wood, Norton, and Sarkar are not so much interested in novelty as in the polar opposite, the *status quo*, which they are highly disinclined to see upset. One might speculate that these advocates of biodiversity-as-novelty-generator trip over their view that creativity is acceptable so long as it does not alter the current, particular biodiverse state of affairs whose peculiar mix of kinds in various biological categories (not necessarily their diversity) more or less satisfactorily meets human desires and needs for resources and services.

6.8   Biodiversity as Font of Knowledge

Biodiversity is often cited as the subject of biological study and therefore as a rich source of human knowledge. This suggestion already came up (in the preceding section) by way of Sarkar's views on "novelty". More famous is E.O. Wilson's trope of the "Great Encyclopedia of Life", which was inspired by the words – now ubiquitous in discussions about biodiversity – of his friend and colleague, the chemical ecologist Thomas Eisner (1982). Eisner was writing less about knowledge, considered abstractly as a human good, than with the relish of a genetic engineer (at a Monsanto Symposium) about the genes that he would like to see diced and spliced into genetically improved models of organisms:

> As a consequence of recent advances in genetic engineering, [a biological species] must be viewed... as a depository of genes that are potentially transferable. A species is not merely a hard-bound volume of the library of nature. It is a loose-leaf book, whose individual pages, the genes, might be available for selective transfer and modification of other species.[97]

Wilson steers Eisner's vision of biodiversity as raw material for biotechnological bounty towards what some might regard as higher ground. He does this by borrowing heavily from the Shannon-Wiener information-theoretical tradition of measuring biodiversity (touched on in Sect. 4.1.2, Measures and indexes). In this tradition, biological entities – for Shannon-Wiener entropy it is species; here it includes genes as well as species – as bits of information. It is but a short step from there to reimagining these bits into a vision of a fabulous library in which they are the contents. Wilson (1992, 151) asks us

> ... to imagine... that all the diversity of the world were finally revealed and then described, say one page to a species... ... this Great Encyclopedia of Life would occupy 60 meters of library shelf per million species.

He asserts (Wilson 2002, 131) that

> Each species... offers an endless bounty of knowledge... It is a living library.

---

[97] Wilson (1992, 302) also quotes this passage, but he (1992, 381) cites the wrong paper and does not reveal the genetic engineering context in which the passage occurs.

And so on. This vision haunts almost every one of Wilson's writings about biodiversity. It also appears in the writings of Holmes Rolston (1988, 98–99), who closely mirrors Eisner's characterization of organisms as constituting a "genetic library". From these origins, the vision appears to have achieved a life of its own, whereby it regularly finds its way into other work, typically without reference to its originators.[98] As one example among many, the philosopher Jeremy Bendik-Keymer (2011, 15), drawing on Rolston, picks up and imaginatively embellishes the library trope:

> If a species is like a book in the library of life, an evolutionary story that is distinct and unique, the genus to which it belongs is a genre of books, for instance, books of sonnets. Its family is then a kind of literature, e.g., poetry –in other words, a section of the library, not just a book in a stack. If, for instance, a species went extinct, but much of its family remained, then the Earth would still have a considerable record of the evolutionary achievement that allowed species such as it to exist. We might miss this particular book of poems, but we would still have poems. Species like it could continue to evolve, since the family would be intact. But if the family goes extinct, then the chance of continuing evolution along anything like the species's line is gone. In other words, by our mass extinction voiding sections of nature's library, we are erasing whole areas of evolution.

Unfortunately, neither Wilson, nor Bendik-Keymer, nor any of the many who have utilized this trope on behalf of biodiversity have given serious consideration to whether or not it is capable of the normative lifting for which it is pressed into service.[99]

Surely some caution is in order insofar as the basis for the trope's power lies in the assumption that all knowledge constitutes a good. One should beware of granting, without qualification or reservation, that any knowledge is worth pursuing or that it is worth pursuing at any expense. The repugnant aim of a project to determine how, most cost-effectively, to torture people is a serious reason not to pursue it. The vacuous aim of a project to count the precise number of pushups that an individual *Sceloporus occidentalis* (western fence lizard) performs over the course of its lifetime (in contrast to studying the role that this signaling behavior plays in a western fence lizard's life) should suffice to raise questions about its worthiness. Still, it seems safe to say that no one would seriously question the pursuit of biological knowledge generally or knowledge of biodiversity in particular on grounds of repugnant purpose, vacuousness, or any similar obvious objection.

It is certainly true that the biological world as it exists today, and specifically, the diversity of kinds in the world right now, is largely unexplored. Focusing just on species diversity, only around 1.5 million species or so have been documented. Few among these have received extensive study. By conservative estimates, the 1.5 million

---

[98] Perhaps another indicator of how uncritically the trope is often tossed out is that some authors acknowledge Eisner by way of Wilson's incorrect citation.

[99] In this context, Bendik-Keymer offers a logically separate argument having to do with the familiarity of the existing contingent of organisms. I will not address that argument here except to note one obvious weakness, which derives from the fact that the vast majority of organisms are not known and therefore completely unfamiliar.

known species are less (and most likely, considerably less) than one-sixth of the actual total of 10 million or perhaps many more.

However, the untapped bits of knowledge locked up in these organisms is but one type of knowledge that one can hope to have about the biological world. Much, too, could be learned from a vastly changed biological world that contained a significantly different set of species with significantly different population sizes (abundances). Minimally, this alternative would provide an unparalleled perspective on biological systems, which now are necessarily viewed (advances in paleoecology notwithstanding) largely through the highly biased lens of their current and very recent states. It is highly likely that the new perspective would immediately suggest new relationships in nature that are currently hard to discern. Just as likely, the new state of the world would reveal that some relationships that now seem hard and fast are, in fact, highly circumstantial and evanescent anomalies.

Also, the very processes involved in bringing about such an altered world (of differently composed and even reduced biodiversity) would be a rich source of knowledge that could not be tapped except by observing them unfold. What better way would there be to study previous extinction events, whose geological remove makes their reconstruction – necessarily on the basis of scant and patchy evidence – a complex puzzle whose solution might be forever underdetermined by the evidence? Ecologist Dov Sax and his various colleagues make similar observations with regard to species immigrations and their role in the extinction of "native" species.[100] These are "experiments" that could not otherwise be conducted over such large spatial and temporal scales. They permit scientists to observe ecological and evolutionary processes in action and with unprecedentedly direct access to how and at what rates these processes unfold.

This is not to say that humans should intentionally go about exterminating species in a morbid grand experiment. It is to say that if human behaviors that are an integral part of valuable human projects impose adaptive pressures, and if those adaptive pressures push some organisms towards extinction, remix others, and create yet others via speciation, then this provides an opportunity to develop an understanding of these processes that might not otherwise present itself.

There might be criteria that justify a preference for knowledge that accrues from the study of currently extant organisms frozen in time, in their current assemblages in their current environments, to knowledge that accrues from the study of flux in all these things. But I am not aware of any discussion that even raises the question of this tension, let alone argues that one or another set of criteria constitute a legitimate basis for determining the answer. In short, with all due respect to Shakespeare and Dante, the problem with the library of organisms is that its value for promoting knowledge can be realized in ways that do not include merely trying to keep every single volume on the shelf.

---

[100] According to Sax and his colleagues, the best evidence shows that the role of invaders in extinction is constrained by many factors bearing on the characteristics of the invader and those of the members of the target community. For discussion of the knowledge-advancing role of species invasions, see Sax and Gaines (2008) and Sax et al. (2007).

6.9   Biodiversity Options

As mentioned in Sect. 6.2 (Biodiversity as resource), Daniel Faith (2007, §1 and §3.2) proposes a conservation axiology based on option value. James Maclaurin and Kim Sterelny, both philosophers of biology, pick this idea up where Faith (2007) leaves it off. In fact, they try to run with what they call (Maclaurin and Sterelny 2008, 154) "The Option Value Option", not as an adjunct to other considerations, but as the sole basis for valuing biodiversity.

I focus my discussion of "biodiversity options" by scrutinizing Maclaurin and Sterelny's account.[101] I justify this focus by the fact that, for all the peculiarities of their approach, these authors are among the few who bring to the discussion a solid scientific grounding, who concern themselves primarily with applying option value to biodiversity, and who go beyond the briefest of gestures along the lines of: "We should keep our options open." Their treatment is also a reasonable basis for a general reassessment of "The Option Value Option" – that is, whether or not option value, properly understood, is a suitable candidate for representing some part of biodiversity's value.

Maclaurin and Sterelny (2008, 154) offer a quick definition of option value:

> [Option value] is the additional amount a person would pay for some amenity, over and above its current value in consumption, to maintain the option of having that amenity available for the future, given that the future availability of the amenity (its supply) is uncertain.[102]

Perhaps this definition is a little too quick. At best, it is misleading insofar as it differs in several fundamental respects from standard definitions of "option value" in the primary economic literature.

---

[101] Maclaurin and Sterelny's treatment of option value (Maclaurin and Sterelny 2008, 149–171) is the subject of Chap. 8, "Conservation Biology: The Evaluation Problem". Chapt. 7 on "Conservation Biology: The Measurement Problem" is preparatory material for Chap. 8's exercise in evaluation. In toto, this treatment of value – their version of *What's So Good about Biodiversity* – occupies over one-fifth of their book, whose remainder is devoted to a detailed scrutiny of their titular definitional question, *What is Biodiversity?*

I am not centrally concerned with how these authors answer the "what" question. However, I believe their treatment of the "value" question is influenced by and very likely partly derailed by building the value into the "what". According them (Maclaurin and Sterelny 2008, 174), the concept of biodiversity is based on species richness (so far, so good), which "… has to be elaborated in different ways for different biological purposes". With "biological purpose", Maclaurin and Sterelny are not referring to some "creative design". Rather, they quite straightforwardly mean, "the purpose of a biological investigator trying to establish some ecological relationship". But this, in turn, means that biodiversity is whatever a biological researcher would like it to mean, so that an ecological relationship established in the research can be said to be about biodiversity. While this in no way trivializes the research or its results, it does seem to trivialize the concept of biodiversity as the basis for a sweeping norm.

[102] Although Maclaurin and Sterelny omit quotation marks and a citation, their definition of option value is, word for word, that of van Kooten and Bulte (2000, 295).

First, it is couched in terms of a premium over "[an amenity's] current value in consumption". Although some economists presume that this "premium" must be positive, others show how, in fact, it can be negative, positive, or undefined; and moreover, that it can be negative even with risk-averse individuals. Calling an amount of money that you would demand in compensation "a premium", which implies an amount you might be willing to pay, prejudices the discussion. Second, the "premium" is, in fact, relative to expected consumer surplus. Expected consumer surplus is not (as supposed in the above definition) the "current value", but rather the *expected* value of future consumption.[103] Third, their definition implies that "the current value in consumption" is a kind of single, fixed price tag. But this is not at all the correct picture. Option value is really defined by reference to expected consumer surplus. Expected surplus is not a single price tag attached to an entity. Rather, it is a statistic computed from values that vary in various possible realized states of the world, taking into account the probability with which each state might be realized. Fourth, their definition, which characterizes the situation as one in which "the future availability of the amenity (its supply) is uncertain", obscures, if not falsifies, the working assumption of most definitions of option value. According to them, the supply is entirely *certain* because determined by the dichotomous choice to conserve or not to conserve. Supply is (assumed) assured by conservation; it is (assumed) zero in the absence of conservation.[104] Fifth, the definition fails to properly attend to the kind of uncertainty that does give rise to option value, according to many economists. That is uncertainty in a consumer's demand for the good, which might or might not be conserved. For better or for worse, in some possible states of the world, the consumer might not want it at all.

Maclaurin and Sterelny give an equally quick definition of quasi-option value.[105] Their acknowledgment of this separate category of economic value is a credit to their discussion because quasi-option value more rarely escapes the confines of specialized economic treatment. But while quasi-option value has a conceptual common ancestor with "garden-variety" option value,[106] it differs fundamentally in explicitly incorporating into the decision model an intertemporal framework, an

---

[103] Economists routinely *translate* the expected consumer surplus into the *present value* of expected surplus by applying the "social discount rate". This discount rate is an ethical hornet's nest that a lack of space advises against touching here. But even the present value of expected consumer surplus is the (expected) value of *future* consumption – albeit in current dollar terms.

[104] Though as mentioned further on in the main text, some treatments of option value do allow that the supply of a resource (such as biodiversity) is not fixed by a choice of development path.

[105] This is the term that Arrow and Fisher (1974, 315) used to introduce the concept. However, other economists, such as W. Michael Hanemann, point out that what Arrow and Fisher call "quasi-option value" is an alternative interpretation of what Weisbrod (1964) originally called "option value". So Hanemann (and some others) use the term "option value" to discuss what Arrow and Fisher and most other economists call "quasi-option value". *Caveat lector.*

[106] Weisbrod (1964) gave rise to the literature that covers all forms of option and quasi-option value.

expectation of acquiring new information, and some notion of "irreversibility".[107] Unfortunately, the authors' Rumsfeldian characterization of this (Maclaurin and Sterelny 2008, 156)[108] – that quasi-option value requires ignorance, not too much ignorance, not too little ignorance, but just the right amount – is rather far removed from the actual theory of quasi-option value that one encounters in the economics literature.

In short, it appears likely that Maclaurin and Sterelny use the terms "option value" and "quasi-option value" to discuss something rather different from what economists discuss in these terms. Divergence from strict economic theory is also evident in the comfort these authors express (Maclaurin and Sterelny 2008, 154) in collapsing garden-variety option value and quasi-option value into a single concept despite the fact that the two notions are based on very different models possessing very different properties. The end result is an unfortunate situation that is nonetheless emblematic of option value-based arguments for biodiversity: If one sticks to definitions of "option value" and "quasi-option value" that fall within the compass of what one finds in the technical economic literature, then there is little reason to think that biodiversity would have some positive amount of it. It is only by first misconstruing what these terms mean, but tacitly assuming an equivalence of these meanings with standard economic ones, that Maclaurin and Sterelny's argument joins many others in achieving some initial appearance of credibility.

This state of considerable confusion is not entirely unsurprising, and the responsibility for it does not originate with Maclaurin and Sterelny. The concept of option value is actually a collection of diverse concepts; and they elude unified understanding even within the field of economics.[109] Though in a minority, some economists (for example, Freeman 1986, 163; Hanemann 1984, 14) dispute whether it is a legitimate, distinct, and useful category of economic benefit – contending that it is

---

[107] See, for example, Hanemann (1984) for a discussion of this. Unfortunately, as previously noted, Hanemann adds to the general confusion by insisting on using the term "option value" to refer to what Arrow and Fisher (1974) originally called "quasi-option value".

[108] I have in mind the theory of epistemic categories propounded by United States Defense Secretary Donald Rumsfeld in a Department of Defense News Briefing on February 12, 2002:

   … as we know, there are known knowns; there are things we know we know. We also know there are known unknowns; that is to say, we know there are things we do not know. But there are also unknown unknowns – the ones we don't know we don't know. [from the official transcript, http://www.defense.gov/transcripts/transcript.aspx?transcriptid=2636]

[109] This is not only my assessment, but also the assessment of the community of economic theorists. The debate about which (if any) conceptions of option value are actually equivalent, which are more general, and which better capture something real and useful in economic theory and practice are ongoing and apparently still unresolved. Some participants in the debate even question whether option value, under any reasonable interpretation, has standing as a legitimate, independent category of economic value. See, for example, Cory and Saliba (1987), whose "Requiem for Option Value" sounds the dirge for the concept of option value, at least as a component of natural value. Hanemann (1984, 14) comes to much the same conclusion.

   In addition to the problem of finding a convincing theoretical basis for option value, it appears that there is no good independent way to measure it. Of course, someone persuaded that option value were a theoretical chimera would not be surprised by the seeming impossibility of assessing its magnitude.

merely an artifact of different ways of computing net benefits. Other economists more specifically find that this suite of concepts has limited application to environmental conservation and is more or less uninteresting in this domain. Even among those economists who agree that option value is "real" and it that it has some general application to nature and the environment, there is disagreement about its most basic properties. For example, my exposition of the concept accords with many economists' conception, according to which option value can be negative (even for a risk-averse individual); but other economists deny that this is possible.

Also, many if not most definitions of option value in the economics literature defy rather than embody the intuitions of non-economists such as Maclaurin and Sterelny. For example, conservationist arguments routinely presume that option value is awarded for avoiding the risk of not conserving. But as economists define it, option value is not so much an expression of risk aversion as it is a choice between different ways of distributing risk. In fact, it is easy to see how circumstances can make conservation the risky choice. Finally, and relating to this last point, it seems that option value, in common with other categories of economic value, has no particular characteristic tendency to favor environmental (or biodiversity) conservation over environmental (or biodiversity) destruction.

In sum, there is considerable evidence to suspect that Maclaurin and Sterelny's nomination of option value to carry the banner of biodiversity's value is based on serious misconceptions and is seriously misplaced. The seeming unawareness of what it actually takes to demonstrate positive option value for biodiversity is especially concerning. Therefore, this section departs from other parts of the book by devoting some detailed attention to what is really involved in the underlying economic concepts of option value and quasi-option value. This should give some better idea about what sort of argument is required in order to establish that either kind of value attaches to biodiversity. And it should make plain how far short of meeting these requirements claims about the option value of biodiversity fall.[110]

First, I explicate the concept of garden-variety option value, henceforth (for the most part) simply "option value". I then show why skepticism concerning its application to the conservation of biodiversity is justified. I repeat this two-part exercise for quasi-option value.[111] In between, I briefly comment on the epistemological vocabulary that enters into these economic discussions.

---

[110] I am aware of no other similar discussion, accessible to non-professional economists, in the environmental and conservation literature.

[111] I must forewarn the reader that my presentation barely scrapes the surface of the conceptual issues that concern option value and quasi-option value. The literature on these two concepts is enormous. Large, too, are the gaps between the accounts of professional economists of how, exactly to define and use it. The economist Richard Bishop (1986, 134), one of the most distinguished contributors to the option value debate, observes that the discussion of option value is "often very technical and confusing". This is a grand understatement. Compounding the problem is the fact that economists sometimes are not the most lucid writers and that they do not adopt a uniform vocabulary or formulation for even the most basic concepts. A full understanding of the controversies requires facility in the concepts and tools of microeconomics, which this book cannot hope to introduce. As a consequence, I cannot provide a general survey or even a broadly balanced analysis.

6.9.1 Option Value and Conservation

As discussed in Sect. 2.1.2.1 (Consequentialism), neoclassical economics posits that satisfying human preferences, no matter their object or reasons (if any) for being held, is what ultimately matters. So it must be for option value, a term that the estimable economist Burton Weisbrod (1964) introduced in a seminal paper. He proceeded by way of a story that is rather odd (to my way of thinking), but which nonetheless captured the imagination of many economists. The story concerns a park run as a business by a private concern.[112] Despite being run with all possible efficiencies, the business is a losing proposition based on the people who actually show up to visit. Moreover, there is every indication that it will remain a loser. Despite this economically gloomy picture, Weisbrod grasps for an economic straw that would yet justify preserving the park. He (Weisbrod 1964, 472) finds it in

> ... the existence of people who anticipate purchasing the commodity (visiting the park) at some time in the future, but who, in fact, never will purchase (visit) it. ...they will be willing to pay something for the option to consume the commodity in the future. This "option value" should influence the decision of whether or not to close the park and turn it to an alternative use.

In other words, Weisbrod suggests that in the park's "option value", we might find a legitimate basis for boosting its total economic value. This hitherto hidden component of economic value is ensconced in the preferences of persons who anticipate visiting the park, though (as Weisbrod allows) they might (or indeed, actually) never do so. These preferences occur in the context of an ability to ensure the certainty of this environmental good's supply in the face of what economists call "uncertainty in demand".[113, 114]

Some context for option value as a category of economic value is provided by relating it to existence value, mentioned at the end of Sect. 2.1.2.1. Although existence value, unlike option value, does not depend on uncertainty, in another basic way, these two categories of economic value are nevertheless kissing cousins.

Therefore the restricted goal of my discussion is to suggest how fragile the concepts are by pulling on a well selected few of the main discussion threads connected to them. For those wishing to plow deeper into this fertile ground, I recommend beginning with Weisbrod (1964) for the seminal idea, Arrow and Fisher (1974) for the original and lucid explication of quasi-option value, and any of the several papers on the subject by W. Michael Hanemann (see references, including Fisher and Hanemann (1986)), who is unusually clear in his thinking and writing on this subject.

[112] Fortunately, this description does *not* really apply to Sequoia National Park, despite the fact that Weisbrod (1964, 471) finds this conceit "useful for the... exposition".

[113] The literature on option value is extraordinarily convoluted on the relationship of uncertainty in demand and uncertainty in supply. I attempt a high-wire balancing act that synthesizes considerations relating to uncertainties in both domains.

[114] This interpretation – in terms of garden-variety option value – is just one of two main schools of interpreting Weisbrod's suggestive but vague story. As already noted, Arrow and Fisher (1974), Hanemann (1984), and Fisher and Hanemann (1986) see quasi-option value as an alternative interpretation.

This is so especially on Weisbrod's formulation, according to which those who anticipate visiting might (or as he says, in fact) never actually do so. Option value and existence value are both types of "passive use" or "nonuse" value, in the parlance of modern economics.[115] In other words, they are both passive ways of "consuming" a resource. In Weisbrod's story, visiting the park is the predominantly non-rivalrous and non-exclusive way of "consuming"[116] the park. This contrasts with its rivalrous and exclusive consumption as a source of timber or granite. When I "consume" an environmental good for its existence or option value, I do not diminish its value to you (consumption in this way is non-rivalrous). Furthermore, there is not a way to exclude you from enjoying the good for its existence or option value. Nor can you be kept from doing this more or less simultaneously with me (the resource is non-excludable). For Weisbrod, it appears that the major respect in which option value differs from existence value is that the former has some pretense of being about a desire (however uncertain) for eventual consumption (even when, *a la* Weisbrod, that desire is never actually consummated), while the latter does not. From this perspective, a non-economist might conceive of existence value as the limiting case of option value when the latter's *un*certainty about demand for a resource fades into the *certainty* of a kind of consumption that is confined to human imagination.

I now continue with a somewhat technical explication of Weisbrod's intuition in the context of a conservation problem, which focuses on a place that can undergo more or less of the kind of development that can affect biodiversity.[117] The ecological euphemism for this is "habitat conversion". The wetland is to be paved over for condominiums. The forest is to be cut down to build those condos. The river is to be dammed (transforming a segment of free-running water into a lake) to provide electricity to the condos. These kinds of conversions oust populations of long-time nonhuman residents. Let's assume that this means that these projects decrease biodiversity. An alternative to pursuing these sorts of development projects, which would realize their economic value in a way that is both rivalrous and excludable, is a conservation project that might possibly preserve the biodiversity.

Let's presume that biodiversity can be quantified and suppose that a quantity $q = Q$ of biodiversity is at stake.[118] The quantity actually conserved can be $0 \leq q \leq Q$. Suppose further that a certain conservation project P will ensure, *with certainty*,

---

[115] This classification predominates but is not universal among economists. See, for example, Vining and Weimar (1998, 322).

[116] I follow economists in using the word "consume" to indicate all uses and even, according to standard economic usage, non-uses.

[117] I am not aware of any other analysis of garden-variety option value with a like focus. However, there have been attempts to apply quasi-option value to biodiversity in the economics literature – for example, Fisher and Hanemann (1986).

[118] Economics lives by this kind of quantified representation, which I model closely on what one encounters in the technical literature. As the discussion in Sect. 4.1.2 (Measures and indexes) there is a major danger in doing this. The choice of what is measured and the weighting of various measurements build in their own set of value assumptions, which thereby evade scrutiny as value judgments.

that q = Q. On the other hand, not pursuing the conservation project (~P) makes it equally certain that q = 0. It is doubtful that these assumptions reflect the actual workings of the world. But they are part of a typical framework for defining option value.

The framework assumptions mentioned so far amount to asserting that there is no uncertainty in supply. To P or not to P (that is, to P or to ~P) – that is the dichotomous question. The answer to it entirely determines q as either Q or 0.

This certainty in supply is part of most economic models of option value. However, option value cannot arise without uncertainty. Many models of it focus entirely on uncertainty in demand; some smaller number incorporate awareness that supply, too, can be uncertain.[119] In the context of our biodiversity conservation problem, demand uncertainty means that certain features of various possible states s of the world can influence Ms. Consumer's *demand* for some level of biodiversity q. These features include[120]:

- Her income (y). In some possible states of the future world, Ms. Consumer loses her job or she loses her life savings in bad investments. Her income plummets, which causes her demand for any level of biodiversity $q > 0$ (along with her demand for most other goods) to likewise plummet. On the other hand, she might win the lottery.

- The prices p_i for various goods i that are complements for the consumption of q – such as the cost of transportation to where q can be "consumed". A fourfold increase in the price of fuel to get to the park might put a significant dent in Ms. Consumer's demand.

- The prices p_j for various goods j that are economic substitutes for q, such as the cost of satisfyingly realistic holographic presentations of wildlife or even goods that compete for the consumer's delight. Deep discounts at the local boutique spa might make getting a facial massage and aromatherapy far more attractive than trying to get up close and personal with q in the personae of mosquitoes, skunks, and bears. Or medications that might derive from the creatures spared by a conservation project might, by the time that they are developed and marketed, be more expensive than alternatives that are born of rational drug design.

- Conditions c_k that affect level of utility that the envisioned consumption of q will actually afford Ms. Consumer. It is possible that she will find Weisbrod's park thronging with rude, rowdy, music-blaring yahoos, outnumbered only by the mosquitoes feasting on every inch of exposed (and even unexposed) human flesh. That might take a bite out of her demand, too. Even aside from these obnoxious possibilities, her terrible sense of direction might result in her getting lost and lead to a frightening, even near-death experience. Or by the time of her actual visit, her bad and worsening knees might prevent her from getting out of her car. On the other hand, by the time of her prospective visit, park officials might have

---

[119] Freeman (1986, 154–155) is as clear as any economist in allowing for both supply and demand uncertainty.

[120] This list is drawn from various sources, including Plummer and Hartman (1986), Freeman (1986), and Hanemann (1984).

clamped down on rowdy miscreants, sprayed for mosquitoes, and constructed trail signs that make it virtually impossible for a literate person to get lost; and successful knee surgery might have made her more mobile than ever.

On the other hand, if the primary value of the biodiversity lies in its potential pharmacological value, then this condition of being the source of a valuable drug dominates its level of utility.

I encapsulate these ideas by saying that the conditions c_k are ones that affect the "quality" of biodiversity.

- Ms. Consumer's very preferences (reflected in the shape of her utility function u()) might change even if the above-listed factors remain constant. She might change her mind about just how much she likes biodiversity. She might come around to thinking that she is much happier not exposing herself to the possibility of getting lost. She might decide that the risk of severe sunburn, poison oak rashes, and insect bites is too much to bear. Or, on second thought, the prospect of making do without running hot water and flush toilets is just too awful to contemplate. This change of heart is not difficult to understand. Even now, most of her friends incredulously ask why she would voluntarily submit to such indignities when, with far less effort and similar expense, she could go to the spa for that aromatherapy and a facial massage. In short, Ms. Consumer's preferences might themselves change and be different in different possible states of the future world.

Subsequent states of the world might vary in any one or more of these conditions. In the economic model for option value, a possible state s of the world is essentially just a possible state of demand. Combining the prices of complement and substitute goods, the composition of states is expressed by:

$$s = \left(y, < p_1, \ldots, p_i, \ldots p_n >, < c_1, \ldots, c_k, \ldots c_m > \right)$$

for some collection of prices p_i on n complementary and substitute goods, and some other collection of m demand-influencing ("quality") conditions c_k.

Suppose that u() is a utility function that expresses Ms. Consumer's preferences as a function of factors that might influence her demand. Suppose further that y_q is the net of income and costs associated with the quantity q of biodiversity. Since the model supposes that the choice of doing or not doing the conservation project entirely determines the quantity of biodiversity q to be either Q or 0, respectively, y_q is just the net cost/income of P or ~P, assumed to be independent of s. That is:

$$y_Q = y_P$$

and

$$y_0 = y_{\sim P}$$

Note that a lot is hidden inside y_q. In particular, preserving q=Q entails the opportunity costs of forgoing all the benefits of development that q=0 would permit. Reserving the stock of a resource for future consumption often precludes its immediate and possibly wealth-compounding investment in alternate lines of development – those condos, for example. These potential benefits foregone constitute the conservation project's opportunity cost. Though it is not the only factor that affects the value of y_p, this alone could diminish y_p (or increase y_{-p}) and it could even make y_p strongly negative.

In this framework, one can characterize the utility U to Ms. Consumer of a certain state s of the world, which influences her demand for quantity q of biodiversity, as:

$$U = u\left(y_q, s \mid q\right) \qquad \text{where '|' means "given the following condition(s)"}$$

As s has been defined, uncertainty about which state s of the world Ms. Consumer will find herself in gives rise to uncertainty in her demand for q. It is good to keep in mind that this is a simplification, which ignores the less frequently considered source of uncertainty about her demand for q: u() itself might not be fixed (last in the list of demand-affecting factors above).

Now suppose that each state s occurs with probability π_s such that $\Sigma\pi_s = 1$. Note that this apparently mundane condition implies a startlingly strong epistemic claim: The set of all relevant alternative states is known and so is the probability of occurrence for each state in this complete set. Continue to assume that the supply of biodiversity is certainly Q with the conservation project P and it is certainly 0 without it (~P). Then the expected utility EU_p for Ms. Consumer with P is the sum of the utilities for each state s, weighted by that state's probability of occurrence:

$$EU_p = \sum \pi_s u\left(y_p, s \mid q = Q\right)$$

On the other hand, her expected utility EU_{-p} in the absence of the project (~P) is:

$$EU_{-p} = \sum \pi_s u\left(y_{-p}, s \mid q = 0\right)$$

So much for the preliminaries. To see how uncertainty figures in creating option value, let's start by assuming that there is none – and particularly, that there is no uncertainty in demand. This is equivalent to saying both that Ms. Consumer's utility function u() is fixed and that u() is constant-valued in s. That is, Ms. Consumer's demand for any level q of biodiversity, like its supply (though contingent on the choice of P or ~P), is fixed and certain:

$$U_p = u\left(y_p, s \mid q = Q\right) \quad \text{for every state s}$$

$$U_{-p} = u\left(y_{-p}, s \mid q = 0\right) \quad \text{for every state s}$$

In this case, the *expected* utility EU in each case is trivially identical to the utility for any, arbitrarily selected state:

$$EU_p = U_p = u\left(y_p, s \mid q = Q\right) \quad \text{for every state s}$$

$$EU_{-p} = U_{-p} = u\left(y_{-p}, s \mid q = 0\right) \quad \text{for every state s}$$

because (per the tentative assumption of demand certainty) the utility for any state s is the same as that for any other state.

What would our certain demander pay up front – that is, *ex ante* – for the conservation project? Assuming that she is an individual of the species *Homo economicus* – that is, she is a preference maximizer who evaluates any set of alternatives by doing a cost-benefit analysis according to the rules of neoclassical economics – she knows the price. That price is the "option price" OP, which satisfies:

$$u\left(y_P - OP, s \,\middle|\, q = Q\right) = u\left(y_{-P}, s \,\middle|\, q = 0\right) \quad \text{independent of state s}$$

OP is the answer to a single question that spans the entire set of possible states s. Beware: we still have not gotten to "option *value*".[121]

Whether OP is positive, negative, or zero depends on the shape of Ms. Consumer's utility function $u()$.[122] That is, it depends on whether $u()$ is higher or lower when the amount of biodiversity is some nonzero quantity Q versus 0. Ms. Consumer might prefer a more biodiverse world. In that case, $U_P > U_{-P}$, which means that she would pay for the conservation project (OP > 0). But nothing *a priori* justifies assuming that her preference is for a more biodiverse world. Nor does any contingent matter of fact about the world have this universal implication. And in fact, she might strongly prefer the benefits of development and regard a nonzero level of biodiversity as an obstacle to that more desirable state of affairs. In that case, $U_P < U_{-P}$, and she would *demand payment* to allow the nonzero amount Q of biodiversity (OP < 0). Of course, if $U_P = U_{-P}$, then she will not be inclined to lobby either for or against conservation.[123]

The option price OP is one way to price the conservation project P. There is an alternative approach to pricing the project, which hinges on the answer to a different

---

[121] Unfortunately, in one of a number of terminological confusions that plague this topic, what Weisbrod calls "option value" in his paper is now called "option *price*" by most, if not all economists, who follow the usage of Krutilla et al. (1972). I follow this now-more-common usage.

[122] As Plummer and Hartman (1986) help to explain further along in the main text, this concept of option price differs in significant respects from the concept more familiar from options in stock and other real markets. See Note 130.

[123] I cannot account for the statement in van Kooten and Bulte (2000, 296) that "OP... is always positive", unless these authors are fixated on the narrow sense of option that applies to stocks. Otherwise, their claim would seem to require *a priori* justification of a proposition that is a contingency and that can tip either way. Empirically, one finds that development (no conservation) sometimes *is* preferred. In such a case, a consumer would demand payment for allowing the conservation project. That is, the option price would be negative. None of this depends on how risk-averse or risk-welcoming the consumer might be.

I detail this because, as previously mentioned, van Kooten and Bulte (2000) appear to be a primary economic reference for Maclaurin and Sterelny (2008). I have not encountered another explicit statement of this erroneous assumption. But it is possible that this assumption is nevertheless implicitly made by those who argue for the (positive) option value of the natural world. If it is (erroneously) supposed that there is no downside (negative option value), then the case basing conservation decisions on option value will (illegitimately) appear much more palatable.

question: The alternative imagines Ms. Consumer dwelling separately in each of the possible states s of the world. For each state, it asks her as part of that state, what amount would she be willing to pay (or demand in payment) for enjoying (or suffering with) a quantity q=Q (versus q=0) of biodiversity in it. The number of questions is equal to the cardinality of the complete set of possible states s. The answers to these questions jointly produce an *ex post* (versus the previously described *ex ante*) valuation in the sense that each separate valuation is contingent on one particular state being realized and Ms. Consumer's projecting herself into it. This amount satisfies an equation that looks suspiciously like the one for option price:

$$u\left(y_P - CS_s \mid s, q = Q\right) = u\left(y_{-P} \mid s, q = 0\right) \quad \text{for a given state s}$$

CS_s is what economists call the "consumer surplus" for state s.[124]

Of course, so long as one presumes that the value of the utility function u() itself is fixed and that its value does not vary from one state s of the world to another, Ms. Consumer will pay (or demand payment of) the same amount CS_s in any state s in order to arrange for (or tolerate) q=Q. This resembles the situation for OP. However, OP is the answer to a single, *ex ante* question, and is therefore the same for each state *by definition*. By definition, the answer to the option price question is found by asking Ms. Consumer what she is willing to pay up front for the conservation project, independent of state. In contrast, each CS_s is the answer to an *ex post* question that is contingent on state s being realized.

So far, CS_s only happens to be the same in each state as a consequence of the tentative initial assumption about a matter of fact – that no difference in any possible state of the world affects Ms. Consumer's demand. This assumption entails that it doesn't matter which state s she projects herself into: Her consumer surplus will be the same in every one. As a consequence, the probability π_s with which each state occurs is irrelevant and the *expected* consumer surplus ECS is identical to CS_s for any state s:

$$ECS \equiv \sum_s \pi_s CS_s = CS_s \quad \text{for any state s}$$

Obviously, under the conditions so far specified – which make Ms. Consumer's demand for biodiversity certain and the supply of biodiversity also certain – the equations that define CS_s and OP make them identical for all s. Therefore, the expected consumer surplus ECS and option price OP are identical, too.

I am finally in a position to state the definition of "option value". Option value is (after all that complexity) quite simply the option price OP less the expected consumer surplus ECS.[125] Since OP and ECS are identical under the assumptions so far made, they entail (under those assumptions) that the option value of biodiversity is nil:

$$OV \equiv OP - ECS = 0$$

---

[124] While this definition might look somewhat unfamiliar, it encapsulates similar (though notationally different) definitions in, for example, Plummer and Hartman (1986, 458), Freeman (1986, 160), and Vining and Weimar (1998, 327).

[125] This definition of "option value" is widely accepted among economists.

I have already shown that option price can be either positive, negative, or zero – that is, indeterminate in sign. Option value can be any of these, too. But clearly, something must be uncertain for option value to be nonzero.

Suppose that some combination of the demand-affecting factors listed above makes Ms. Consumer's demand for biodiversity depend on which state of the world is actually realized, so that generally, $CS_{s1} \neq CS_{s2}$ for different states s1 and s2. In this case, it might be thought that paying the option price up front is a kind of insurance that reduces risk. But this is not your normal insurance, in which one knows with near certainty the undesirability of, or negative demand for, the state of the world for which the insurance payout compensates.

Paying an option price OP up front that is greater than the expected consumer surplus ECS is not insurance, but rather a matter of trading one uncertainty for another. Of course, by paying OP, regardless of which state s of the world obtains, Ms. Consumer avoids the consumer surplus lottery. She might be so inclined by considering that the lottery would typically require her to pay a very high price for a level $q = Q$ of biodiversity in a state of the world in which she finds that this level of biodiversity highly desirable.

But things might turn out in a way that makes the high level of $q = Q$ of biodiversity highly *un*desirable for her. That, in turn, makes her choice to not enter the consumer surplus lottery itself a bet with its own risks – for she might wind up paying a lot for something that, as it turns out, is repugnant to her. Specifically, paying OP is a bet that the state of the world will turn out to be one in which her demand for biodiversity is, in fact, high. In this case, the price she pays *ex ante* for a high level of biodiversity $q = Q$ is lower than she would have paid *ex post*. It is also a bet that Ms. Consumer will *not* end up in another state of the world in which biodiversity $q = 0$ is more desirable to her than $q = Q$. Viewing things retrospectively from that possible world, she will see (*ex post*) that she erred by not *demanding* payment for allowing a conservation project that ensured $q = Q$. In this alternative world of unwanted biodiversity, Ms. Consumer would regret having paid a positive (option) price for it. In other words, paying the option price up front is also a gamble. Like any gamble, it can be lost.

I now show informally, by way of example, how option value can be negative or indeterminate. In other words, the difference between OP and ECS is not so much a matter of risk aversion as it is a matter of preferring one kind of risk to another. In fact, the examples illustrate how risk aversion can cut either way.

Initially, consider four states that vary in just one demand-altering variable – namely, the "quality" (as perceived by a consumer) of the quantity $q = Q$ of biodiversity.[126] Suppose that there are two "components" to this quality. One component has

---

[126] The choice of a model that contains more than two possible states is significant. There is a substantial body of theoretical work based on a two-state model that suffices to make most of the points that are salient for my discussion. This theory demonstrates how option value can be negative or positive – depending on a consumer's risk preferences and on the shape of her utility curve. My treatment leapfrogs over this work to a more realistic multi-state model, which also illustrates how option price and therefore option value might be neither positive nor negative, but indeterminate.

Table 6.1 A hypothetical set of states reflecting uncertain demand for biodiversity based solely on the uncertain "quality" of a given level of biodiversity

State (s)	s1	s2	s3	s4
Probability (π_s)	0.19999	0.40001	0.39999	0.00001
Surplus (CS_s)	−700	−10	360	10,000

to do with the experience of actual encounters with various creatures.[127] The other has to do with the possibility of Q amount of biodiversity yielding some very valuable drug.[128] Table 6.1 shows the four states, the probability that each will occur, and the consumer surplus for the conservation project in each state.

In this example, a 60% supermajority of the time, the consumer surplus for conservation is negative (states s1 and s2). Moreover, there is a one-in-five chance that the surplus is very negative (s1). However, there is also a good but less-than-even chance of a substantial positive surplus (s3). And there is a tiny chance – 20,000 times less likely than the very bad outcome of s1 – of a stupendously large positive surplus (s4). The story behind this example could be: In s1, development (for which the biodiversity is sacrificed) turns out to be an excellent investment and pays a handsome dividend. In s2, development does okay and better than conservation. In s3, people find that they really like to be around the mosquitoes, skunks, and bears. And in s4, a conserved plant is found to produce an alkaloid that, without side effects, provides immunity to leishmaniasis. The numbers in Table 6.1 are cooked to yield $ECS \approx 0$, just because zero is an easy number to think about. But the basic story arguably reflects possible real-world economic probabilities and valuations (within the limits of economic modeling).

To find the option value of Q amount of biodiversity, one needs to find the fixed (state-independent) option price OP that is willingly paid with the knowledge of which states are possible and the knowledge of how likely it is for each state to occur, but without knowing which state does, in fact, obtain. In this example, since $ECS = 0$ and $OV = OP - ECS$, option value $OV = OP$. As a consequence, the question of whether $OV > 0$ has the same answer as the question of whether $OP > 0$. This lets us focus on the question of whether or not, under these conditions, a consumer would willingly pay *anything* up front for the conservation project. Such willingness would confer a positive option value on Q.

So what is the option price (which is also the option value) in this case? Plainly, most of the time (states s1 and s2 combined), the quantity Q of biodiversity is *not* demanded. In fact, there is a substantial chance (in s1) that Q is considered a very bad thing. Given this, would a consumer be willing to pay something for the slight chance at a jackpot payoff (s4), or even a fairly good, but less-than-even-chance at very good outcome (s3)?

---

[127] The "experiential value" of biodiversity is separately examined in Sect. 6.11.

[128] Biodiversity as pharmacopoeia is explored in Sect. 6.5.1.

Does aversion to risk push the price into positive territory or the opposite? One way to understand risk aversion is in terms of its equivalence to the diminishing marginal utility of income. A risk-neutral consumer is indifferent to the marginal utility of her income. But a risk-averse consumer prefers to spend income under conditions in which her income is high and her income's marginal utility is low.[129] This equivalence does not (yet) come into play because for now, I am supposing that only the "quality" of biodiversity affects demand for it. However, taking the equivalence seriously, one can see that a person's risk aversion alone does not dimininish the attraction of betting on a jackpot that is is extremely unlikely to be collected (s4). Being risk-averse comes into play only if a chance at the jackpot means substantial exposure to another, very bad outcome. That is exactly the situation in the example at hand. One must assume that a risk-averse person is just the kind of person who might decline a shot at s4 because it only comes with a substantial exposure to the very bad outcome in s2.

In a circumstance such as this, with an expected surplus of zero, would a person averse to risk be inclined to pay some positive sum in order to avoid the consumer surplus lottery? I believe that there is no determinate answer. It is plausible to think that a consumer might *demand* payment to endure exposure to the most likely outcome, which is negative, especially when there is a one-in-five chance of the outcome being very negative. Perhaps it is only a risk-*seeking* individual who, under these conditions, would pay for a chance to win the s4 jackpot.

Up to this point, I have focused on uncertainty in demand. The choice between a state-independent *ex ante* payment and payment *ex post* becomes even more murky with the introduction of uncertainty in supply. In the real world, a conservation project will not ensure a level $q=Q$ of biodiversity. Nor is $q=0$ certain in the absence of a conservation project. Rather, conservation will merely shift to the right the distribution of probabilities for q, such that $0 \leq q \leq Q$. In general, it seems that this uncertainty in the supply that remains even after a conservation choice has been made can only lead to an increased reluctance to commit funds up front for conservation. Though much more can be said about supply uncertainty, I only briefly return to it when considering income as another source of demand uncertainty.

So far, I have supposed that the "quality" of biodiversity entirely determines demand for it, as specified in connection with Table 6.1. But further complications can and are likely to arise from other factors that affect demand. For example, the price of a complement good could be a major factor. Suppose that the demand for the drug that emerges in s4 is affected by the affordability of health insurance, which might be the only or primary means for affording access to the drug. If the price of insurance were so high as to routinely make it unaffordable, then demand for the drug might diminish

---

[129] For example, Freeman (1984, 3) takes risk-aversion in state s to be equivalent to the condition that

$\partial u_s / \partial y_s < 0$

Table 6.2 A hypothetical set of states reflecting uncertain demand for biodiversity as in Table 6.1, but also taking into account the cost of a complement good

State (s)	s1	s2	s3	s4
Probability (π_s)	0.19999	0.40001	0.39999	0.00001
Surplus (CS_s)	−700	−10	360	50

to miniscule levels. That would similarly diminish consumer surplus for level Q of biodiversity in s4. In this case, Table 6.1 might become Table 6.2.

Because the likelihood of s4 is so small, the expected consumer surplus ECS remains the same as it is for the Table 6.1 example – zero. But now, even a consumer fixated on the jackpot in the scenario for Table 6.1 does not have this temptation to motivate her to pay to ensure that q=Q.

The consumer's expected income might also have a gross effect on her willingness to pay a price *ex ante* from (perhaps) a lesser income and when no benefit is ensured, versus paying *ex post* from a greater income and when benefit for payment is ensured. Suppose that biodiversity is a normal good – that is, a good for which (at a constant price) demand increases with increasing income. Suppose also that income is likely to increase with no conservation as the result of the investment in development. Finally, suppose that some level of biodiversity is likely to remain even in the absence of a conservation project (supply uncertainty). Under these conditions, option value is likely to be *negative* for a risk-averse person and positive for a risk seeker. The world of higher income is a world of higher demand for any level q of biodiversity. At higher levels of income, the marginal utility of income is lower. A risk-averse person who prefers to spend dollars with lower marginal utility will prefer the choice of only having to pay for the good of some level q of biodiversity in a world in which she actually demands it; and then, preferably with low marginal utility dollars.

What conclusion can be drawn from all this? Economists Mark Plummer and Richard Hartman (1986, 464–465) arrive at a fair summary in their attempt to find a "big tent" version of option value:

> … "option value" has very little to do with the value of an option.[130] Instead, if a price change is proposed under uncertainty [a source of demand uncertainty that I have not put in my examples], option value is a measure of the premium or discount a consumer is willing to pay or accept to purchase the price change by making a constant payment of $\overline{S}$ [the expected consumer surplus, which I have designated 'ECS'] rather than a payment of S(T) in each state [which I have designated 'CS$_s$'] of the world. Although both methods have the same effect on expected wealth, the variation in S(T) [i.e. CS$_s$] may provide the consumer with additional benefits or harm (relative to the constant payment of $\overline{S}$ [ECS]) because the payment varies as the marginal utility of income varies.

---

[130] Options on publicly traded stock differ from options on environmental goods in other obvious respects. For the former, only the option premium (representing the option value) is due up front. The option confers a right to exercise without the obligation to actually take that action. Therefore, the option might or might not be exercised later; full payment of the option price is due only if the option is exercised. The model of option value for environmental goods typically requires up-front payment of some more substantial portion – and sometimes all – of the option price. That price is presumed to include the cost of the resource-conserving project.

> In essence, option value is a measure of the value of one institution for diversifying risk relative to another… …faced with a choice between two payment methods with the same expected dollar value, a risk-averse individual will choose that method which has greater success in diversifying the risk faced by the consumer…
> This reasoning is very different from that originally envisioned by Weisbrod.

My treatment of the subject minimally supports a similar conclusion regarding option value as it applies to biodiversity: First, in reasonably realistic conditions involving more than two possible contingencies (possible states of the world), the option value of biodiversity, not unlikely, is indeterminate. Second, even if risk aversion is rational and warranted (a proposition that requires independent justification), this does not imply that the option value of biodiversity is positive. To put the previous point in another way, the redistribution of risks afforded by a state-independent and fixed *ex ante* payment has risks of its own. The end result can just as well be an *increase* in risk (favoring immediate "consumption" of biodiversity via development) as it can be a decrease in risk.

Third, the dependence of option value – either positive or negative – on risk preference makes it all the more difficult to reasonably invoke it as a justification for some choice. It is hard to imagine any decisive reason for why one attitude towards risk might be more rational or more justified than other attitudes. Even if a preponderance of persons adopt one attitude in preference to the others, this does not constitute a *justification*. And in fact, there will be some distribution of risk-related preferences – ranging from risk-avoiding through risk-indifference to risk-loving – in society at large. From the economic viewpoint, the sign and size of option value will depend on this variable distribution. Even in a case where risk aversion entails a positive option value, if there is no bias in society towards risk aversion, then there is no net premium for added risk.

But even if one were mistakenly inclined to suppose that a predominant preference for risk aversion in a population underwrites a moral imperative to honor such a preference, the case for positive option value remains in jeopardy. That is because, as a matter of fact, aversion to a risk diminishes as more and more persons bear it. Even for a significant risk, the aggregate of the premiums goes to zero as the risk is distributed.[131] This consideration can reduce to irrelevance risk considerations in making a decision – even when the decision affects a population whose individuals are generally risk-averse in their own personal decision-making.

It is undoubtedly a relative strength of quasi-option value that it does not build in any assumptions about risk preferences. At least in this way, it is less problematic than garden-variety option value.

One final point, which saliently bears on the application of option value to environmental or natural goods, directly relates back to its basic definition in terms of the difference between option price and expected consumer surplus. When it is positive, option value represents a premium over the expected consumer surplus.

---

[131] This is the famous Arrow-Lind Theorem, due to (surprise) Arrow and Lind (1970).

Sometimes a willingness to pay this premium for a conservation project requires a risk-avoiding preference; other times it requires a risk-seeking preference. Either way, the fact that it is a premium over the expected surplus should be a source of discomfort for an environmentalist: It means that biodiversity must do *more* than compensate for the expected surplus of the conservation project. It must somehow redistribute uncertainties in a way that justifies the additional premium represented by the option value. In other words, option value places an additional burden on environmental or natural goods to show their conservation-worthiness.

6.9.2 *Risk, Uncertainty and Ignorance*

The discussion of option value in the preceding subsection and the discussion of quasi-option value in the next subsection adopt a usage of "risk", "risk aversion", "uncertainty", and (in the case of quasi-option value) "ignorance" that is customary in the economics literature. Unfortunately, this usage levels important epistemic distinctions, which are critical in other, related discussions – particularly discussions of the Precautionary Principle.[132]

Here is one (European Environment Agency 2001, 170, Box 16.1) possible way to restore a proper perspective, which brings those distinctions back into relief:

...there is the familiar condition of **risk**, as formally defined in probability theory. This is where all possible outcomes are known in advance and where their relative likelihood can be adequately expressed as probabilities. Where this condition prevails, risk assessment is a valid technique...

Under the condition of **uncertainty**, as formally defined, the adequate empirical or theoretical basis for assigning probabilities to outcomes does not exist. This may be because of the novelty of the activities concerned, or because of complexity or variability in their contexts. Either way, conventional risk assessment is too narrow in scope to be adequate for application under conditions of uncertainty...

Many case studies... involve examples where... appraisal laboured not only under a lack of certainty as to the likelihood of different outcomes, but where some of the possibilities themselves remained unknown. Here, decision-making is faced with the continual prospect of surprise. This is the condition formally known as **ignorance**. Even more than uncertainty, this underscores the need for a healthy humility over the sufficiency of the available scientific knowledge and, crucially, for an institutional capacity for open reflection on the quality and utility of available bodies of knowledge...

Once it is acknowledged that the likelihood of certain outcomes may not be fully quantifiable, or where certain other possibilities may remain entirely unaddressed, then uncertainty and ignorance, rather than mere risk characterise the situation. The adoption of robust, transparent and accountable approaches towards the various aspects of risk, uncertainty and ignorance can be identified as one crucial means of regaining public confidence in regulatory decision-making. [bold in the original]

It is useful to see how these epistemic categories relate to the previous (and subsequent) discussion of economic value. Cost-benefit analysis in general, and the

---

[132] See Sect. 6.3 (Biodiversity as service provider) for a discussion of the various elements that enter into Precautionary Principles.

concept of option value in particular, are built on the notion of expected net surplus and related notion of "expected" values. Expected values are assessed on the basis of some high degree of epistemic assurance concerning what states of the world can possibly obtain and with what respective probabilities. To put this in another way, the most clearly legitimate domain of economics is the domain of decision-making under *risk*, according to the above set of definitions. Insofar as this epistemic condition is not satisfied – because uncertainty or ignorance (both defined as above) is involved – economics wanders out into territory where its base assumptions do not apply.

In light of this, the entire discussion of option value in the preceding subsection should be read as a discussion of economic value under standard conditions of economic risk, not uncertainty, strictly understood. Part of this point's importance will become apparent in the discussion of quasi-option value, which follows.

6.9.3   Quasi-option Value and Conservation

Quasi-option value attempts to go beyond the domain of decision-making under risk (strictly understood) to decisions made under uncertainty or ignorance or both (again, strictly understood). In this respect, it represents a substantial extension of the framework for economic analysis. There are several other respects in which the defining framework for quasi-option value radically departs from the framework for cost-benefit analysis generally, and from the framework for garden-variety option value particularly.

I highlight these distinguishing characteristics of the quasi-option value framework in developing an account of how it applies to a conservation problem that resembles the one that focused the discussion in Sect. 6.9.1 (Option value and conservation).[133] Once again, the problem is to choose whether to conserve or to develop some place that harbors some amount of biodiversity.

Time element. First, the decision model explicitly incorporates a time element that is fundamentally absent from the framework for garden-variety option value.[134] I adopt the classic formulation (of quasi-option value), in which there are two decision points separated by a period of time, period 1. Period 2 is a time period that follows the second decision point. One can easily imagine how this decision model could be extended to more than two decision points.

Irreversibility. Second, the model applies only to development that is assumed to be "irreversible".[135] "Irreversibility" is a slippery concept. Here, it suffices to suppose that the quasi-option framework applies only in cases where, in some sense,

---

[133] I follow the seminal formulation of Arrow and Fisher (1974), as further articulated by Hanemann (1984), Fisher and Hanemann (1986), and Hanemann (1989).

[134] Hanemann (1984) and elsewhere emphasizes its temporal element as a distinguishing feature of quasi-option value.

[135] The "irreversibility" assumption for quasi-option value is typically introduced by way of an informal example. For Arrow and Fisher (1974), it is the building of the Hells Canyon Dam on the Snake River on the Oregon/Idaho border in the United States. It is never carefully defined. Most explications of quasi-option value beg the question of what "irreversible" means by stating that it means "reversal would be extremely costly" (Vining and Weimer 1998) or something similar.

development in period 1 cannot be undone in period 2. That is, it applies only when one can assume that development is forever. This assumption is itself tied to specific assumptions about costs and benefits in period 2, which I visit momentarily.

The irreversibility requirement for quasi-option value precludes its application to some decision problems. But the specific concern here is with the conservation of biodiversity. Is development, as it affects the conservation of biodiversity, irreversible?

It is plausible to suppose that more development means less area for at least some species; and that the species-area relationship makes it plausible to conclude that less area means lower levels of biodiversity.[136] So let us grant the premise that development leads to extinctions or extirpations of local populations. Let's also grant for this discussion the stronger claim that "lower level of biodiversity in a developed area" means "*permanently or irreversibly* lower level of biodiversity". Then, insofar as development permanently reduces biodiversity, it passes the irreversibility test.

In sum, the working assumption is that development in period 1 reduces biodiversity. This, in turn, entails forgoing for all time the benefits (and costs) that might otherwise have accrued to the amount of biodiversity that the development permanently removed. These conditions jointly constitute the "irreversibility" of development and they are elaborated below, under **Costs and benefits of information and irreversibility**.

Information. The role of information in reducing uncertainty (strictly understood) is explicitly recognized. It is assumed, quite plausibly, that more information is available to inform decisions made at the second of the two decision points. By the time of that second decision, the benefits and costs experienced or uncovered as a consequence of the first decision are better known. New benefits and new costs might be uncovered. Also, better estimates might be found for the likelihood of realizing any benefit or cost (whether newly discovered or not). In other words, information gathered between the two decision points can change the computation of the expected value of development versus that of conservation.

Couched in the epistemic vocabulary of the immediately preceding subsection, quasi-option value is a creative suggestion for how to frame cost-benefit analyses (based on conditions of known risk) under conditions of true uncertainty (when the probabilities are not well known) and ignorance (when the range of possibilities is not well known).

Controversy surrounds the important question of whether or not information critical for making the best second decision can derive from development undertaken in period 1. It is unclear why this is at all controversial. Surely, sometimes, something can be learned about the net benefits of investing in a development proposal

---

Unfortunately, "irreversible" admits multiple meanings; running them together can and probably does undermine clear discussion.

For a valuable start at picking apart the concept of irreversibility, see Manson (2007).

[136] This crude analysis is the basis for much reasoning about extinction rates, which He and Hubbell (2011) have recently called into question. Insofar as this principle is an unsure basis for diversity estimates, it removes one plausible basis for thinking that quasi-option value analysis applies to the real world.

by undertaking some part of it. This implies that at least some learning is or can be endogenous to the decision problem. It also implies that the quasi-option value framework might well justify undertaking development in period 1 specifically in order to provide information that could inform a more nearly optimal choice for period 2 – a choice that might also prove to be more nearly optimal for both periods taken together.

Therefore, it is somewhat baffling that a number of theorists presume that the information that informs decision 2 must concern only the existence or magnitude of *conservation benefits*. And furthermore, this information is presumed to flow from research that is exogenous to the decision problem. In cases that concern the conservation of biodiversity, the rather weak rationale for these presumptions (Fisher and Hanemann 1986, 178) is that relevant information about the properties of indigenous species will not come from developing their habitat; rather the information will flow from research that is undertaken quite apart from the development decision and the framework for making it. But this argument ignores the fact that engaging in even partial development can help to assess how much benefit the development might actually provide as well as the magnitude of costs that might become apparent as a result of some partial failure to conserve.

The question of whether and to what extent endogenous information plays a role in the framework for quasi-option value is significant. That is because observation of how actual development in period 1 goes could play a significant role in informing the second (and any subsequent) decisions regarding how much to develop. Therefore, this information could be the basis for deciding at the first decision point to proceed in period 1 with some development (as opposed to none) or with a greater rather than a lesser amount. However, the answer to this question might have less effect on the basic shape of the quasi-option value proposition than some other considerations that I consider shortly.

Risk preferences. Quasi-option value radically parts with (garden-variety) option value in being completely divorced from questions of risk preference. This is a good thing insofar as it is hard to find a general justification for either a preference to avoid risk or to seek it.

Arrow and Fisher (1974, 318) acknowledge that "there is something of the 'feel' of risk aversion... by a restriction of reversibility." But wherever that "feel" comes from, the inducement to utilize newly acquired knowledge does not appear to depend on whether individuals prefer to avoid risks or seek them. In comparison to those hard-to-justify preferences, the rational basis for incorporating into the next decision knowledge acquired after the last one has the "feel" of solid footing.

Costs and benefits of information and irreversibility. One succinct characterization of quasi-option value is that it is the value or expected value of information that is conditioned on less, rather than more, "irreversible" development.[137]

---

[137] See the discussion of irreversibility by Fisher and Hanemann (1986, 179). An even more general definition of quasi-option value along these lines might be one that makes it the difference in the expected value of information along different development paths – with the conservation path regarded as an alternative , "biodiversity development" path.

The plausibility of this proposition rests on combined presuppositions regarding irreversibility and information that frame quasi-option value.

First, the framework presumes that development (itself) is forever in the sense that if undertaken, the model disallows "undevelopment". It also presumes that the costs and benefits that arise from development are never subject to revision. Once attained, benefits continue to register as beneficial for all time; and similarly for costs. On the other hand, so long as development is not undertaken, the *possibility* of doing so later is also presumed to remain for all time.

One or another of these assumptions might give pause. But however plausible they are as independent propositions, they jointly entail that any period 1 cost or benefit, which derives from period 1 development, extend to period 2 where they apply at precisely the same levels. Then during period 2, period, these period 1 costs and benefits combine with those that derive from any additional (period 2) development.

Unfortunately, this pleasingly neat cumulative model skates past several uncongenial facts: Some investments yield compound returns whose greatest benefits are realized only long after the initial investment. Second, some investments yield unexpected benefits. And third, investments sometimes lead to other, otherwise unrealizable and even previously unknown investment opportunities. The model also overlooks the frequently encountered real-world circumstance that some investments have a limited window of opportunity. The decision not to develop now is sometimes practically or effectively reversible. All too often, though, the required alignment of will with resources occurs just once and for all time.

Second, this separation of what is separate – between the irreversibility of development (on the one hand) and the irreversibility of its costs and benefits (on the other hand) – can be applied to conservation, too. While conservation itself might be thought to be reversible – because development is always an option – its costs (for example) might not always be so. This is not just due to permanently forgone opportunities in development, which are connected to the first presupposition. For example, if, as seems likely, a more biodiverse state of a world in which humans also manage to insinuate themselves into every nook and cranny is more likely to be one with a world of frequent and severe human pandemics, then the loss in human life and morbidity might be the sort of compound loss that is never truly recouped.

Third are some surprising and unjustifiably asymmetrical presuppositions concerning exactly what information newly available at the second decision point will reveal about costs and benefits for period 2. The entire focus is on previously unknown benefits of biodiversity and its conservation, not on any previously unknown costs of keeping it around. As Maclaurin and Sterelny (2008, 154) put it:

> … as our knowledge improves (and our circumstances change) we will come to discover new ways in which species can be valuable.

Routinely ignored is a similar possibility that newly acquired knowledge – for example about how complex zoonotic disease systems work – might uncover new ways in which conserved species might come back to bite us people.

It is also routinely presumed that new information will *not* reveal ways to attain "lost" benefits by other means. For example, if it is thought that biodiversity is the future's pharmacopoeia, then we currently respiring decision-makers must consider the question of how long we can justify the conservation choice that lets us wait to uncover this benefit. This requires some realistic assessment of expectations for exogenously acquiring the knowledge to derive the same benefit within this timeframe in an alternative way – for example, by dint of fast-developing technologies for rational drug design.

In other words, whatever support quasi-option value gives to conservation relies on a particular and apparently biased view of what kind of information might be forthcoming, combined with the assumption that conservation, unlike development, does not also have "irreversible" consequences.

Time element again. The question of what kind of information a decision-maker should expect to acquire within a set timeframe has a twin – perhaps the most vexing question of all – which has to do with setting that timeframe in the first place: How long is long enough to wait for the second decision point? This question is, in turn, closely aligned with the question of what are justifiable expectations for getting information – any information – that would increase the chances of making a more nearly optimal decision (at the next decision point) for all time?

Insofar as the focus is on exogenous research, there typically is, and should be, a time limit on doing more studies and undertaking more research. This is especially true when (as in biodiversity as pharmacopoeia), the odds of finding the sought value in conserved biodiversity are known to be extraordinarily slim. It is even more true when clear and legitimate benefits from development can be identified and known with relative certainty. Moreover, when development benefits are ones that meet basic human needs – say, by turning over an enclave of nature to farmers who can feed themselves from it – there might be a strong moral case for cutting off research quickly.

It should be apparent from this exposition that quasi-option value is not really a "component" or category of economic value in the way that say, existence value or bequest value or (sometimes) garden-variety option value is claimed to be.[138] The strain of trying to cast it in this way can be felt as a kind of paradox.

Suppose (as I suggest in Note 137) that quasi-option value is cast as the difference between the expected value of future information that is available on a "conservation" development path versus the expected value of future information available on a "development" development path. Suppose that (in some scenario) quasi-option value exists, that it can be computed, and that the expected value of future information with conservation is greater than that for development. That is, suppose that the quasi-option value is non-zero.[139] Computing the expected value of

---

[138] This point of view is not original with me. Some economists say much the same thing, for example Vining and Weimer (1998, 331).

[139] Unlike garden-variety option value, quasi-option value cannot be negative. Therefore, non-zero implies positive.

information along each development path requires (in the customary way for this computation) knowing exactly what pieces of information to expect and with what likelihood. If one makes the assumption of only exogenous information, then by the definition of "exogenous", this is the same for each development path. What differs between the development paths (by assumption) is how this information will benefit decision-making (that is, facilitate a more nearly optimal decision). For example, it will be known, with probability π_c , that conserved biodiversity will yield a cure for cancer. But if we really have this information, then quasi-option value collapses into a run-of-the-mill computation of the net expected value of conservation versus development. There is no need for a special category of economic value. *A fortiori*, there is no need for a separate, special computation of it. Which contradicts the intuition that quasi-option value says something that cannot otherwise be said about economic value.

This means that quasi-option value is better regarded as a rule for how best to conduct cost-benefit analyses as a multi-stage decision process. The rule directs a decision maker not to ignore the possibility that information newly available only at the second decision point will show that the optimal amount of development for *both* periods 1 and 2 is less than that already undertaken in period 1. For if this possibility is realized and if development cannot be reversed, then the decision maker will have failed to find the optimal cost-benefit solution for the two periods combined.

The meaning of quasi-option value is essentially just this framework for decision-making. For my treatment of it, there remains only to add the main formal results based on this model. The crux of these results (Fisher and Hanemann 1986, 177) is

> ... that optimal first-period use of the area [to be conserved or developed] is *less* likely to be full development ($d_1 = 1$) [where d_1 means "the amount of development in period 1"] when it is possible to learn about the benefits precluded than when it is not. [italics in the original]

However, it is important to note that this result obtains only under two fairly restrictive assumptions, which the authors carefully spell out. One assumption is that all learning is exogenous to the decision problem. I have remarked that this is open to serious question. The other has to do with the linear dependence of net value on the degree of development. Built into this assumption of linearity assumption are others about the relationship of development to biodiversity and the relationship of biodiversity to value. So far as I can tell, there is no *a priori* reason to believe that any of these relationships are linear. Nor is there any empirical evidence for this, despite the fact that the incremental model of biodiversity (Sect. 5.1.1, The incremental model) is routinely adopted.

As Arrow and Fisher (1974, 319) put it, "... the effect of irreversibility is to reduce the benefits [of development], which are then balanced against costs in the usual way." This is because – again as Arrow and Fisher (1974, 317) say – "Given an ability to learn from experience, underinvestment can be remedied before the second period, whereas mistaken overinvestment cannot, the consequences persisting in effect for all time."

Let's pull back to gain perspective. I think it useful to view quasi-option value, first and foremost, as a strategy to avoid a certain kind of opportunity cost – namely, the cost connected with permanently forfeiting the use of future information in ways that might (or then again, might not) be beneficial. The circumstances are fairly specialized in two respects: First, the *forfeiture* is not a forfeiture of the information itself. Rather, it is a forfeiture of whatever benefit there might be (and there might be no benefit) from having a larger set of development paths (here including conservation as a possible development path) from which to choose. Second, the *permanence* of the forfeiture is a direct consequence of the irreversibility requirement. It entails the permanent removal of development path alternatives – specifically those that involve less development for both periods combined than would be optimal for period 1 alone.

From this perspective, it seems critical to ask whether these two specializing circumstances warrant special consideration under the rubric of quasi-option value. If one focuses on the primary thrust of quasi-option value as a kind of opportunity cost avoidance strategy, then it seems that special consideration is *not* warranted. With each decision comes *some* opportunity cost. And surely there is at least an important *moral* sense in which no action or its consequences can be reversed. Any action that significantly benefits or harms other persons is irreversible in the sense that the benefit has been conferred and enjoyed, the harm done and suffered, and the actors have behaved well or not so well.

None of these things can be undone. We moral agents are chagrined by a person's enjoyment of benefits ill gotten because, even though we can later punish the beneficiary for her methods of obtaining them, her enjoyment of them at the time can never be erased. We are equally discomfited by the pain of someone who suffers undeserved harm or misfortune, even though we can later relieve her suffering by empathizing with her and perhaps by punishing her offenders. Neither the undeserved benefit nor the undeserved harm is ever really reversed. These considerations, no matter how obvious, are hard to capture in an economic framework in which any benefit can be removed by imposing a price; and any harm is just a price that can be reimbursed.

Furthermore, an irreducible element of human decision-making generally is that every decision is made despite the inevitability of better information that might guide a better choice later. A convincing case often can be made for more study and research. But there is also always a case, which gets increasingly compelling with time, to decide and to act. Working within the framework of quasi-option value entails a salient hazard: One might be tempted to think that it justifies an indefinite postponement of any decision that might feature a permanent removal of some development path from the collection of alternatives. But this would be a bad mistake in practical reasoning. It cannot be permissible to indefinitely postpone a decision that nonetheless could be made more optimally in light of still unavailable, but always possibly forthcoming, information.

One might object to this line of reasoning by saying that it is possible to fix the time of the second decision point. But this objection begs the question of what principles justify abandoning study and research at one particular time rather than another.

For instance, so long as new information material to the decision has not yet arrived in period 1, no condition material to the framework has changed; and in particular, there is no more justification for settling on the time of the second decision point at that point than there was at the first decision point.

It simply won't do to say, with Maclaurin and Sterelny (2008, 154), that "as our knowledge improves… we will come to discover new ways in which species will be valuable." I can (and, alas, do) say much the same thing about the accumulating junk stored in my garage: If I wait long enough, surely I'll find some use for it. But I recognize that this is but a poor rationalization. It is not a justification for failing to clean out that space and put it to some obviously good use.

So long as there is no answer, let alone a convincing answer, to the question of how to set the decision points, it seems that the framework for quasi-option value is seriously incomplete. Given that it is a seriously incomplete guide to decision-making, I think that we are well advised to use it with considerable caution.

6.9.4   Specific Claims About the Option Value of Biodiversity

I now turn to Maclaurin and Sterelny's discussion of how they think biodiversity's value is primarily bound to its option value.

These authors take but one paragraph to introduce quasi-option value together with garden-variety option value. This introductory paragraph (Maclaurin and Sterelny 2008, 154) already gives substantial cause for concern when its conclusion declares that these two very different conceptions can be considered of a piece. This initial concern is amplified by their subsequent discussion, which veers back and forth between the two, but mostly loses touch with the requirements of both.

On one hand, there is the authors' admonition to "hedge our bets, insuring against unpleasant surprises" – a risk avoidance maneuver that bears only on garden-variety option value. There is also their concern about "ignorance of our own future preferences". This appears to be an acknowledgment that the shape of personal utility functions can change over time – one way (mentioned in Sect. 6.9.1, Option value and conservation) in which demand uncertainty can enter the (garden-variety) option value picture.

On the other hand, there is the surprising claim (Maclaurin and Sterelny 2008, 156) that "the option-value approach to conservation biology depends on our being ignorant, but not too ignorant" – a consideration that appears related to the value of information. This seems to be something about quasi-option value, but not garden-variety option value. Unfortunately, as my review of these two concepts in the preceding subsections should make clear, little can be coherently said about garden-variety option value within the framework of quasi-option value; and vice versa.

But mostly – and sadly, in this their treatment is representative – Maclaurin and Sterelny's remarks have no clearly identifiable connection to either framework. For example, their suggestion (Maclaurin and Sterelny 2008, 155) that "The solution is to focus not on mere possibilities but on probabilities" appears to state a general, necessary condition for computing expected values in any standard cost-benefit analysis: This calculation weights the costs and benefits of each possible outcome by the likelihood of its occurrence; so both the outcome set and the probabilities must be known with some high degree of confidence. But this has nothing specifically to do with whether or not option value of either flavor figures in the analysis.

Maclaurin and Sterelny (2008, 149–157) argue that the case for the value of biodiversity must rest on the case for its option value. They present three possible ways, which they call "cases", to consider biodiversity for its option value. Whether or not they regard this survey to be exhaustive, it is clear that the authors do regard their proffered cases as exemplars and therefore their strongest evidence that option value is the key to the value of biodiversity. I therefore turn to examining these cases. Unfortunately, my examination confirms the premonitions for a less than sanguine result.

6.9.4.1 Phylogeny

In this first "case" (Maclaurin and Sterelny 2008, §8.5), the thesis under examination is that biodiversity, understood as phylogenetic diversity, has option value. The discussion by Maclaurin and Sterelny is fraught with difficulties quite aside from its problematic invocation of option value. I include these "extracurricular" problems in my discussion both because they typify what one encounters in appeals to option value and because they are too basic to be ignored.

One should expect that Maclaurin and Sterelny would launch the economic calculus of either option value or quasi-option value, incorporating phylogenetic diversity into their calculations. Phylogenetic diversity seems particularly well suited for this role, for these authors follow Daniel Faith's conception of it as a measure (of "feature" diversity, as explained in Sect. 3.3.2.1, Features). Measures are exactly the sorts of things that fit comfortably into the calculus of economics.

In the case of garden-variety option value, one should expect to see discussion of uncertainty in demand for phylogenetic diversity, computations of expected consumer surplus, and comparisons of *ex ante* versus *ex post* prices under varying circumstances. In the case of quasi-option value, one should expect to see discussion of timeframes for development and the costs and benefits of information about development – in both the standard economic sense and in the sense of "articulating a phylogenetic trajectory".

Unfortunately, absolutely none of these requisite elements are to be found. It is impossible to say whether Maclauren and Sterelny's economic reasoning is valid or not, because there is no line of economic reasoning at all that might lead to the conclusion that some positive option (or quasi-option) value attaches to phylogenetic diversity (or phylogenetic "development"). Maclauren and Sterelny (2008, 157) skip the argument, assume the conclusion – the most basic circularity fallacy (discussed in Sect. 2.2.5, Circularity fallacies or begging the question) – and assert that any speciation and any condition that encourages speciation is good insofar as it explores distinct "evolutionary possibilities":

> … in an explicitly conservation biology setting, confining our discussion to sexually reproducing organisms, and to reproductive isolation… [o]ption value explains the importance… of reproductive isolation… [This is because] Speciation allows daughter species to diverge radically in morphology, physiology, ecology, and behavior from their stem. For these reasons many people think of option value as mandating the preservation of species… every [one of which] represents a new and potentially important trajectory in a space of evolutionary possibility.[140]

With no argument concerning option value to examine, I could end my own discussion (which was supposed to be about the "case" for "phylogenetic option value") here. But it is instructive to follow along with Maclaurin and Sterelny to see how their discussion, like many others ostensibly about option value, devolves into a quite different one that begins to look like one that I've already remarked upon.

As noted just above, Maclaurin and Sterelny (2008, 139–142) adopt Faith's conception of the phylogenetic distinctiveness of one organism relative to another as the Google-mapped, shortest phylogenetic distance from one to the other. The passage just cited indicates that they also follow Faith in the dubious assumption that this phylogenetic distance is principally useful for measuring not (as one might have supposed from the name) some notion of evolutionary relatedness, but phenotypical or morphological distinctiveness. Whether justified or not, this conceptual leap leads Maclaurin and Sterelny to what they consider to be the key questions. These are: whether "all speciation represent the same amount of option value" and whether "some evolutionary trajectories represent more option value than others".

At first, it seems as though these authors (Maclaurin and Sterelny 2008, 158) might be concerned to answer their questions in a way that reprises the theme of Faith's notion that some species are more equal than others by virtue of the distinctiveness of their traits. Woe to the creature in a phylogenetically crowded neighborhood, which on that account is presumed to be morphologically and phenotypically near-identical to phylogenetic near-neighbors. To use their own examples, the authors' answers to their questions bode ill for a creature such as *Percina tanasi* (the snail darter), which has the misfortune of having evolved within a rather crowded evolutionary neighborhood – a neighborhood that does, indeed, happen to define a correspondingly compact and crowded morphospace of darters (at the genus level, which are part of the extremely speciose family of perch-like fish). On the other hand, their answers (Maclaurin and Sterelny 2008, 163) smile on *Rhynochetos jubatus* (the kagu). And sure enough, its morphospace has lots of elbowroom – presumably all vacant and available for future "evolutionary exploration".

Unfortunately, the use of Faith's notion of phylogenetic distinctiveness as a stand-in or measure for the distinctiveness of morphology or traits immediately puts this discussion on shaky foundations for the reasons first mentioned in Sect. 3.3.2.1

---

[140] The foundation of their premise in distinct phenotypes is more explicit on p. 163, where they insist (again with no supporting argument) that it is critical to "attempt to represent phenotype distinctiveness".

(Features). Over the course of evolutionary history, very distant lineages have repeatedly "found" the same morphologies and other traits. There appear to be a limited number of adaptive "tricks"; different phylogenies often stumble upon the same ones – stretching from a small number of different body structures and symmetries up to the most dramatic cases of full-blown convergent evolution. A distinct phylogeny does not at all guarantee a distinct morphology or phenotype.

However, at some point, Maclaurin and Sterelny switch their pitch. They stop talking about the distinctiveness of an "evolutionary trajectory" as having a value that corresponds to the distinctiveness of its morphological *destination* in the tree of life. Rather, it emerges after further discussion that most value lies in species that are an especially promising *starting point* for a new trajectory – their "evolutionary plasticity" or "evolutionary potential" to explore uncrowded morphospaces. In fact, Maclaurin and Sterelny write as though these starting points are equivalent to "options". And at least in their mind, this connects to option value.

Maclauren and Sterelny (2008, 158) then veer back and forth between their two notions – on the one hand, biodiversity value as plasticity and potential, and on the other hand, biodiversity value as an actually traversed trajectory – as if they were equivalent. For they go back to the latter, insisting "that we should conserve as representative a sample of evolutionary history as possible." The authors seem to be only vaguely aware that their implicit suggestion that these two things are equivalent incorporates another unfounded assumption – that what has the most evolutionary potential also is an important representative of evolutionary history.[141]

This last assumption is simply false. There are many well-known instances where the phylogenetic isolation and phenotypical distinctiveness of a creature are the result of the dying off of nearby sister branches in the tree of life. The demise of most of an extant organism's near relatives can be strong evidence that it shares with its extinct cousins a lack of robust evolutionary potential or that it lacks an ability to adapt to changing conditions (adaptive plasticity). The phylogenetic isolation of a creature such as the tuatara is an indication that it and its now-extinct relatives are relatively incapable of striking out on a new adaptive course. Stephen Jay Gould (1996, 72) makes this point by reference to *Equus caballus*, the modern horse – "a remnant of a remnant", as he characterizes this species. It is, as he says, a remnant of steady perissodactyl (not just horse) decline in contrast to equally steady ascent of artiodactyls as the dominant representatives of modern macrofauna. *E. caballus* is an isolated twig at the tip of the tree of life where once was some quite bushy foliage. This certainly leaves a lot of horsey morphospace vacant. But given that that space was vacated by failed relatives, it is likely to remain unexplored in the evolutionary future.

---

[141] At one point, Maclaurin and Sterelny (2008, 163) acknowledge that the snail darter might, in fact, take off on "an evolutionary trajectory… that will… make it of enormous consequence for our own future projects." The rest of their discussion does not take this point seriously. And they do not acknowledge the worse problem, discussed in the main text, that evolutionary trajectories that rest in isolated regions of the tree of life are often that way precisely for lack of further evolutionary potential.

The authors (Maclaurin and Sterelny 2008, 163) dig themselves yet deeper into this factual hole by claiming that more recently evolved species are less "evolutionarily plastic" – and are on that account less valuable than those that are the product of an ancient speciation event. But disregarding the normative conclusion (for which no argument is offered), this proposition is also false. Some organisms are evolving, sometimes very rapidly – (as remarked in Sect. 6.7, Biodiversity as value generator) before our very eyes and likely behind our backs – as opposed to those "fossil organisms" that stopped evolving relatively long ago. The demonstrated ability of the modern quick-change artists, in contrast to the demonstrated incompetence of evolutionary stick-in-the-muds, to adapt and evolve under current conditions is strong evidence of their continued capacity to further respond to ecological change.

Let's return to the normative question. Some creatures might occupy a relatively uncrowded morphospace. Some creatures might, by some yet-to-be-specified measure, have greater "evolutionary potential" than others. Some other creatures might more "effectively represent evolutionary history". But the confusions, back-and-forths, and flaws in Maclaurin and Sterelny's narrative, which conflates all these things – distinctiveness of "traits", evolutionary longevity, and evolutionary plasticity – are largely irrelevant to the question of why any of these properties should be considered the basis for judging one organism more valuable than another. So far as I can tell, all of these properties are normatively irrelevant; and Maclaurin and Sterelny offer no reason to make us think otherwise. These properties are even more irrelevant (if that is possible) to the question of whether or not any kind of option value is involved. One will search Maclaurin and Sterelny's discussion in vain for even a hint of awareness of option value rudiments such as as *ex ante* versus *ex post* evaluation (for garden-variety option value) and the influence that new information might have on successive decision points (for quasi-option value).

Maclaurin and Sterelny, like Faith and others writing in a similar vein, do not answer the question of normative justification because they do not even ask it. And if they answer the question of how option value is connected with norms for biodiversity, then that answer is inscrutable. Maclaurin and Sterelny (2008, 163) give us only this hint about what they have in mind:

> We have imperfect knowledge of threats and opportunities the world will bring to us, and we have imperfect knowledge of how our own preferences will change over time.

The first part of this compound thought (regarding knowledge) might be relevant to quasi-option value; the second part (regarding demand uncertainty due to changes in utility functions) might be relevant to garden-variety option value. But the authors are mute on the crucial question of *how* these considerations are relevant. Furthermore, they flow from two entirely different conceptual frameworks. If, as seems possible, they are trying to combine them, it is hard to imagine what this admixture amounts to. Again they leave only a hint (Maclaurin and Sterelny 2008, 163):

> A diverse, adaptable, evolutionarily plastic biosphere is like individual health. It is fuel for success for our projects, both collective and individual.

Here, as in so many of the various attempts to link biodiversity to something valuable, the discussion comes around to some notion of "health". Unfortunately, the authors offer no argument as to why or in what respects an evolutionarily plastic biosphere is like individual health. Nor do they help us understand why and how phenotypical diversity fuels successful projects. Neither proposition is self-evident, and there is strong evidence against both generalizations: The plasticity of infectious organisms is perhaps the greatest threat to human well-being. And a small number of phenotypes associated with a startlingly small number of creatures often account for the lion's share of any category of benefit – for example the overwhelming dominance of a small number of organisms as sources of medicine. As I shall show in Sect. 6.9.4.3 (Ecological option value), Maclaurin and Sterelny themselves provide support for excluding most creatures from having a role in "ecological option value".

6.9.4.2 Bioprospecting

Maclaurin and Sterelny's discussion of bioprospecting (Maclaurin and Sterelny 2008, §8.6) takes up some lines of reasoning that Sect. 6.5.1 (Biodiversity as pharmacopoeia) examines. These authors once again neglect to present an argument. Nor can one find premises from which an argument might be constructed on their behalf. They nonetheless assert their conclusion (Maclaurin and Sterelny 2008, 167), that:

> … bioprospecting option value will weigh phylogenetically distinctive species much more heavily than those from speciose clades.

One can speculate that the authors believe that more "speciose clades" are less likely to manufacture the right kinds of bioactive molecules. But these authors present no evidence for this proposition and I am not aware of any evidence for it. The fact that some of the most common pharmaceuticals are derived from organisms such as sheep (a member of the mammal-dominating order Artiodactyla), common molds, and bacteria bluntly contradicts it.

One can also speculate that the authors believe that the more removed the lineage of an organism from that of any other organism, the more likely the organism is to produce useful bioactive substances. Once again, I am not aware of any evidence whatever for this; Maclaurin and Sterelny offer none.

Or one can speculate that the authors think that bioprospecting is like the lottery and that the winning lottery numbers are imprinted, not on distinct species, but on distinct lineages. But there is substantial evidence that bioactive molecules are often concentrated in groups of creatures that occupy nearby phylogenetic perches in the tree of life. One example is (from Smith and Jones 2004, 7841) the ">180 pumiliotoins found in virtually all anurans". This sort of evidence makes it clear that finding a bioactive molecule is not at all like a random draw from a phylogenetic deck of cards in which one would wish to have as many winning cards as possible.

Rather, the phylogenetic deck is stacked. The production of bioactive molecules is largely a matter of adaptive necessity of a particular kind of lifestyle – a need that might commonly be shared by every species in a speciose clade; but a need that is completely absent from distant and distinct clades.

In sum, it is hard to avoid the impression that Maclaurin and Sterelny fail to provide a premise or an argument for the simple reason that no defensible premise exists for an argument that might lead to their conclusion, quoted above.

Certainly nothing in their discussion of "bioprospecting option value" relates it to the framework for garden-variety option value. There is no discussion of what kinds of demand (or supply) uncertainties are in play, how these affect the difference between expected consumer surplus and option price, or how this difference varies in sign or size according to risk preference. Nor does anything in their discussion relate to the framework for quasi-option value. There is no mention of the decision framework, including the tradeoffs involved in postponing a development decision in the hope of obtaining new information that relates to the costs and benefits of possible bioprospecting outcomes and the likelihood of these outcomes.

In short, the authors do not discuss anything having to do with option value. What they offer are some considerations that are at home in a standard cost-benefit analysis in which option value (of either kind) plays no role. A case for conserving biodiversity on this basis requires evidence that the expected value of conserving is greater than the expected value of developing. That evidence, which I consider in Sect. 6.5.1, is tenuous.

It is remarkable that in the end, Maclaurin and Sterelny themselves give a dim assessment of the expected value of biodiversity from bioprospecting. They grant that there is a dismally low payoff to bioprospecting; that rational design of drugs is more efficient; that, even in the natural world, "second source" species are highly likely; that "big pharma" has largely abandoned bioprospecting for the sound economic reasons that maintain corporate profits; and that there is a very high potential for "enhancing the chemical diversity of an organism by adding to it a gene coding for alien enzymatic activity". All of this argues for assigning vanishingly small likelihoods to a better outcome, and in particular, a better pharmaceutical inventory, in a world with conserved biodiversity. It argues that, at least on these (standard cost-benefit) economic grounds, development at the expense of biodiversity is almost surely justified.

In an even more remarkable concluding twist, the authors disown the case for bioprospecting option value. This move, they say, is prompted by their belief that "ecological option value" can step into the breech to assume the burden of justifying biodiversity-as-option value. I now turn to this last "case" of option value.

6.9.4.3 Ecological Option Value

Unfortunately, there appears to be no more reason to wager that "ecological option value" (Maclaurin and Sterelny 2008, §8.7) will carry the day for biodiversity's value than that the previously examined "cases" of "option value" will do so.

The "option value" in the phrase "ecological option value" belies the fact that yet again in this "case", nothing in Maclaurin and Sterelny's discussion connects to the framework for either garden-variety option value or quasi-option value.

Instead, Maclaurin and Sterelny hitch their nominal option value wagon to Paul and Anne Ehrlich's original precautionary argument whose later and more sophisticated incarnation is discussed towards the end of Sect. 6.3 (Biodiversity as service provider). Their discussion (Maclaurin and Sterelny 2008, 168) takes as premises, first, that stable ecosystem function is a good, and second, that changes in the way ecosystems work are a bad to be avoided, and that the consequence of "the removal of any species from an ecosystem risks a domino effect, leading to wholesale species loss and ecosystem breakdown." The discussion then concludes that a very small class of animals – mostly large, furry ones with (as they say) "high metabolic demands" and "slow reproductive rates" – are far more likely than others to be the critical protectors of the *status quo*.

Though after looking at the other "cases" of "option value" it might no longer seem so odd, it truly is odd that, again, nothing said here has anything to do with option value. However, several more odd qualities in this particular discussion are worthy of brief additional comment. First, it appears that the argument is not so much pitched at conserving biodiversity as it is at preserving a small and select cadre of creatures. This position is vulnerable to being flipped on its head and considered an argument for how dispensable most creatures (for most creatures are not furry megafauna) are – that is, an argument *against* biodiversity rather than for it.

The second oddity, already mentioned, is the discussion's turn to loosely precautionary reasoning. This divagation into precautionary territory is apparently unintentional. For the authors explicitly promote reasoning about option value as far more tractable than precautionary reasoning, emphasizing their desire to distance themselves from the latter (Maclaurin and Sterelny 2008, 185, note 7):

> … because option-value reasoning is tied to some future assessment, albeit rough, it does not depend on the decidedly controversial "precautionary principle," beloved of green politics.

Of course, the fact that Maclaurin and Sterelny conflate their precautionary reasoning with "option-value reasoning" is not itself an argument against some version of the (precautionary) reasoning at which they vaguely gesture. But because they are unaware of having strayed into precautionary territory, they have not thought to do any of the basic preparatory work for making one's way in that tricky terrain – including choosing and justifying choices for the key ingredients (sketched in Sect. 6.3) that go into any plausible Precautionary Principle. At most, one could say that they narrow the threat of harm to the uncertain fate of large furry mammals. But this is a long way from characterizing the nature of the harm, its seriousness, and the uncertainties involved; and on the basis of these considerations, justifying "appropriate" precautionary responses.

The end result is that, not only do these authors offer no argument for biodiversity's value in terms of option value, they also fail to provide any viable precautionary argument. As a consequence, they wind up offering no argument at all for the value of biodiversity.

6.10   Biodiversity as Transformative

Taking a cue from some early work by Bryan Norton (1987), Sahotra Sarkar (2005, 81–87) develops the idea that biodiversity has "transformative value". Sarkar (2005, 82) defines this notion in terms that connect it directly with the nomenclature of standard neoclassical economics – as "the ability to transform our demand values". Demand values are what most economists would call "preferences". Economists tend to focus on just those demand values (preferences) that are expressed in market transactions, which (as remarked in Note 28 in Chap. 2) can be either real or imaginary. Sarkar (2005, 80) expresses some discomfort with this – asking, for example, what the market value of freedom could possibly be. As a quick cure for his malaise, he labels such values as "intangible", summarily sets such values aside, and then forges ahead with his theory.

According to that theory, to a first approximation, biodiversity has value because it can change human preferences. Sarkar is quite aware of two hazards in this position, which he calls, respectively, "the boundary problem" (Sarkar 2005, 95–96) and "the directionality problem" (Sarkar 2005, 96–98). These problems are different perspectives on the stubborn fact that a person's preferences might be changed so that she prefers the trivial to the important, the banal to the original, the excessively narrow to the sweepingly full spectrum, and even the bad to the good. The solution to these hazards, according to Sarkar (2005, 102), is to allow only transformative value that is also "systematic":

> An entity has "systematic" transformative value if we have reasons for giving it this value other than the mere fact that some individual stood transformed by it: we have a generalizable account of how it acquired such a value.

In other words, if one can offer a theory for why some thing X causes values to be transformed and the theory generalizes to many people whose values have been transformed by X, then the value of X can be regarded as "systematic" in the required sense. It passes muster according to Sarkar's theory.

Unfortunately, Sarkar's test does not and cannot do the vetting job that he assigns to it. There are good psychological theories for why a person in circumstances that make her the witness of the systematic abuse of another generally transforms the witness' values regarding trust. Saliently in western consumptive society, there are excellent theories for how commercial marketing can and does transforms consumers' values so that they prefer and acquire objects, many of which do not benefit them and which sadly sponsor much of the biodiversity-reducing "economic development" that Sarkar and other conservationists are concerned to forestall. It appears that some transformative values can be as environmentally destructive as others can be environmentally benign or beneficial.

But are these consumptive transformations "systematic" in the required sense? Unequivocally, yes. Modern neoclassical economic theory offers one of the most fully developed and systematic theories of value acquisition, which purports to shows how people, according to one model of rationality, come to prefer (for example) subdivisions to the wetland they replace. Modern marketing theory has come to an eerily sophisticated understanding of how to apply this theory. And that theory is routinely and systematically applied to transform the "demand values" of persons who, as a consequence, "voluntarily" adopt preferences to consume that which profits the corporation behind the marketing.

One might question many of the precepts on which neoclassical economics is built – saliently including its characterization of *Homo economicus* as a fully functioning rational person. But this does not disqualify that theory as non-systematic in Sarkar's sense. Sarkar says nothing to dissociate his theory from this standard version of economics. And in fact, his theory of transformative value appears to be built squarely atop it.

But there is more to Sarkar's failure to adequately qualify transformative value. For his theory does not fail just because it borrows its theoretical structure and content from economics and marketing, which notoriously can transform demand values in perverse ways. It fails more fundamentally because of a lacuna in reasoning. The mere fact that a demand or preference comes about via the systematic working of some state of affairs – whether via the omnipresence (and possibly omnipotence) of marketing or biodiversity – can never be held as an endorsement of that state of affairs. A newly acquired preference might or might not be worthy of satisfaction; its object must be assessed on its own merits. The "systematic" character of its genesis is not a legitimate criterion for making this normative distinction.

6.11   The Experiential Value of Biodiversity

In the course of making his argument for the transformative value of biodiversity, Sarkar repeatedly mentions "the experience of biodiversity". This appears for him (for example Sarkar 2005, 82, 83, 96) to have its own value – insofar as Sarkar seems to regard it as the source of "good" transformations. This notion of "the experience of biodiversity" seems ill-considered. As I observed in this book's Preliminaries, biodiversity is an extraordinarily abstract concept. It is so abstract that it is hard to say how, if at all, a person can have "experiences" of it.

When one goes out on a walk in the woods, one finds oneself encountering and having an experience of individual plants and animals – "objects" that instantiate substantial universals (in the simple ontology of Fig. 3.1). One can plausibly say that one experiences the graceful needled foliage of a neck-craning individual woody plant growing by a brook *as* an eastern hemlock – the particular kind of tree that it is; or that one experiences a particular, mellifluously whistled song *as* a Rose-breasted Grosbeak – the particular kind of feathered creature that one surmises is the star performer. Moreover, individuals of some kinds might be so numerous that the individuals quickly become anonymous and our experience of them fades into an "experience" of their kind. And as one's realization grows of the variety of these encountered kinds, one might hold on to this variety or diversity of kinds as a major or even dominating contribution to an experience of an eastern U.S. hardwood forest.

But is this an experience of *biodiversity*? Consider first a single species: Is the experience of hearing a Rose-breasted Grosbeak burst forth in song an experience of the species *Pheúcticus ludoviciánus*? Now consider several species that enter into a single moment. Is the experience of the disembodied grosbeak's song emanating from somewhere in a mixed wood of *Tsuga canadensis* (eastern hemlock), *Pinus strobus* (eastern white pine), *Acer saccharum* (sugar maple), *Betula papyrifera* (white birch), and *Fraxinus americana* (American ash) an experience of species diversity? Now consider a disjointed sequence of moments. When later on, one spots a Wood Thrush, and then a Cedar Waxwing, does one then have an experience of avian diversity? When one comes upon the unexpected yellow birch, does this enhance one's experience of tree diversity? What if "later on" is 30 years later? What if some other person, not the original observer, has the later experience?

Perhaps it is okay to say that these are experiences of biodiversity. But if they are, it is only in some very extended sense of "experience" that is stretched well beyond its customary meaning to include abstract ways of knowing, categorizing, comparing, and analyzing the world. This would be a sense "experience" that Plato could happily endorse – experience that "sees" not just the mere Platonic shadow of a particular eastern hemlock, but eastern hemlockness itself; experience that "hears" not just the shadowy sing-song tune of a particular concertizing Rose-breasted Grosbeak, but Rose-breasted Grosbeakness itself. Bringing this up a level to diversity, it would involve experience of speciosity itself, not merely experience of this species and that, and certainly not merely experience of particular barky and feathery things.

I am put in mind of Elaine Scarry's provocative discussion of justice and the value of beauty in grasping *that* abstraction. She notes (Scarry 1999, 101) that

> … the symmetry, equality, and self-sameness of the sky [taken as something beautiful] are present to the senses, whereas the symmetry, equality, and self-sameness of the just-social arrangements are not [because, among other reasons] it is dispersed out over too large a field (an entire town or entire country), and because it consists of innumerable actions, almost none of which are occurring simultaneously.

This suggests how one might legitimately say that a person is "experienced" in (matters of) biodiversity. However, this is not the same thing as to say that that person "experiences" biodiversity. More likely, that person has had a great many disjoint experiences – in the normal, non-extended sense – that involve a great variety of organisms. It is through the accumulation of these experiences, combined with a reflective understanding that integrates them, that one comes to "have experience" in the great variety of life forms. One might venture to characterize this as "experiencing" biodiversity. But this is clearly a figurative stretch of the normal sense of "experience" to something related but distinct. Less figuratively, one would characterize this as gaining an appreciation of biodiversity.

A plenitude of experiences (in the non-extended sense) might be acquired by a single person over the course of a lifetime. Someone such as E.O. Wilson who actively seeks them out might amass an impressive store of them over time – each new experience informed by, and building on, the knowledge and understanding acquired from the totality of previous ones – eventually yielding a deep-seated appreciation of life's variety. Wilson's appreciation of biodiversity – his experience in this matter – might be similar to the lifelong experience (or appreciation) of justice by a judge who, in various jurisdictions, has overseen a just and equal application of the law in many different cases that present many different circumstances bearing on their just disposition.

In fact, this extended sense of "experience" need not be restricted to a single person. It could be jointly "had" by a community of communicating observers whose individual experiences can be pooled to construct a collective picture, which all members of the community can then assemble from these various, initially disconnected pieces. In other words, the extended sense of "experience", unlike the non-extended one, admits of disjoint experiencers as well as disjoint experiences had by a single experiencer.

Clearly, these constructions of extended "experience" are at a far remove from objects that are sensible to a person at any moment in place and time – even objects-experienced-as-a-kind (for example, the feathered being-as-Rose-breasted Grosbeak). In addition to one person's non-extended experience, the extended "experience" draws on that person's memories. It draws, too, on the experiences and memories of others, conveyed in symbolic form, perhaps over the course of hundreds of years. It could even include theoretical extrapolations that let us say such things as that humankind has had "experience" of the more than 10 million species, despite having collectively encountered and catalogued just 1.5 million or so over the past few centuries.

A person certainly does experience individual animals and plants, and even many of them, and many different kinds of them in one scene at one time or in the course of a day's travel; and one is commonly aware of not just these individual organisms, but of their variety. That was part of the experience of the unseen Rose-breasted Grosbeak singing from some perch in the rich mix of trees in a mixed coniferous and hardwood forest. It is this kind of "experience" that is central to the "experience of biodiversity" in Sarkar's favored example of a neotropical forest (Sarkar 2005, 82). So let's return to the key question for this particular "experience".

Is such an experience an experience of *biodiversity*? Perhaps, but only in an extended sense in which one might "experience" criminal justice in a particular trial – upon witnessing the handing down of a sentence that seems proportional to the crime committed, given an accumulated knowledge of other, similar crimes committed in similar circumstances, disjoint in space and time, for which one might or might not have had the direct experience of the sentencing; and given some even more abstract understanding of principles of criminal justice. We might be said to have that experience, but only by virtue of evaluating what is now before us in light of a complex web of prior experiences, along with an even more complex web of abstract concepts that have to do with the nature of justice.

However, this modestly extended sense of "experience" is apparently not what Sarkar has in mind. His thinking is based on a strict dichotomy – between "direct" and "indirect" experience. He grants that some "experience" of biodiversity might be "indirect" – by which he appears to mean "theoretical knowledge". This is the sort of "experience" that comes from computing quadratic entropies or computing phylogenetically minimum-length paths to connect two points in the tree of life. But he insists (Sarkar 2005, 82–83 and in other, similar passages) that all such "indirect experience" of biodiversity is firmly rooted in the "direct experience" of biodiversity "which brings about a transformation of our demand values". He views "direct experience" to be the sort of thing that grabs us by the sensory organs, that tugs at our viscera, that dazzles us, and thereby transforms us. It is the stuff of revelatory vision – the antipode of a quadratic entropy computation.

Sarkar is driven to this dramatic but implausible conception of the "direct experience of biodiversity" precisely because of the transformative role that he requires of it. That role demands that this phrase be invested with direct and visceral qualities because, according to him, its value depends on selling us something. That is, it depends on its acting on us just like the cleverest of demand-transforming commercial marketing campaigns. Advertisements do not succeed by means of theory far removed from experience. Quadratic entropy cannot sell biodiversity.

It is possible that by insisting on this conception of the "direct experience of biodiversity", Sarkar is again led astray by the recurring error of mistaking the diversity of kinds for the qualities of some particular kinds in a diverse collection; or worse, mistaking diversity for the qualities of some particular individuals in a particular place. In that woodland walk, it is the towering presence of certain specimens of eastern white pine and eastern hemlock that most powerfully insinuate themselves into a person's experience. So too does some sense of the place – the feeling that it conveys through its quality of light and sound. This, one might suspect, is a consequence of its particular assemblage of plants and animals. But then the person walking through the woods is directly experiencing those qualities of light and dark and sound, not the diversity of organisms that collectively conspire to create those qualities.

These points are not lost on environmental organizations, which understand that some particular kinds – often the big furry animals that attract Maclaurin and Sterelny – are good "salesmammals" for natural value. But in these cases, it is not the diversity of kinds, let alone the direct experience of the diversity of kinds that is the selling point – that is, the source of demand-transforming power. Rather, it is some sort of experience of individuals of a "charismatic" kind – say, an encounter with a tiger; or an experience of a tiger as a tiger; or, at one level of indirection removed, a conjuring in the imagination of The Tiger – that is, the conception of a being that constitutes an exemplary tiger.

This criticism, of course, does not preclude the possibility that a more "indirect experience" of biodiversity might, after long and sober reflection, eventually unveil its supposed value. This is not the wakeup punch in the gut that Sarkar looks to as instrument of transformation. But I take I take it to be a succinct description of the other, less-than-compelling theories of biodiversity value covered in this chapter, which depend on some line of causal-utilitarian or precautionary reasoning, or a theory of biodiversity as a constituent of a good human life.

6.12 Biodiversity as the Natural Order

The value of biodiversity as implementing a "natural order" of things is inextricably intertwined with the just-so model of biodiversity value. Much of what I wish to say about this view of biodiversity's value is presented in Sect. 5.1.4 – The just-so model of How biodiversity relates to its value. But the remarkable prevalence and persistence of this theory of environmental value merit some additional remarks here.

The notion that the *actual* order of things is also a *natural* order permeates several venerable philosophical and religious traditions. Leibniz's version of this notion, which Sect. 2.3.2 (The value of diversity in general) visits, is unusual in pressing a view that very directly ties the value of this natural order to some great, underlying variety or diversity. Leibniz did not imagine that this actual diversity might vary from time to time. For him, the diversity of the natural world was once and for all time set by its Creator. The distinct kinds make possible the world's perfection, which consists in the natural ordering of these kinds with respect to each other.

Putting aside the static quality of Leibniz's vision of diversity, one can see how closely aligned it is with a major current of modern thinking about why biodiversity is valuable. In the updated version of the natural order story (one that admits geological and evolutionary change), the actual order is seen to have *deviated* from the natural order, which only existed at some prehuman or (on some accounts) later prehistorical or historical juncture; and the natural order could only have persisted in the absence of human interference after that. Still according to the latter-day perfectionist view, biodiversity is valuable to the extent that it exemplifies this state of perfection or one derived from it without anthropogenic inflection. Human activities can do nothing to improve on the perfect state, which, after all, was already perfect (in the relevant, natural sense); their effects can only be adulterating, except insofar as they deflect the state of nature back towards the imaginary state defined by a history uninflected by humans after the prescribed juncture of perfection. To the extent that humans cause biodiversity to deviate from the natural biological order, they compromise the natural good of the world. This sentiment is the spine of the position that holds up exotic creatures as aliens, less desirable than natives.[142]

Of course, the notion of a *natural* order carries with it the burden of understanding what makes a particular order "natural", why a natural order is a good (or at least better than an "unnatural" order), and how it should guide human choices and actions. There is not a little irony in tying biodiversity to this burden, for the notion of "wilderness" also largely hinges on some similar notion of natural perfection, which famously, humans can only "trammel", if they are not careful.[143] Hitched to the same post, the debate about the value of biodiversity plunges into the depths of the same tangled debates that once consumed the value of wilderness – debates that precipitated that latter concept's widespread supplanting by biodiversity as emblematic of natural value.

By common consensus, wilderness, conceived as a place "untrammeled by man", appears to have vanished almost completely from the planet. But so has biodiversity – insofar as it is conceived as an expression of natural order in the biological world. As remarked in Sect. 5.1.4, there is essentially no ecosystem on the planet – considered

---

[142] This is not to say that the perfection of natives versus the imperfection of exotics is the *only* salvo that is routinely lobbed at the latter. It *is* to say that some substantial consternation about exotics has to do with human assistance in introducing them to their new neighborhoods, whose biodiversity is thereby rendered less "natural". I am fully aware of another claim that is now routinely made – to the effect that exotics decrease biodiversity *simpliciter*. But support for this claim as a general thesis is tenuous unless "biodiversity" is understood to mean something like "*natural* biodiversity".

[143] This is a reference to the definition of "wilderness" in the 1964 Wilderness Act in the United States.

as including its diverse community of organisms – that is un*affected* by a wide range of human activities. If the "natural order" is one in which (according to one prevalent suggestion) humans have not heavily *interfered* with the "natural progression"of the world, then this order is largely or perhaps entirely absent in the current state of the world. This makes it difficult to find grounds for thinking that the *current* order, which is evidently already a deviation from the natural order, should be defended, conserved, or even restored to some very recent state – a state that will inevitably still be marred by human-induced deviations from the natural order, which (on most perfectionist accounts) already have a history centuries or millennia long.[144]

Current-day thinkers tend to substitute the term "naturalness" for "natural order", but to all appearances, their philosophy sprouts from the same roots as that of Aquinas and Leibniz. The verbs "affect" and "interfere" (italicized in the preceding paragraphs) figure prominently in characterizations of how human influence on the "natural" world have compromised its naturalness. By the lights of some thinkers, the extreme degree to which humans have affected and interfered with nature is thought to have sufficiently destroyed the natural order so as to entail "the end of nature".[145]

The environmental philosopher Dale Jamieson accepts the basic framework in which the "end of nature" claim is made. But he believes that the rumored "end of nature" is greatly, or at least somewhat, exaggerated. While setting aside the question of how much the quality of naturalness *explains* about nature's value, he (Jamieson 2008, 164) regards naturalness to be "a matter of degree", and the degree of nature's naturalness to be an axiological barometer – that is, a salient property or characteristic that is a sound basis for judging the degree to which nature has value as nature (in contrast to its value as resource, for example). Finally, the degree of naturalness, according to Jamieson, is a function of how human influence is involved in the causal chain that led to a temporary terminus in the current state of the world. In this significant respect, he is a latter-day exponent of the perfectionist tradition of valuing naturalness,[146] which makes it important to understand the extent to which pieces of the current, actual world exemplify this property.

Jamieson suggests that assessing how much naturalness is justifiably attributable to a place must take into account what he regards as two distinct kinds of human influence, because according to him (as I explain below), these have differing effects on naturalness. To this end, Jamieson (2008, 163) tries to drive a wedge

> …between X affecting Y, and Y being a product of X. I may affect your decision about what to study in many ways, for example by providing you with information or advice that you may or may not take into account. This, however, is quite different from the case in which your decision about what to study is a product of my influence.
>
> Consider the following example. Human action affects the length of the growing season in the Great Lakes region of North America, but the fact that there are zebra mussels in the

---

[144] This point stands even before any additional support it receives in the much-discussed conundrum that modern restoration practices constitute yet another means of human interference.

[145] This phrase is ensconced in, and made famous by, the title of Bill McKibben's book, *The End of Nature*.

[146] Jamieson himself is likely to balk at association with perfectionism.

Great Lakes is a product of human influence. They were transported there by ships, and deposited along with ballast water. The distinction between these two cases (the human impact on the length of the growing season and human impact on the presence of zebra mussels in the Great Lakes) has intuitive force (or so I hope)...

Unfortunately, there are significant gaps in the logic of Jamieson's proposal. First (initially putting aside the validity of the distinction), he neglects to state explicitly *why* he thinks the alleged distinction between affecting and producing makes a difference for nature's naturalness, or for determining whether or not "the end of nature" is at hand. As he grants (Jamieson 2008, 163), if the end of nature is not at hand, this must be so despite the fact that "human influence is so pervasive that no part of nature remains untouched." Or, to put this in the terms of modern ecology, if we are not at the end of nature, this is so despite the fact that there might not remain a single biome on the planet that is not an anthropogenic biome (Ellis and Ramankutty 2008).[147]

Jamieson does not explicitly state the purpose of his distinction "between X affecting Y, and Y being a product of X". But it seems clear that he believes that it serves to separate two kinds of human influences, which differ in their effect on nature's naturalness. On the one hand, humans can *affect* nature and do so without thereby affecting its state of being natural. On the other hand, any change in nature that humans *produce* is, by virtue of that, an unnatural state, or at least a state of reduced naturalness. Establishing the relevance of this difference to the issue that motivates Jamieson's discussion – of whether or not nature is already at its end – requires one final claim of an empirical nature. This premise, which I supply on his behalf, is: The changes that humans *produce* are restricted in scope – sufficiently so (compared to changes due to people merely *affecting* nature) that the end of nature is some way off.

The second gap in Jamieson's logic is located in the alleged distinction itself rather than in its alleged implications for naturalness. As he says, Jamieson "hopes" that his readers will feel the intuitive force behind it. I suppose that he also hopes that his readers will feel the intuitive appeal of the distinction's force in bifurcating human influences on nature, along the lines I have just suggested. But he offers no supporting argument; I do not feel the intuitive force that Jamieson hopes I will feel; and no matter how hard I look, I do not see any light between X affecting Y (on the one hand) and Y being a product of X's influence (on the other).[148]

One might think that the clues for finding the light will be found in the two contrasted cases (briefly mentioned above, in the first quotation from Jamieson) of person-person interaction, which Jamieson offers as models for the two contrasted cases of human-ecosystem interaction. But I think that there is more in the person-person model to derail the analysis of the human-ecosystem interaction than to keep it on track. The interhuman interaction concerns a decision, of possibly great personal

---

[147] As the main text explains, the very point of Jamieson's distinction is to enable him to say that although both forms of human influence result in anthropogenic biomes, only one reduces a biome's naturalness. See also Sect. 5.3 (The moral force of biodiversity).

[148] Jamieson uses "product of X" and "product of X's influence" interchangeably. The latter phrase appears at Jamieson (2008, 164).

significance, that one person (in this case, a student) makes as (one initially supposes) an autonomous rational agent. Quite clearly, the manner in which Jamieson exerts influence on that agent is an important moral consideration. Offering guidance that is not based on his own personal interest in the choice is permissible; threats are not. But none of this carries over to influencing the environment because, quite simply, it is not an autonomous agent that makes decisions that matter to it.[149]

The person-person model also goes off track considered apart from its role as a model for human-environment interactions. I believe that whatever difference there is between the two interactions hinges on tacitly transforming the student from a fully autonomous agent in the first case, in which Jamieson's role is merely "providing… information that [she] may or may not take into account", into someone or something whose freedom to decide is severely truncated in the second case, where the decision is no longer hers, but rather "a product of [Jamieson's] influence". Perhaps the student is really now a robot, which Jamieson has programmed with a certain decision procedure that yields the decision that he "recommends". Or less dramatically, perhaps the student is still a person but Jamieson has given her a drug that temporarily disables or seriously impairs her ability to deliberate in a way that properly balances all her own interests. Or perhaps Jamieson's influence on the student is more subtly coercive, by dint of his position of relative power over the student. If one resists this tacit transformation and presumes that the student remains as fully autonomous and rational in the second ("product of") case as in the first ("affected by") one, then the opening between these two cases collapses. Insofar as it is natural to say that her decision about what to study is a *product* of Jamieson's influence, it is equally natural to say that is because he has *affected* her decision.[150]

Nor does intention appear to provide an opening between "affecting" and "being a product of". In Jamieson's two human-environment interactions, both the shortening of the growing season and the presence of the zebra mussels in the Great Lakes are causally attributable, in part, to human actions and patterns of behavior. But no one *intended* to shorten the growing season; and no one intended to transport zebra mussels into the Great Lakes. Both are inadvertent and unplanned side effects of actions taken with intentions that had nothing to do with those effects. In both cases the lack of intention is clear because awareness of these specific effects and even of the possibility of their occurring – arguably a necessary, but not a sufficient condition for their being intended – emerged only after the fact.

---

[149] Jamieson disagrees with this assessment, saying "It isn't obvious to me that autonomy can't be ascribed to nature." [personal correspondence] See also Jamieson (2008, 166) and Jamieson (2010, §6, "Respect for Nature"). I briefly return to this topic at the end of this section. Section 8.2.6 (What appropriate fit is not) reconsiders the position taken by Jamieson and others by way of contrast with my own, antithetical view: Insofar as autonomy can be ascribed to nature, it is devoid of key elements that ground moral respect for other *persons*' autonomy.

[150] As I have already pointed out, Jamieson (2008, 164) states that "the distinction between X affecting Y, and Y being a product of X's influence, is undeniably value and a matter of degree." But this does not cut against my point, which is that, if X and Y are chosen in a uniform way (both fully autonomous rational agents, or not), then the distinction loses its ability to do the logical work for which Jamieson employs it.

Perhaps "product" is supposed to connote "a salient result of actions or behavior". But this interpretation does not help Jamieson, either. Both the shortened growing season and the presence of zebra mussels in the Great Lakes surely are products in this sense.

Maybe Jamieson just picked unfortunate examples, and a more felicitous choice would tease out some relevance to the distinction of intent and salient results. Suppose that the case of *producing* a Great Lakes full of zebra mussels were replaced by that of constructing a condominium in a wetland area. Condos, unlike the relocation of zebra mussels, *are* produced in the primary sense of "made or fabricated". Their fabrication is the product of human planning and design, and therefore (unlike the transplanting of zebra mussels) intentional. Directly tied to the intentional production of the condos is the salient result consisting in the wetland's transformation. Perhaps this combination of intent and salient result justifies saying that, while the condo project does not *affect* the wetlands, its transformation of the place is a *product* of human influence.

But this new example does not really salvage Jamieson's goal of distinguishing those human influences that undermine naturalness from those that do not. While no one would dispute that the condos were produced and that *they* are not natural, that was never in dispute; Jamieson's proposed wedge is superfluous in making that determination. So far as intention is concerned, no one would dispute that the condos required careful planning and design: The intent was to build condos. But one must take care to distinguish that intention from a full accounting of the salient consequences of acting with that intention. The peculiar (intentional) nature of intention is such that, when doing A ends up (also) realizing B, a person who intended to do A need not also have intended to do B. That is certainly true when she had no idea that B would come to pass as a consequence of her doing A. But it is also true when she knew B to be an inevitable consequence of A; and it is true even when she intended to do A *by* doing B. In the first case, she might acknowledge that she knew that B would follow from A. In the second case, she might acknowledge that she did B precisely in order to do A. In both cases, she might be held responsible for B. But in all these cases, she might also truthfully maintain that it was only A that she intended. So it is reasonable to question whether an intention to *produce* a condominium complex is also an intention to obliterate the wetland that it displaces.

Granted, one might say that the intentional production of condominiums (also) *produced* a transformation of the wetland – whether or not that transformation was intended. However, with equal clarity and cogency, one might say that the wetland was *affected* – in fact, greatly affected and perhaps even destroyed – by the development project. This latter characterization – in terms of the effects of human activities – is common in the ecological literature, which frequently speaks of how humans have *affected* virtually every biome on the planet by various means. These means include behaviors, which, on the one hand, have resulted in changes in climate and growing seasons in the Great Lakes region among others, and on the other hand have resulted in bringing creatures such as zebra mussels to places such as the Great Lakes heretofore unvisited by them.

Where else can the opening between "being affected by" and "being a product of" be found? What else could it consist in? On the one hand, why can't one just as easily say (*pace* Jamieson) that the changing length of the growing season in the Great Lakes region is a product of human influence as say (along with Jamieson) that the domination of zebra mussels over other mollusks in the Great Lakes is a product of human influence? On the other hand, why can't one just as easily say (*pace* Jamieson) that human behavior has affected the Great Lakes aquatic ecosystem as (with Jamieson) that human behavior affected the growing season in that region? So far as I can tell, all these cases depict states of affairs that are partly the *product* of human influence. Which is to say that all these cases depict states of affairs that have been *affected* by human actions and characteristic behaviors. Which is to say that human actions and behavior have played some causal role – a role that is, in fact, salient both for the growing season and for the zebra mussels' newfound homes.

Perhaps Jamieson is just trying to say what I suggested at the outset – that the world's naturalness – the degree to which it approximates the natural order – varies inversely as the quality, power, or degree of human causal involvement. And perhaps one should take his definitions of "produced by" and "affected by" as stipulations (not necessarily concordant with common usage) according to which the former has a higher quality, degree, and power in its deleterious effect on naturalness than the latter.[151] Still, it is hard to see what illumination this thesis throws on the value of nature. That is partly because "degree of causal involvement" is, at best, a tricky philosophical knot to unravel. This is evident from Jamieson's own cases: It is highly doubtful that there exists any straightforward or incontrovertible basis for comparing the degree of human causal involvement in the shortened growing season versus the degree of causal involvement in a hitchhiking organism establishing itself in a new locale. At the highest level of causal description, both were caused (in part) by characteristic human behavior connected with economic pursuits. But even if one could somehow disentangle degrees of human causal involvement here, there, and everywhere, why would a lesser degree of naturalness (understood in terms of degree of human causal involvement) be more perfect (and presumably more valuable on account of that) than a greater degree? Aquinas and Leibniz had an answer to this. So far as I can tell, Jamieson and latter-day perfectionists do not.

It is important to understand that the conundrum of determining "the true natural order of the world" is not one that has emerged just recently as the result of industrial-age developments (in several senses of that word). The preponderance of evidence now makes it clear that *H. sapiens* profoundly transformed every place it invaded, starting with that species' initial exodus from southwestern Africa 70,000 or so years ago. As of 1,500 years ago with the occupation of New Zealand (and Madagascar probably 500 years before that), "every place it invaded" had

---

[151] This interpretation is not obvious in the ordinary meanings of "affecting" and "is a product of" or in Jamieson's exposition. But Jamieson says as much in personal correspondence: "As I use the terms, 'affecting' implies that there are other, more powerful causal influences; 'is a product of' implies that this was the only or most powerful causal influence."

become essentially "every habitable place", with the exception of a few isolated islands. By now, those too are no longer untouched by human activities.

The primary influence on ecosystems has often been exerted from the top – by the direct killing of megafauna by hunters and their eviction by agriculturalists. In a survey by marine ecologist Jeremy Jackson (2006, 28):

> Food webs were severely disrupted by the loss of almost all of the top predators and megafaunal herbivores, and patterns of vegetation changed greatly in response. The vast herds of bison in North America were partly an artifact of the elimination of all other large herbivores that had competed with bison to graze.[152] Seed dispersal and distribution patterns of Neotropical plants with large, armored seeds, whose germination depended on consumption by large frugivores, were greatly altered by the extinction of 15 genera of large herbivores; a pattern that was apparently reversed at the eleventh hour by the arrival of goats, cattle, and pigs four centuries ago. And perhaps most spectacularly, the elimination of large herbivores in Australia resulted in the accumulation of vast amounts of uneaten vegetation, which was vulnerable to wildfires, which in turn transformed much of the vegetation to arid scrub, controlled more by fires than by herbivores.

Even these dramatic human-wrought transformations do not fully convey how dim is the prospect for finding in this morass a "natural order" in the sense of "being unaffected by humankind" or "not a product, in significant degree, of human activities". In response to all these anthropogenic perturbations and multiple other non-anthropogenic forces, all of which interacted in cascading patterns impossible to unravel, virtually every ecosystem on the planet took a "development path" (to borrow the handy phrase from economics) utterly different in both small and gross detail from the path that it would have taken in the absence of human influence. Some of the more recent (industrial-age) human impingements – particularly those on the biogeo-chemical cycles of, for example, nitrogen and carbon – have had a rapid, globally sweeping, and deep penetration that significantly affects both terrestrial and aquatic (including marine) biota, and most likely every biotic as well as abiotic interaction worldwide. But the sort of changes that Jackson describes, though enacted over longer, millennial timeframes, had already erased any vestigial hope of reconstructing a "natural order" in the sense under consideration. And as Mann (2011) recounts, the remixing and reconstitution of ecosystems that followed on Columbus' 1492 voyage to the New World rivaled, if it did not exceed, more recent remixings and reconstitutions.

As a consequence, even if one makes the category mistake of confusing the diversity of kinds for the particular identified kinds that assemble themselves into the particular communities in the particular habitats to form the particular ecosystems

---

[152] The stupendous size of the bison herds was likely also a consequence of the disappearance of bison predators. The last of the predators to vanish were the human ones – Native Americans, who in previous millennia had hunted the competing, nonhuman predators of bison to extinction. In the early sixteenth century – before epidemics decimated Native Americans – Hernando de Soto apparently did not see a single bison while stumbling around the southeast on out to the Mississippi, an area which he was the first European to see. Perhaps it was de Soto who sowed the germs of the destruction of the previously resident Native American people in that region. By the late seventeenth century, Robert de LaSalle, paddling down the Mississippi, observed that bison were omnipresent on the plains along the river. For an intriguing discussion of this, see the chapter on "The Artificial Wilderness" in Mann (2005).

that one can observe today or anytime since the beginning of human prehistory, it is hard to understand how any state of the world could qualify as the natural order or even be characterized in terms of its deviation from the natural order. The question – of how much the actual order (evaluated at any point in time after the prescribed juncture of last perfection) deviates from the natural order (understood as the terminus of an imaginary development path from that juncture taken in the absence of *H. sapiens*) – is most likely incoherent. But even if it were meaningful, we humans are utterly impotent to imagine in even the most general terms the world to which that development path would have led. That reason alone – the practical impossibility of adopting that imagined world either as an action-guiding norm or even as an action-valuing norm – is reason enough to disqualify it.

Finally, even if humankind had, or could have, or could *conceivably* have a science powerful enough to produce a picture of this "what if" world, this picture would be utterly irrelevant for distinguishing good behavior from bad. For such concepts as right action and good character necessarily apply to human moral agents confronted with decisions about how to act in this other, very different, well-peopled, current world that, for better or for worse, they actually inhabit. Morality has to do with norms that moral agents can be reasonably expected to see as applying to acting in the real world in which they find themselves. By contrast, morality has little to do with some alternative world without moral agents – including one without all those trammeling people. Norms that are centrally grounded on some such fiction literally make no moral sense. This leads to the conclusion that norms based on some ideal of natural order are not just practically impossible to adopt, they are morally irrelevant.

At this point, perfectionists tend to adopt a backpedaling strategy in order to salvage their approach. They suggest that, while there is no *perfectly* natural order, there are (as Jamieson believes) *degrees* of naturalness – that is, degrees of approximating natural perfection – and we humans should strive to achieve higher degrees. Two considerations make it difficult for this strategy to salvage credibility for the perfectionist view. First, the norm or standard that is said to admit of degrees of attainment is, as I have just argued, morally irrelevant. It is hard to see how this is remedied – that is, how an irrelevant norm could possibly be made relevant – by declaring that it admits of degrees.

Degrees of naturalness are sometimes defined by association with (or non-anthropogenic derivation from) states of the world at various earlier times. The gold standard for "naturalness" is affixed to some sufficiently early time, with naturalness becoming more and more alloyed with the compounding of human-induced deflections over time. Many choices present themselves for the gold standard: before the Pleistocene extinctions, before the deforestation of most of the world's forested lands, before the flora and fauna of most of the world's islands were dramatically shuffled by human activities and those of their companion plants and animals, before Columbus' journey, or before the introduction of vast quantities of reactive nitrogen into virtually every environment on the planet. Then again, you or I might dearly wish the time to be set (or reset) to that of our childhood world, which we yearn to repossess.

But this suggestion for setting a standard of naturalness by historical reference does little but illustrate something important about norms: The choice of a normative standard for anything cannot be arbitrary. It must be based on some set of endorsable principles that can be seen as applying to a broad range of choices, not just this one. It is very difficult to imagine any credible principle that could be used to justify the selection of one point in history over another… except by going back to a criterion relating to human influence or its absence. In that case, the successive historical references are reduced to mere tokens that, with the advance of time, mark successively more heavy and pervasive trammeling, and successively more compounding of human-induced effects successively more intertwined. That criterion, I have already argued, offers no hope of a solution. But the requirement for generally endorsable principles adds yet another reason why it is futile to build a norm out of the relative absence of human-induced effects: The absence of human influence – whether expressed by reference to some point in time or not – is singularly unpromising as a basis for any such principle because some of the most highly prized things are uncompromisingly human creations. Therefore, human influence alone cannot be held to be a normatively tainting influence.

Second, the attempt to salvage the notion of naturalness by emphasizing that it admits of degrees only serves to emphasize the moral impotence of the underlying conception. "Degrees of naturalness" seems utterly powerless to adjudicate between real and even imagined alternatives for biodiversity. Would a North America with Asian elephants tromping around as surrogates for the elephantids that existed on the continent 10,000 years ago be more or less naturally biodiverse than one without them? Coyotes have, with systematic zeal, expanded into the void left by exterminated wolves and mountain lions. Would a world in which, with equally systematic zeal, they were shot be more or less natural than one in which coyotes are everywhere? In a world in which essentially no free-flowing rivers remain, is it more or less natural to weed out creatures that thrive in the dammed legacy of sluggish riparian systems and reservoirs, while coddling the creatures whose previous habitat has vanished *in situ*?

Incoherence, practically insurmountable epistemic demands, an absence of generally endorsable principles, moral irrelevance, moral impotence: individually and collectively, these reasons suffice for abandoning the natural order approach to biodiversity's value. But these are not the only reasons. For "the natural order" to connect with "the good of nature", there must be some reason to persuade us that the natural order matters – that it is good in a way that commands moral respect. Even if (among other things) the "natural order" were a coherent concept and even if it were morally relevant (in the limited sense of offering a coherently characterizable norm that *could* apply to actual persons acting in today's actual world), there remains the question of whether it *should* apply. No matter how many iterations of fractionation, distillation, and refinement they undergo, the notions of human "product" and "effect" offer little hope as the keys to an answer.

The notion of a natural order builds on a particular and peculiar model of value, which is not peculiar to Jamieson's treatment of it. I comment briefly on it here, as preface to a more comprehensive treatment in the last chapter of this book. What makes this axiological model worthy of comment is that it typifies and sets the terms for many more general discussions of natural value, not just those concerning natural order. These terms, I think, are largely responsible for bedeviling an understanding of nature's value.

First, as I have observed, the model for natural order is a state model – one that is saliently oriented towards achieving a certain state of perfection conceived in terms of some notable absence of human influence. It is no small problem that such a state cannot, as a matter of fact, be achieved, approximated, or even imagined. There is no getting around the fact that nature has been enormously, incalculably, and unimaginably affected as a product of human activities.

This first characteristic, which has to do with the nature of a human-inhabited world, is an empirical one. The second characteristic is conceptual: According to the natural order state model, virtually all human actions and behavior that impinge on nature *by definition* cannot have anything but the deleterious effect of making the actual state of the world diverge from natural perfection. In the context of a perfectionist state model, there is something of the theory of original sin in this precept, which views as inherently damaging the living of a characteristically human life.[153] The third characteristic is that an implicit reference to historical memory is curled up inside the state model. That is, the state of perfection – or of the relative *im*perfection that is a deviation from it – is underdetermined by the state of affairs observed at any instant in time. One cannot solve for the degree of perfection without bringing into the equation the causal history of what one observes.[154] For example, and saliently, there is no general principle whereby one can distinguish between an exotic and a native organism by merely observing the current state of an ecosystem. The exotic, considered as an organism that doesn't belong, cannot be fingered without reference to observed past states (including, most obviously, a previous record of the exotic organism's absence), or by reference to observed events – most saliently human actions and activities that introduced the creature by something other than "natural" means.[155]

---

[153] Of course, the theory goes on to urge that, although people necessarily sin, they ought to arrange to sin a little less.

Also, for this brief discussion, I but briefly touch on the roles of both conservation and restoration biologists in further influencing the course of the nature. Conservation biologists commonly perceive their mission as one to slow, or ideally, to stop the perceived deviation from the natural order. This is even true of to ecosystem service conservationists insofar as they deviate from a strictly economic argument to a position that defines the "natural order" in terms of some gold standard of service provision. Restoration biologists, at least in their admirably simple, initial understanding of their mission, perceive their mission as one of making history loop back on itself at an earlier point of lesser deviation. Making a case for the good of these kinds of nature-influencing projects versus the bad of other human-engendered changes these projects are supposed to rectify must deal with an obvious incongruence. See below in the main text. Chapter 8 further explores this conundrum.

[154] Of course, this might cause some to question whether this is, after all, a state model.

[155] Mark Sagoff brings home this point in Sagoff (1999). See, for example, his discussion of mute versus trumpeter swans in the Chesapeake Bay region. Even an analysis of genomes, which might link two distantly disjoint populations and implicate humans as the likely intermediary, must rely on an historical analysis of the populations and rates of change at DNA loci. See, for example, Alter et al. (2007).

A fourth characteristic applies to many, but not all, conceptions of a perfect natural order. This is a principle of permanence, which says that any human-induced deviation from perfection cannot be undone, only further compounded.[156] This is what Bill McKibben (1989, 210) calls "the permanent stamp of man". There is a tension between this principle and the conception – of inducing nature to loop back to a previous historical state – that some restoration biologists have of their mission. The tension is relieved by reference to some common-sense notion of "undoing" what was done or "putting back" what was there. But any relief in contemplating the possibility of restoring nature's perfection is temporary. For putting aside the giant leap from "putting back" in everyday life (for example, restoring to its original, perfectly judged position the potted plant that I inadvertently moved when I leaned on its container) to resurrecting an historical state of an ecosystem, the restoration must still drag along the ball of defiling human influence at the end of its historical causal chain. And the last few links of human influence, comprising the restorative reshaping, are no less significant than those that led up to its previously imperfect state.

I close this section by expanding on some earlier remarks (in connection with Jamieson's attempt to distinguish "affected by" from "being a product of") about how the natural order ties into what is called the "autonomy" of nature. This topic is important because the word "autonomy" implicitly carries the normative load of an entitlement. Those who connect the natural order to autonomy are none too clear about what this connection consists in. One might start, simply enough, with the thought that in the natural order of things, nature is autonomous. If "autonomy" with regard to nature means merely "unaffected by humans" or "not a human product", then the proposition that perfectly ordered nature is autonomous is underwhelmingly true, for it follows in a straightforward way from the perfectionist definition of the natural order. But in this case, "autonomy" fails to confer any entitlement, for it merely translates a matter of fact about the perceived absence of human influence.

On the other hand, suppose that "autonomy" has the sense in which it is customarily attributed to a person, whose autonomy moral agents are obligated to respect. This sense of "autonomy" entails that the autonomous subject have an interest in exercising the freedom to pursue an existence that that subject conceives as good for herself. As applied to the natural world, this notion of morally respect-worthy autonomy would entail that nature have similar entitlement to pursue its existence on its own terms. But since nature has no conception of a good for itself, let alone an interest in pursuing such a conception, it seems certain that nature is not autonomous in this sense, no matter how influenced (or not) by people. As a consequence, there is nothing to object to – on the grounds of "respect for autonomy" – about humans anointing themselves "masters" of a natural world that they are determined to "dominate". I return to the topic of autonomy in Sect. 8.2.6 (What appropriate fit is not), where I distinguish my own views about nature's value from those for whom nature's autonomy is foundational for its value.

---

[156] Restoration biologists who take their mission to loop back to a previous historical state obviously do not fully embrace this principle.

It is also unclear whether autonomous nature is always the natural order. Whether or not that is so would seem to depend on whether the autonomy is expressed by or exercised by something that is truly natural. If kudzu in the American southeast is considered unnatural, then its expression of autonomy will not pass muster as part of the natural order. Some similar objection might be raised to exclude Eduardo Kac's biotopes (discussed in Sect. 5.3, The moral force of biodiversity). But this requirement mires the proposition in circularity: Whatever it is that expresses its autonomy is not the natural order unless its naturalness can be affirmed on the basis of assessing its autonomy. In other words, the naturalness of the starting point forever remains in doubt.[157]

6.13 Other Value-Influencing Factors

I end this chapter with a survey of some other factors that have been said to affect the value of biodiversity, while not necessarily being claimed to constitute the core of its value.

6.13.1 Viability and Endangerment

Sahotra Sarkar says that he is primarily concerned with attaching biodiversity value to places. By this, he does not so much mean that biodiversity is a two-argument function, which takes "place" as an argument alongside "diversity", as that a place presents a conservable quantum of biodiversity. In Sarkar's formulation of this thought, a place's biodiversity value cannot be assessed from (what he calls) its biodiversity *content* alone. Rather (still using his vocabulary) its value is a function of the *viability* of that content in that place (Sarkar 2005, 173–178; 2002, 136). It is unsurprising and uncontroversial that, according to Sarkar, a place with no biodiversity content has no biodiversity value. But it does seem surprising that, according to his viability requirement, the most biodiverse place on the planet also has absolutely

---

[157] One of several variations on the theme of "respect for autonomy" is "respect for nature's health". The kernel of the theme is a slide from some characteristic of the human condition that commands moral consideration to a homonymous characteristic attributable to nature. Unfortunately, in the contextual shift from people to nature, all of the morally compelling considerations are left behind. Nature is simply not capable of autonomy or of being enslaved in anything like the sense in which human autonomy demands moral respect and human slavery demands moral condemnation. Nor is nature capable of having an interest in being "healthy" in a way that a person understands is a prerequisite for living her life fully and well. What is left in the nature context is a word – "autonomy" or "health" – which is a normatively suggestive but otherwise hollow shell. Ecosystem health insinuates itself at several points in the next two chapters. For an excellent discussion that largely aligns with my views about ecosystem health (but not about nature's autonomy), see Jamieson (2002, 213–224).

no biodiversity value – if its complement of biodiversity is doomed. This view defies the common perception that the value of something commands increased consideration with the realization that it might soon be lost. It is valued more, and it is appropriate to value it more, on account of the short time left in which its treasures might be regarded in any way except as fond and fading memories.

Some sense can be made of Sarkar's position if construed as dealing with the kind of triage that is a central concern of biomedical ethics and, indeed, that is a central concern of Sarkar's own practice of conservation biology. A triage protocol assumes that it is possible to classify "patients" according to their expected response to "treatment". It is especially concerned to identify patients that will not survive, despite all efforts to the save them.[158] But if this analogy is what guides Sarkar's viability doctrine, then it exposes some major logical and axiological shortcomings in that doctrine.

First, in the sorts of extraordinary circumstances that call for triage, there are compelling grounds (based most obviously, but not solely, on consequentialist principles) for letting valuable patients succumb without attempting to save them.[159] But this does not somehow demote the value of those patients, let alone reduce their value to zero. Rather, it increases the tragedy of their demise and imminent loss, and the tragedy of an inability to forestall this loss. To assign value to some thing or being based on good or bad fortune outside the valued entity's control is not just an error; it is a morally disturbing error. It is the sort of move that gives false comfort to the compulsive rationalizer. She need not confront the great value of what, in bad fortune beyond her control, is lost. Instead, she can maintain that, by the time it is about to vanish, what is inevitably lost has essentially no value whatever. In essence, Sarkar's seemingly innocent axiology writes off rather than acknowledges a morally horrible circumstance: Sometimes, we find ourselves in the position of having to sacrifice something of the greatest imaginable value (human lives) in order to retain a precarious grip on a still-valuable, though diminished world.

Sarkar would probably respond by saying that he is simply focused on right action, whose rightness, he would say, is both judged retrospectively and revealed beforehand by his optimizing, aggregating consequentialist calculus. In this calculus, a species' value (by virtue of its contribution to biodiversity) is conceived not in a

---

[158] Medical metaphors generally, and the triage metaphor in particular, are not unfamiliar to the discipline of conservation biology. Its founding statements, for example, by Michael Soulé (1985), self-consciously conceive conservation biology as the discipline that restores "health" to its natural world "patients". This, I believe, is the origin of now common talk about "ecosystem health".

[159] Medical triage protocols are generally regarded, not as a more or less uniform "prescription for action" under any pressing circumstance, but rather as a class of protocols that vary according to context. Hospital emergency rooms, pandemics, and battlefields are significantly different contexts that call for different protocols. Battlefield triage protocol often rests on a four-way classification of patients in which those who have no chance of survival are in a class whose members might receive no attention at all. However, medical triage protocols for other contexts, such as emergency rooms and pandemics, often do allocate some care (beyond palliative care) to "doomed" patients. My purposes are not served by trying to split these differences.

single dimension, but as the two-dimensional "value area", which is the integral of the value curve traced over time. To say that a species is doomed is to say that its value curve is expected to shortly plunge to zero, whereafter it contributes nothing of value. This understanding warrants ignoring such a species, from the viewpoint of a practical actor (a conservation biologist) who is supposed to work for the broadly conceived good of biodiversity at large and in the long term. That broadly conceived good is the sum of value areas under the value curves for all species over time. The greatest contribution to it comes from species whose value curve extends on indefinitely, or at least for a good long while.

In other words, Sarkar might appeal to the principles of aggregating consequentialist logic that underlie his evaluative practice. But in this context, it is important to recall that, for Sarkar, his operationalization of biodiversity is one and the same as biodiversity. What his algorithm fails to select is excluded from consideration because no other, independent basis is capable of justifying inclusion. In light of this, to be excluded from consideration is to be devalued.

This brings up a second disturbing implication of Sarkar's axiology. It appears to sanction the view that, insofar as humanly controllable activities and behaviors are known to have already contributed to pushing some organism to the brink, a sense of relief should set in. For at that juncture, little will be lost in the final, fateful stroke that pushes the ill-fated organism over. This implication is made more disturbing by the likelihood that at some level of social organization, the institutions that sanction and fund conservation according to Sarkar's precepts are the same as those that sanction and fund those humanly controllable activities and behaviors that have contributed to pushing some organism to the brink. Under these circumstances, one might think that it is singularly important *not* to devalue a place or its resident "chunk" of biodiversity whose imminent total demise will be ensured by continuing insults.[160]

Sarkar's discussion is troubling in a third way, which has to do with his proposed criteria for segregating triage "patients" into disjoint classes.[161] His preferred patient-classifying tool is what conservation biologists call "Population Viability Analysis" (PVA). By Sarkar's own admission, there is a radical lack of scientific consensus on how to use PVA in conservation, which is an unsurprising consequence of radical disagreement on what it even means. In their utter opacity, the various definitions of PVA are reminiscent of the definitions of scalar "indexes" of biodiversity itself. Researchers who propose various competing formulas for computing PVA routinely fail to connect their computation's scalar output to some independently characterizable condition or property in the real world. In its stead is typically a vague declaration that it satisfies an intuition, which apparently varies considerably from researcher to researcher. This kind of vagueness is open to manipulation (whether conscious or unconscious) and constitutes a formidable hazard. It is a moral hazard insofar as it operates in the realm of moral decision-making.

---

[160] The discussion at the end of this subsection revisits this source of malaise with triage from a somewhat different perspective.

[161] As previously noted (in Note 159), the classification of patients is a critical part of any triage protocol.

This is not the end of the troubles with Sarkar's use of viability as a value determinant. As it turns out, his views on viability are in sharp discord with other tenets of his axiology. Logic dictates that if *viability* enhances or is essential to value of biodiversity, then *endangerment* militates *against* the value of biodiversity. Generally, the more something is endangered, the less viable it is; and conversely. Sarkar defies this logic. According to him, (for example, Sarkar 2002, 147) despite diminishing biodiversity's value, its endangerment *adds* to the mandate to conserve it. And for him, the mandate to conserve is equivalent to the attribution of value. In other words, his axiology rests on principles that are essentially self-contradictory.

Of course, if Sarkar asserts both A and not-A in support of his proposition concerning how environmental value attaches to biodiversity, this misfortune in logic entails nothing about the proposition's truth value. But it does contribute to the impression of a general inattention to drawing out the implications. Most of all, it increases the aforementioned hazard that whatever "reasons" are forthcoming will be reasons of convenience rather than carefully considered ones.

Sarkar's proposal that viability is a necessary condition for biodiversity to have value toes the line of yet another contradiction. This second incongruence concerns conjoining viability with rarity and endemism (geographical rarity) as signal attributes of biodiversity value. Even if an organism has a viable population, if that population exists in just one place on the planet, then its vulnerability to extinction is elevated.[162] As a consequence, one would think that narrowly established endemics would be at the bottom rather than at the top of Sarkar's triage list, so far as the implications of his thinking about viability are concerned. Sarkar ignores this implication of his stance on viability.

My discussion of viability and endangerment has focused on Sarkar's treatment of this topic because his is one of the few in which there is an extended attempt to rationalize the inclusion of these factors in a biodiversity-valuing ethic. But Sarkar's views and his seeming unawareness of their contradictions and other flaws are widespread. He is not alone in viewing endangerment as a key element in biodiversity value. More widely known than Sarkar's discussion of the "Viability Problem" is the famous definition of "biodiversity hotspot" due to Myers et al. (and described in Sect. 4.2.2.1, Geographical rarity). The notion of "hotspot" embraces both rarity (in the form of endemism) and endangerment (in the form of a history of impingement that has already drastically reduced local vegetation) as reasons to conserve biodiversity.[163] It is unclear whether these authors intend to justify these suggestions

---

[162] Most ecologists believe that this threat derives mainly from the ongoing human "conversion" of land. However, direct land conversion is not the only threat. Even an impenetrable physical barrier that circumscribes the tiny spot where a geographically rare organism lives is not an ecological barrier against such factors as changes in climate and the deposition of nitrogen.

[163] The alert reader will notice that this is something of a cheat on the definition of biodiversity as a state of affairs. I discuss causal history as a possible value-altering element in Sect. 6.13.3 (Causal factors and history). It is also useful to recall (from Note 20 in Chap. 4) that the endemism that Myers et al. consider is restricted to that of plants and vertebrates. The latter group of creatures is the subphylum of the phylum Chordata that is the most species-impoverished of the commonly compared groups on the planet. On the other hand, there is strong evidence for a high correlation of insects – from the most species-rich phylum Arthopoda – with plants.

on the grounds that rarity and endangerment are key elements of the value of a place's biodiversity. If this is not the intended normative grounding, then there appears to be none. Therefore, for the sake of this discussion, I presume that this is the intended normative grounding for their suggestions. But then those suggestions are very hard to defend.

Myers et al. do not create the tension (or flat-out contradiction) that comes from Sarkar's playing on both sides of the viability/endangerment court. But the idea that endangerment adds to biodiversity *value* does not need the disagreeable company of conflicting ideas for it to lead to another conundrum, which arises from its application to a theory of right action.[164] As I previously observed, embracing the notion that endangerment *decreases* value seems to entail also endorsing the reprehensible attitude that little or nothing is lost by ignoring that which is endangered. The notion that endangerment *increases* value has the equally uncomfortable consequence of seeming to endorse behavior that increases value via promoting endangerment. If hotspots are what really count, then other things being equal, it seems as though conservationists should do what they can to create more of them. One can imagine eliminating dispersed populations of widespread species to the point that they are endemic to a single, last remaining place where their heightened contribution to biodiversity then makes them merit our assistance in making its last stand.

Of course, this suggestion is absurd. The obvious response to it is that the doctrine of enhancing value through endangerment is intended to apply only to triage "patients" that are selected beforehand according to other criteria that those who must perform the triage cannot control. Compare: there was a bar brawl and the ambulance brought the injured to the ER. The doctrine is not supposed to justify standing outside the ER and stabbing random passers-by.

The problem with this response is that it ignores disqualifying disanalogies between the triage of biodiverse places-*qua*-patients and the triage of human patients. First, if all humans are presumed equally entitled to their health and life, then medical triage need not say anything about how this human life is more valuable than that other one. Both are presumed equal.[165] In contrast, a rational biodiversity triage protocol presumes significant inequalities of places with respect to their biodiversity value. Moreover, it is advisable to avoid stipulating that a place is "entitled" to its biodiversity. For on what grounds could a place be said to have such a right? Whatever those grounds might be, it is safe to assume that they are very different from the grounds that justifies a human's entitlement to health and life. Finally, the proposal to create more hotspots is the opposite of stabbing people – if one takes seriously the notion that hotspots really are more valuable places by virtue of the endangered condition of their biodiversity. Just as a place's value might be

---

[164] Myers et al. (2000) prefer the term "threatened" to Sarkar's "endangerment". As used by these two sets of authors, the two terms appear to be synonyms.

[165] I acknowledge that a principle of utter equality of all human lives is not universally granted. Some would say that a young life has greater value than an old one; that a life endangered by the foolish choices of its owner is less entitled to be lived than the life of a person who did not so contribute to her compromised condition. To the extent that human equality it is not granted, the force of the disanalogy diminishes.

thought to be enhanced by the cultivation of a garden there, another spot might be more highly prized by virtue of its being a hotspot. By contrast, no one could legitimately claim that stabbing a person changes her value; and in particular, no one would claim that a stabbed person (the flawed analog of the hotspot) is more valuable than an unstabbed one.

Myers et al. provide just the two aforementioned criteria for assessing a place's value as a hotspot. There is the endemism of vertebrates and plants, and there is endangerment – the latter implicitly determined by induction from the amount of primary vegetation already removed. I have already touched on the difficulties of calling on endangerment to make the case for value and will not recite that verse here.

Setting aside endangerment leaves just endemism – and the endemism of a relatively small group of organism (plants and vertebrates), at that – to assess biodiversity value. As I observed in Sect. 4.2.2.1 (Geographical rarity), endemism is geographical rarity and it is simply a conceptual error to conflate rarity of any kind with diversity. So if the endemism of a species logically entails a disproportionately large contribution to biodiversity value, then this contribution must come in through a logical back door that is not legitimately part of the concept of biodiversity itself. It is also possible that those who value endemic biodiversity presume that it contingently correlated with efficiency in conservation. That is, they might think that as a matter of contingent fact, by conserving endemics one conserves more for less cost and effort; in other words, endemic species are the key to a good heuristic for the species set cover problem. But this presumption is highly questionable: A species' high degree of endemism might entail its high degree of vulnerability to unpreventable perturbations in the specialized conditions (for example, the temperature ranges of its climate or the biogeochemistry of its habitat) that make life for that species' individuals possible. If so, this would make it a bad bet for the investment of conservation resources. And, as a "marker" for high biodiversity, there seems to be no reason to privilege endemism over direct assessments of the actual diversity of species (or the diversity of kinds in any other category). Even if, as Myers et al. seem to suppose, by felicitous happenstance or perhaps by an as-yet uncovered causal connection, hotspots turned out to be extremely species-diverse places, the value of the biodiversity (or species diversity) in those places would then properly derive from the fact that they contain lots of different species, not because the species are endemic to those places or because they are endangered.

In summary, it appears that both legs holding up Myers' case in support of the proposition – that a place's biodiversity is more valuable by virtue of that place being in a hotspot – give way. I now return briefly to the notion of efficiency in conservation as somehow undergirding judgments of biodiversity value.

6.13.2 Efficiency

In the preceding subsection, I discussed the triage-like orientation that seems to be the context for the role of viability or endangerment in some conceptions of biodiversity value. In this subsection, I explore the possible derivation of this view from neoclassical economic analysis in general, and cost-benefit analysis (CBA) and (economic) efficiency in particular.

Sarkar (2005, 160) draws the connection to economic efficiency quite plainly:

> The critical consideration is that of economy (usually called 'efficiency' in the biological literature…): the representation of as many different and as a high a concentration of individual biodiversity surrogates as possible in the least number of places.

This characterization pushes consideration of biodiversity value to the very center of the neoclassical economics stage, which has to do with the efficient allocation of scarce resources. The allocation of each place is a considerable cost; and so one should strive to allocate places to biodiversity as efficiently as possible. In that way (and only in that way) can the greatest possible biodiversity benefit – conceived as the greatest possible amount of biodiversity – be achieved at the lowest possible cost.

Compare: there is the value, to me, of acquiring the stratospherically priced season tickets for my local major league baseball team. Of course, in contemplating whether or not to acquire them, I do not consider just the value of my experience of live ball games. I also think of the costs, which include that of the car that I might have to forego to attend the games (which necessarily includes the requisite $8 beer and other baseball food), and the fact that my son might have to forego a college education.[166] In considering these costs, I am now doing a CBA in which the benefit of witnessing baseball – live and well played – are weighed against its various costs (in the form of benefits foregone). But I do not confuse the value of a foregone car or of my son's education, which enter into those costs, with the value of experiencing baseball. That latter value would not change if my neighbor, about to move out of the area and unable to use his remaining season tickets, bestowed them on me *gratis*.

In short, the value of experiencing live baseball is something that I can get a grip on without reference to its cost. I can even see how some baseball experiences might have greater value than others. For example, there is more than one major league baseball team where I live and ticket prices for the two teams are not equal. As a matter of fact, the quality of my experience will differ between the two teams because one team plays the game with greater and more dedicated enthusiasm and in a more exciting style than the other. But this has nothing to do with which team's tickets are more dear.

When they talk about efficiency, it is unclear whether Sarkar and Myers et al. really take themselves to be addressing the value of biodiversity. If they are, then their discussion is based on a confusion, which conflates the value of biodiversity (a presumed benefit) with the costs of acquiring it and with the question of its most efficient acquisition. One cost of a biodiverse place might be its cost as real estate – valued for the development of something other than biodiversity. This might vary from place to place, just as the season baseball tickets vary from team to team. But this has little or no bearing on the value of the resident biodiversity, which one can safely assume varies

---

[166] Prices for seats at major league games apparently have risen to this dizzying height. See Mallozzi (2009). Fortunately for my son, he has graduated college and need no longer worry about my being blinded by my love for baseball.

more or less independently of real estate values – much as the baseball ticket prices are no sure reflection of the quality of the baseball experience that they afford. The fact that a tract of biodiverse land cannot be acquired except (in some cases) at enormous expense has little to do with the value of its biodiversity.

Furthermore, the determination of economic efficiency is customarily a determination of the relative efficiency of various alternatives. This requires the computation and comparison of the economic value that each of several possible development paths is expected to realize. Couched in current dollar terms, this is the expected net present value of development along each of these paths. In this calculation, greater viability, which increases the likelihood of survival, increases value while endangerment has the opposite effect. Myers' et al. (and Sarkar, who, as I observed in the preceding subsection, also plays on the opposite side of this court) seem unaware of the devastating consequence of this consideration for the development path that attempts to preserve hotspots. The very definition of "hotspot" entails the precarious condition of its biodiversity, which in turn entails a substantially depressed expected net present value.

So in the end, it seems that conflating biodiversity value with the efficiency of acquiring it does not produce the result that Myers et al. might have hoped for. Considerations of economic efficiency appear to argue against preserving hotpots rather than for preserving them. I take that unexpected and unwelcome result to be a call to revisit the initial confusion on its own terms. To say with these biologists that the value of biodiversity varies inversely as the cost of the real estate on which it resides is not correct and not useful, and is itself an act that tends to devalue what they are trying to value.

Sarkar volunteers an observation that illuminates one of the hazards of failing to maintain a clear separation between the value of diversity on the one hand, and on the other hand, the plethora of other considerations that might enter into a conservation decision using CBA, or any other evaluation scheme, for that matter. He points out that his algorithm (sketched in Sect. 3.3.4.1, Place (again)) for iteratively accumulating a collection of biodiverse places (the places that, by his precepts, humanity *ought* to conserve) is capable of assigning radically differing biodiversity values to any *one* place – depending on an arbitrary starting point defined by the arbitrarily selected initial subset of places.[167] This initial selection is arbitrary in the sense that it has no axiological justification. It defines the initial condition for an algorithm that is itself arbitrary in the same sense: While it obviously must satisfy some intuition for Sarkar, he offers as chief justification that it is practical to implement. This does not qualify as sound axiological reasoning.

In the end, this scheme falls far short of qualifying as a principled system of evaluation. That dismal assessment is made all the more clear by reference to a similarly unprincipled system of criminal "justice". No one would consider satisfactory a system

---

[167] This casual acceptance of varying value assignments must be distinguished from the radically different and eminently sensible observation that, as experience in assigning values grows, heightened practical wisdom might justify a re-evaluation of things previously evaluated with less experience.

that assigned guilt or innocence to a sequence of defendants according to an algorithm justified by its ease of computation, starting with a first case decided by the flip of a coin. Yet this is an exact analog to Sarkar's principles of "biodiversity justice".

It is tempting to resist this unfortunate conclusion by retreating back to the triage metaphor. In this spirit, it might be suggested that the sequence of choices does, in fact, matter – because it affects evaluatively relevant aspects of the situation on which the next triage choice is made. One is not in a position to determine – in a single, sweeping, all-knowing, godlike gesture – what is of greatest biodiversity value everywhere on the planet. Instead, one must plunge into the gritty business of trying to make the first decision, then the next, at every step, trying to save what most adds to the accumulating total value of biodiversity. A model of economic market valuations might be thought to incorporate this sequential quality. This is easily observed on the stock market ticker, where there is an incremental adjustment in market value for every transaction that is done "at the margin". More generally, the price commanded today by a commodity can float upward as demand and purchases make its supply on the market dwindle. One might say that it is no different for biodiversity. With each biodiversity transaction, one tries to buy the most with what remains to pay for it.

Unfortunately, two stubborn problems undermine this retrenchment, along triage lines, to legitimize the arbitrarily initiated, sequence-dependent assignment of biodiversity values. First, the credibility of this move is undermined by a problem that often infects consequentialist-inflected thinking: That is, it systematically blurs the distinction between a procedure for deciding how to act rightly, and a justification of the rightness of an action (however decided upon) according to some independent principle. This point pushes beyond the previous point that the triage rules themselves appear to have no principled grounding. The new point is that, even using a well-justified decision principle (which biodiversity triage is not), it is possible to arrive at a wrong decision – a decision that results in a world with less value rather than more. To put this in another way, some independent set of principles that assess the rightness of action is required to also assess the quality of any decision-guiding principle that purports to arrive, in the ponderance of cases, at a right action. No such independent set of principles accompanies biodiversity triage.

Second, insofar as the move back to triage is based on the analogy of incremental price adjustments along a series of market transaction decisions, it ignores a critical difference between the running of real markets and the running of Sarkar's place-selection algorithm. The difference is that real markets are real. For that matter, so are the shifting conditions and classification of triage patients in an ER. The transactions in a real market and the treatment of patients in an ER are actual sequences of events. This is not true of the grand hotspot scheme due to Myers et al. Nor is it true of Sarkar's algorithm, though this might initially be obscured by the fact that it spits out first this place, then that. But these are just partial results obtained before the algorithm runs to completion. In fact, Sarkar's algorithm runs to completion "on paper" and produces value assignments to form a conservation plan – which is a prelude to taking any action in the real world. Like any algorithm, it is just a sequence of steps – a recipe. But the recipe creates or purports to create, as output, a single map of biodiversity values – in a single, sweeping, all-knowing, godlike running of the algorithm.[168] As a consequence, the major support this approach derives from the analogy of a sequence of market transactions, in which each is separately evaluated in light of its predecessors, is ill gotten.

I do not have the space to further pursue this discussion of conservation triage and its relationship to "conservation efficiency". So at this point, I must refer back to the questions raised in Sect. 6.13.1 (Viability and endangerment) about whether it is reasonable to think that key guiding principles for biomedical triage also apply to the evaluation of biodiversity. But there is a final point to be made quite aside from the question of whether the triage metaphor is valid. If the object of conservationists really is to cram as many species as possible into as little and as inexpensive a space as possible, then an honest assessment of *all* means to accomplish this end is called for. Recent science has yielded some nascent understanding of which means merit consideration. For example (see Fridley et al. 2007, 4, 6), it appears that habitats with high positive Native Exotic Richness Relationships (NERR's) are ones that invite efficient packing. I cannot think of a legitimate reason why those who promote biodiversity place efficiency would not therefore want to promote the "assisted migration" (further discussed in Sect. 7.2.1) of species into such places – as a cheaper, leaner means to the end that they seek. A good case can also be made for the greater efficiency of several of the other biodiversity-enhancing suggestions in the list in Sect. 5.3 (The moral force of biodiversity).

6.13.3 Causal Factors and History[169]

In Sect. 6.12 (Biodiversity as the natural order), I discussed how the causal influence of humankind on nature is sometimes thought to degrade the natural order of things. I return to causation in this subsection, but without that previous discussion's burden of relating it to a perfectionist vision of the natural world. Here I explore the human causal contribution to a harmful state of affairs saliently characterized by some undesirable state of biodiversity. But I now understand this in terms that transcend tarnishing a perfect state of nature.

So far as biodiversity is concerned, causal influence is typically assigned at the collective level – focusing on the behavior of human societies rather than the actions of one or a few human individuals – and regarded as historically persistent.

---

[168] In saying this, I'm not denying the obvious ability to re-run the algorithm at a later time; and I'm sure that Sarkar has this in mind. But a rerun merely improves, or purports to improve true knowledge of the global value map. It is the algorithm's best understanding, at the time of that latest run, of where in the world biodiversity value lies.

[169] This discussion owes much to perceptive comments by Jeffrey Lockwood, who uncovered some weaknesses in an earlier draft. It also owes to a spirited discussion with Jeremy Bendik-Keymer about the role of intent in moral responsibility.

This is of moral interest because the degree and nature of a causal role in inflicting harm can entail something about responsibility for the harm – most notably something bearing on the degree of blameworthiness for it. It can also have a role in determining what obligations there might be to mitigate the harm – to "right the wrong", according to some principle of corrective justice. Something like this picture seems to be bound into the frequently encountered claim that humankind is responsible for "harm to biodiversity".

The precise way in which causal responsibility figures into moral responsibility for any harmful state of affairs is subject to several ethical provisos. These are especially important – and tricky – for something as causally complex as the biodiverse state of the world. But the complexity of the causal role is just the start of the moral complexity. The degree and nature of moral responsibility also depends on what those with a causal role are actually aiming to do. Bearing on this question of aim or intent, but also constituting independent considerations, are: what the best experts know about the impacts of human behavior and activities, what people generally and reasonably can be expected to understand about this, what people are capable of doing to address their impacts, and even what reasonably can be expected of people's moral sensibility in such matters.[170] The thorny question of how to understand the notion of *collective* or group responsibility also intrudes. Group responsibility stands alongside group rights as one of the more vexed topics in moral theory, which unfortunately places it outside the scope of my discussion.

Finally and most fundamentally, the causal intricacies of human involvement in how a complex state of affairs comes to obtain has little moral interest unless the coming to be of that state of affairs causes or itself constitutes a harm of some kind. Few would deny that for a change in the biodiverse state of affairs to achieve moral interest requires good reasons for thinking that, not only did humans have a hand in that change, but that as a consequence of their causal role, either biodiversity itself is harmed, or other harms ensue from the change. The ubiquitous catchphrases "harm to the environment" and "harm to biodiversity" are often used so indiscriminately that even the identity of the sufferer of harm is left uncertain; this makes a complete mystery of the exact nature of the harm (born by whichever subject suffers it). The moral proposition is further obscured by routine omission of all the considerations in the preceding paragraph, even though every one is critical to a clear understanding of how causal responsibility bears on moral responsibility.

When it comes to the aim or intent of human behavior that influences the biodiverse state of the world, the presumptions that various thinkers adopt span the broadest imaginable spectrum. At one extreme, there are those, such as Jeremy Bendik-Keymer (2010; 2012), who suggest that there is *no* intent behind humanity's causal role in what he (2011, 15) calls a "disintegrating of – here a deliberately poetic term – nature's cornucopia *as we have known it*" [italics in the original]. According to him, this is the result of utter thoughtlessness or "wantonness" – quite

---

[170] See, for example, Kawall (2010) for a discussion of the epistemic element in responsibility from a virtue ethics point of view.

literally, something more or less completely out of mind.[171] At the other extreme are charges that humanity is engaged in a "war against nature", that "Homo sapiens… [is] waging a war against Nature"[172] and that such a place as Hawai'i is (Wilson 2002, 43) "a killing field of biodiversity". The trope of "waging war" is a starting point for a sequence of tropes, which circulates through "the wounds on the land"[173] – the aim and expected result of warfare's assaults, against which Nature is said to be making its "last stand" (Wilson 2002, 42). It is easy to see this as the expression, from Nature's point of view, of the proposition that – either through utterly unthinking behavior, or through monstrously sociopathic behavior characterized by satisfaction in waging a cruel war – humanity is reducing the world to biotic rubble.

Typically absent in accounts at both extremes is careful attention to many of the critical points – up to and saliently including a convincing account of where, exactly, the harm lies – that would serve to anchor the thread of a moral argument.[174] The issue of intent is a crucial one in assessing the degree of moral responsibility and in assessing how reprehensible is some action or pattern of actions leading up to some harm. But the issue of harm is logically prior. Establishing that harm is actually done is a prerequisite for a causal-historical argument for any kind of moral

---

[171] Bendik-Keymer's use of the term "wanton" is prone to misinterpretation. While "wanton" commonly suggests maliciousness (which necessarily involves some kind of deliberateness), Bendik-Keymer (2012, 450) explicitly excludes that element from his usage, which focuses solely on the element of thoughtlessness. Bendik-Keymer (2012, 450) attempts to address the "puzzle" of "how people can produce wanton consequences without deliberately doing so." But given his usage of "wanton", there is no puzzle about the possibility of this: It seems clear that thoughtless behavior (in Bendik-Keymer's sense of "behavior by actors who do not consider salient consequences of their actions") might produce consequences that those actors did not (deliberately) aim at (because they didn't even think of them).

[172] The first cited phrase is Rachel Carson's call to arms, from an April 13, 1963 broadcast of *C.B.S. Reports* on *The Silent Spring of Rachel Carson*:

Man is a part of nature, and his war against nature is inevitably a war against himself.

This statement echoes *Silent Spring*'s use of the word "war" to describe human impacts on the natural world (Carson 1962, 7, 93, 99).

Carson's call to arms is taken up by Dave Foreman. The second cited phrase is part of his presentation of the position of the Rewilding Institute, http://www.rewilding.org/thesixthgreatextinction.htm. Its Conservation Fellows include such prominent conservation biologists as Michael Soulé and David Maehr, though these scientists are not on record as explicitly endorsing this way of expressing the Institute's purpose.

[173] This is Michael Soulé's characterization of nature's state, in his address, "NATURE'S ASPIRIN: A CURE FOR MANY OF NATURE'S ILLS", delivered to the Western Conservation Summit in January 2009 (available at http://www.michaelsoule.com/, with text at http://www.facebook.com/note.php?note_id=204698819568168). Soulé is one of the more prominent of myriad writers who avail themselves of the trope of "nature's many wounds" which are inflicted by many acts of "wounding nature".

[174] Bendik-Keymer attempts to avoid the issue of whether or not harm is done when a species goes extinct. He does avoid the 'h'-word, but only by way of locutions that rely on it for their normative impact – as, for example, when he (2012, 450) says: "Wanton behavior is given to destruction by the way its thoughtlessness loses touch with what matters in our world and so tends to damage it." To "damage" something is to harm it or to harm those with an interest in its intactness.

responsibility. If collective harm is established, and only if it is established, one can forcefully argue that humanity has an obligation to mitigate that harm – at the very least by reigning in the instruments that have given rise to and perpetuated the harmful conditions. But if no harm is connected with those conditions, then causal responsibility (of any sort) has no moral significance – for causal responsibility by itself does not entail anything about whether or not those conditions constitute or cause harm. This might seem to be an obvious point, but the number of narratives that ignore it counsels for making it explicit.

That the presence of humans and their activities has altered the natural world in general and its biodiverse state in particular is not in dispute. But in itself, the difference in biodiversity – between the actual, human-inhabited world and some imagined world devoid of either a human presence or of the presence of humans who nevertheless do not engage in the characteristically human activities that are causally relevant – is not a reason for believing that it constitutes a harm. This is true even if that gap is correctly characterized as largely a matter of humans being instrumental in diminishing biodiversity – for example, by playing a role in the "premature" extinction of various species.

In other words, a separate reason must be supplied to establish that the human-induced differences are harmful – because causal responsibility for those differences is neutral with respect to whether or not they are harmful. To establish harm generally requires that a subject that suffers the harm be identified. A harm is a harm *to* someone or something, or to some group of persons or things. When it comes to biodiversity, there are two classes of theory on this. The first presumes that the harmed parties are current or future humans or both. The harm consists in depriving us currently respiring humans and our yet-to-be respiring descendents of such things as ecosystem services, mankind's pharmacopoeia, some preeminently valuable knowledge, or some of the many other goods that are supposed to flow from the world's biodiversity. But on the evidence examined in this chapter, none of these things have the kind of value that is claimed for them.[175]

A different take on who or what is harmed arises from the common use of the phrase "harming biodiversity" in a context – such as "the war on Nature" – that suggests that its literal meaning is intended: As a consequence of "losses in biodiversity", biodiversity itself is harmed. After all, if you cut off my arm, I will suffer a harm as a result of its loss and you will have harmed me. This grants *literary* license to say, more generally, that a loss of some part *of* X is a harm *to* X. That includes, for example, X = a car, which "suffers" the loss of a hubcap; or X = the set of all cars on the road, which suffers the loss of mine, which just broke down. But in these cases, the "suffering" is a matter of moral indifference and the harm to the car when I remove its hubcaps does not warrant moral condemnation of my action. Much the same can be

---

[175] I realize that it is impossible to prove the nonexistence of a successful argument for biodiversity's value. I also understand (and have previously noted) that some philosophers prefer to avoid the notion of harm altogether. I revisit the latter issue, particularly with respect to virtue ethics, in Sect. 8.2.6 (What appropriate fit is not).

said of biodiversity, when it "suffers" a loss or when it is "harmed". Biodiversity – that entity, which is commonly measured by the number of species – is simply not capable of suffering in any morally interesting sense. For example, it obviously is not sentient and so cannot suffer in the sense of experiencing pain. Just as obviously, it cannot suffer from the thwarting of its projects, because biodiversity has no project to thwart.

It's clear that merely establishing a human causal-historical relationship with biodiversity cannot, by itself, ground a theory of biodiversity. These difficulties – of identifying a harm, which in turn, requires identifying a subject suffering the harm – cast doubt on whether a causal-historical theory can even clear the preliminary hurdles. But they are not the only hurdles, as consideration of the other, above-mentioned moral provisos reveals. Let's focus attention on the matter of intention, starting with the position at one extreme – that humankind is guilty of the kind of vicious behavior that would be involved in "waging a war on Nature".[176]

Consider: If something is done that in fact constitutes a harm, and if it is done "viciously" – in the sense of "intending cruel or violent harm" – then these factors significantly color how one views the moral burden of causally responsible parties. In the broad moral scheme of things, the nature of the intent varies. Other factors, all varying partly, but not completely independently of intent, intertwine with intent to produce a final moral assessment of culpability. One of those factors is how directly the harm is a consequence of actions and behavior. The more tenuous and convoluted the causal chain and the more that causal factors beyond human control dominate, the weaker the case for moral responsibility is generally supposed to be. Another variable, related to the last, registers the capability of agents to perceive the causal connection to the harmful consequences and to recognize that these consequences do indeed constitute a harm. Furthermore, benefits that accrue to the harming agent are commonly thought to substantially increase how reprehensible is the harm inflicted. At the most extreme, an agent's well-calibrated *aim* to benefit herself by inflicting harm on others boosts the warrant for regarding this behavior as vicious.

At the other extreme, if causing a harm is an unintended, inadvertent, and relatively indirect consequence of unknowing and oblivious behavior whose harmful conse-quences have no obvious benefit to the perpetrators, then the tie between causation and responsibility generally remains very weak, though it still might justify an obliga-tion for compensating action. The tie is strengthened to the extent that (among other possible factors) specific intent, knowledge of the harmful consequences, and a view to gaining benefits that flow directly from the harm creep back into the situation.

One path through this thicket is marked by categorical signposts, which signal how intent and benefits combine to progressively strengthen the case for tying moral responsibility to causal responsibility:

1. **Unintended harm.** The causal agents act and behave to attain some benefit, but they are not (fully) aware that their action or behavior is harmful. If some harm

---

[176] The analysis that follows is meant to "lead by example". It is intended to suggest some issues that merit but are rarely given sober moral assessment rather than to be a complete working out of those issues.

is done, the agents might not perceive it as a harm. Or some consequence might register as harmful in some small or insignificant degree, but absent an appreciation of its full and possibly great significance to the party harmed. In short, though harm is done, no (or little) harm is intended.

In criminal law, some cases of reckless endangerment fall into this category. However, these cases typically require that the harm be foreseeable. My definition is broader and includes cases in which even the ability to foresee cannot reasonably be expected.

2. **Collateral damage.** The causal agents are (fully) aware that among the consequences of their actions are very likely unwanted harmful ones. These unwanted harms are viewed as unavoidable in attaining a benefit, but they are not directly instrumental in attaining that benefit. This corresponds to what, in recent, military-derived parlance, is called "collateral damage". It is less clear that a harm of this kind is completely *un*intended, for the action aimed at the benefit is intended and the attendant harm is expected. Perhaps one could express this unclarity by saying that the harm is only "indirectly" intended.

   Most commonly, a case of collateral damage is not a case of willful rapacity. That is because the harm is not a means to the end of a benefit; nor does the harm itself constitute any part of a benefit. However, these exclusions do not entirely preclude rapacious behavior in which the harm, though unwanted, is seen as justified by the benefit. Picture the bank robber who "must" shoot the teller poised to trip the alarm. Even the act of scamming an elderly couple out of their pension can be viewed in similar fashion: The scammer derives no great benefit from the emotional and financial distress that she sadly "must" inflict on her gullible victims. The benefit of the scam is the money transferred from scammed to scammer. Of course, the scammer also can be said to treat the elderly couple as a (mere) means to her end. In that respect, the scam example might be said to belong to the following category.

3. **Means to benefits.** The causal agents understand their acts to be harmful, but additionally regard their acts of harming to be a means or part of a means for attaining certain benefits, though the infliction of harm in itself does not constitute a benefit. Thus, it is by means of using a hostage as a human shield that the desperate criminal protects herself and makes good her escape. But placing the hostage in harm's way has no benefit independent of that. The desperada would gladly don her invisible cloak, had she remembered to bring it along for the heist.

   This is a step beyond "indirect" intention. The use of a harmful means to a beneficial end makes it plausible to say that, in contrast to collateral damage, the harm is intended in a direct way, qualified only by the fact that the focus and motivating intent is to attain the beneficial ends. It is also plausible to say that the good of such ill-gotten benefits are tainted in some serious way by the nature of the means employed to attain them.

4. **Constitutes a benefit.** The act of harming is not just a *means* to a benefit, but *constitutes* the benefit or part thereof, and the causal agents are (fully) cognizant of this; that is, they understand that the harm to others amounts to a benefit for themselves.

This constitutive relationship gives us leave to say that the harm is as much the focus of intent as the benefit that it constitutes. Such cases are the ones most prone to justifiable characterization as vicious.

Fortunately, this kind of viciousness is probably quite rare. In the realm of criminal activity, it is the psychotic stuff of premeditated rape and murder for their own sake.

Let's now return to this question regarding biodiversity: To what extent can human causal influence be said to be vicious, insofar as it results in the loss of biodiversity? Let's grant what seems highly doubtful in light of this chapter – that loss of biodiversity constitutes a harm. Still, the only reasonable answer seems to be: almost not at all.

Down through history, most human behavior affecting biodiversity pretty clearly falls into the category (1) of "unintended harm" in the list above. It is true that for at least 70,000 years, human activities and behavior have had a major causal influence on the "development path" for biodiversity. The terrifically effective predation of megafauna – by both humans and their invited and uninvited companions – over the course of the human diaspora was undoubtedly unmindful. People also undoubtedly noticed disappearances of their prey as well as the depredations by such animals as pigs and rats that accompanied or infiltrated their communities in various peregrinations. But they adapted as best they could to any changes that required adaptation and continued going about the business of surviving and being human. In the current millennium, the most dramatic remixing and reconstituting of which organisms live where – which heavily influenced which ones survived and which ones did not – followed close on the heals of Columbus' 1492 visit to the New World. This first giant step towards recreating an ecological Pangaea was entirely "out of mind" of anyone at the time. Indeed, it was out of mind and human comprehension until the late twentieth century (Mann 2011).

Only very recently have some of these circumstances, which affect the nature and degree of moral responsibility, changed. The nature of the impacts of characteristic human behaviors and activities that have pervaded human history down to the present time are finally coming to be understood at some very coarse level. Probably – and improbably, because the story is now so well known – the very first recorded extinction in which human activity was understood to have played a role was that of the *Raphus cucullatus* – the dodo – which vanished from its last residence on the island of Mauritius sometime in the seventeenth century. Four centuries would pass until it became common knowledge (at least among those with their nose in the subject) that for 70,000 years or so, *H. sapiens* as a species (or as a global society or as a collection of societies or as a collection of individuals) has had a significant, ongoing, causal role in bringing about the current, general, biodiverse state of affairs – whether regarded for better, for worse, or neither.

General *moral* consciousness – the awareness that a causal role in species extinction could possibly have moral implications – is at least as recent. Although it should be taken for what it is – a data sample of two – Bill Bryson (2003, 476) recounts how two American birding enthusiasts – who one would think were among the most

sensitively attuned in their appreciation of birds – reacted on their sighting of
Vermivora bachmanii (Bachman's warbler), at a time when it was thought to be
possibly extinct:

> Perhaps nothing speaks more vividly for the strangeness of the times than the fate of the
> lovely little Bachman's warbler. A native of the southern United States, the warbler was
> famous for its unusually thrilling song, but its population numbers, never robust, gradually
> dwindled until by the 1930s the warbler vanished altogether and went unseen for many
> years. Then in 1939, by happy coincidence two separate birding enthusiasts, in widely sepa-
> rated locations, came across lone survivors just two days apart. They both shot the birds,
> and that was the last that was ever seen of Bachman's warblers.

Bryson goes on to recount similar stories that make it clear that this sort of attitude was not confined to the United States. The last nesting pair of *Pinguinus impennis*
(great auk) was likely the two birds killed by four Icelandic fishermen in 1844.
The fishermen viewed these creatures as bundles of very valuable feathers, not as
the last individuals of a valued species. The final, fatal blow to both the dodo and the
great auk followed on a long history of hunting the birds for their eggs, skin, fat,
flesh, and feathers both by humans and their animal traveling companions, including the invited pigs and crab-eating macaques along with the uninvited rats.

But even the late-developing consciousness of human implication in the extinction of one or more species is not yet a consciousness of a human-induced reshaping
or general diminution of *biodiversity*. And there is a yet wider gap to an awareness
that human-induced changes in biodiversity might warrant moral consideration.
Even if one grants that the extinction of a species is also a diminution of biodiversity
(something that might seem assured only if one considers species richness to be
biodiversity's ultimate measure), whether or not it also diminishes the *value* of biodiversity or in some other way constitutes a harm is a question that can only be
answered by reference to some model (in Chap. 5, The calculus of biodiversity
value) or theory (in this chapter) of biodiversity value.

The anecdotes that I have related are representative of a more general body of
anthropological evidence that gives little reason to think that, at least until the latter
part of the twentieth century, much if any human influence on biodiversity goes
beyond category (1), "unintended harm" – the weakest basis for moral responsibility. Within that category, not even the decisive requirement for reckless endangerment – that the perpetrators should be reasonably expected to foresee the harm – is
met. *A fortiori*, nothing approaching viciousness can be ascribed to the behavior of
humans with regard to biodiversity up until very recently.

An over-hasty look at environmental history before the late twentieth century might
initially leave a different impression. For example, some attempt was made to rein in
the slaughter of the great auk before the final species-fatal killings. But in fact, this
was nothing more nor less than a regulatory gambit in a regional economic dispute:
It expressed Newfoundlanders' regard of themselves as sole owner of this feathery
economic resource and resentment of New Englanders' plunder of it. Moreover, the
economic resource was understood strictly in terms of a single species of bird, not in
terms of biodiversity. To cite another example – of a type of a phenomenon that has
pervaded human history: The effect on biodiversity of uninvited stowaways (such as

the rats that accompanied humans in their diasporas and peregrinations) can barely achieve the level of "unintended harm" – insofar as a case can be made that it was, as that category specifies, an unintended, inadvertent, and relatively indirect consequence of unknowing and oblivious behavior whose harmful consequences had no obvious benefit to the human perpetrators. This behavior does not even rise to the level of reckless endangerment, because there can be no reasonable expectation that (up to a few decades ago) the human wanderers could have anticipated any harm – even presuming that a case can be made for regarding the resulting changes in biodiversity as harmful.

But what about human behavior in the latter part of the twentieth century and later, when some understanding of patterns of biodiversity change and the connection of these to human activities and behavior have become apparent? I believe that the shift in focus to more or less the current time does little to shift which moral categories apply.

At the top of the most direct, current-day sources of human-induced changes in biodiversity stands human appropriation of land for uses that involve its radical transformation. The reasons why people appropriate turf typically have nothing to do with biodiversity and everything to do with such things as growing food, building buildings and infrastructure, and providing transport. Needless to say, every hectare of land that goes under the plow, under the foundation of a structure, or under the asphalt running up to it becomes quite a different place. So does a wide swath of surrounding land. And this change in environment is reflected in which collections of organisms do well or not so well there. In one of the more comprehensive and credible surveys that shed light on this, Wallace Erickson and his colleagues (2005) estimate that buildings and other infrastructure are responsible for the vast majority of the 500 million to 1 billion bird deaths annually in the United States alone. Human transportation continues in the grand tradition of the Columbian exchange, reshuffling the biotic deck by providing transportation for non-ticket-paying, nonhuman stowaways to places that they would not likely have visited otherwise or quite so soon. Roads and road traffic kill all manner of creatures that attempt crossings. They also are ecological walls, despite an unimpressive presence in the vertical dimension. They substantially alter the mix of organisms and hence the biodiverse state of affairs in their vicinity, not just by subdividing previously undivided habitat, but also by creating habitat edges, whose asphalt boundary creates distinctive ecological conditions quite unlike either the previously contiguous expanse or the interior of the new subdivisions.[177] Merely turning on the lights turns off the lights for some number of creatures. For example, on the Hawaiian island of Kaua'i, the threatened *Puffinus auricularis newelli* (Newell's shearwater) is threatened partly because (Mitchell et al. 2005):

> Street and resort lights, especially in coastal regions, disorient fledglings causing them
> to eventually fall to the ground exhausted or increase their chance of colliding with an

---

[177] Forman et al. (2003) provide an introduction to a growing literature on road ecology – a discipline mostly unknown to all but those with a Ph.D. in the field. I note in passing that there's no denying that, in creating these new kinds of habitat, the diversity of habitats is increased.

artificial structure (i.e., fallout). Once on the ground, fledglings are unable to fly and thousands are killed annually by cars, cats, and dogs or die because of starvation or dehydration.[178]

There is no more direct way to alter the biodiverse state of a place than to "repurpose" it. Yet few people are aware – or can reasonably be expected to be aware – of something as recondite as the ecology of roads and roadkill. And the appropriation of land for farming and structures, as well as our facilitating the travels of invited and surreptitiously hitchhiking plants, animals, and microbes, are little-changed, millennia-long-standing aspects of how humans have operated in the world and continue to do so.

I am not defending willful ignorance; I am not suggesting that there is not sometimes a moral obligation to acquire certain kinds of knowledge that are requisite for acting well in the world; nor am I less than enthusiastic about educating ordinary citizens about what their style of living on the planet entails for biodiversity and the natural world generally. I am certainly not defending moral permissiveness, morality by majority vote, or morality by long-standing habit. And, as I argue in Chap. 8, I think that there is a compelling case for thinking that there is something of possibly great normative significance in the great magnification of effects by the greatly increased human numbers and the vastly greater efficiency with which humans pursue their biodiversity-transforming activities.

However, I question whether there are any reasonable grounds for believing that any of these current-day activities rise much above the level of category (1), "unintended harm" – Bendik-Keymer's notion of profound thoughtlessness – again, if indeed, human-induced change in biodiverse states of affairs does constitute a harm. For none of these activities does it appear that altering the biodiverse state of affairs itself constitutes a benefit (category (4)). Nor is that alteration a means to any benefit (category (3)); for while building a road might change biodiversity, bulldozing, not the change in biodiversity is the instrument for accomplishing that end. It is often even difficult to press the charge that these activities are pursued under the dark cloud of a cold calculation of collateral damage (category (2)) when the calculators do not regard the effects on biodiversity to be "damage". Although now, in the twenty-first century, one can legitimately ask whether this might be a matter of reprehensible negligence in moral education, changes to biodiversity simply still do not generally rise to the level of individual or social consciousness as a behavior-guiding norm. At least they do not unless and until a directly noticeable and notably unpleasant effect for humans is perceived to be tied to such changes.

Viciousness simply does not to enter into the picture. I do not think that you, my reader, are acting viciously for your role – and you do have a role, at least as a

---

[178] This is an extremely common phenomenon. The authors of this report apply this identical commentary to *Pterodroma sandwichensis* (Hawaiian petrel), *Puffinus pacificus* (wedge-tailed shearwater), and *Oceanodroma castro* (band-rumped storm petrel). And their concern about the effects of artificial lighting extends to other avian and even non-avian species, such as *Eretmochelys imbricata* (Hawksbill sea turtle). Other reports reiterate the same challenge to yet other species in other places – for example, St. Kilda (in the Outer Hebrides off the coast of Scotland) and the Canary Islands.

purchasing consumer – in appropriating and transforming the land on which the building in which you dwell sits, for your role in the presence of the habitat-dividing road that runs up to your front door, or for your role in appropriating and transforming the land that produces the bounty on your dinner table. Nor do I think that society at large is acting viciously by sanctioning and facilitating this individual behavior of yours, which extends to all members of society collectively.

I have focused on human transformation of the landscape for its outsize contribution to changes in the state of biodiversity. But of course, many other human activities and behaviors also affect that state. The effects of some, such as hunting, are as direct as those of land transformation. The effects of most activities tend to be less direct, though not necessarily minor. Among that majority are, for example, the production of chemicals that find their way into various organisms and the escape of fertilizers to places – for example, rivers, lakes, and oceans – that they are not intended to fertilize. But for the most part, these seem to have no special quality that bolsters their qualifications for anything beyond a weak moral tie to their human causation. If the relative indirectness of their causation has any bearing, it militates towards weaker rather than stronger moral significance.

In short, a sober assessment of the extent to which intent and benefit tie moral responsibility to causal responsibility for changes in biodiversity turns up meager grounds for thinking that this nexus has itself undergone significant change through the great swath of human prehistory down to the present. The Pleistocene megafauna of North America, the great auk, the passenger pigeon all succumbed to human activities that were focused on providing human goods, not on anything remotely resembling vicious intent. There really is little to distinguish these cases morally from an eagerness to develop the Port of Anchorage on the broad backs of belugas.[179]

In continuing a long-standing pattern of human behavior, which includes relative unawareness that it has any broad environmental significance and little sense that this behavior rises to the level of causing a harm and therefore warrants moral consideration, there is little to distinguish the case of those belugas from most any other, similar case in the last 70,000 years. However, two parallel shifts in human thinking have recently had more than a little significance for how biodiversity is regarded. First is the nascent understanding, vested mostly in a relatively small number of scientists and environmentalists, of how the effects of human behavior and activities on individual species enter into the larger biodiversity picture. The second is the rise of some strongly held, strongly promoted, but I would say, weakly justified convictions on just what that picture *should* look like. This is what is variously called, "biodiversity management", "biodiversity development", or what I call "the biodiversity project", which is the subject of scrutiny in Sect. 8.1 (The disvalue of the biodiversity project).

The biodiversity project largely comprises novel experiments in species manipulation. With its emphasis on saving a shortlist of species on their way to oblivion, it includes captive breeding and the introduction of populations of creatures regarded

---

[179] See Sect. 6.3 (Biodiversity as service provider) for an account of the Cook Inlet belugas.

as surrogates for previous residents. On the opposite side of the spectrum are programs to control and extirpate "pest" species. For the world's most unwanted, "specicide" (the programmatic elimination of a species), even "genicide" (the programmatic elimination of an entire genus) – for example of all *Anopheles* species of mosquitoes – is part of the program.[180] Sometimes one and the same species, such as the *Canis lupus* (grey wolf), lurches back and forth between the poles of being the reviled target of specicide for its depredations and heroic embodiment of ecosystem salvation... for its depredations. One might think that differences in geographical, ecological, and historical context might justify shifts in the status of that creature (and others) as hero or villain. But credible principles undergirding the moral relevance of these contextual elements elude coherent formulation. This kind of Janus-faced attitude, which divides a small part of the biotic world into the wanted and the unwanted, and makes the vast majority of organisms an afterthought or, more likely, lost in a sea of thoughtlessness, poses a problem of consistency for such value models as the incremental one, discussed in Sect. 5.1.1 regarding The incremental model. Serious as that problem is in itself, it has far greater significance as a symptom of a general failure to address the pivotal question of how causal responsibility for the current, biodiverse state of affairs morally sanctions *any* kind of biodiversity project.

I leave a more thorough exploration of this crucial question to Sect. 8.1.2 (Responsibility for nature).[181] However, I wish to mention a few points, which connect the current discussion of deriving moral responsibility from causal responsibility to the question of determining whether, by virtue of humanity's role in changing the state of biodiversity, humanity inflicts some harm. The first point has to do with just how precarious is the understanding of the underlying causal responsibility. This understanding commonly derives from a vague gesture at a difference in states of biodiversity – the difference in biodiversity between our actual, human-inhabited world and some imagined world without us.

The first term of this difference is or should be a matter of empirical investigation. But every scientist who works on biodiversity acknowledges how difficult it is to characterize the biodiversity in our current, actual world. This is reflected in radical ignorance of even the simplest of characterizations – in terms of the number of species of living organisms that inhabit the planet. Current estimates vary by more than an order of magnitude. Even on a far smaller scale, marine biologist John Spicer (2006, 2–4) lyrically recounts the impossibility of knowing "who" lives at Wembury Bay – Spicer's "back yard" and perhaps as intensively studied a place as

---

[180] See, for example, Judson (2003). Judson's suggestion might shock those not inoculated with the biodiversity project serum, but it is entirely congruent with a much larger body of scientific, environmental, and "conservationist" writing within that project. Not atypical is its narrow focus, which excludes from view the consequences of annihilating an entire genus of insect – for example, on bats and the other insectivorous creatures for whom they are a dietary staple.

[181] One might also ask whether "biodiversity management" constitutes vicious behavior. I argue in Sect. 8.1 (The disvalue of the biodiversity project) that it is more likely that its ends are either incoherent, or coherent and lacking any sound justifying reasons; and as a means, it is ill informed, unjustifiably self-confident, and imprudent. But not vicious.

any on the planet. Still, one might plausibly say that, although current ignorance of biodiversity in our actual world is profound, one can expect scientists to make steady, if slow, progress in peeling back that ignorance.

The second term of the difference is quite another matter.[182] Characterization or evaluation of the biodiverse state of an alternative world devoid of human influence (after some prescribed juncture) or very differently influenced by humans seems an utter impossibility. One need only consider the multitude of specific changes that people have wrought directly, and which have initiated a corresponding multitude of cascading causal threads, which have woven themselves into causal webs that cannot possibly be unraveled. In every place where people have dwelled, species have been shuffled, forests cut down, watercourses altered. Every place of human habitation has been changed in its hydrology, biogeochemistry, and fire regime. There is simply no unraveling any of this, let alone all of it.

That consideration is huge; yet it is dwarfed by another: People have transformed the planet in ways that fundamentally change the terms on which any organism must struggle to live and procreate *anywhere* on this planet – not just where people have built cities, roads, or cut down a forest to farm the land. That is most evident in such changes as those in the planet's climate, the ocean's acidification, and the human-produced deluge of reactive nitrogen into the land and the planet's fresh- and salt-water bodies.[183] For an organism – even one well removed from most doings of humanity – to survive anywhere in this world is a very different proposition from what it would have been for it to survive in a world in which *H. sapiens* had never emerged, or had not been so populous, whatever that world might have looked like. This, I believe, necessarily confounds any attempt to construct a coherent norm, one that is based on how organisms would live in a nonhuman, or pre-human, or minimally human world whose biotic and abiotic structure would require a way of life very different from what the actual, current world demands of any organism here, now. As a consequence, even if human causal responsibility for the current, transformed biodiverse state of affairs were regarded as morally reprehensible, and even if (contrary to fact) it were possible to conjure up a reasonable picture of what the biodiversity of the world would have been in the absence of those human-induced transformations, there is no biological possibility of (let alone practical credibility in) posing that picture as the norm for what the biodiversity of the current world ought to be.

One can begin to get an idea of the problem by considering the question of what obligations people might have to save organisms whose way of life no longer works well in the human-altered planet on which they now live. Do people have an obligation to reverse those life-framing alterations? It seems impossible to maintain that position, for reasons that I have already suggested: First of all, it is impossible to know what that reversal could consist in, given the impossibility of unweaving the

---

[182] Here I reprise some themes from Sect. 6.12 (Biodiversity as the natural order), which expressed concern for the prospects of finding some credible characterization of the natural order of things.

[183] Air currents transport and deposit atmospheric nitrogen on the land; watercourses carry nitrogen to any body of water into which they flow.

causal web. Even it were possible to define "reversal", the radical transformational nature of the human influence on the systems involved and the broad spatial extent of the resulting transformations through a long history of startlingly persistent human behaviors and activities would make reversing them an impossibility. One might think that this problem could be ameliorated by aiming at some more recent restoration target – say, within the range of a human generation's memory or the generation before that, or before Columbus and his successors began the process of re-creating an ecological Pangaea. But the justification for shooting at any such a target requires some as yet unspecified principle. It is hard to see how any principle could be anything other than arbitrary.

Even if reinventing the biotic and abiotic conditions for living an earth-dwelling life were not impossible, leveraging causal responsibility in an attempt to establish a norm would still be problematic. That is because if an organism is struggling now, one likely cause is that it finds itself in conditions that are a consequence of a very long and complex history of characteristic human activities and projects. While we now might see ourselves as having a choice to modify some of those activities and projects or to cease and desist from some of them entirely, acting on that choice would not effect any significant change in the conditions that already obtain and that currently define the terms for most every organism's struggle for existence. In other words, what humankind can do doesn't achieve what is required to justify doing it.

On the other hand, what humankind cannot do it cannot be expected to do: The more ambitious project of actually redefining the terms for existence on this planet collides with generally accepted presuppositions concerning moral duties – that they are special and sufficiently specific to circumstances in a way that makes it possible and practical for moral agents to focus their moral attention, to make sensible decisions, and to act within the compass of capabilities that they can be reasonably presumed to have. Those capabilities do not, for example, plausibly extend to the saving of most every struggling species. But what about doing what *can* be done to "manage" the survival of struggling plants and animals whose ways of life no longer work in the current world? What about putting what few we can in zoos or arboretums? There are many reasons why, I believe, these are very bad ideas. But for this discussion, I will say only that any grounds for the responsibility to do this are seriously or entirely undercut by the fact that such heroic efforts at best achieve a temporary and brief extension of "life" for a vanishingly small number of the "living dead". And the cost of that temporary reprieve for these few is often a life that those living-dead individuals must live, which is in significant ways incongruent with the way of life for their kind.

It might be countered that when the decision is made to lay the concrete for a parking lot over the last known remaining individuals of some species of plant, or to shoot the last one or two individuals of some animal that has the misfortune of being viewed as "game", then the capabilities (for restraint) are not in doubt and the responsibility is clear. But this argument focuses far too narrowly on the last proximate act leading up to extinction. The moral significance of that act is arguably dwarfed by a longer history of anthropogenic stage-setting, which made the final act an anticlimactic denouement – the inevitable, because the only possible, resolution of the plot. One can imagine some minor variations in the details of exactly how the final act is played. But the conclusion is always the same – whether the result of a rash but intentional act, or the result of one last and fatal inadvertent insult.

The validity of the points that I am making here does not hinge on the culpability of people acting 70,000 years ago – or any time since then when awareness of both the facts of the matter and their moral implications, if any, could not be reasonably expected. It suffices that essentially all the biomes of the world are now anthropogenic biomes, and that the success, or failure, or state of being on the brink of failure, of any organism is directly related to whether or not it thrives in whichever anthropogenic biomes it calls "home".[184]

To put this entire discussion in perspective, I reiterate that many of the various stubborn barriers to understanding what causal responsibility for the current biodiverse state of affairs entails for moral responsibility are built into a state-based model, which is at the heart of the biodiversity project.

---

[184] See Sect. 5.3 (The moral force of biodiversity) for a brief explanation of the concept of "anthropogenic biome" or "anthrome", as proposed by Ellis and Ramankutty (2008).

Chapter 7
Some Inconvenient Implications

The preceding chapter diligently followed many and varied twists and turns in many and varied arguments on behalf of the proposition that biodiversity deserves preeminent attention as what is preeminently valuable about the natural world. The risk of such a detailed examination is that of leaving the impression that the many and varied failures in arguing the case for biodiversity's value are due to many and varied missteps that are are largely independent of each other. Along with that goes the risk of leaving the impression that at least some of these missteps are capable of rehabilitation by means of more careful reformulation.

That, I believe, would be an incorrect impression. This chapter tries to forestall it by pulling back from the microcosmic churn of swings and roundabouts related to specific arguments. By means of a few selected topics, it examines how a program that identifies certain states of biodiversity to be emblematic of natural value and for that reason, undertakes to bring those states into existence insofar as that is possible,[1] regularly finds itself burdened with certain uncomfortable and inconvenient implications; and how this leads it to seek new outlets of expressing nature's value, which paradoxically sweep biodiversity aside as a primary consideration. The chapter again takes up the persistent questions (already touched upon in Sect. 6.7, Biodiversity as value generator) about how the timeframe of changes relating to biodiversity affect the question of its value. It closes by examining, or really revisiting, a possible complaint regarding this book's critique – that despite all efforts to the contrary, it misconstrues the biodiversity program in some way. In one version, the complaint is that when biodiversity advocates say that, for example, biodiversity is good for human health, they do not really mean what those words seem to mean. In particular, the complaint is that, in such a claim (about biodiversity's promotion of good human health) "biodiversity" really means something rather different from what biodiversity is taken to mean in Chaps. 3

---

[1] This is what I call "the biodiversity project" in Chap. 8.

D.S. Maier, *What's So Good About Biodiversity?: A Call for Better Reasoning About Nature's Value*, DOI 10.1007/978-94-007-3991-8_7,
© Springer Science+Business Media B.V. 2012

(What biodiversity is) and 4 (What biodiversity is not). Maybe it means something like "all and only those particular organisms that tend to promote human health"; and that ends the discussion.

This chapter, in turn, is prelude to the book's final chapter, which shows in a more systematic way why an evaluative program focused on biodiversity is highly unlikely to be able to account for the value of the natural world.

7.1   Value Deflation from Nondiscrimination

It is a familiar story about an exclusive club. When you make the unnerving discovery that the rules of your club require that the riffraff be admitted, then your club does not seem anywhere near so fine a thing as it did before the riffraff actually shows up. That, I think, is part of what happens in setting up and sticking with some reasonable-seeming rules for the inclusion of kinds (for example species) in the biodiversity club. And it is part of what undermines the biodiversity view of natural value.

In other words, value deflation comes not so much from constricting (or really, clarifying) the concept of biodiversity as from expanding it to the full compass of its rules of membership. If one agrees to define "biodiversity" as a matter of the diversity of kinds in biotic and biota-encompassing categories, then one must live with the consequences. In particular, one must admit all kinds within any agreed-upon category without excluding or pretending to exclude kinds based on other agendas. In truth, a diversity-contributing kind might, on some other grounds, just as likely be repugnant as the opposite. But its contribution to diversity does not depend on the kind falling on one side of the divide rather than the other. Obviously, so far as the diversity of species is concerned, human parasites, disease organisms, and their vectors must be granted admission to the species diversity club; and their value contribution to the club is neither less nor more than that of polar bears. Similarly, so far as the diversity of functions is concerned, the fire-inducing, water-reducing, channelizing functions of *Tamarix* are on a par with all other functions. And if some such function as productivity is a good, then it is hard to justify not embracing this valuation when it comes to algal blooms and kudzu.

Of course, one can stipulate that biodiversity comprises just the plants and animals that humans generally like to have around. Ditto for the functions: Any that are perceived not to contribute to such a good as an "ecosystem service" can be proscribed. Then the value of biodiversity becomes obvious... and tautologous: Biodiversity is good because only good things are allowed to be called "biodiversity".

This point is at once trivially obvious and hard to keep in mind. The just-so model of biodiversity value (Sect. 5.1.4, The just-so model), in particular, tends to obscure it. This, I believe, is because whatever attraction this model has hinges on its ability to convince us that the value of the *status quo* is not just positive, but as good as it gets, and what we should be prepared to defend. While evidence and arguments for this position have yet to be produced, it might nonetheless be true. But even if it is true, it must still account for inconvenient kinds – organisms and functions – some of which are, quite literally, scourges of humanity, but nonetheless legitimate parts of the current biodiversity picture.

Those who embrace biodiversity as the foundation of natural goods wind up wrestling with a version of the venerable theological "problem of evil". If one thinks that God is omnipotent, omniscient, and morally perfect, then God knows where there is evil (because omniscient), can eliminate it (because omnipotent), and will eliminate it (because morally perfect). But as a matter of fact, the world appears to contain distressingly large amounts of both evil and suffering. That seriously militates against the existence of such a God. To my knowledge, no one has proposed that biodiversity has God-like characteristics. But the distinctly perfectionistic strain in much biodiversity advocacy (explored in Sect. 6.12, Biodiversity as the natural order) makes it vulnerable to similar puzzlement. If biodiversity is at the heart of the good of nature, then it is at least very puzzling that so much biodiversity is responsible for so much human suffering.

7.2 From the Natural/Artifactual Trap to Nature as Science Project

As noted in this book's Preliminaries, environmentalists self-consciously embraced biodiversity in large measure to avoid dealing with the elusive distinction between the natural and the artifactual. This is the context in which attempts to define wilderness – as the model state of "the natural" – foundered. Chapter 3's exercise in the conceptual clarification of biodiversity respects this foundational attitude and consistently shuns the temptation to give the distinction any role in defining biodiversity.

However, in backing off from this perceived trap, environmentalists and scientists who embrace biodiversity as central to their cause fall off the plateau where the range of common-sense notions of nature dwell and plunge into an abyss where nature reemerges as a science project. This project concerns itself with "scientific" ways to make nature "right". That is, it is framed by a conception of right action with regard to the natural world.

I say "plunge into an abyss" rather than "slide down a slippery slope" because there are none of the usual signs of slippery-slope logic here. Slippery slopes are customarily greased by difficulties in justifying any specific line of demarcation between two polar opposites, which difficulties are supposed to lead to the collapse of *any* meaningful distinction between those extremes. But those who hold biodiversity as a central locus of natural value do not slide down the science-project slope because of diff iculties in finding a brightly marked stop line. Rather, they altogether decline to consider the possibility that an interesting line of demarcation might exist.

I reserve for the next chapter the task of examining how the science project comes around to justifying its philosophy of realizing, as much as possible, some standard of biodiversity. In this section, I merely put the science project on display for what it is. In any event, it appears that many "conservationists" and other environmentalists are entirely prepared to take the plunge. But if my reasoning in Chap. 8 is correct, then the "environment" that remains at the bottom of the abyss is not at all what those who value the natural world have in mind.

In what follows, I divide the free fall into a few logical segments, though proponents of the great plunge tend to intermix them.

7.2.1   *"Assisted Migrations" and Such*

I start the free fall where Sahotra Sarkar starts it – conjuring up a wetland from a grassland. As I have remarked (in Sect. 4.1.1, Wilderness), Sarkar is quite taken by Keoladeo in India. He strongly advocates that the wetland in that park be maintained by welcoming the cattle that mow down the wetland-inhibiting grasses. This, he believes, is justified as a path to the diversity of wetland birds. One can imagine that these birds, too, are happy with this arrangement. The bovines control the vegetation that "naturally" wants to turn the wetland into grassland. To be clear, this place was grassland prior to the cattle's entrance.[2]

Sarkar's suggestion is entirely in the spirit of the various biodiversity-enhancing schemes that I suggest in Sect. 5.3 (The moral force of biodiversity). Like some of the other suggestions in that list, it is an intentional manipulation of a place in order to increase its (local) diversity. It is curious, too, that the intentional introduction of cattle is unequivocally the intentional introduction of an exotic species – both because domesticated and because it did not evolve in this particular wetland. But these are the least of the malaise-inducing considerations, because Sarkar's inclination is open to a number of more serious criticisms.

First, there is good reason to be skeptical that Sarkar's proposed means (the recruitment of bovines) actually achieve the ends – of increasing local species diversity as a whole – which he seems to regard as paramount. He provides anecdotes to persuade us that local *avian* diversity might increase, but even this is not supported by reference to any respectable survey. And certainly, no study shows how the radical transformation of this grassland into wetland really affects the diversity of non-avian species. For example, it is quite doubtful that anyone knows the fate of myriad invertebrate herbivores that make their home in grassland, but not in

---

[2] Sarkar is far from alone in heralding *Bos primigenius* as biodiversity savior in its role of "keystone herbivore". That designation comes from biologist Stuart Weiss. Mentioned in Note 76 in Chapter 6, his single-minded focus on *Euphydryas editha bayensis* (bay checkerspot butterfly) in the San Francisco Bay area of California (Nash 2009), leads to his similarly enthusiastic promotion of bovine intervention. I say "single-minded" in the sense that I cannot find evidence that Weiss' exhortation to "bring in the cows" takes into consideration any effect of the bovines other than as savior of the bay checkerspots.

the wetland. Since arthropods often far outnumber all other species, it is entirely possible that Sarkar's cattle actually decrease overall local species diversity by virtue of decreasing the diversity of arthropods alone.[3]

However, I think it important to focus more on teasing out some implications of endorsing these supposedly biodiversity-increasing means than on the issue of whether or not they actually achieve their ends. Suppose that introducing domesticated breeds of *Bos primigenius* into a place does indeed increase that place's diversity in a category of species that have the happy ability to fly in under their own power. What reason could there be to discriminate against organisms less well endowed in the transportation department? So far as I can tell, there is none. So it seems that, according to Sarkar's principles of "conserving biodiversity" (as he would call it), "conservation projects" should be encouraged to load up trucks, when their goal of beefing up biodiversity requires the recruitment of transportation-challenged plants and animals.

Some "conservationists" would not object to this. Some of them will like the birds that tend to frequent a wetland more than those that prefer a grassland. Keoladeo seems like a nice place for a wetland; and once it's created, there are obviously vacant "ecological niches" – unoccupied wetland apartments that "ought" to be filled. The birds are only too happy to oblige.[4] Indeed, recent ecological research has yielded some nascent understanding of when and how niche filling occurs; so there is some general promise for plotting to bring in the right kinds of less self-obliging transportation-challenged organisms.[5] But consideration of niches adds nothing to a basic principle that is thought to justify facilitating the influx of groups of organisms for diversity's sake. And sometimes, as apparently the case with some amphibians (Tingley et al. 2011), a niche already occupied by a congeneric creature is not only *not* a barrier, but rather a positive indicator that it occupies territory that would be a fine new home for an exotic interloper. Such a creature is effectively a sign that reads "bring in my cousins". According to Sarkar, an aim of "biodiversity conservation" is to obey signs like these.

Some might find fault with Sarkar's position on the technicality that cows "do not belong in a wetland". Whether or not this notion of "belonging" can be made respectable,[6] "conservationists" and restorationists tend to regard the "belonging" requirement as a minimal barrier to gross manipulation. For example, restoration biologists

---

[3] I also touch on this point in Sect. 4.1.1 (Wilderness).

[4] It is interesting to note that very often the "wrong" birds oblige. A salient case is presented by the tidal marsh that is being "recovered" from the salt evaporation ponds that once lined San Francisco Bay. The recovered marsh was "supposed" to be for the benefit of such "desirable" shorebirds as American avocet, black-necked stilt, snowy plover, Forster's tern, and Caspian tern (Ackerman et al. 2010). Instead, California gulls, which conservationists have labeled *avis non grata*, have seized the opportunity to enjoy a repast of the eggs and chicks of the other species. Management of this wetland includes twice-daily hazing of the gulls (Sommer 2011).

[5] See, for example, Fridley et al. for a discussion of niche-related processes. This is the essence of my suggestions (5) and (6) in Sect. 5.3 (The moral force of biodiversity).

[6] I express my doubts in Sect. 8.2.7.1 (Implications for natural value generally and biodiversity in particular).

routinely suggest substituting extant plants and animals for extinct or extirpated ones. The criteria that qualify an organism for this substitution role start with "belongs to the same genus as", goes through "looks a lot like", "plays the same role as", or (as comes up shortly) "plays some role in a new ecosystem", and ends with "is really cool".

But why stop there? Indeed, some number of conservation biologists see no reason to.[7] There appears to be a difference in spin, but no difference in principle, to be found in a class of "conservation" proposals that have their most ambitious expression in The Mother of All Conservation Moves.[8] This is a project that essentially aims to establish a United Van Lines for relocating the biotic world – that is, "assisting the migration" of organisms by scooping up populations from one place, then transporting and depositing them in places that, in a warming climate, more closely resemble cooler times in their old homes.[9]

The choice of phrase "assisted migration" is worthy of close scrutiny, because it is a clear sign of the arbitrary (in the sense of "unprincipled") nature of the thinking behind it. Though that label was affixed by highly trained ecologists, "assisted migration" clearly misapplies the term "migration". That term properly applies to regular seasonal journeys undertaken by many species of fish and birds, and some species of insects and mammals (such as some species of whales, as well as some species of African and North American ungulates). In contrast, The Mother of All Conservation Moves is not migration by any stretch of any imagination. It is a wholesale introduction of exotic organisms – a "human-assisted invasion" (and perhaps a human-assisted exodus, too). It is also worth noting that the "assistance" is broad and cuts through several layers of human society. A project such as this one requires institutionally sponsored design, engineering, and ultimately, implementation. In these significant respects, it is even more an "artifact" of human enterprise than the intentional but unplanned and institutionally uncoordinated, non-designed, and non-engineered introduction of, for example, *Tamarix* spp. in the North American southwest.

However, the idea of "assisted migration" is not new. Humans certainly have engineered (some would say, at least in some cases, "mis-engineered") many such translocations, and have found a number of rationalizations, the latest fashionable one being "biological control". These efforts include, for example, the exotic *Bufo marinus* (the infamous cane toad) brought to Australia in order to predate on the native *Dermolepida albohirtum* (cane beetle), which the recalcitrant amphibian stubbornly refused to do; and currently the exotic *Diorhabda elongata* (the tamarisk leaf beetle) brought to the American southwest in order to prey on the human-introduced and now naturalized *Tamarix* there.

---

[7] Though, as Marris (2008) points out, there is far from universal agreement among conservation biologists about whether conservation (however understood) is served by The Mother of All Conservation Moves, discussed immediately below in the main text.

[8] This moniker is from an online commentary on this topic, available at http://environmentalvalues. blogspot.com/2008/02/unnatural-acts-of-conservation.html.

[9] See, for example, Dean (2008) and Marris (2008), which represent the tip of the conservation literature iceberg, which agonizes about conservation on a warming planet.

The resurrection of such creatures as mammoths and aurochs also has a solid core of the essential characteristics of "assisted migrations". The dramatic appeal of resurrection has partly to do with "migrations" that span a gap in time. Migration in time would certainly require reconsidering what an Elvis species truly is. But a moment's reflection makes it clear that this kind of "migration" is not just a migration in time, but also from one *place* to another. While these creatures have been resting in peace *en masse*, their habitat has also been put to rest, permanently retired to the environmental annals. In the interim, the new place – really, any place of arrival on today's planet – has become a place quite different from the one from which these creatures departed. Resurrecting a species would rehearse the tension-producing device, standard in time machine stories, of getting the coordinates wrong, whereupon the time traveler finds herself in the most unpleasant place imaginable. According to the U.S. Environmental Protection Agency, "An exotic species is a non-native plant or animal deliberately or accidentally introduced into a new habitat."[10] On that definition, the reintroduction of aurochs or mammoths anywhere on the planet would be a human-assisted invasion of an exotic from the past.

Of course, "assisted migrations", both spatial and temporal, could be a means of not just conserving biodiversity, but of boosting local, regional, and even global diversity. As I observed in Sect. 6.7 on Biodiversity as value generator, isolated populations of an immigrating organism can experience new and substantial stresses in its new home, which can and do lead to adaptive phenotypical transformations, which, in turn, can lead to new species. Science will undoubtedly become increasingly clever in understanding which organisms and which circumstances are most likely to yield the desired species-generating effect.[11] I will have more to say about this in Sect. 7.3 on Biodiversity value in human timeframes, below.

I do not claim that every conservation or restoration biologist endorses the sorts of projects that I have described. But the projects I describe are but a small sample of an extensive literature, which recounts a great many like projects; this evidences a great many scientists who do endorse them. However, the more important claim is that there is no visible basis – no credible, let alone generally endorsable principles – for adjudicating disputes concerning which of these projects are worthy of pursuit and which are not. That is, there are no apparent principles for justifying the biodiversity goals that these projects seek to realize. Their success or failure is therefore based on whether their means are considered to have achieved their unjustified goals.[12] As I said at the start of this section, "conservation" as science project is not so much a slippery slope as it is a plunge – a leap of faith followed by free fall.

---

[10] http://www.epa.gov/bioiweb1/aquatic/exotic.html.

[11] See Sect. 6.7 (Biodiversity as value generator) for remarks on how rapidly adaptations can occur and Sect. 7.3 (Biodiversity value in human timeframes) for more reflections on and implications of rapid evolution.

[12] It is also a cause for concern that the biodiversity goals tend to be assessed in terms of a very limited set of plants and animals that are relatively easy to observe or are the center of conservation attention. The vast majority of species fall into groups that conservation biologists and conservation organizations routinely ignore.

7.2.2   All Creatures That on Earth Do Dwell – Wild, Domesticated, and Transgenic

Species diversity is the mainstay component of most accounts of biodiversity, though some dispute this – a controversy discussed in Sect. 3.1.1. The proposition that the diversity of species is valuable in and of itself does not discriminate between domesticated (in the sense of "accustomed to or dependent on human provision and control") and wild. Many of those who feel the strong tug of the natural versus the artifactual are moved to discount those organisms (animals and plants) that have had the misfortune of being bred into existence. They tend to employ the obvious cheat of simply proclaiming that the relevant, diversity value-bearing category is that of *wild* species. So far as I can tell, there is no justification for this proclamation that is both plausible and respectable from a species-egalitarian point of view. It certainly arouses the suspicion that advocates of this move are more interested in the normative force of "wild" than in the normative force of "diversity of species".

Be this as it may, Chap. 3's definition of biodiversity is extensible – allowing for the inclusion of any number of categories of kinds. So if those who are most interested in wild species wish to have "wild species" as a category that excludes domesticated plants and animals, then this does not preclude the inclusion of another category of "domesticated species" diversity, which registers the contribution to diversity of those non-wild organisms. But this "separate but equal" policy accomplishes little aside from obscuring the "equal" contribution to species diversity of organisms in both subcategories. Insofar as a greater diversity of species is desirable, the (human-assisted) creation (via breeding, for example) of a domesticated one should be as great a cause for celebration as the emergence of one in the "wild". By the same token, the extinction of a domesticated organism is as much a blow to species diversity and should be as much a blow to its value (as a species contributing to species diversity) as the extinction of a wild one.

This phenomenon of inventing subcategories for purposes of exclusion is as common in the domain of biodiversity as it was in the Jim Crow South and for the same basic reason – namely, malaise with admitting the equality of kinds within a broader category.[13] Jim Crow sustained segregation on the basis of the category "race" via non-racial categories – such as "owning property" and "being able to read". These latter categories, largely coextensive with the racial one or enforced in a way that made them racially coextensive, deselected the same, non-white populations from the broader category "citizen".

However, this kind of obscuring maneuver seems more difficult to pull off with regard to the broad category "species". It is difficult to find a category coextensive with "domesticated" that, in a more "respectable" way, captures the underlying discriminatory attitude that domesticated plants and animals are somehow tainted by a

---

[13] For the use of a Jim Crow strategy in the different context of identifying the basis for ultimate value, see Sect. 2.1.1 (Concepts and categories of value).

human effort – partly methodical, partly unconscious – to make them accustomed to or dependent on human provision and control. For example, the category "purposefully manipulated" won't do. It casts an unexpectedly wide net, which covers many "wild" plants and creatures, whose purposeful manipulation is often called "wildlife management". Culling, "assisting migrations", captive breeding, and deploying as biological control, among other practices, fall under this category. Retreating to the yet broader category "shaped by people" just makes matters worse, for it is possible that this category encompasses a majority of organisms on the planet today. It is highly likely that most species on the planet already bear the adaptive imprint of human activities – including some number of new species that owe their very existence to the presence of adaptation-inducing conditions that humans have had major responsibility for creating.[14] And in the long term, the ranks of this majority are likely to swell.

Of course, domesticated organisms tend to manifest the strongest of human influences. But there remains the unanswered question of why they count less as species because of this. In what respect does their contribution to biodiversity decrease as their shaping by humans increases? The answer to this question must derive from some theory of how biological diversity creates value; and that theory must justify such a non-egalitarian stance. If, for example, the value comes from a diversity of resource-generating capacities, then this would support the antitheses of the commonly expressed sentiment. The extinction of a domesticated species, which is typically *designed* to serve as a human resource or to service human needs or desires, would generally have a far more serious effect on human welfare than the extinction of a wild organism, which far more rarely serves any human need. Imagine a world without wheat, or corn, or rice, or potatoes, or cattle.[15] In contrast, essentially all people everywhere on the planet would continue living lives as good as they are now without *Ursus maritimus* (polar bear), *Delphinapterus leucas* (beluga whale), or *Eubalaena* spp. (right whales). On the other hand, transgenic, silk-producing goats might offer enormous benefits (Lazaris et al. 2002).

A moment's reflection on the potential for genetic engineering provides another perspective on "assisted migrations" (discussed in Sect. 7.2.1, "Assisted migrations" and such). Such "migrations" are perhaps a relatively crude tool for a job that might be more neatly accomplished by designing species that are "right for the job" – including "right for the place as the climate warms". Why move a species from one place to another just because of its intolerance for heat if one can more conveniently genetically enhance its heat tolerance and almost instantaneously readapt it for the place where it is otherwise already well adapted? I realize that the technology for this general approach is not trivial. But in the face of climate warming, that has not

---

[14] See Sect. 6.7 (Biodiversity as value generator) for a description of several human-induced mechanisms that undoubtedly lead to speciations.

[15] This is obviously true of the agricultural societies that now dominate the world. Pat Musick pointed out to me that it would not be true of now-rare hunter-gatherer societies whose sustenance still derives from wild animals and plants.

stopped many from sounding the call to do just this – at least for domesticated plants and animals. And the technology for human-"assisted migrations", while obviously different, does not seem to be so different so far as the degree of technological challenge is concerned. The more important point is that, on biodiversity grounds, it is hard not to endorse both projects and endorse them both for the same, biodiversity-enhancing reasons.

Which leads into a discussion of the relationship of biodiversity to the places that might be said to be more or less biodiverse, or contribute more or less to biodiversity.

7.2.3 Home Is Where You Make It – In Situ, Ex Situ, and Ecosystem-Engineered

It seems difficult to deny that the contribution of each species to species diversity – whether that species is wild, domesticated, the unintended fruit of human activities, or transgenic – is equal to that of any other. I am not aware of any principle that one could defend as generally endorsable that underwrites the distinctions between these categories as evaluatively relevant to species diversity.[16]

Another distinction defines a dimension orthogonal to the one along which the aforementioned distinctions run. It is the distinction that separates those species that make their contribution to diversity *in situ* (where it was found) versus those that make it *ex situ* (someplace else).[17] As with the categories of "non-wild", existence *ex situ* is sometimes held to be the basis for curtailing the validity of a species' contribution to species diversity. And to a large extent, the *in situ/ex situ* dichotomy seems to be the doppelgänger of wild/non-wild one. A creature or plant is *ex situ* because a conservation biologist brought it to a zoo, aquarium, park, or an arboretum; or introduced it to someplace other than its "historical range".[18] So it is a kept thing (like a domesticated organism) or it is an alien thing – in a place where people have placed it, but not really where it's supposed to be because it would not have been there otherwise.

Something like this seems to weigh on conservation biologists who agonize over how well biodiversity is served by *ex situ* conservation and who therefore covet the *in situ* label as conferring more favor on a conservation project that they already favor for independent reasons. This explains why – as in the effort to preserve coral and that to preserve the razorback sucker, both described below – conservation practice so often defines and applies the *in situ* label with so many degrees of freedom that in the end, it is hard to understand what distinguishes it from *ex situ*.

---

[16] Whether or not some other category of diversity, such as functional diversity or phylogenetic diversity heaps favor upon some species and casts disfavor on others is a separate question.

[17] This distinction features prominently in the conservation biology literature, which warrants its discussion here.

[18] Preservation of genetic material in a seedbank or germplasm bank is also sometimes included within the compass of *ex situ* conservation.

But in cases where conservation *ex situ* is rationalized on some other basis – for example, because thought to be the best practical alternative – the distinction is summarily dropped. The end result is that, while the distinction between *in situ* and *ex situ* is often held to have some great normative force, conservation practice defines and applies these labels with such unlimited freedom that they become labels of convenience. Donlan and his colleagues (2006) do not use the distinction explicitly to characterize their proposal to "rewild" North America, but the force of their proposal is that North America is *in situ* for the creatures that they propose to introduce. Tim Caro (2007, 281, 282) expresses his displeasure with this proposal, in part, by labeling Pleistocene Rewilding as *"ex situ"* conservation akin to zoos. The Mother of All Conservation Moves blurs the distinction in a similar way. Are the plants and animals in their new home *in situ* or *ex situ*?

But one need not look to such grandiose projects that might never be undertaken. Chinks in the *in situ*/*ex situ* distinction show in projects already underway and widely regarded as good, if not exceptional, conservation practice. They are especially evident in projects that involve intensively managing the affairs of a biological system *in situ* according to some spec sheet. When power drills are used to make divots for coral sprigs in coral reefs (Fackler 2009), and when these reefs are vacuumed on a regular basis to rid them of algae, and when sea urchins are planted on them to graze on the algae (The Nature Conservancy 2011), I suppose that the resulting entity still has some claim to being a coral reef *in situ*.[19,20] After all, the coral took up residence there some time ago and has not yet moved; nor have the "conservationists" moved it. But beyond the "natural" location – if that is what is what "keeping the same GPS coordinates" means – it is hard to discern any difference between this and an intensively managed garden that could be sited anywhere that the gardened species might survive with continual, possibly furious, "management".[21]

---

[19] I am not aware of any criticism of the coral vacuuming project from either environmentalist or scientific quarters. It is possible that this is another case where the imprimatur of "fighting aliens" (in this case, the alien algae) suppresses the criticism that some marine biologists have leveled against the use of vacuums in other contexts involving coral. In these, the vacuums are used to harvest coral "rubble" as a source of pharmaceutical calcium. But the reasons for why this sucks appear to also apply to the first use of vacuums – factors such as vacuuming up other "native species" (such as anemones and other invertebrates), destroying critical habitats for other species (such as hiding places for juvenile fish), increasing water turbidity, which decreases light available to and critical for the coral, and finally, inhibiting the coral-building bio-accretion by both sucking up the bioeroded calcium carbonate particles that are required for that, and by disturbing the settlement of coral larvae which are the germs of new coral growth.

[20] The relatively low-tech coral vacuuming-and-sprig-planting effort represents just one among several coral gardening techniques. A higher-tech method (Normile 2009) seeds a coral nursery on custom-made ceramic disks from existing coral. After 18 months of nurturing in their ceramic nursery, the entire nursery – seedlings-*cum*-ceramic-substrate – is cemented to a struggling reef.

[21] Some famous and influential conservation biologists and ecologists, such as Dan Janzen (1998) wholeheartedly embrace the "gardenification" of nature. I return to the theme of gardens and their relationship to the natural world in the final chapter of this book.

Coral gardening, practiced as described in the preceding paragraph, will necessarily be short-lived. The root causes of the corals' demise are the joined increases in the temperature and acidity of the water. These conditions push calcite and aragonite (the two crystalline forms of calcium carbonate that calcareous creatures produce) to the edge of their saturation level. That is the point at which calcification rates for coral (and for other creatures, such as calcareous plankton, which manufacture body parts from inorganic carbon) become zero. Below that point, these creatures quite literally dissolve. But even that does not put coral beyond salvation "*in situ*" (here migrating to the use of scare quotes). What it calls for is technology that effectively encapsulates the coral, combined with the application of some simple inorganic chemistry to remove excess carbonic acid. But at this point, are the corals conserved *in situ*, or are they are now ensconced, *ex situ*, in an "at sea" aquarium, a garden, a snow-globe, or a glass-encased "biotope" that Eduardo Kac might have designed? The basis for making this distinction seems at once elusive and also essentially inconsequential in weighing these alternatives against the one of transplanting the coral – in other words, another "assisted migration".

The prospect of an at-sea coral aquarium is an emblematic example of how *in situ* blurs or devolves into *ex situ*. But it is not isolated. Consider the case (described by Rosner 2010) of *Xyrauchen texanus*, the razorback sucker. Several vestigial populations of these creatures persist in dam-enforced isolation up and down the Colorado River. The most genomically inclusive of these populations is Lake Mohave's. This lake is the product of Davis Dam, completed in 1951 – a habitat totally unlike the Colorado River of the days before that dam. Of course, even at the time of Davis Dam's completion, dams upstream from the Davis Dam site had made the river completely unlike the free-running Colorado of pre-dam days.

X. texanus is a river fish lately become a lake fish. It is unclear whether this species is a native in a now-alien environment or an alien in its native environment or something in-between. Perhaps it inhabits some new category. However that might be, the fish's survival in Lake Mohave now depends on its larvae being plucked out of the Lake in an annual "Razorback Roundup" – to put them out of the voracious reach of recently arrived rainbow trout that lunch on the hatchlings. Hatchlings thus spared an early demise are (literally) ferried to a hatchery where they are nurtured for three years, at which time they achieve a size that gives them a fighting chance in Lake Mohave's newly hostile environment. It is a pity that the razorback suckers cannot understand the irony of their primary tormentor-predator: The trout are raised – for the pleasure of anglers – in the same hatchery as the razorback hatchlings that are saved from being devoured by them.

This kind of story concerning a riverine system is hardly unique. In fact, because of the ubiquity of dams, it might well be more the rule than the exception. Dams utterly transform an environment that was previously ruled by sometimes fast-running rivers with floods and flows that varied enormously by season. Afterwards, engineers gesture at conditions of yore by opening up the sluice gates at various times, by building fish ladders, by revising turbine designs so as not to slice and dice quite so many swimming, scaly creatures. The justification for this effort is truly hard to fathom in terms of championing the value of the natural world. The plants and animals that previously lived *in situ*, but in a very different environment, struggle to survive in what is still – after all the engineered simulations – a thoroughly alien, ecosystem-engineered environment. Moreover, because a radically new set of traits matter vitally in their now-alien environment, the creatures themselves are arguably not exactly of the same kind as their ancestors. Fish for which ladder-climbing capabilities are critical for survival live a life on terms significantly different from fish for which those capabilities are irrelevant.

Lest one think that only aquatic systems are the focus of this kind of "biodiversity management", consider the case of the rare *Dendroica kirtlandii* (Kirtland warbler). Its breeding preference for young jack pines has not served it well after forest managers almost entirely suppressed the fires that used to clear the way for new trees. Nor has its lot been improved by a large influx of cowbirds, which remove the warblers' eggs from their nests and deftly substitute their own, which the warblers then dutifully incubate for their deceptive tormentors. With these realizations were born programs to cut down old jack pines, plant jack pine seedlings, and slaughter cowbirds. A nice management touch is that, while the warblers abandon pines by the time the trees are 20 years old, the management regime leaves them in the ground until they are 50, when they are large enough to have significant commercial value.

Another bird, *Falco peregrinus* (peregrine falcon), took a few steps back from the brink of extinction when organochlorine pesticides such as DDT were banned just a few decades ago. Now, these raptors are welcome in large cities, where the fledging of their young has often become as much of a human effort as an avian one. Human caretakers relocate nests from high up on bridge towers, considered too hazardous for the precarious fledging process. And they see to it that the eyasses that fly down to street level but cannot make it back to a 33rd floor nest by means of their own still-unsteady wing power take the elevator.[22]

Perhaps people are saving *some* creature with efforts such as these. People are therefore, incrementally, preserving some small piece of species diversity. But this account of the matter can seem uncomfortably facile. For what is the value of what is saved in these sorts of ways? Is a vacuumed coral the same creature as the coral that existed before these human ministrations? Can something of value be saved by creating something else rather different from it? A similar question arises for a razorback sucker that is ferried to razorback kindergarten and then ferried back to a lake habitat completely alien to its way of life. Of course, in some ways this new razorback-like creature remains more or less the same creature it has always been. An analysis of its DNA would suffice to show continuity in the genome, even if it also might show some adaptive shifts. But is the human relationship to this creature the same? What does the sameness of that relationship count for? Is the value of the nouveau-razorback diminished because of the radical break in the form that relationship takes?

---

[22] See Note 4 in Chap. 4 for the source of this example.

The answers to these questions seem to have a significant bearing on the value of the creatures concerned and even on the value of nature more generally. And yet they are entirely disconnected from a discussion that is narrowly focused on diversity. The value of the contribution of each of these creatures to diversity is plainly equal. Yet it seems that *something* is diminished, that *something* valuable is lost, in a world in which corals exist because people relate to them largely as coral vacuuming servants; in which razorbacks exist because people relate to them largely as aides who ferry them to and from razorback kindergarten; and in which Kirtland warblers hang on because people muster sufficient ire against cowbirds and sufficient lucre from mature jack pines despite the warblers' unfortunate preference for younger trees. Alongside these examples, one can place many others, including some previously mentioned in this book – for example, the airlifting of *Leptodactylus fallax* ("mountain chicken" frogs) to safety and the deployment of cows to save *Euphydryas editha bayensis* (the bay checkerspot butterfly). It seems that an unremitting and narrow focus on biodiversity (or really, a few, select species) is blind to something very basic and terribly wrong in these pictures. The mere continuation of an organism's lineage might have some value. But it is quite possible that the projects that achieve this are profoundly value draining. The next chapter explores this possibility.

Just in case coral, the razorback sucker, the Kirtland warbler, and the bay checkerspot butterfly don't evoke a felt need to raise these kinds of questions about continuing lineages, I play the polar bear card. While *Ursus maritimus* might not survive much longer *in situ* – in its increasingly alien, iceless Arctic habitat – the creature might persist for a while longer in zoos – such as New York City's Central Park Zoo. There is Gus (Kifner 1994). In some ways, Gus is a fine specimen of Polar Bear. He certainly is physically fit and, as reported in 1994, keeps himself that way by endlessly and mechanically swimming a tight figure-8 pattern in his pool. Apparently this was his way of dealing with confinement to a 5,000-square-foot enclosure, which, even apart from its miniscule size for such a creature, is unrecognizable as a place for polar bears. The Zoo describes it in this way:

> The Polar bear exhibit is a rocky expanse containing a *waterfall*, some ice, and 90,000 gallons of *freshwater*. The water is *10 feet deep* and the temperature isn't regulated, but adjusts with the seasons between *45-75 degrees F*. Each bear has a den with *air conditioning* to help get through the hottest days. [italics added][23]

This might be equivalent to a person's perpetual confinement to a 50-square-foot enclosure – a standard size for a jail cell. Concerned about Gus' repetitive behavior, the zoo hired a psychiatrist to help him deal with his "neurosis".

There is no doubt that Gus is a container of polar bear genes – just as the rounded-up razorback sucker kindergarteners are containers of razorback genes, and so on. In his role as polar bear gene vessel, Gus undoubtedly contributes to both species and genomic diversity.[24] Perhaps there was some hope that he would pass his genes

---

[23] http://www.centralpark.com/pages/central-park-zoo/polar-bears.html. My italics emphasize the novel features of the polar bear's zoo habitat.

[24] This theme echoes one for which Dale Jamieson (2002, 172–173, 185, 187) offers a very convincing exposition.

onto another generation of polar bears, with the cooperation of his consorts Ida and Lilly (who was euthanized in 2004). However, polar bears are notoriously disinclined to procreate in captivity; Gus and his consorts never produced offspring and the prospects appear dim.

In Gus' defense, one might still argue that this is a time when the animals are succumbing to an Arctic environment that is increasingly hostile to their lifestyle. People must do what they can to help prolong this magnificent expression of nature. But can a polar bear that is perceived to need psychiatric help be said to contribute to the value of nature – or even the value of biodiversity? I would suggest that the answer to this question does not lie far from the answer to another question: For whom was the psychiatrist hired? Zoo visitors, whose entrance offerings keep the Zoo open, do not like to see such mechanical, repetitive behavior.

Those who defend biodiversity as a primary pillar of natural value might try to deflect the source of discomfort in considering these questions from Gus. They might say that any discomfort about Gus is really a reflection of the loss in diversity of habitats, considered as integral to biodiversity. Specifically, they might claim that even if species diversity is only maintained *ex situ*, the diversity of habitats is diminished, and along with it, overall biodiversity. This, it might be said, adequately accounts for the feeling that something is lost in having polar bears confined in close quarters as exhibition specimens.

But this defense rides on a false presumption. In fact, it does not seem that habitats are suffering at all, so far as their diversity is concerned. The habitat of vacuumed coral or of an *in situ* coral aquarium-in-a-bubble, the lake-*cum*-kindergarten habitat of razorback suckers, and jail cell-sized habitats for polar bears might not be "natural" according to a definition of that term that clings to a principle of biodiversity as natural order (Sect. 6.12, Biodiversity as the natural order). But the increased role of people in defining them is irrelevant to the fact that these habitats are unique and therefore make a unique contribution to habitat diversity. No such habitats existed prior to their recent engineering and construction. And there is no obviously good reason – on grounds of diversity – to exclude these recently established habitats from an accounting of the diversity of habitats overall. "No good reason" does not inhibit the invention of Jim Crow rules – like the ones that try to segregate domesticated organisms from the wild – that disqualify some habitats from the diversity accounting on similarly arbitrary grounds.

And in fact, a great many conservation and restoration biologists embrace this principle. They make no apologies for ignoring the taint of the artifactual and in assigning themselves the mission of creating more, different habitats. They think themselves already sufficiently knowledgeable and capable of designing environments for the habitation of many nonhuman creatures – just as, for example, city planners are capable of designing cities for the habitation of human creatures. And they are impressed with the prospects for advancing these capabilities and increasingly putting them to work on behalf of diversity. Reinventing themselves as ecosystem or habitat engineers, they promote the creative design of new ecosystems.

To be clear: ecosystem engineers and managers design and implement projects to achieve particular ecosystem objectives. This is *not* the ecological *science* that comes from study of ecosystems that are newly emerging as the result of inadvertent human influences. It is this latter kind of science that Michael Soulé (1990, 235) seems to have had in mind in his 1990 vision of a new field of "recombinant ecology" or "mixoecology". So too, Ellis and Ramankutty (2008), in their proposal for a science of "anthropogenic biomes" or "anthromes". But it is a quite different matter to proclaim that it is permissible, advisable, or obligatory to *engineer new ecosystems* – that is, to claim that right action with respect to ecosystems entails the active and creative design of new spaces and places as habitats for specific biotic mixes, which mixes are part and parcel of the overall design.

What are the particular ecosystem objectives that ecosystem engineers propose to aim for? Unfortunately, to the degree that design goals are made clear, it is clear that they have no justification based on credible or generally endorsable principles. This general phenomenon is systematically explored in Sect. 8.1.3 (The ends of the biodiversity project). Here, I bring it into view via one example – namely, appeals to "ecosystem health". Ecologist Young Choi (2007, 35), for example, suggests that

> Future-oriented restoration should focus on ecosystem functions rather than recomposition of species or the cosmetics of landscape surface. If a person lost one of her or his legs, the physician would put a prosthetic leg [sic]. The primary purpose of the prosthetic is to rehabilitate the function of leg rather than to recompose original flesh and bones. Physicians call this process rehabilitation, not restoration.

Choi makes a glaring yet terribly common and commonly condoned error by ignoring critical elements of disanalogy in his analogy to human health[25]: A human amputee might well have an interest in restoring her ambulatory function. But no ecosystem has any such interest. Whatever "ecosystem health" is, interests in it appear to be entirely human. Humans, too, are by definition the only parties with an interest in identifying some small, select group of ecosystem functions as the valuable properties that should be promoted, while (by dint of environmental Jim Crow segregation rules) excluding others.

Choi's statement cannot be tossed aside as exceptional; it is representative of a multitude of similar ones. One possible way to make sense of it is that it is a metalepsis in which "ecosystem health" really means "ecosystems that make people physically well and economically well-off". This reading also explains why talk about "ecosystem health" more often than not bleeds into a dissertation on the economics of "ecosystem goods and services", without much of an explanation of how these two apparently very different things are related. The mystery is dispelled if one supposes that Choi and like-minded ecosystem engineers presume that, by definition, "ecosystem health" and "ecosystem goods and services" are essentially one and the same thing. In that case, "ecosystem health" is, or is equivalent to, "economic health". This helps to explain why talking about paraplegic ecosystems serves no clarifying purpose.

Carrying the "ecosystem health" rubric to its logical extreme, proponents of the North American rewilding project propose that we ought to "reinvigorate" (Donlan

---

[25] The problems discussed here connected with "ecosystem health" parallel those discussed in Sect. 6.12 (Biodiversity as the natural order) regarding the autonomy of ecosystems.

et al. 2006, 674) wild places, evaluated by "holistic measures of ecosystem health" (Donlan et al. 2006, 672). Without revealing what this end consists in, they jump to the means for achieving it. On this basis, they enthusiastically endorse the substitution of one night heron for another (species of) night heron, one tortoise for another, and then, with an extra burst of enthusiasm, just about any creature for any other extirpated or extinct creature that can be conceived as performing "similar functions". All told, the criteria for "sufficient similarity" are so loose that it appears that few substitutions are proscribed – a seed-disperser for a seed-disperser, a predator for a predator, and a megafaunal specimen for another that now only haunts museums because the semblance is in some unspecified regard sufficient to generate (Donlan et al. 2006, 665) the "excitement" of "actively restoring evolutionary processes".

Some conservation biologists dismiss the North American rewilders' proposals as outside the mainstream of conservation biology. But it is hard to see why that is so merely on the basis of the principles from which they derive. The next chapter argues that the rewilders' basic principles are entirely mainstream. The North American rewilders are "wilder" mainly by virtue of more clearly recognizing their implications and by being more inclined to voice them with candor.

Another reading of "ecosystem health" and its role in justifying the engineering of ecosystems unites it via shared preamble to other, seemingly independent considerations. This preamble points to the undeniable fact of enormous and irreversible human-induced factors that have dramatically modified ecosystems planetwide. I have previously mentioned some of the sweeping and systematic factors, such as the swift change in climate and the dramatic influx of reactive nitrogen into many environments. From this solidly factual point of departure, a sense of the "natural order" as the healthy order of things takes over: The current set of ecosystems, as a whole, have broken loose from any original, or "natural" design principle. They have fallen into disrepair – that is, become "unhealthy" – because the haphazard and *un*designed consequences of human activities have made them deviate from the principles of natural design. This state of ecosystems – now untethered from any "natural" point of reference – places proper responsibility for further "evolving" their design squarely on human shoulders. This gives ecosystem engineers more or less free creative license for new designs.

The reasoning often takes one final step, which circles back to the interpretation of "ecosystem health" as equivalent to "economic health". This step is taken by way of portraying the essential design choices as constrained by a dichotomy: Surely (the reasoning goes), human designers should design ecosystems to suit human ends and, more specifically, human desires; for the (only) alternative is to let ecosystems run amok – careening off in unpredictable directions that are likely not conducive to yielding what people want.

The just-described line of reasoning underpins the "voided warranty" model of responsibility for nature – one of two models of responsibility for nature that Sect. 8.1.2 (Responsibility for nature) explores at length. In a *Discovery News* interview, Erle Ellis, a progenitor of the idea of anthropogenic biomes or anthromes, clearly and succinctly sums up the basic idea behind the voided warranty model: According to it, humankind has screwed around with the natural world to the point where the original specification no longer applies. Therefore (the logic goes), people now own the design and the responsibility falls on us to draw up our own specification. From there, it is but a short step to the suggestion that properly fulfilling this responsibility requires a design that aims at satisfying human desires – the ultimate arbiter of value in the human economy:

> The biosphere… [is] a system that's already been transformed by human activity… [So] we are responsible for the way that nature behaves now… and in the future. If we want to live in an environment that is desirable for all of us, it's up to all of us to make that happen.[26]

And from there, it is but another short step to the suggestion, now so prevalent, that ecosystem design is properly the design of ecosystems that realize ecosystem goods and services (Sects. 6.2 (Biodiversity as resource) and 6.3 (Biodiversity as service provider)).

The major point of this discussion is that the biodiversity narrative has wandered into economic territory by way of ecosystem health; and this leads to an economic constraint on ecosystem design that places little or no constraint on species diversity or biodiversity more generally. But it should be pointed out (and several sections of Chap. 6 do emphatically point out) that the effect on the diversity of species of making this consideration preeminent is likely to be enormously deleterious. For the entourage of organisms needed to satisfy this constraint is almost certainly small. Of course, that can be remedied by an arbitrary decision: Whether justified or not, add species diversity as a goal of ecosystem design alongside the one of serving humans.

Echoing suggestion (7) in Sect. 5.3 for enhancing biodiversity, that is exactly what some conservation biologists propose to do – at least, at the start of their narrative. For example, Harris et al. (2006, 174) urge that the "target" of ecological engineering is the

> … judicious creation (and, where possible, restoration) of habitats of species designed to conserve and protect these assemblages in new areas made appropriate by changing biophysical conditions and to restore natural capital, and therefore ecosystem goods and services.

On the grounds of better serving genetic diversity (and echoing suggestion (5) in Sect. 5.3), they go on to urge that the design of new habitats should not just consider "local material",[27] when deciding which organisms to toss into them. But in the end, even they find that this diversity is desired only insofar it serves economic interests. The authors (Harris et al. 2006, 173) predict that:

> Restorationists will almost certainly rely more heavily in the future on the notion of "restoring natural capital" and restoration of delivery of ecosystem goods and services delivered by the ecosystem under consideration and not simply on those metrics based on the numbers and arrangement of the biota.

---

[26] See http://www.youtube.com/watch%3Fv=sTUOHMkGa0Q. The part about responsibility comes from the very end of the piece.

[27] The phrase "local material" appears to be a device for dodging the phrase "native species". That latter phrase goes hand in glove with "exotic species" or the lightning rod term "aliens". In this more normatively charged vocabulary, Harris et al. are proposing to expedite alien invasions.

It is clear from their further discussion that the "will" in "restorationists will rely… on… the restoration… of ecosystem goods and services" is not a mere reporting or prediction of fact. Rather, it is an urgent recommendation to adopt an action-guiding principle.

Margaret Palmer, with her sizable group of distinguished collaborators (2004, 1252), echoes Harris and company. According to them, a major thrust of the "new ecology" is found in

> … "designing" ecosystems [in a way that] goes beyond restoring a system to a past state, which may or may not be possible. It suggests creating a well-functioning community of organisms that optimizes the ecological services available from the coupled natural-human ecosystems… [These include] systems consciously created de novo where other alternatives are not possible. The latter are synthetic systems consciously created to achieve ecological, social, and/or economic goals… Such systems are not necessarily based on historical views of ecological structure at a given location… Instead, systems may be designed… by means of a blend of technological innovations, coupled with novel mixtures of native species, that favor specific ecosystem functions.[28]

One can try to blunt the force of this apparent recommendation for freestyle creation of ecosystems by clinging to the conditional phrase "where other alternatives are not possible". But these authors are advisedly pessimistic about the possibility of re-recreating historical tableaux under conditions that so drastically differ from historical precedent that impossibility slips into inconceivability. No other alternative that satisfies their "possibility" condition is on offer. The conclusion is obvious: In their view, there is no reason not to define restoration as "anything that optimizes the ecological services".

Let's take stock of how the ideal of biodiversity arrived at this postmodern picture of neo-conservation and restoration. At least one part of the route follows the idea of creating preserves for individual organisms that might be the last preserves of certain specific genomes. Focusing on this element of biodiversity simultaneously defocuses the habitats that individual organisms occupy. All at once, snow globes, zoos, and the like blur into the undifferentiated space that makes greater diversity of species and their genomes possible. Another part of the route runs along the narrative that engineered nouveau ecosystems themselves do not diminish, but rather enhance, biological diversity by virtue of contributing to ecosystem diversity. This logic thus holds out the attractive prospect of achieving greater biodiversity overall.

One can easily imagine additional benefits that accrue from the engineering approach to solving the natural value problem for biodiversity. Having themselves designed and constructed the habitats, engineers might be better equipped to manage and maintain them than when the design specifications are "given" and sketchy. For as things stand now, the design specs are known only to The Creator of one's theology, or only insofar as they can be reverse-engineered. This points to automated management as the final goal of a technologically achieved biodiverse world.

---

[28] It is difficult to fathom what these authors mean by "native species" in this passage.

Moreover, insofar as the organisms inhabiting what in Chap. 8 I shall call a "biogeoengineered" habitat are bioengineered with just those biogeoengineered environments in mind, one might reasonably expect them to be better equipped to live in them than those organisms still at the mercy of the vagaries of adaptive selection. In this significant respect, the bioengineered entourage would be far superior to the current entourage, which must try to adapt – the old-fashioned way – to the rapidly changing environments that humans are engineering anyway. Of course, as things now stand, engineers tend to engineer environments with mainly or only people in mind. As a consequence, their designs often exceed the adaptive capacity of other, long-time biotic residents. Therefore, neo-conservation does require a shift in attention towards accommodating them.

But there are significant reasons to be hesitant to embrace a world of brave new ecosystems. First, the conclusion of "greater biodiversity overall" is almost certainly false on any reasonable understanding of "biodiversity". I have no doubt that human designers are capable of achieving greater ecosystem diversity when cut loose from any other design goal. But in reality, that goal unfailingly transmutes itself into something quite different, most notably "optimizing ecological services". Thus reincarnated, the goal is almost certainly incongruent with preserving a diversity of species, for few species do any respectable amount of work to earn their preservation. And it seems at least as dubious that, by optimizing services in ecosystems, the abiotic diversity (that is, the diversity of physical and chemical environmental characteristics apart from living organisms – relating to light, temperature, topography, soil, hydrology, and air) of those systems will thereby be increased. Efficiency in all other domains typically leads to standardization. There is no evidence whatever that would lead one to believe that it breaks differently for ecosystems.

The second reason for hesitation is more basic, more profound, and independent of the actual effects on biodiversity. It has to do with the question of whether or not a world with even an enormous diversity of snow-globe ecosystems is a world in which the value of nature is realized. I leave my inquiry into that topic for Chap. 8.

7.3   Biodiversity Value in Human Timeframes

Biologists fret over current losses in biodiversity and particularly global species diversity because they feel that these losses cannot be made up. By now, I hope, the ambiguity in this statement is transparent. In the absence of resurrection science (Judson 2008b), it is trivially true that an extinct species will not be, well, resurrected. But the *diversity* of today's kinds (of species) is entirely capable of resurrection. The diversity that a particular set of kinds can achieve is also achievable by any number of other, even very different sets of kinds. Diversity, if that is the true holder of value, can be maintained, enhanced, or resurrected by different and even disjoint sets of kinds.

However, there is another sense in which the claim about not being able to make up for biodiversity losses might be true. It might be said, and often *is* said, that biodiversity – strictly understood in terms of the diversity of kinds in biotic and biota-encompassing categories without regard to which particular kinds make up the diverse collection – will not recover in a human-meaningful timeframe. This concern can be construed as legitimately about biodiversity in a way that does not trade on conflating diversity with how it is constituted. Furthermore, as Sect. 6.7 (Biodiversity as value generator) observed, this proposition is supported by current scientific understanding, according to which species diversity typically appears to take at least 1 million years and perhaps 5–10 million years to recover from a major extinction event.[29] That is a long stretch of time. Five million years encompasses more than 150,000 human generations. Even the still-contested evidence for a more compact recovery from the Permian extinction (dubbed "The Mother of Mass Extinctions" by Douglas Erwin (1993, Chapter IX)), which might reduce it to a "mere" 1 million years, would span more than 30,000 human generations. Overall, 1–10 million years quite likely constitutes a major stretch of the run of *H. sapiens* as a species on this planet. Most large mammals (such as we humans) tend to dwell in the lower end of that range.

The whacking down of value connected with biodiversity – no matter for how long – might suffice to upset an intrinsic value theorist who is a strong objectivist about the value of diversity. Even putting aside the value of a species as a species, for such a person, the winking out of each successive species, which makes its contribution to the world's diversity, is concomitantly a winking out some quantum of the natural value that previously illuminated the world.

But so far as I can tell, a quite different rationale undergirds most soundings of the long-time-to-recover alarm. According to it, the demise of a species still takes a bite out of biodiversity's natural value. However, this depression in value hinges on the projection of the actual presence of many generations of humans valuers who would suffer the resulting deprivation. In this section, I assume this metaphysically less adventuresome view – that biodiversity value is contingent on valuing people actually being around – to consider whether or not the alarm is a false one.[30]

Even if the alarm about timeframes for recovery from extinction events is not entirely false, several major difficulties substantially dampen it. The first is that, if the value of biodiversity depends on there being people around, it is just as important for those people to retain a capacity to value biodiversity. There is good reason to believe that, in a world that goes on for even a little while with significantly

---

[29] In addition to Sect. 6.7, see Ehrlich et al. (1997, 101).

[30] See Sect. 2.1.1 (Concepts and categories of value) for the background discussion of objective versus subjective value. My background assumption will not sit well among those with a predilection for strong objectivist metaethics – most commonly held by hard-core intrinsic value theorists such as Holmes Rolston III. According to such a view, value inheres in objects independent of human (mis)judgment or valuing behavior. This grounds a strong objection to any proposition that might entail that the value of the natural world suddenly burst forth – in an anthropogenic "big bang" of value – with the grand arrival of humans. Nor will it sit well with metaethical subjectivists of the ilk that are happy to depict a value landscape *sans* actual people, but only by defining value in terms of what value people *would* assign, *if* they were present.

diminished biodiversity, this capacity will be lost – perhaps within a few generations. A common countering gambit poses the question: what would future persons think of us, if we currently respiring persons were complicit in reducing biodiversity? But this move merely deflects attention back on current persons. Insofar as this question strikes a vein of sensibility, it is the sensibility of some (unfortunately, probably small) number of us who are living now; it is not the sensibility of those future persons who are supposedly harmed, wronged, or deprived.

Suppose that the behavior of currently respiring humans is part of the causal explanation for diminishing biodiversity – whether that diminishing effect is unintentional or understood as collateral damage of otherwise valuable projects. Then the values embodied in pursuing those projects – as well as the goods that they produce – are likely to be transmitted into the future together with the altered state of biodiversity. Inheriting those values and lacking direct contact with a world that is different either in its state of biodiversity or in human regard for biodiversity, future persons will lack the basis, which some of us now alive still take for granted, for developing different values with respect to biodiversity or nature generally.

This likely incapacity of future persons' finding anything of great value in a more biodiverse world suffices to counter the future persons' retrospective blame gambit. But additionally, not only will future persons likely be devoid of this evaluative capacity and devoid of any basis for developing it, even its miraculous reinstatement would not matter. For it will be entirely *reasonable*, given the circumstances that form their future world and their future perspective on it, for future persons to regard a valuing attitude towards biodiversity or nature more generally as foolishness. This is the sad conclusion of the following narrative[31]:

A child is almost totally accepting of unvarying aspects of the environment in which she grows up. Her developing mind transmutes the "what is" that frames her life and informs the development of her evaluative capacities into "what is the norm". No matter how much you and I might deprecate the quality of that child's experience, the child herself has no access to this critical perspective, which would require her to depart from the familiar – her norm. The obstacles to this access are not merely physical and in the limitations of her experience, but emotional. A child needs permission and encouragement to enter the emotional space in which it is permissible to evaluate some aspect of her world in a comparative, counterfactual, or critical way. In short, access to a critically evaluative perspective is not something innately present in the human animal. Rather, it is something that people develop and learn through experience of alternatives, through experience in judging between them, and with encouraging guidance in doing these things. In the persistent absence of these requisite conditions for developing her evaluative capabilities, an evalua-tively impotent child becomes an evaluatively impotent adult.

In these respects, there is nothing special about values connected with biodiver-sity or the natural world. One should expect that a generation of children who

---

[31] This narrative is adapted from a wide-ranging essay, available online at http://environmentalvalues. blogspot.com/2008/08/future-persons-future-values-ignorance.html.

become adults in a world characterized by the consistent and persistent refusal to regard any part of the world as off limits for the pursuit of human projects will embrace the values embodied in pursuing those projects without that constraint; and they will embrace the goods that those projects yield. They will have no capacity to do otherwise. In particular, they will have no capacity to think the worse of their world – including the utilization of a previous generation's natural world in its making. The bases for feeling deprived, harmed, or wronged will quickly attenuate and vanish entirely just a few links down the generational chain.

Of course, this future generation will have historical accounts, which will provide a view of today from their several generations' remove. These accounts will make them aware of a natural world and attitudes of old that made some of their ancestors think it worthwhile to wall a part of it off from the domain of human enterprise and spare some amount of that past world's biodiversity. But there is little reason to suppose anything other than that such accounts – of a world and attitudes so remote from their own – will strike those inhabitants of a future time as endearingly quaint, at best. More likely, reports of a bygone eagerness to forgo the projects that nevertheless were pursued to create the future world valued by those future inhabitants will strike them as a kind of folly that, much to their relief, did not prevail. Regret seems highly unlikely. At most, one could imagine the kind of casual nostalgia that is an acknowledgment of the fact that the "we" of any generation lives in its own time, not in a previous one. It would be not unlike the feeling one might have about not having been present to witness or receive first news of Hannibal crossing the Alps. There could be no real sense of loss in being cut off from this – no sense of being cheated out of something that one has a right to have – for one does not have a right to anything that is not even a possibility.

Without any basis for feeling or believing that they have been cheated out of an entitlement, without the loss of anything connected to any experience available to them, and without even the capacity or possible means to acquire a capacity to understand living a human life in a possible world in which they do not happen to live, a shrug of the future shoulders seems both far more likely and far more appropriate than an accusation that we-now acted in a morally reprehensible fashion for our role in creating the actual the world in which they actually live. It is important to emphasize that feelings and beliefs of (at worst) indifference held by future persons regarding the behavior of current persons, will not be grossly inappropriate. They will not be at fault for overlooking some terrible moral misdeeds, as we-now would be if we exhibited indifference to our ancestors' practice of slavery. On the contrary, their likely charitable attitude towards us and our behavior will be quite reasonable. The norms for their world will be built around the circumstances of human life in that world. These circumstances will be infused by the many goods whose creation was made possible by our-now willingness to forgo (or ignore) any value that would have required a different relationship to the natural world – one in which not all of it were always jealously eyed as the domain of human enterprise and development.

All told, these considerations suggest that persons just a small number of generations from now will thank rather than revile us-now for our willingness to make incursions into the natural world on their behalf.[32] If some of us-now characterize these incursions in terms of destruction, it is only we currently respiring persons who do so. Future persons would not be so inclined. Unable to recognize goods that accrue to living in a world that includes a natural world, they would characterize our behavior in terms of creation – the creation and passing on of the goods that are actually present in their life. Contrary to the claims of those who project the sensibilities of our own time onto the future, future generations would feel an enormous sense of community with us precisely for so assiduously creating the world and circumstances familiar to them and in which they strive to live their life as best they can.

This narrative touches on two fundamental truths about the nature of values – including natural values – and their trajectory through time.[33] The first is that a person's capacities to make judgments about value – including well-considered ones – are developed and shaped by her experience in the world in which she actually lives, including experience in her practice of making those judgments. The second is that a person's best-reasoned judgments are not about timeless abstractions that exist apart from the business of living a human life in the circumstances in which that person finds herself. Rather, they are highly sensitive to those circumstances, which partly determine what goods are available and can possibly be made available to her.

These two truths are connected to the formation of preferences. Political philosopher Debra Satz (2010, 69) observes that "Our preferences... do not come from nowhere." Satz makes this point in the context of challenging the thesis that the preferences that a person brings to a market transaction are a reliable barometer of respectable values. This thesis fails to take into account the fact that preferences are shaped by many factors, including such ones as "tradition, peer pressure, and social context", and even markets – factors that "may not reflect what is really important for us" and that make them less than worthy of satisfaction. The broader version of Satz's point is that *all* preferences – including ones that *are* of respectable provenance and *are* worthy of satisfaction – do not come from nowhere. Those preferences can have a significance for how we live our life only insofar as, having been shaped by the circumstances in which we live that life, they will play a real, practical role in a life lived in those circumstances. And insofar as preferences that are well grounded in the circumstances of our life inform our best value judgments, the same is true about value.

To put this point in another way, values, including those that attach to the natural world, are not exogenously given. However, the basic shape of the world in which we persons find ourselves is an exogenous force that shapes and bounds the set of

---

[32] In saying this, I do not wish to join Bill McKibben in supposing that human activities are bringing the planet to the "end of nature" because they are destroying the one and only state of Nature that is natural. McKibben's view differs quite profoundly from my understanding of Natural value starting from people (Chap. 8), in ways that I discuss in Sect. 8.2.6 (What appropriate fit is not).

[33] The theme of temporally determined circumstances strongly and legitimately shaping what is important in living a good human life in those circumstances resurfaces in Sect. 8.2.7.1.

possibilities that we can imagine for how best to pursue our life. Powerful, too, is the force of the values that previous generations hand down to later generations. These are not just expressed in the actual shape of the world as a given generation experiences it, but also in the norms that infuse the stories about why previous generations shaped the world in the way they did. In these two ways, previous generations' values, which a later generation has no hand in creating, strongly influence that later generation's values.

Let's get back to the difficulties with the alarmist narrative surrounding extinction events. A second major difficulty with it has to do with a serious disanalogy between past major extinctions and the current one – supposing with some prominent biologists that a major one is actually underway. While past extinctions were not facilitated by people, nor were the recoveries from them. The opposite is now true – and it is as true as much for the recovery end of things as it is for the extinction end. Insofar as the concern about recovery is based on a perception that humans are impotent to positively influence the state of diversity, its basis is open to question. That is because this gloomy assessment overlooks some potentially substantial capabilities that people have for quickly and substantially contributing to biodiversity. And just as for extinctions, some major part of this capability to effect adaptive change and speciations is exerted even without our intending to do so.

Quite possibly, the *least* of the intentional capabilities is the human capacity to *conserve* the diverse kinds that currently exist, particularly those that totter on the brink of extinction because the world in which they now live has become inhospitable to their way of life and possibly unalterably so. This makes nearly moot the question of how likely it is for the forces that brought about those unalterable changes to abate. They have already done their work, which abatement will not undo. And in any event, it seems highly unlikely that people will suddenly decide to complicate their pursuit of what they regard as their own welfare, which has been and continues to fuel most changes.

Of course, people are certainly capable of heroic salvation of some small number of organisms that capture their attention. The range of schemes and technologies that conservationists marshal on behalf of such efforts is enormously impressive. But again: all too often, a species teeters on the brink because the human-transformed world as it stands and in which the organism finds itself some time ago stopped satisfying some of its basic lifestyle requirements. As a result of development projects – such as the damming of a river and the introduction of game species in the case of the razorback suckers – the chemistry is no longer right, nor the hydrology, nor the temperature; and the older denizens have a limited capacity to compete and defend themselves against their new neighbors. It is easy to treat this set of unfavorable conditions as the initial conditions for a new engineering and management project, whose solution, concocted from clever technology and attentive management, keeps the old-timers slogging on for awhile. But not for more than a very short time.

Such projects as the one that keeps the razorback sucker going are biodiversity performance art. As exhilarating as it is to behold, it is hard to imagine that the technological performers will be able to keep it up for very long or for very many organisms in very many places where an old home has become a truly alien environment. At the end of the previous subsection, I mentioned genetic engineering as offering the promise of more permanently installing the requisite adaptive capabilities in various organisms. But it is unclear whether this kind of wizardry adapts old organisms so much as it creates new ones for the new world.

It seems that far and away the most promising strategy is to focus on "enhancing" biodiversity by bidding farewell to the failing and welcoming a new class of of organisms with far better chances for survival in the modern world. In Sect. 5.3 (The moral force of biodiversity), I give a list of promising means to do this – methods that are already at the disposal of humankind, or not far out of reach. Aside from applying our increasing gene-manipulation know-how, that list emphasizes the fact that we are aware of several different sorts of conditions that people can manipulate to encourage speciation. There is reason to be sanguine about prospects of getting good at this with practice – perhaps good enough to attain speciation rates that, like rates of extinction, exceed historical norms. That, by definition, would be the salvation of biodiversity, or at least, species diversity.

Moreover, the results of speciation projects need not await the far future. There are strong reasons to believe that they could be designed and managed so that new species emerge with great rapidity. For example, it is known that invertebrates and other groups of organisms that have a short reproductive cycle can exhibit adaptive changes with startling suddenness when abruptly exposed to new stresses. Section 6.7 (Biodiversity as value generator) provided several examples of this. The genomes of even megafaunal species such as *H. sapiens* can undergo rapid, significant adaptive changes under stress. The evidence says that the human residents of the Tibetan Plateau diverged significantly from their lowland parent population when subjected to the stresses of high altitude. This divergence took only about 2,750 years (Yi et al. 2010, 77).[34] And this appears (Wade 2010b) to be a quite general phenomenon of the human diaspora: Human populations that experienced the stress of environments quite different from those of their African origin have had more of their genes subjected to selection than the stay-at-home Africans.

Growing evidence points to a general mechanism for evolutionary alacrity. It appears that much evolutionary change occurs by means of "soft sweeps", which involve a mixing and matching of existing variations in a genome – in contrast to "hard sweeps", which must wait for a rare favorable mutation to sweep through a population. In a study involving *Drosophila*, the investigators (Burke et al. 2010, 1) remark:

> Standing variation is theoretically predicted to lead to rapid evolution in novel environments, and case studies of ecologically relevant genes bear out this prediction.[35]

This mechanism might well account for the rapidity of adaptive changes in humans over the course of their diaspora out of Africa (Pritchard et al. 2010).

---

[34] See also Wade (2010a), which refers to the work of Yi et al.

[35] See also Wade (2010c), which refers to the work of Burke et al. (2010).

There are many ways to apply new stresses and each of them suggests a different project. One project might be the systematic introduction of "exotic" species – or, viewed from the other side, systematically subjecting organisms to "exotic" environments. This takes advantage of the fact that the stresses on an immigrant arriving in an alien environment can be the catalyst for adaptive changes that put the immigrant on the path to becoming a species distinct from that of its parent population. Moreover, new immigrants can and often do hybridize with cousins in their newly adapted environs. A hybrid that constitutes a viable new species can emerge in a single generation.[36] Finally, it is becoming increasingly evident that hybrids occur with fair frequency in nature – even in the absence of human facilitation. Hybridization provides an especially fast route to "finding" an adaptive advantage. Remarking on various plant and animal hybrids found in nature, biologist Sean Carroll (2010) observes,

> One lesson from the sunflowers appears to be that hybrids may succeed if they can exploit a different niche from their parents. The same phenomenon has been discovered in animal hybrids.

Currently, we observe these speciation phenomena relating to "exotics" in circumstances that are a matter of happenstance. A project whose goal it is to enhance biodiversity would set about creating these circumstances in a carefully planned way. The concept of "invasive surgical strikes" could serve biodiversity well.[37]

The suggestion for programs to "enhance" biodiversity can be expected to provoke controversy. But as I previously noted (Note 44 in Chap. 6), no discernible controversy surrounds some such programs, which are ongoing. And more importantly, I do not know of any objection that is grounded in a principled consideration of biodiversity or its alleged value such as might flow from some one of the theories examined in Chap. 6.

Nor can an objection find easy refuge in any of the models of biodiversity value discussed in Sect. 5.1 (How biodiversity relates to its value). It is easy to see that none of those models provide support for a preference to focus entirely or even preferentially on conserving those organisms that constitute the current biodiverse horde. And none provide a basis for shunning a program of "invasive surgical strikes" to enhance biodiversity. This might be most obviously true for the incremental model. It counts up or down, without prejudice for the identity of what is counted (as consistency demands, according to egalitarian principles). And according to it, counting up is always good. When it comes to the threshold model, an increase of any kind takes biodiversity away from any supposed threshold. And the kinds that together might form a quantum in the quantum model also hide

---

[36] See again Sects. 6.7 (Biodiversity as value generator) and 5.3 (The moral force of biodiversity), and the discussion of immigrants in Sect. 6.3 (Biodiversity as service provider).

[37] I understand that the possible effect of a new immigrant on existing citizens must also be assessed. But this does not militate against the potential of an assisted immigration program. A great majority of introductions do not result in extirpating organisms already resident.

behind a veil of anonymity. None of these models offer grounds for preferring some means of increase or means of maintaining constant biodiversity (despite flux in membership) over another. As a consequence, none provide a basis for shunning such programs as "invasive surgical strikes" to accomplish these goals.

Even the just-so model, on a "weak" interpretation, can be seen to endorse a program of biodiversity enhancement when its goal is the conservation of biodiversity, properly understood. On this interpretation, what is "just so" is not each and every specific organism in its "proper" place as a constituent of the current biodiverse horde, but rather the level of biodiversity. This allows for an exchange of kinds that maintains the "just so" level.

Of course, the conservation of the kinds themselves is one means to the end of biodiversity conservation. And the achievement of that end by this means is easier to recognize than the same effect attained by way of the churning produced by a balance of ongoing and perhaps accelerating losses and gains. With a static kind of conservation, a constancy (and so conservation) of diversity is seen in the constancy of composition – that is, in the familiar and recognizable "faces" (kinds) that continue to stick around. A flux in the composition – some losses, some gains – makes the overall level harder to gauge; it raises concern of gaps after losses and before gains; and it raises anxiety about novelty in the elements and their assemblages. Newcomers tend to be regarded with suspicion. But it is hard to see how that prejudicial attitude can legitimately be said to diminish a newcomer's contribution to diversity, when that is an actual effect of its arrival.

Another problem in perception has to do with how new speciation "events" are viewed, along with the adaptive changes that lead up to them. The typical sense of being underwhelmed by a speciation is the polar opposite of the potentially great drama in the demise of a commonly observed, large, or "charismatic" species. As biologist Olivia Judson (2008a) remarks, the signs of adaptation and speciation tend to appear "small" and undramatic to the human observer, however large and dramatic the change for the organism itself – the altered size of a bird's beak, a plant's phenological shift in flowering, even a lizard's installation of cecal valves to revamp its intestinal plumbing. Aside from the human tendency to shrug off those few cases that come to light, there is far less awareness of speciation than there is of extinction in the first place. For even the most dedicated evolutionary scientist, detecting speciation is a difficult labor of intense, extended, and selectively focused study. So it is not surprising that few are willing or able to devote a career to such studies. Even if the current situation on the planet falls short of a great speciation event, as I have previously suggested, it is possible that a lot of adaptation and speciation is going on behind everyone's back or under our very noses.

Of course, a similar claim about ignorance of extinctions is also plausible. And in fact this claim is routinely made – reflecting the relatively intense scrutiny of extinction in the context of biodiversity, in contrast to adaptation and speciation. There likely are many extinctions of which humankind is unaware largely because it is hard to be aware of the disappearance of something that was invisible before it disappeared. A veritable chorus of ecologists, conservation biologists, and environmentalists have echoed E.O. Wilson's oft-voiced complaint (for example, expressed in Wilson 2002, 100), that our woefully inadequate grasp of what organisms exist makes it safe to assume that some are slipping away unnoticed because they have never been noticed in habitats that are being transformed in ways that would make it difficult for these unknowns to persist. It is therefore remarkable that no one, to my knowledge, has recognized that a similar speculative logic applies to speciations. These, too, might well exceed those actually noticed based on what is known about mechanisms for speciation.

My central point is served by even the weakest conclusion warranted by the evidence about major adaptive shifts up to and including speciation: Speciations *can* happen, as evidenced by the fact that they *do* happen; they happen at rates that are well within historical human timeframes; and sometimes they happen within one or a small number of human generations. Moreover, there is a nascent and growing understanding of how these facts could be placed in service of the goal of increasing biodiversity, should that goal be thought worthy of our attention; and most views about how biodiversity is valuable quite directly lead to the conclusion that that goal *is* both worthy of attention and worth the effort to achieve.

For those who (implicitly or explicitly) embrace a threshold model of biodiversity value (Sect. 5.1.3, The threshold model), the long timeframe argument would seem to have an appeal that is limited by a realistic assessment of where the threshold is. On this model, pruning biodiversity back towards the threshold entails no direct loss in value – only perhaps an increased risk that the threshold will be crossed. As Martin Jenkins (2003, 1177) remarks in his discussion of this issue, humans have benefited enormously from

> … [converting] some 1.5 billion ha … of land area to highly productive, managed, and generally low-diversity systems under agriculture. Even with regard to indirect ecological services, such as carbon sequestration, regulation of water flow, and soil retention, it seems that there are few cases in which these cannot adequately be provided by managed, generally low-diversity, systems.

Jenkins' remarks give reason for confidence in the geo-historical record as solid inductive evidence that no biodiversity threshold is near. The touchstone of biodiversity is, by common consent, species diversity; and no matter how painful it is for some to acknowledge, the truth is that the planet still hovers near its all-time global high, despite reductions in which *H. sapiens* almost certainly had some causal role, significantly starting something over 70,000 years ago when that bipedal species emerged as a transforming force on the near side of the bottleneck that almost pinched it off.[38] Even though some species have succumbed since then, even though some have emerged since then, and even if the rates for either or both species appearance and disappearance are unusually high by geological standards, the resulting changes to overall species diversity are likely extremely small with respect to the totality of the diverse contingent. And even in the midst of this flux, the set of currently extant species is overwhelmingly the set of species that dwelled

---

[38] See Note 59 in Chap. 6 for references that present the evidence that the current "high tide" in diversity exists at most taxonomic levels.

on the earth 60 millennia ago, despite the fact that the place looks a lot different now. Moreover (as Jenkins observes), people appear to have adapted themselves to whatever changes have occurred. These considerations militate against a belief that some biodiversity threshold is anywhere close by.

A final point concerns still-shaky speculations about rates of species diminishment.[39] Even aside from legitimate questions about estimated current rates and how they compare to geo-historical norms, the final significance of those rates depends on their additional and independent speculative extrapolation into the future. That extrapolation seems to require the concomitant extrapolation of the conditions that produce the rates – saliently, that organisms continue to confront conditions and pressures not previously within their experience. That, in turn, seems to require the assumption that the conditions and pressures themselves continue to change in ways that never allow the contingent of organisms extant at any one time to fully realize the survival benefits of adaptations to previously encountered conditions and pressures. I am not aware that these questions – which must take into account the nature of future human projects and whether they continuously spawn novel anthropogenic biomes – have been previously raised or addressed. If biodiversity is indeed diminishing, then one can imagine that the rates of diminishment will tail off as a consequence of the relative stability of the anthropogenic pressures that spurred a previously higher rate.

At this point, it is best to step back from thinking about any particular model or theory for how biodiversity's value varies with biodiversity. A discussion that narrows its focus to diversity inventories, the extent to which the inventories change, and their rates of change measured in in human timeframes distorts the overall importance of these things. As I have previously remarked, one must consider just how pervasively humans have changed the rulebook – encoded in largely unalterably changed biotic and abiotic conditions – for what works biologically and what does not.[40] Dominant among the causes for various species' struggles is that the new rulebook does not favor them. But tossing out that rulebook is not just beyond human willpower; it is beyond human capability. It is not a real option. Insofar as this impossibility is what is entailed by talk about "saving biodiversity", saving biodiversity courts not just irrelevance, but folly.

The illusion that "saving biodiversity" makes sense, I think, is fed by fixating on the perseverance of some small number of familiar elements in isolation from the

---

[39] Estimates for current rates of extinction and their comparison to geo-historical norms are still highly uncertain for a number of reasons. These include gross (two orders of magnitude) uncertainty in the number of extant species, the validity of extracting norms for extinction rates based on a very biased geological record, questions regarding the validity of extrapolations based on the Species Area Relationship (He and Hubbell 2011), and (as I have been keen to emphasize) the absence of any accounting for rates of speciation driven by the same adaptive pressures that are a salient factor – now as ever – in pushing some extant species towards extinction.

[40] The polar bear's fast-vanishing habitat is the marquee example of the phenomenon that affects many less well-known though more abundant organisms.

broader context whose salience is thereby lost. That broader context sets the conditions for how life must be lived on this planet as it now exists, not as it existed sometime in the past. There is a strong temptation to think about some familiar species: if only we could help them in their struggle to persevere, even if for just a few more years. But these familiar kinds struggle under quite unfamiliar burdens. It is very unlikely that their burdens can be relieved in any systematic or lasting way, even if that is the conservationist's dream.

At the same time, whether or not people *choose* to launch biodiversity-enhancing projects, we are very likely doing just that, even if we don't know it. The unfamiliar burdens on most organisms will almost certainly give rise to unfamiliar adaptations, unfamiliar species, and unfamiliar relationships of living things to each other and to their unfamiliar environs. Which seems not at all bad, if biodiversity is truly supposed to be at the core of nature's value.

7.4 Leaving What Is Not Biodiversity Out of It

One of the most difficult problems in getting a handle on the vast literature on biodiversity is figuring out what its various authors mean by "biodiversity" or a "biodiverse state of the world" and what they mean in claiming biodiversity to be a salient good of the natural world. Two chapters (What biodiversity is, What biodiversity is not) of this book are dedicated to getting a handle on this problem; that material served to construct a semantic guard rail, which (as observed in Chap. 6) various and sundry lines of subsequent argument frequently hop and consequently lose their logical way. One pervasive formal failure reasons from premises that relate to the beneficial actions and interactions of a small number of specific organisms in a very particular assemblage to a conclusion that asserts that these actions and interactions exemplify the goodness of biodiversity. Arguments for Biodiversity as safeguard against infection (Sect. 6.5.2), for example, very typically take this form. All in all, this leaves little doubt that there is pervasive confusion about what biodiversity is, what's so good about it, and perhaps even about whether there is any difference in meaning between the state description and normative claims about it: One plausible explanation for much of the confusion is that it is underlain by the mistaken belief that "biodiversity" *means* "that which is good about nature".

I wish to address two dismissive objections to this book's negative assessment of the evaluative program for biodiversity, which, I believe, stem from these sorts of confusions about biodiversity and its goodness. The first objection bluntly denies the normative nature of the claim that this book attributes to much of the literature on biodiversity: According to this complaint, it is simply a mistake to think that a great proportion of scientific and environmental literature on biodiversity is centrally concerned to judge that biodiversity is a great good and at the core of nature's goodness; it is equally a mistake to think that this literature's aim is to promote the position that the basis of right action with regard to nature is to promote biodiversity. In other words, this objection asks us to think that those who – paper after paper, book after book, conservation campaign after conservation campaign – seem to be holding up biodiversity as the core of natural goodness and who seem to be exhorting us to take action in order to promote the good of biodiversity are, in fact, not saying what they appear to be saying.

An objection along these lines is so astonishing that it is hard to know how to respond. Fortunately, I believe that a response is safely left to the Pope John Paul II and to E.O. Wilson – both of whom say with unusual clarity that there is a moral imperative to conserve biodiversity. It is safely left to the large cadre of ecologists who have devoted their careers to studying how biodiversity correlates with human goods. It is safely left to the even larger cadre of conservation and restoration biologists who justify their many and varied projects for the good that will accrue to biodiversity. And it is safely left to the world's major conservation organizations, which portray themselves as biodiversity's billion-dollar saviors.

The second objection does not deny the undeniable normative thrust of the vast biodiversity literature. Rather, it charges that this book construes biodiversity so narrowly that it necessarily thins out any value that might be attributed to it. This complaint expresses a sense that I have cheated. According to it, by some sleight of hand so deft that no one seems able to identify or articulate it, I have clipped biodiversity into a bonsai version unrecognizable as its real self – a version that is never allowed to bear the value fruit that is so overwhelmingly palpable to believers in the good of biodiversity. Thus, the complaint runs, the book achieves a semantic victory that is otherwise hollow.

On the face of it, this second objection appears as lacking in sense as the first one. The definition that has guided my exploration of biodiversity is the definition that I develop in Chap. 3 (What biodiversity is), which is explicitly designed to be as broad and as broadly sympathetic to the biodiversity program[41] as any that I am aware of or can imagine. That definition, which admits diversity of kinds in any and all biotic and biota-encompassing categories, is a generalization and expansion of the ones in common use by biodiversity experts in the biodiversity program. One would think that, if there is any value to be found in biodiversity, it is very likely to be found under this very, very large umbrella.

But also, it is hard to imagine making the umbrella more expansive without courting the triviality of "all that we mean by, and value in, nature". I am not aware of any respectable scientist who thinks that her research regarding biodiversity is research into either one of these things. Nor am I aware that any of those who propose and argue for the various theories of biodiversity value presented in Chap. 6 believe that, in fact, they needn't have spent their valuable time on this. There is no need to convince the world that biodiversity is valuable for human health when it is given that biodiversity *means* all that we value in nature. And I would like to think that anyone who seriously discusses biodiversity takes herself to be discussing something about diversity. But "all that we mean by, and value in, nature" seems to

---

[41] I use this term as I did at the outset of this chapter, to mean the program that identifies certain states of biodiversity to be emblematic of natural value and for that reason, undertakes to bring those states into existence insofar as that is possible. See Note 1.

have little or nothing to do with the contribution or role of diversity in things, let alone the role of biological diversity.

It is easier to understand this second objection as concerning a different kind of narrowness. I do maintain a strict and (one might complain) narrow insistence that, however it is construed, "biodiversity" mean *some*thing about the natural world; and that it mean *one* thing that has something to do with diversity. By "*some*thing about the natural world", I mean some distinctive natural property, no matter how richly conceived and complexly structured, that applies to identifiable parts of the world and to the world at large. This definition leaves open the question of whether biodiversity is a good or a bad. It denies that "biodiversity" *means* or includes in its meaning "that which is valuable in nature" or "that in nature which is worthy of approval". That semantic equivalence would trivialize any value claim made on biodiversity's behalf. If someone nevertheless insists that this is what "biodiversity" means, then one can acknowledge her eccentric definition and allow that when she says that biodiversity is valuable, she is saying something that is indubitably true by virtue of being a tautology. For the rest of us, connecting biodiversity to value still requires justifying reasons.

By "*one* thing that has something to do with diversity", I mean that when the value of biodiversity is at stake, I do not tolerate switching the meaning of biodiversity. Most saliently I do not tolerate using "biodiversity" in a way that has nothing to do with diversity. Nor do I abide slides in meaning from factual premise to evaluative conclusion. If there were evidence that, in some particular environment, cows afforded protection against malaria, then I do not allow that this means or even is evidence for the proposition that "biodiversity protects human health". The proposition is about cows, organisms in the genus *Plasmodium*, and mosquitoes. As it is presented, it either has nothing to do with the diversity of anything; or if it does, it fails to take into account the diversity of the disease organisms and its vectors, which compromise human health. Putting this succinctly: to mistake "cow" for "biodiversity" is, at best, grotesque synecdoche.

A complaint of arbitrary narrowness might alternatively be interpreted as misgivings with my refusal to admit, in support of biodiversity's value, much of what some find valuable in the natural world. In this case, the charge is that I have unacceptably narrowed my conception of natural value so as to exclude whatever value one might attribute to a sentient creature on account of its sentience, or to any organism on account of its being a subject of a (nonhuman) life, or to any particular "community" composed of such organisms on account of their being thrown together in the way that they were, or even to any particular species as a species. But the required premise for this complaint is, once again, the unacceptably distended definition of "biodiversity" that makes it all that is valued in nature. For it is only with this premise that I can be criticized for omitting something from under biodiversity's umbrella that does, indeed, have natural value. Once that false premise is discarded, it is obvious that my views about what biodiversity is, and about what its value is or is not, do not preclude me from considering these other possible sources of natural value on their own merits. To put this in another way: There might be credible grounds for finding great value in the natural world, which have little to do with nature's biodiversity.

If that turns out to be true, then this is one more reason to think that, not only does biodiversity not mean "all that we… value in nature", but that this normative proposition is contingently false.

I cannot emphasize too strongly that each of the above-mentioned propositions regarding value in the natural world – based on sentience, being a subject of a life, natural communities, and so on – is worthy of consideration. Each has a history of lively debate, which has produced distinctive arguments, conundrums, and a range of contested conclusions. I also cannot too strongly emphasize that weighing them down with baggage from the biodiversity morass serves none of these separate debates well. Worst of all, the considerations most critical for good answers to the question of the value of an individual organism and that of an individual species have little to do with biodiversity. As a consequence, when they are couched in terms of "biodiversity", the salient points are obscured or lost entirely. If the value of an individual nonhuman organism lies truly in its role as a subject of life, then that is entirely lost in its portrayal as "a component of the world's biodiversity". And the most important considerations for why some particular species might be valuable as a species are lost when those considerations are buried several layers deep under the rubric of "biodiversity management".

I need to be especially careful about one special case – namely, the value of a species. I have remarked just above and many times before that the value of a species as a species is distinct from its value in contributing to species diversity or as a collective vessel for genomic diversity. However, confusion can arise because there are some obvious connections between the value of species diversity and the value of a species as a species. Most notably, they kick up some of the same issues. And salient among them is the question of whether all species are deserving of equal moral consideration, whether it is right to sacrifice some for the benefit of others, and whether or not the good of species generally entails a duty to create more. But this partial overlap of questions is not sufficient to justify burdening arguments in the debate about the value of a particular (or any) species as a species with the merits of arguments about the good of biological (or even just species) diversity.

A great deal of the biodiversity literature obscures this point with its own characteristic linguistic sleights of hand. For example, a species becomes "a component of biodiversity". Proposals to weight the odds in favor of one species at the expense of another are disguised as questions of "biodiversity management". Phrases such as these grease the slide into discussions of the value of particular species without acknowledging that this is the real topic of concern. If there is some value attached to *Euphydryas editha bayensis* (the bay checkerspot butterfly), then that value might have something to do with the value of the bay checkerspot *as a species*, which has value like any other kind in this category. Or the butterfly's value might have something to do with its being the particular creature that it is. But that, again, would bear on its value as a species, not as an anonymous cipher that boosts the biodiversity census. Or it might have something to do with the value of all the individual living beings that happen to be creatures of that kind or of any other. But that would have to do with the value of individual living things. If any of these considerations account for the bay checkerspot's value, then the loss of that species would incur the irreplaceable loss of some of that value. Other butterfly species might evolve, with their own unique characteristics and unique individual butterflies. These would have value, too. But it would be a different value. It would be separate and apart, and not a substitute for the value of the bay checkerspot and all the individual insects of that species.

None of these considerations, which apply to a butterfly species or to butterfly individuals, apply to the butterfly's contribution to biodiversity. Unlike a specific, distinct, and unique kind such as a species, biodiversity by its very nature admits substitution within its "components", properly understood as categories that admit a diversity of kinds. The biological diversity of the world would be arguably intact, were a new species of insect to emerge and take over the bay checkerspot's niche. Ironically, this fungibility is viewed by many biodiversity advocates as a distinctive advantage of their evaluative focus, for it gives humans a great deal of freedom to engineer and manage the planet to better suit human needs and desires – all without affecting biodiversity or its value. This is the basis for thinking, with Maclaurin and Sterelny (Sect. 6.9.4.1, Phylogeny), that once you've seen one snail darter, not only have you seen them all, but also you've seen all perch, which, in turn, is tantamount to seeing all bony fish.[42] By their lights, most any snail darter could sub for *Percina tanasi*. Importantly, the logic of a single substitution is not invalidated by the first one made. It remains valid for an indefinite number of substitutions. In the end, biological diversity can be achieved in any number of ways that do not rely on the continuation of even a single one of the familiar plants or animals or other organisms that inhabit the earth today.

In sum, there is no good reason to suppose that this book has, by means of a semantic sleight of hand, made biodiversity's value disappear before its readers' eyes. It might be more accurate to say that, by trying for formulate nature's value so emphatically in terms of its biodiversity, biodiversity proponents have dismissed or obscured a great many other promising ways to find value in the natural world.

7.5 Biodiversity Value *Post Mortem*

What should one make of the "inconvenient implications" of the biodiversity program to account for and better realize the value of nature? Chapter 6 and all the material leading up it portrays the program as consistently failing to gain any credible purchase on its own terms. I would say that the roundabouts and elements of discomfort and confusion presented in the current chapter are symptoms of a more systematic failure. They are strong signals that biodiversity is, in some fundamental way, unsuitable for shouldering the normative load that so many environmentalists, conservationists, and scientists have asked it to carry.

---

[42] See Sect. 3.3.2.1 (Features) for mention of perch and other, especially bushy parts of the tree of life.

If the biodiversity program is unsalvageable, one reasonable reaction is to abandon it and rejoin the longer-running programs to get at natural value through the value of individual living things or of individual species as the species that they are. However, these other programs are not without their own, formidable problems; and so one might be inclined to make one last attempt to rehabilitate the biodiversity program.

But how could that possibly be done? The obvious route is to once and for all, directly confront the great bugaboo that William Cowper (Sect. 2.3.2, The value of diversity in general) put us in mind of: Some kinds of diversity are good, but others are bad; and some are very, very bad. So it is for biodiversity. Therefore, one must take care to identify just "the right kind" of biodiversity – the kind that does, indeed have great value. But what is "the right kind"? Is it the natural kind, or the wild kind, or the kind with some species but not others? Perhaps it is the kind of biodiversity that is embodied in all ecological systems as we know them and as they function today; or perhaps yesterday; or how about a century ago; or before the Columbian exchange; or a millennium ago; or before the human diaspora out of Africa. But all these answers are different ways to lurch towards some version of the just-so model of biodiversity value. And so this route runs once again into the brick wall of finding credible principles for why some particular *status quo* should be held worthy of a norm. It is hard to see any reason why one should think that the prospects for any other salvaging project would fare better.

The idea of wilderness, which was largely emblematic of environmental value in the pre-biodiversity days of environmental thinking, foundered on vexations in finding a non-arbitrary conception that made wilderness seem worthy of being highly prized. Then, in the mid-1980s, the neologism "biodiversity" was invented and a parade of highly respected figures, led most publicly and insistently by E.O. Wilson, pasted it on their environmental banner, which they then waved at everything that one might suppose underlies or constitutes the good of nature – "free" services, medicines, meeting genetically deep-rooted biophilic needs, and so on. But in the end, it seems that in the hasty retreat from the Scylla of wilderness as the emblem of environmental value flew straight into biodiversity's Charybdis.

When given a clear meaning based on clearly defined biological categories, the overwhelming evidence is that even the most generous conception of biodiversity fails to establish it as a good of the natural world. What is worse, the illusion that the good of nature saliently consists in its biodiversity fuels a number of uncomfortable implications, some of which this chapter puts on display. Worse yet, those who continue to advocate for the biodiversity program seem immune to the discomfort of the prospect of their very real projects transforming the the natural world into creations of their own imagination – creations in which most of the rest of us will see little of nature and little of natural value. The tragedy of this situation takes its final, disastrous turn by encouraging the adoption of a biodiversity program by those who primarily view the natural world as an *impediment* to realizing their projects, except insofar as nature is a resource that fuels economic development and gain. In the biodiversity program, those with an interest in such projects find the perfect, environmentalist-approved cover for exploiting the natural world without undue impediment and without any genuine regard for its natural value – all in the name of natural value.

Thus, the more that biodiversity is valued as emblematic of natural value and the more that a conception of right action is based on it, the less of nature there will be to value. I believe that the converse also holds: The more that something is held to be naturally valuable, the less it tends to be connected with biodiversity, *per se*. Or at rate, the less it is connected with any clearly defined and consistently used notion of diversity, such as the one that this book adopts. This radical decoupling of biodiversity from natural value is a consequence of the fact that the idea of biodiversity as the basis for nature's value is bound to particular way framing nature's value that is singularly unpromising. Chapter 8 attempts a systematic and theoretical examination of why this is so. Before moving to that concluding discussion, and without trying to anticipate the principles that it develops, I wish to push back from the detailed evaluation of the many failed arguments examined in the preceding chapter and conclude this chapter by returning to a theme in this book's Preliminaries – with some much higher-level, *un*detailed, but still pre-theoretical reflections on why the hoped-for goldmine in biodiversity value never materializes.

One might think that, however tarnished it seems to be, the idea of biodiversity could yet glitter if only one were more clever in picking the right categories (species, function, phylogeny), or if only one were more adept in distinguishing the distinct and diverse kinds within these categories. But I think that these potential difficulties have little to do with the elusiveness of biodiversity's value. True, even its most central category of "biological species" is a little ragged around the edges. The notion of functional kinds is a lot more than a little ragged, and in this case, different choices in the principles according to which one individuates functions can lead to dramatically different views of the value of functional diversity. But I do not think that any of this raggedness necessarily leads to critical failure.

Nor need biodiversity be sidetracked by the chimera of *the* right measure or index that mashes all categories of biodiversity into a single number. This, I believe, is largely nonsense when it is proposed as a representation of value, but there is reason to hope that clear thinking might ultimately prevail and recognize that it is, indeed, nonsense. Nor is the elusiveness of biodiversity's value the result of undeniable limitations and deficiencies in the nascent biological sciences that bear on biodiversity. Scientific shortcomings do throw a monkey wrench into some lines of reasoning. This is most notable in trying to characterize conditions that might justify applying some form of the Precautionary Principle (touched on in Sect. 6.4, Biodiversity as (human) life sustainer). But even this, I think, is not central to the basic failure of biodiversity to anchor the values that might attach to the natural world.

To understand the sources of failure, I think it more promising to look to three other, general difficulties that attach to setting norms for biodiversity. These are (i) the great remove of biodiversity from common, sensible experience, (ii) its extreme abstractness as a state of affairs that is subsequently set up as a target to be realized by the actions of moral (individual and social institutional) actors, and (iii) its great remove not just from sensible experience, but also from the moral and social experience of human valuers. The last source of failure is the most critical in its undoing; Sect. 8.1 (The disvalue of the biodiversity project) expands on it and gives it a principled grounding. Here, it suffices to suggest that biodiversity as a norm is morally abstract by virtue of its complete abstraction from the attitudes, behaviors, and actions of persons both individually and collectively, and in its abstraction from the "environments", broadly construed, that make their lives meaningful and worthwhile.

But at least in part, that last critical source of failure likely derives from the first two. The discussions – of biodiversity's great remove from ordinary experience (Sect. 6.11) and its conceptual abstraction as a Cautionary sign (Sect. 2.3) – provide the hints needed to discern them. The second (abstractness as a concept at least three degrees removed from real objects in the world) might entail the first (remove from common, sensible experience), but the converse does not necessarily hold: Not all things removed from common, sensible experience are particularly abstract. Most likely, few people – not even the human co-residents of *Batrisodes texanus* in Travis county, Texas – will ever have direct experience of individuals of that other, non-voting Texan. But at least the individual creatures of that disenfranchised species are no more abstract than any other individual object in the real world.

As I remarked in Sect. 6.11 (The experiential value of biodiversity), biodiversity is not composed of particular objects, creatures, or places of the real world; it is not even of one or another particular kind of object, creature, or place; and – despite the fact that much discussion about biodiversity devolves into discussion of species diversity – it is not even the diversity of kinds in one category. Rather, it is composed of the diversity of kinds in multiple different and complexly inter-related categories. It is at least three layers of abstraction removed from the things, creatures, and places with which we interact as environmentalists, biologists, and moral agents. There is no sensible entity or even simple collection of entities that any person encounters in the world that *is* biodiversity or that in any straightforward way stands in for it.[43]

Biodiversity's remove from experience is not just a matter of its scientific complexity or the abstruseness that creeps in from such obfuscating constructs as quadratic entropy. Few things are so scientifically complex or so abstruse as the planet's climate and changes in it – that other signal environmental issue of today. Climate phenomena can be fully comprehended and evaluated by at most a few hundred highly trained climatologists worldwide. But at the same time, shifts in the climate are fully palpable to the senses in terms of the common, daily experience of weather, including record-setting temperatures, droughts, floods, and phenological shifts.[44]

---

[43] It might be thought that this is precisely the role played by charismatic species such as polar bear. But, as I argue in Sect. 4.2.1 (Charisma and cultural symbolism), this claim is based on a false representation of species diversity.

Also as discussed in Sect. 6.11, the remove from ordinary experience does not necessarily mean that we cannot "experience" biodiversity in some very extended sense of that term.

[44] Of course, the basis for this sense of the climate through experience of weather is a common-sense understanding of how that weather fits into the bigger climate picture.

There is no truly analogous experience of biodiversity. Even when it comes to the diversity of species, no one person, not even the most highly trained and industrious biologist, is equipped to understand what collection of species exists. That is so even in highly confined and closely studied areas.[45] Of course, over the course of time, even the ecologically naïve might become aware of the absence of a previously noticed organism – provided that it is sufficiently large and conspicuous to be easily noticed by people – in some particular, familiar-because-oft-frequented place. But that says nothing about the *diversity* of organisms there – only that the kinds of conspicuous organisms might be in flux. It typically takes some greater amount of time, expertise, and intense investigation to detect the immigration or emergence of other, and sometimes much less easily noticed organisms.

The possibilities for experiencing biodiversity in an extended sense are even more limited than those for justice, which Sect. 6.11 used as an aid for reflection on this topic. With justice, at least, it is possible to point to particular experiences of witnessing the unjust treatment of one person by another – by dint of some well-known and widely accepted norms which make it possible to identify the unjustness of at least the most extreme actions. It is hard and perhaps impossible to find any norms that help us with biodiversity in this way. Just-so models of biodiversity value might initially seem to provide a norm in the form of the *status quo*. But as I have previously observed, the road to justifying any *status quo* for biodiversity is a road to nowhere.

This leaves the third source of elusiveness, which, I think, most decisively puts biodiversity – considered as a primary bearer of nature's value – beyond the effective reach of persons as valuing beings. Value, I think, must be conceived in terms of relationships involving human valuers and something that they value. With biodiversity, one starts with the "something valued" – as a recognizable and even scientifically computable summary characteristic or characterization of nature. But this starting point is radically detached from any sort of valuing relationships that people and societies actually enter into with the natural world and which, I will argue in the next chapter, give the natural world its value. The problem with biodiversity is that it is neither a cause, nor an effect, nor constitutive of any such relationship. Rather, the state of biodiversity cuts indiscriminately across circumstances in which nature is valued and those in which it is not. As a consequence, a blinkered focus on biodiversity cannot possibly account for natural value.

This is a radical critique. It does not merely say that it is difficult for persons or societies to determine how to behave in order to achieve, in some maximizing or satisfying way, some adequately biodiverse state of the world. It does not even say that such an achievement would be worthy, if it were practically possible; but sadly it is not. The thesis of practical impossibility is an empirical one, subject to disproof by some ingenious biodiversity project that defies initial expectations to ultimately achieve success.

---

[45] In the context of Sect. 6.11 again, I have already mentioned John Spicer's (2006, 2–4) lyrical account of how practically impossible it is to know "who" lives at Wembury Bay, despite intense study.

My critique is not subject to that sort of empirical disproof. It says that no biodiversity project can be coherently characterized as "ultimately successful" in promoting the value of the natural world. On the other hand, there is ample evidence – from all the actual proposals for pursing a biodiversity program – that coherence is elusive. And worse, if it can be achieved, that is possible only at the cost of a program that would be enormously destructive of natural value.

This final and decisive failing of the biodiversity strategy for characterizing natural value is also the key to a more fruitful way of thinking about it. My suggestion, which I develop in the next chapter, is quite simple in concept: Start from the other end of the valuing relationship. Start with people and their societies, and try to understand the nature of our valuable relationship to the natural world. Starting with valuing agents rather than going on a treasure hunt for things valued surely has at least one enormous advantage: Very likely, what is discovered about our valuing transactions will be morally relevant.[46] This suggestion, in effect, is a suggestion to return to the unspecified "moral reasons" that such distinguished partisans as Paul Ehrlich and E.O. Wilson only gesture at (Sect. 6.1, Unspecified "moral reasons"). I think that it returns to the *heart* of the matter, and to Wordsworth's poetic inspiration for this book.

When one starts with human valuers and asks how the natural world might be valuable to them as the natural world, one does not find biodiversity, understood in any reasonable way, on the receiving end of their evaluations. The various maneuvers to "hack" biodiversity into something that it is not very likely reflect an intuitive recognition of this truth – that biodiversity is not the core of what people value as nature.

---

[46] I borrow the term "transaction" from Jamieson (2002, 234), where he conceives of an act of valuing as a transaction between valuer and thing valued. It has an unfortunate tendency to suggest "market transaction" or business deal. But one should push aside that specialized connotation to embrace its broader meaning of "an interaction".

Chapter 8
Natural Value Starting from People

> *Enough of Science and of Art,*
> *Close up those barren leaves;*
> *Come forth, and bring with you a heart*
> *That watches and receives.*

– final verse of "The Tables Turned" (1798),
William Wordsworth

This book has examined in detail a set of views that coalesce around the idea that biodiversity is a preeminent handle for the value of nature. Far from armchair philosophy, this idea, coupled with several others closely associated with it, has come to dominate the justification of and motivation for very real projects now routinely undertaken by conservationists, restorationists, and environmental organizations on behalf of the natural world. The basic precept is that by promoting biodiversity, natural value is preserved and even created. In this chapter, I call the collection of these projects "the biodiversity project".

One way to summarize the preceding chapters' analysis is to say that there are literally no good reasons, and certainly no decisive reasons, to espouse the biodiversity project. The evidence and reasoning on offer from the biodiversity project's advocates provide no credible grounds to suppose that nature-as-biodiversity is particularly valuable. To the contrary, what they have to say often lends greater credence to the supposition that the biodiversity project might terribly compromise whatever natural value there is in nature.

Getting to this point has required engaging with the biodiversity project on its own terms, following the twists and turns of the best of its many intricate threads of argument. It has required grappling with (among other things) puzzles about the significance of biodiversity tallies, the relationship of biodiversity to "ecosystem function and services", "good biodiversity" versus "bad biodiversity" or just "not so very important biodiversity", and even heuristics for solving the NP-complete set

D.S. Maier, *What's So Good About Biodiversity?: A Call for Better Reasoning About Nature's Value*, DOI 10.1007/978-94-007-3991-8_8,
© Springer Science+Business Media B.V. 2012

cover problem.[1] My discussion has shown how, along every thread of reasoning, the logic gives way, or the evidence; or the reasoning breaks free of any credible and generally endorsable justifying principle; or it yields impossible-to-adjudicate internal contradictions; or it urges seriously counterintuitive and unacceptable actions. I could leave matters there, along with the conclusion that this dismal record warrants an urgent recommendation to abandon the biodiversity project.

But there is more to the biodiversity project's failure than an absence of credible principles that can be honestly endorsed without producing irresolvable paradoxes or recommendations for action that apparently guarantee a diminishment of nature's value. There is more to it than transgressions of logic, evidence, consistency, and coherence. Nor can the project's failure be accounted for specifically and solely as a matter of the promiscuous transmigration of the meaning of "biodiversity" from one protean form to another. One should expect that at least one thread of argument would have been saved from itself by this book's disciplined insistence on sticking with a well-characterized, scientifically informed, yet generous conception of biodiversity. Surely, if a theory of natural value is capable of hitting any target, it should be able to hit this broad and nonmoving side of the value barn.

The "more to the biodiversity project's failure", I believe, has to do with the normative and metaethical framework within which the biodiversity project poses questions of value. Otherwise put, it entails the presuppositions that circumscribe the set of possible answers. The failure of this framework retrospectively lends to the consistent failure of each and every one of a large collection of multifarious arguments a quality of inevitability. This renders as relatively insignificant distractions the various ways in which the various twists and turns lead to dead ends. Therefore, it is to this flawed framework that I now turn my attention.

My argument proceeds in two stages. Section 8.1 presents the first stage, which teases out and probes from several different perspectives the normative implications of the biodiversity project's founding normative assumption – namely, the proposition that the value of the natural world supervenes on its instantiation of certain properties of biodiversity.[2] This is a "state-based" view of value, which is plausible and attractive for evaluating certain things. Some states of the economy are better than others. Some states of a person's health can usually be said to be better than others. So it is, again, for states of my automobile. But I argue that it is neither plausible nor attractive to think that the value of the natural world hinges on its exemplifying certain biodiverse states and not others. Nor do I think that it is plausible or attractive to think that arranging to bring "good" or "better" biodiverse states into existence can possibly contribute to natural value.

Note that this presentation argues that it is the very framing of the biodiversity project that makes it a nonstarter. This is part of what I called my "radical critique" at the end of the previous chapter. It is a far more penetrating and devastating critique than one that merely observes that each of its best supporting arguments so

---

[1] See Note 1 in Chap. 1 for a prominent proposal to operationalize the value of biodiversity in terms of solutions to the set cover problem.

[2] "Instantiation" is used in the sense illustrated by Fig. 3.1 and explained in Sect. 3.1.

far proposed happens to fail due to insufficient attention to matters of fact, logic, and principle. The new, radical critique says that the biodiversity project frames the question of how the natural world is valuable in such a way as to essentially *guarantee* that no argument on its behalf can succeed.

In other words, I urge that we should seriously doubt the prospect that, perhaps someday, a biodiversity project advocate will have the long-awaited brilliant stroke that finally answers the question, "What is so valuable about nature?" in terms of biodiversity, in a way that is both morally compelling and viscerally motivating, and that, for once, slaloms down a course guided by principles that promote generally endorsable ends without crashing into obstacles of fact, logic, or paradox. Rather, we should fully expect more desultory meanderings, more irresolvable puzzles, more counterintuitive conclusions, and more misplaced emphasis. Unfortunately, there is reason to believe that, despite this dismal prognosis, the biodiversity project will not easily succumb from the natural causes that by all rights should be the death of it. That is because, however dysfunctional its framework, its advocacy is sustained by serious moral hazards, which maintain an extraordinarily tenacious grip on its advocates. I call this suite of bleak implications the "disvalue" of the biodiversity project.

This grim assessment of the biodiversity project's prospects argues for a tectonic shift in thinking about nature's value. Section 8.2 presents the second stage of this chapter's argument by offering the requisite seismic charge in the form of a theory of what I call "appropriate fit". This theory frames natural value in a way that differs quite radically from the way it is framed by considering its biodiversity. My approach incorporates some elements that might seem at home in other frameworks – including environmental virtue ethics, Paul Taylor's "respect for nature", and one particular strand (while eschewing many others) of Bill McKibben's thinking about what nature means to people. It shares a general perspective that is largely compatible and sympathetic with these and other proposals. However, despite these sundry resemblances, I believe that my framework makes use of other, substantially different elements. And where it recycles old ones, a theory of natural value according to appropriate fit selectively cobbles them together in a fresh and, I believe, more compelling way.[3]

The proposal rests on straightforward, intuitively appealing, and (I believe) compelling principles. That is partly because it follows the prescription written out at the end of the previous chapter. It does not start out by dissecting nature and grading its various analyzed states. Rather, it brings human valuers into the evaluative picture from the very start. In particular, it identifies the most basic ways in which people actually relate to the natural world in a valuing way. These ways of valuing nature I call the "core of natural value".

This resulting framework slices through assorted Gordian knots that tie up the biodiversity project. Unlike that project, the framework of appropriate fit has some considerable power to account, in a rich and satisfying way, for widespread and

---

[3] Section 8.2.6 (What appropriate fit is not) calls out significant respects in which a theory of appropriate fit distinguishes itself from these and other theories.

deeply held intuitions about the unique qualities of nature's value. Finally, while standard, biodiversity project-style thinking dissects and tends to eviscerate nature – leaving humankind's relationship to it off to the side except as an economic afterthought – framing that relationship in terms of appropriate fit gives some plausible basis for understanding its breadth, depth, and visceral grip on people.

8.1   The *Dis*value of the Biodiversity Project

This section examines some of the reasons and moral psychology behind the systematic dysfunction of the biodiversity project. To start, Sect. 8.1.1 (Biodiversity and the state of nature) characterizes the biodiversity project in terms of how it locates the value of the natural world as flowing from states that are characterized in terms of their biodiversity. Section 8.1.2 (Responsibility for nature) then explores how this view of the locus of natural value encourages the adoption of highly questionable models of responsibility for the natural world. These models leverage the intuitive attractiveness of such rules as "you break it, you own it" and "you break the warranty, it's your problem". Though their underlying intuitions derive from inapt analogies, they nonetheless underwrite the narrative of many environmental pragmatists and conservation biologists, among others. Next Sect. 8.1.3 (The ends of the biodiversity project), focuses on the project's endemic and epidemic confusion about its biodiversity-characterized ends – its seeming lack of credible principles that adjudicate between competing views about just which biodiverse states have the greatest claims on us and why. Section 8.1.4 (Biogeo-engineering value into nature) then characterizes what I call "biogeoengineering projects", which are seen as comprising a principal form of "right" action, entailed by the biodiversity theory of natural value. In other words, these projects are regarded as a salient means of creating a "good" or "valuable" natural world, characterized in terms of its biodiversity-deconstructed bio-parts. Examination of these claims shows them to be highly problematic. Section 8.1.5 (A metaethical gloss on biogeoengineering) is an excursion into the metaethics behind the observed difficulties faced by the central "state of biodiversity" plank of the biodiversity project's value platform. Section 8.1.6 (Lesser evil arguments) highlights arguments that play a role in the moral psychology of biogeoengineering by obscuring salient moral context. Section 8.1.7 (Moral corruption – a symptomology) then probes this moral psychology – particularly assorted corrupting influences that promote the disinclination to confront and even notice systematic failures of practical reasoning. Finally, Section 8.1.8 (A parting look at the biodiversity project), returns to the enterprise of biogeoengineering bio-inventories. This time, the focus is on biogeoengineering as supposedly "right action" – that is, as a characteristic behavior reflecting certain attitudes towards the natural world. I suggest that these attitudes are antithetical to those that embody natural value, regarded as a key element of "appropriate fit" of people in the world. This observation is the springboard into Sect. 8.2's topic – Natural value as uniquely valuable human relationships.

8.1.1   Biodiversity and the State of Nature

Chapter 3 (What biodiversity is) suggested a structured conception of biodiversity whose building blocks are various categories defined by biological science. Or more precisely, the building blocks comprise the diversity of kinds within those categories. This conception is both congruent with most sensible usage, and maximally congenial to the evaluative thrust of the biodiversity project: As I remarked in presenting this way of understanding biodiversity, it makes good initial sense of the many and varied actual claims that are made regarding biodiversity's value. This distinguishes it from other interpretations (such as one that builds goodness into the definition of the term), which render those claims circular or incomprehensible. Also, because there are few restrictions on choice of categories or how they might be combined for evaluative purposes, my conception provides the broadest and most flexible of substrates for attaching value to the natural world, insofar as that value stems from biodiversity.

In this subsection, I make some general observations about the evaluative program of the biodiversity project, which is best understood in terms of evaluating these structured, biodiverse states of the world. My observations rise above the level of the specific details – of precisely which categories whose diversity of kinds, in which weighted combinations, enter into consideration – that differentiate various views of biodiversity assessment. Therefore they should cut across all reasonable views, held by all reasonable partisans of the biodiversity project, of how to characterize and assess the biodiverse state of the world.

While the details that separate those differing views do not matter for my discussion, the fact that all these views take these details to be critically important *does* matter. So does the manner in which the details are taken under consideration. Specifically, a salient characteristic of the biodiversity project is that it locates value in nature within a highly stylized portrayal of nature as a bio-parts warehouse, as a collection of bio-properties (or biogeochemical properties), as a processing plant, or as some combination of these. The graphic tools of stylization come from science. This is what, as biodiversity project advocates say, makes the project "science based". The assessment of the biodiversity is an assessment of nature portrayed as a scientifically ascertained and scientifically characterized inventory of various categories of natural stuff, some structured combinations of which are said to engender scientifically ascertained and characterized properties and processing capabilities.[4]

So understood, nothing within the biodiversity project dictates how normative assessments of its scientifically stylized pictures should be done. But as much of

---

[4] In Sect. 8.1.3 (The ends of the biodiversity project), I widen the discussion somewhat to encompass "scientific" characterizations of "good" nature that appeal to ecosystem health, integrity, and vibrancy. There, I also consider characterizations that involve approximations to historical ideals. Historical ideals essentially come back around to proper bio-parts, in proper proportions, located in their proper bio-places. Such ideal states as "healthy" ones are also sometimes viewed as reducible to bio-warehouse specifications; though (as I discuss in Sect. 8.1.5), sometimes not.

this book's discussion shows, the project strongly gravitates towards economics. This is entirely unsurprising, for no portrayal of nature could be more congenial to the tools of economic analysis. Sliced and diced into commodified units of trade in the marketplace – either real markets, ones conjured up by exercises in contingent evaluation, or ones reconstructed through (for example) hedonic proxies – the natural world slips quite comfortably under the economist's lens. That lens is supposed to bring into clear view the role of the stuff, properties, and processing of nature in the human economy. Under the black light of economic analysis, some biodiversity project advocates see the stuff of nature glow as a warehouse of inventoried "natural capital" and its properties and processing capabilities luminesce as a more or less efficient engine for "ecosystem services". This is why it seems to many environmentalists that economics is the perfect evaluative overlay for a science-based inventory of nature – a supposedly "scientific" and "objective" evaluative adjunct to a supposedly "scientific" and "objective" accounting for nature's riches.

But although economic valuation looms large in the biodiversity project, nothing in my analysis of its evaluative heart ultimately depends on coupling the valuation to economics. One can strip away the economic calculus and substitute for it another evaluative overlay.[5] Better yet, one can view the evaluative overlay in a mathematically abstract manner – as a natural value function, which one might describe in this way: Take a state of the natural world; then slice and dice it into a collection of measured, inventoried, and indexed parts, properties, and processes. The natural value function maps these "components", constitutive of this state of the natural world – "ecosystem components", in the lingo of modern ecology – into a greater or lesser amount of (natural) value.[6, 7] The biodiversity project reimagines the natural world in terms of the umbrella property of biodiversity and subsumes all of the above-mentioned "components" under that rubric. This leads to a final simplification whereby the natural value function is understood to operate over the single domain of biodiversity.

Actual assessment means applying some preferred natural value function to its argument of the biodiverse state from place to place in order to produce a value-topographical map. This map shows value peaks, valleys, ridgelines, and low-lying

---

[5] For example, the overlay can be one that assesses the degree of approximation of a state of nature to some historical standard for bio-parts, in certain proportions, in certain bio-places.

[6] This common and facile use of the term "component" is not congruent with mine in my definition of "biodiversity". In that definitional context, "component" refers to a category whose diversity of biotic or biota-encompassing kinds contributes to biodiversity overall.

[7] It is worth making special note of the fact that the economist's natural value function is essentially scalar-valued and therefore its values form a totally ordered set. This results in a stunning axiological simplification, which sustains the fiction of *Homo economicus* – the economically rational human – who is capable of applying simple comparisons of less then, more than, or equality to every pair of valued things. Economic valuation requires that this amenability to pairwise "tradeoffs" capture the essential quality of values of all flavors – natural and otherwise. The deeper implications of this are discussed in Sect. 8.2.3 ("Living from" nature, uniqueness, and modal robustness).

bogs of value into which the biodiversity evaluation function maps the sliced and diced "ecosystem components" or "components of biodiversity" that constitute some biodiverse state of the world. This makes it apparent that the natural value topography can be changed by changing the biodiverse state of the world. Therefore, at the most abstract level, the biodiversity project's goal can be characterized as imagining and effecting changes that realize a biodiverse state that yields a more favorable value topography – for example, filling in low-lying bogs of value while shoring up crumbling value peaks.

My account of the biodiversity project's evaluative framework is intended to evoke just how abstract an affair it is. One might be struck (and discomfited) by how detached from this account is anything evocative of the natural world. But the evaluative function eventually has to make contact with nature because the values cannot be manipulated directly. They are accessible only through arguments of the natural value function – that is, through the "natural" objects (known as "components of biodiversity") in the right biodiverse combinations and arrangements, which constitute the biodiverse states in which the values reside. Therefore, scientific expertise is required to figure out how best to realize a desirable value topography by manipulating the sliced and diced natural "components" that also constitute the manipulable elements in the domain of the natural value function. The language in which biodiversity project advocates express this will be familiar to those familiar with the biodiversity and conservation literature; it is typically couched in terms of "management". It urges that we "manage" an ecosystem for this or that "component" – that we "manage" for this or that species, "manage" for this or that property, or "manage" for this or that process.

Most saliently and sweepingly, the biodiversity project urges that we "manage" ecosystems for biodiversity – to achieve some biodiverse state of the world that is regarded as "better" than others. "Better" usually translates to "a state of the world that is 'more' biodiverse". Such a recommendation is quite straightforwardly understood in terms of a natural value function – economic or otherwise. It would show how, according to its computation, "managing for X" builds mountains of natural value while avoiding the creation of low-lying value bogs. Sometimes, as when X is some species or other, "managing for X" can mean "culling or even ridding the world of X". But most notably, broadly, emblematically, and especially for biodiversity, it means "promoting it as an end". That is, X is held up as an important end of conservation and biodiversity is held to be the all-embracing stand-in for all X – that is, all ends – that are allied with natural value.

In the remainder of Sect. 8.1, I tease out what a picture of natural value along these lines entails for whatever normative claims it stakes out. Four salient intertwining elements thread their way through this discussion. They jointly delineate what is at once most characteristic of and most troublesome about the biodiversity-based theory of nature's value.

The first element is this value portrait's "state-based" character, which attaches value to particular states of the world. This naturally leads it to detailed characterizations – in terms of component element and jointly produced properties – of which states of the natural world are supposedly better than others. Even aside from bringing biodiversity into the state characterization, this view of the source of natural value engenders difficult-to-defend views of such normative concerns as "responsibility for nature" and why the biodiversity project is supposed to be an appropriate way to fulfill this responsibility.

It is illuminating and perhaps surprising to notice that this same general view of nature and its value undergirds the view, most famously promulgated by Bill McKibben (1989), that we are at *The End of Nature*. This lament is typically tied to such observations as that the state of today's world in no way resembles its state some 70,000 or perhaps fewer years ago, before humans starting leaving their indelible mark on it. If one truly believes that "the natural" and whatever value that attaches to it is a matter of its achieving some previous and irrecoverable state, then there's no doubt that we are confronting the end of nature. This same belief might also make it seem incontrovertible that "loss of nature" – to the point of almost losing it entirely – calls for a certain form of action to avoid the actual, final demise. This form of action is the fourth element, which I characterize below.

I have already mentioned the second salient feature of this peculiar axiology in connection with its congeniality to economic analysis. That feature is its deconstructive nature. Some might find this an odd description, for phrases such as "global biodiversity" have a sweep that can appear to swamp the kind of intricate accounting and computing that lie behind it. And there's no doubt that some proponents of the biodiversity project veer back and forth between grand gesture and the microvivisection of nature into an inventory of its many and intricately related "components" (in the sense of "bio-parts"), which are accounted for in terms of biodiversity. But in the biodiversity project's best, detailed, and published arguments, it is the world of sliced and diced components, properties, and processes that dominate.

This deconstruction of nature into its myriad components in multiple categories undoubtedly makes for fascinating science. In fact, it is the stuff of biodiversity science, which attempts to examine how the parts in the world's natural parts inventories combine and interact as "working parts" to produce its various observable properties and ways of functioning. There is nothing wrong in that. The problematic step is to identify "that which is valuable in the natural world" with a sliced and diced characterization of its biodiversity. At that point, nature blurs into the working parts inventories, and its value cannot be seen except as that which supervenes on its various detached inventory of pieces. The question of what variety and inventory level of X's constitutes a good biodiverse state of the world and therefore a good state of nature seems to eviscerate the value of the natural world by trying to load much of that value, piece by piece, into its various inventoried parts *as inventoried parts*.

The third element is the notable absence of people – as valuing subjects – from the foreground of the biodiversity project's picture entitled *Valued Nature*. Valuing subjects do appear within the picture's frame, but only in the role of parties collectively responsible for deforming what used to be a very good state of nature into a very bad state; and also as actors responsible for rectifying this situation by removing some of these deformities. The persons within the picture exhibit no other behaviors and are depicted as devoid of the expression of all other attitudes, which could possibly depict valuing transactions[8] with nature.[9] Alongside *Valued Nature* hangs a small companion piece, entitled *Subject Valuing Nature*. In it, people appear only as the acknowledging beneficiaries of the bounty that flows from what *Valued Nature* portrays as a "good" state of nature.

Which leads directly into a fourth aspect of the biodiversity project's view of natural value. This has to do with the form of action that it encourages. In fact, the biodiversity project's practitioners emphasize that it *entails* an obligation to respond to nature in a characteristic way. Most saliently, if we are near or right "at the end of nature" as a consequence of humans transforming the world into something unrecognizable, then the job of humanity is to recover or reassemble the world's lost former state as best as it can manage. Seen through the lens of biodiversity science, the objective of the biodiversity project takes on the appearance of recreating "good" or "better" biodiverse states of the natural world.

Partisans of the biodiversity project urge that these states are to be achieved by properly adjusting and recombining the world's inventories of bio-parts in creative ways: Restock depleted parts, destroy overstocked parts, wrestle the "right" parts into the "right" combinations, get rid of the "wrong" parts where they don't belong, and break up "wrong" combinations. This view of how humanity should go about creating (or recreating) and maintaining a "good" natural world trips over the third feature. There is a remarkably limited consideration of valuing subjects in arriving at these formulations of right action – except, of course, at the point where they are expected to act. Most notably absent is any consideration of what might be the most basic moral test of the biodiversity project: Are the attitudes and behaviors exemplified by engaging in such a project ones that can truly said to be valuing of the natural world?

Suppose for a moment that some set of actions urged by the biodiversity project were incontestably well intentioned. Suppose that there were no controversy over what biodiverse states are the really valuable ones. Suppose as well the absence of serious questions regarding the appropriate means to achieve those supposedly uncontroversially valuable biodiverse states – taking into account (among other things) human capabilities in manipulating the natural world and the epistemic capabilities required to foresee the results of human impingements on nature. Would the answer to the above question then change?

Within the standard evaluative framing of the biodiversity project, neither this question nor any relevant to answering it are even asked. In the remainder of this section and then in the next, I give this question careful and detailed attention. I answer it in the negative. But I do this by arguing for a much stronger proposition: The behaviors that are supposed to carry out the biodiversity project's mission are not just failed attempts to realize natural value. Rather, they are decidedly antithetical to a valuing relationship to nature.

---

[8] See Note 46 in Chap. 7 for an explanation of my use of this term, which owes to Dale Jamieson.

[9] In the next subsection, I examine at least one view associated with a strand of environmental pragmatism that might initially seem to run contrary to this characterization. But I argue there that, insofar as this view hews to any principle other than that it is good to make people feel good, it presumes that certain states of nature are particularly worthy of resurrection.

8.1.2 *Responsibility for Nature*

One can begin the task of exposing the disvalue of the biodiversity project by delving into how it exposits human "responsibility for nature". As I remarked in the preceding subsection, apart from inserting people into the picture as the beneficiaries of nature's bounty, this is the most notable entry point for people in the biodiversity project's portrayal of valuable nature. I shall show that the biodiversity project's notion of responsibility is informed by, and reciprocally reinforces, its state model for "good" nature. The result is a strikingly inapt notion of responsibility for the natural world.

I consider two models of responsibility commonly on offer. The first is what I call the "broken nature" model; the second (already encountered in Sect. 7.2.3), the "voided warranty" model. Because the biodiversity project does not have exclusive rights to the state-based conception of "good" nature, it is not surprising that the connection of this conception to responsibility for nature inhabits thinking that originates from outside the biodiversity project. Such an expression of the broken nature model serves as the counterpoint for my exposition of it. On the other hand, my exposition of the voided warranty model stems from its expression – as a "scientific" assessment of responsibility – from within the biodiversity project.

Pruned down to essentials, the train of thought for both models goes through these stations: The state of nature can be intact… or not. It is said that, in fact, it is currently not intact. Nature is broken; and, people are the vandalizing culprits. Or if nature is not broken, at least people have made it deviate so badly from original design specifications that those specs no longer apply. On account of their causal responsibility – for either breaking nature or at least voiding its warranty – people come to "own" the natural world.[10] That is, people are held to own nature in much the same way that their ownership of any sufficiently valuable possession incurs responsibilities for repair or (outside-of-warranty) maintenance.

On a variant reading, the responsibility is not so much propriety as *fiduciary*, although it likewise entails the responsibility for nature's care and management. Unfortunately, critical elements of the fiduciary model are rarely fleshed out. The identity of the beneficiary, the terms of the trust, and how the beneficiary enters into the trust are all left a mystery. Because the fiduciary model is, to my knowledge, so incompletely developed, I stick with the simpler owner model, though the fiduciary one might also serve with some further development.[11]

It is the notion of the natural world as precious, owned property with norms for intactness that I regard as radically inapt. Yet it serves as a principal basis on which the biodiversity project elaborates the details of specific human responsibilities to

---

[10] It is possible that this narrative implicitly presumes from the very start that persons have a proprietary interest in nature, perhaps by virtue of the kind of creatures-in-the-world they are. In this case, the conclusion is already contained in the premise. This does not matter for my discussion, which tries to understand the logic of the arguments for responsibility that are actually presented.

[11] However, I revisit the concept of a fiduciary trust in Sect. 8.1.7 as a way of characterizing a form of moral corruption.

nature. Insofar as the relationship of people to nature is *not* one of owner to a prized possession with a spec for proper assembly and functioning, it should not surprise that the responsibilities that one ordinarily supposes to attach to such a relationship make no sense with respect to the natural world. But the fact that this model of responsibility does not make much sense is obscured – along with other relevant moral context – by the manner in which project advocates present it. It is, they say, a dichotomous choice: Either we humans can take responsibility for the current non-intact state of the natural world; or we can shun this responsibility. Of course, the second choice seems repugnant. After all, "we" are fine, upstanding, responsible sorts of people. But the first choice – of wrestling nature back into an intact state – can seem quite unattractive, too. The questions it raises about the behavior – sometimes involving radical rearrangement of things – that supposedly fulfills the responsibility can make that alternative unpalatable, too. This is so even from within a state-based conception, when the aim of these efforts seems to have taken flight from any sober conception of nature.

Insofar as neither choice presents a welcome alternative, this is a choice between two "evils". This overarching consideration frames the choice and imposes a special kind of logic on it. According to that logic, we humans – who are in a position to choose – are obligated to choose the lesser one of the two evils. But this tacitly removes from the table the alternative of firmly rejecting the account of responsibility that appears to force the choice in the first place. And (as I will argue) rejecting that account of responsibility *is* the correct choice. I say more about the obfuscating effects of lesser evil arguments in Sect. 8.1.6. In Sect. 8.1.7, I discuss the symptoms of moral corruption, which aids and abets this type of argument as the result of an environment in which those who ply it are prone to biases stemming from their own, morally irrelevant stake in the matter.

To this point, my account of the notion of human responsibility for nature as the responsibility of an owner for precious, breakable property or property no longer operational according to original specifications has glossed over one critical piece of logic. Some normative principle is needed to tie the effects on the natural world for which humans are *causally* responsible to *moral* responsibility for nature's care and management. People are causally responsible for many changes in the world that do not incur moral burdens. Merely causing something to undergo a change does not normally entail an obligation to care for and maintain the thing changed. The basics of everyday human life are full of examples. When I repaint my turquoise kitchen orange, I do not thereby incur any special responsibility for it. When I change a carrot by eating it, I have determined its fate; but that does not underwrite any special obligation towards it or carrotkind. There is no norm, let alone a norm that rises to the level of moral consideration, that makes the initial state one that I ought to have respected and failed to do so in causing it to change. Moreover, in neither of these everyday cases nor in a plethora more do I have a contract, explicit or implicit, direct or indirect – let alone a morally binding contract – with the object that I change. Nor do I have a contract of any such kind with any agent who has some morally compelling stake in the object or in its pre-changed state.

So *some* connecting logic is required to distinguish the natural world from the multitude of other things whose state can be changed with*out* incurring any responsibility for them. That is, some normative principle must be invoked to bridge the gap from "cause of change" to "obligation to care for thing changed" when the "thing" is the natural world. For the biodiversity project and allied views, that principle derives directly from the state model of nature, which is assumed to admit intact states versus states that are not intact. Intactness can be interpreted in terms of not being broken – that is, having a structure and associated set of functions that are the "right" structure and that function the way they ought to function. Or it can derive from a model in which the norm for nature is a state in which OEM parts, authorized by the original design specifications, remain properly installed in their proper place.[12] The requisite normative principle hinges on the normative force loaded into such terms as "intact", "broken", "OEM", or their equivalents. Those who subscribe to the first conception of intactness claim that causal responsibility for changing nature in certain ways constitutes destroying its proper structure and (partly because of this) making it work improperly. This authorizes application of a "you break it, you own it" rule to bridge the logical gap. Those who subscribe to the second conception of intactness claim that causal responsibility for certain kinds of changes to nature constitutes unauthorized meddling, which invalidates the original design. This authorizes invoking the "you break the terms of the warranty, it's your problem to maintain" rule. Despite the difference between their respective bridging rules, the two different models wind up with a similar proprietary notion of responsibility.

The first route to responsibility based on the "broken nature" model emphasizes the element of corrective justice. A principle of corrective justice says that persons have the duty to repair the wrongful losses that their conduct causes. When it is nature that suffers "wrongful loss", "repair" tends to assume a restorative tenor. It involves the moral requirement, already discussed, to assume the burden of correcting nature – righting the wrong of breaking it by making nature whole again. That is, it involves restoring it to its "rightful", unbroken state.

A second route to responsibility (also within the "broken nature" model) is via retributive justice. This form of justice is supposed to demand that offenders perform gestures of reconciliation and that they reform their character via therapy or acts of redemption. Any program of retributive justice must supply and justify criteria according to which an act qualifies as either reconciling or redeeming. Unfortunately, when it comes to retributive justice for the natural world, this requirement is not met. Once again, the framing of acts of retribution for wronging nature distracts from this critical deficiency by directing attention elsewhere. In this case, focus tends to be entirely on a psychological speculation for which a few anecdotes are offered as evidence: The speculation is that retributive acts, typically comprising small-scale gestures of "unbreaking" nature, are psychologically welded to the notion of responsibility to serve retributive justice in the minds of some persons

---

[12] 'OEM' is an acronym for the nearly-as-opaque phrase "Original Equipment Manufacturer". Its meaning comes from industry and varies from industry to industry. But the root definition of "OEM part" is "a replacement part made by the maker of the original". While the conservation and environmentalist literature does not use "OEM", as the main text recounts, this acronym precisely captures one major vein of thought about the state of nature and how human responsibility for it arises.

who, by performing them, feel that they fulfill that responsibility.[13] In other words, the narrative anecdotally observes that some people feel as though certain acts are retributive offerings. With no intervening logic, it concludes that therefore, they *are* retributive offerings. But no thread of reasoning shows how the small-scale gestures of "unbreaking" nature that arouse the feelings of self-satisfaction actually fit into a respectable theory of retributive justice. And so there is no reason to think that these feelings rise above the delusion that people cling to in order to feel better about themselves.

This might justify the conclusion that acts of reconciliation and acts with a therapeutic or redemptive quality with respect to nature are of secondary interest when set beside the primary task of getting the "real" job – of "unbreaking" nature done. And it might lead one to appraise the rightness of these acts on the same basis as the "corrective" actions towards which they gesture. This appraisal would come down to the crippling criticism that there seems to be no principled way to hold up any particular state of nature as the unbroken ideal that humans should strive to realize. That includes states defined by historical ideals, which are often held to rationalize gestures of historical re-creation.

However, advocates of acts of reconciliation and acts with a therapeutic or redemptive quality hold them to be important in their own right, independent of their efficacy in "unbreaking" nature. I think there is something to that. However unjustified in terms of their effects on nature and whatever confusions there are in rationalizing them, the impetus for these acts encapsulates a nugget of wisdom. The nugget is that what is important is not so much "unbreaking" nature, but "unbreaking" the human relationship to nature. It is unfortunate that this particular approach is so poorly judged. A closer look at its missteps is nevertheless worth taking, for the light they shed on the difficulties of bringing the natural world into the domain of human responsibility.

This view of responsibility for nature – in which restoration plays a reconciliatory, therapeutic, or redemptive role – features in the "environmental pragmatist" philosophy of Andrew Light, for whom it undergirds what he calls "ecological citizenship".[14] The notion of "ecological citizenship" is difficult to summarize briefly because it seems to be many and varied things. Among other things, it is the communitarian idea (Light 2006, 177) "that the larger community to which the ethical citizen has obligations is inclusive of the local natural environment as well as other people."[15] Propping up his broadly communitarian conception of persons-in-their-environment

---

[13] Of course, as I insist just below in the main text, it is relevant that the acts offered in the name of retribution are gestures towards (what the actors consider) *corrective* justice. In this respect, the account of responsibility for nature from retributive justice does not stray far from the account that stems from corrective justice.

[14] The terms "reconciliation", "therapy", and "redemption" are not Light's, but I believe that they accurately reflect the spirit of his thought and illuminate its essential contours.

[15] This is one of several pieces of Light's account of "ecological citizenship". Other pieces appear in Light (2006) and in Light (2002; 2003b).

is the notion that people can build better communities as ecological citizens – by banding together and participating in activities to promote a set of environmental goods. On Light's accounting, these goods are quite disparate, aside from the vague connection of each of them to "the environment" or to "nature". One signal good is "environmental sustainability". Light does not explicitly say what he means by that phrase. But his separate mention of watersheds (Light 2003b, 56) and this use of New York City as a central point of reference suggest that he has in mind New York City's source of potable water. Also, more than once, he airs a concern for "sustainable consumption". Taken together, these considerations suggest that, by "sustainability", Light means something like "securing the resources required to support urban populations". Another, quite different, environmental good in Light's disparate collection is the (Light 2002, 169; 2006, 178) "ecological viability" of a city park such as New York City's Central Park. Finally, in yet another category of environmental goods, are restoration projects such as the Chicago Wilderness program, aimed at restoring oak savannah around Chicago (Light 2002, 160, 164, 168, 170; 2006, 172, 180; 2007a, 8, 20, 22). It is especially in connection with this latter type of project that Light's appeals to individual reconciliation, therapy, and redemption as salient in moral justification occur and might appear to make sense.[16]

According to Light (2000, 64–65), restoration is a way to heal our human characters, which have been corrupted by a relationship to nature in which we have done it egregious harms, such as placing it in bondage. As he puts it (Light 2000, 66), acts of restoration are to be viewed as arranging "for… nature to free itself from the shackles we have previously placed upon it." Light (2000, 65, 66) suggests that these acts serve as a means to reconcile our past transgressions against nature and as a sort of character therapy, which alters our relationship to nature so as to curtail future transgressions: They are a way "to interact with the flora and fauna" so as to "increase bonds of care that people will have with nonrestored nature", and to make it "less likely to allow [nature] to be harmed further."[17]

On top of his narrative of past abuse of nature coupled with reconciliation and reform of our relationship with it, Light layers a narrative about damage to our own character. For example, he (Light 2000, 62) writes:

> … it seems that Katz is right in assuming that somehow our actions toward nature morally implicate us in a particular way. In the *same* sense, when we morally mistreat another human, we not only harm them but harm ourselves (by diminishing our character, by implicating ourselves in evil…). [italics added]

---

[16] Light (2002, 153; 2006, 172) also talks about "restorations in [New York City's] Central Park". But it is hard to know what "restoration of nature" might mean for this massive engineering project. See Sect. 8.2.5 (Natural value outside its core) for a description of Central Park's origins in Frederick Law Olmsted's imagination.

[17] This last phrase is sufficiently important to Light that he repeats it *verbatim* in Light (Light 2003b, 59) and Light (2007a, 18). It appears that Light intends the phrase to apply to restoration projects generally, including ones (such as the unstraightening of the Kissimmee River) whose size and scope, in Light's judgment, make the involvement of individual citizens problematic.

Immediately afterwards, Light affirms his conception of a certain moral equivalence between persons and nature, saying that "doing right by nature will have the *same* reciprocal effect of morally implicating us in a positive value as occurs when we do right by other persons." [italics added] All told, this vein of thought suggests that in Light's view, when it comes to nature, what most needs restoration is our character. In my expository terms, Light seems to be saying that the diminution of our character requires therapy; and acts of restoration (or nature) have the requisite therapeutic power. In a corresponding way, that we have been "implicating ourselves in evil" requires the purging of a devil in us; and acts of restoration are the appropriate sanctifying rituals.[18] One can see how this connects back to the notion of reconciliation by looking to the model of truth and reconciliation commissions, which are justified based on their therapeutic and sanctifying value.

Light's narrative conforms to the template that I sketched above by emphasizing the importance of reconciliation, therapy, and redemption: It conveys the distinct impression that the actual effects of restoration on nature are considerably less important than its therapeutic effects on the persons who restore. But restoration (conceived as a way of acting on nature) plays a central role – as the archetypal form of character therapy. Light therefore feels obliged to defend restoration against the well-known and oft-aired charges that it embodies a vice of "faking nature"[19] or a vice of arrogant domination.[20] He (Light 2003b, 62n13; 2007a, 19) also is concerned to convince his readers that valuing nature need not carry the metaphysical burden (or (Light 2006, 174–175) motivational impotence) of attributing intrinsic value (on any of its meanings) to nature itself or that it be capable of reciprocating as if it were itself a valuing subject.

I have no trouble granting that acts of restoration need not embody either of the vices that concern Light. And I concur with him in thinking that acts of valuing nature need not carry with them difficult metaphysical burdens. But the mere absence of these particular deficiencies hardly sustains the strong normative claim that Light (2000, 64, 65) wishes to make – namely that restoration is a salient means, worthy of general endorsement, for people "to form a substantive normative relationship with", in succession: a person, "the land around me", and finally, nature. His answer to his own question (Light 2000, 64), "Can restoration help engender… a positive normative relationship with nature?" is "… that it can".

---

[18] This is the only place in Light's corpus, of which I am aware, which assumes a distinctly theological overtone. However, it is important to include mention of it for its nontrivial role, in Light's narrative, of reinforcing and being reinforced by its secular counterpart (of therapy).

[19] The phrase "faking nature" does not originate with Light. It is the title of Robert Elliot's seminal 1982 paper and of his 1997 book on this topic. Elliot breathed a life into that phrase, which subsequently has been lived in many philosophical papers concerning restoration.

[20] These potential objections to restoration dominate the discussion in Light (2000). They feature prominently in Light (2007a), too. Light is specifically concerned with Eric Katz's views, expressed in various papers, including Katz (1992). I distinguish my own views from Katz's in Sect. 8.2.6 (What appropriate fit is not).

One can tentatively join Light in supposing that acts of restoration connect people to nature in *some* way, though (as even he says) the evidence for this is largely anecdotal. But this is not the central issue. What escapes Light's notice is that what is centrally at stake cannot be determined by measuring what, as a matter of fact, people report feeling about their "connection" to nature after participating with like-minded companions in something they call or are told to call "a restoration project". Rather, what is at stake is whether or not restorations model the *right* kinds of connections by being model acts of valuing nature. Even if restorations are not infused with the vices of fakery or hubris, they might still be entirely inappropriate ways of acting on, or with respect to, the natural world.

For his justification of restoration as a model way to connect with nature, it is legitimate to first expect from Light an independent characterization of the norms for this connection – what Light (2007a, 19) calls the "normative ecological relationship" between persons and nature. In addition, it is legitimate to demand persuasive grounds for thinking that, according to these independently characterized norms for the relationship, restoration therapy is an exemplification-*worthy* expression of it. Unfortunately, Light's discussion meets neither of these requirements.

Instead, one finds unhelpful analogies, hypothetical and real anecdotes, and undefended claims for the therapeutic value of restoration. Looming behind Light's treatment is the proposition that he repeats in one paper after another: People are less likely to harm nature when, having made a connection with it, they care about it.[21] Light (2000, 62–64) tries to make this proposition plausible by way of discussing his relationship with another person. But on the face of it, this analogy is devoid of the key analogous elements that are morally relevant. "Harm", in its morally relevant application to other persons, means "acting contrary to their interests and will". But nature at large has no interests; nor does it have a will of its own to act against. If, as Light says, we have placed shackles on nature, this does not obviously have the salient moral implications that shackling a person would have for her ability to lead an autonomous life. So if he insists on calling human effects on nature "harmful", then he must acknowledge that this is a sense of "harmful" that does not obviously qualify for moral consideration. At least, it doesn't qualify on the basis of his broad analogy to other persons.

Light (2007a, 10–17) also presents a series of hypothetical anecdotes that involve the smashing of various fragile objects for "no reason", as he (Light 2007a, 15) says. He points out that when some person (who could be the smasher herself) stands in an especially meaningful relationship to the object, its smashing can indirectly harm that person. Light later argues that people are less likely to "smash" a restoration that they themselves have created. This, he suggests (Light 2007a, 20) engenders "normative ecological relationships" that extend beyond the restorations. In terms of his "smashing" examples, Light seems to be saying that an increased disinclination to "smash" a restoration that one has helped to create would lead to a similar

---

[21] It is plausible to think that this proposition entails that people are *more* likely to harm nature when, for *lack* of a connection with it, they do *not* care about it.

disinclination to "smash" nature generally. But insofar as we can say that people "smash" nature, there is almost always some good reason. It might be a good place to build structures for dwelling or transport; or it might be a good place to grow food crops. Moreover, restorations of nature – unlike imagined restorations of the smashed watches and glasses in Light's examples – often purport to restore a place-in-time with which neither the restorers nor anyone else alive ever had a meaningful relationship. Included in this class of restorations is Light's oft-featured exemplar of the Chicago Wilderness program, which seeks to mimic a time when fires were set rather than suppressed by the ecosystem managers in charge before Europeans took over that role.

Aside from the morally irrelevant analogies and hypothetical anecdotes, Light (2007a, 22, 21) leans on "anecdotal evidence" that people form "important relationships with the restorations that they participate in helping to produce."[22] According to him (Light 2007a, 22), restorations also connect one participant with others. If these two propositions were true, it would not be surprising on grounds that have nothing to do with restorations – for people often form an attachment to things that they make, while group activities often are fun and help to develop cooperative relationships between the participants as they reinforce their mutually held beliefs. But this says nothing about whether or not those beliefs have any morally respectable foundation. Popular "team-building" exercises demonstrate that the goals of the group need not rise above the level of the frivolous or be morally ambiguous (for example, corporate profits). Light adds to these kinds of generic good feelings a restoration's special therapeutic effect. This effect arises from the restoration participants' belief that it helps to absolve them and perhaps humanity generally from previous malfeasance or the sins of previous evil practices. They believe that their restorative gestures purify their characters. It would be unsurprising for these beliefs to yield very good feelings, indeed.[23]

---

[22] It noteworthy how detached this statement is from any consideration of the value of *nature*. Of course, Light could say that his concern *is* radically detached from any concern for nature and that his "normative ecological relationships" only apply to human restorative creations. That would be consistent with the idea of redemptive restoration. But then his norms fall well short of their promised application (Light 2007a, 2, 3, 8, 10) to "natural systems".

[23] Light (2007a, 21) briefly mentions ecosystem services as another possible reason for valuing nature restorations and he brings it up elsewhere (Light 2006, 172, 178). But whatever the merits of this rationale for manipulating nature, it applies to something like the New York City watershed rather than the Chicago Wilderness. The assurance of the ecosystem service of providing potable water for New York requires managing defecating cattle, sewage runoff, and the quantity of toxic chemicals that lawn-owners use within the watershed. This seems a lot more like cow- and toxin-management than nature restoration. Restoration of nature or not, this particular effort is 300 miles away and far out of sight of the New Yorkers who benefit from appropriating the watershed's water-cleaning service. And even if the watercourses went right through Times Square (as the Great Kill once did, according to Manahatta Project Director Eric Sanderson), it is hard to see how fencing cows away from them could be central to fostering attitudes that bond people to nature in the way that Light's theory of natural value requires.

Light also tosses in ecosystem health as a reason to be respectful of natural systems (Light 2006, 172, 179; 2007a, 3, 5). I discuss the normative impotence of this rationale later in this subsection and elsewhere (Sect. 7.2.3, Home is where you make it – in situ, ex situ, and ecosystem-engineered).

However, while good feelings might be generally good for people to have, a good feeling-inducing activity is not morally justified merely on the basis of that description. Some things that make people feel good or that bind them together can range from frivolous to mildly inappropriate to utterly abhorrent in some respects. This is not an improbability, but rather a common occurrence, which arises in commonly encountered conditions, such as misinformation, self-deception, misjudgment, and bad reasoning. I might feel good about restoring my relationship with a lapsed friend by accommodating her desires for addictive substances. Her psyche might be substantially bolstered, too. However, such an action on my part would be reprehensible, despite the fact that I might dearly desire her friendship. A group might bond by virtue of developing their common hatred of other groups based on race, ethnicity, or gender.[24]

Therapeutic or redemptive restoration focuses on human attitudes and behaviors with respect to the natural world. It focuses on restoring human relationships with nature – something that seems possible to wrap one's mind around – rather than focusing on restoring nature itself – whatever that might mean. There is, as I said in introducing this conception of human responsibility for nature, a vital germ of wisdom in this focus – one that engenders my own proposal for the core of natural value in Sect. 8.2. However, this germ cannot flourish if it merely serves to convince people, because they so dearly *want* to be convinced, that their acts of restoration make amends for what they imagine to be previously inflicted wrongs, or that these acts thereby set things right for themselves and for nature, and for their relationship with nature.

In short, the desire for a mending of things and the desire for therapeutic redemption do not, by themselves, justify any means of realizing a feeling that things are mended. Some convincing argument, based on credible and generally endorsable principles, is required to show that the feeling achieved by performing the prescribed redemptive and reconciling gestures is, in fact, justified and not a "false" feeling. In the absence such an argument, one is justified in suspecting that redemptive restoration, despite its good intentions, falls into the trap of guilt-relieving self-deception.

Moreover, whatever one thinks of the appropriateness of the end of satisfying a desire to experience redemption and reconciliation with respect to nature, one needs some independent reason to suppose that the means for achieving this desire are themselves appropriate. These means, as I have observed, are in the main presumed to be gestures towards historical re-creations. It seems that their appropriateness might depend, at least in part, on whether it is appropriate to go about the business of re-creating states of nature that were lost to world some time ago. In short, the question of appropriateness brings us to the view of responsibility as a matter of corrective justice.

At this point, this view of responsibility for nature encounters yet more trouble along the same lines as that encountered by a theory of Biodiversity as the natural order (Sect. 6.12) and which figured in Sect. 7.5's Biodiversity value post mortem: What reason,

---

[24] Light thinks that he has a response to such extremes, which I examine in the next subsection. But my argument here can comfortably put aside their consideration and rest content with behavior that is merely wrong-headed without having any underlying malice.

supported by principles that people should be willing to endorse generally and out-side this narrow application, is there to believe that acting so as to make a place, now, resemble its former self at some certain point in the past in certain selected respects is not a collection of arbitrary choices? Why that time? Why those respects? And even if there were reasonable grounds for regarding some particular past state of nature as the lost Eden, what grounds would justify a willingness to ignore the overwhelming number of other respects in which the re-creation will inevitably dif-fer from this historic ideal? Most fundamentally, why is it good to mimic anything in the past that cannot possibly be convincingly mimicked in innumerable respects?

The science of ecology reports that in some very fundamental ways, the environ-mental contexts that framed past states of the natural world and that framed life as those worlds knew it simply no longer exist. There is no prospect whatever of rein-stating them: These changes are as irreversible as any. The relatively recent and highly publicized changes in climate, which are coupled with dramatic alterations in the carbon cycle, constitute perhaps the most talked-about framing element. But there are others; and jointly, substantial changes in them guarantee that no modern world can be made to resemble the world's former self.

Questions with even greater sweep present themselves. Considered generally and apart from the business of historical recreations of nature, mimicry is often amusing and good mimicry artful. But what elevates nature re-creations to the level of decisive importance when it comes to nature? And why should one regard arbitrary and, of necessity, woefully incomplete mimicry to be an appropriate way to fulfill responsibilities to nature? Even if we humans we have sinned against nature, what makes us think that this is an appropriate way of doing penance?

One might be tempted to think that the justificatory barrenness of historical re-creation is a malady that infects just this particular proposed means of executing corrective justice for harms done to the natural world. But that is not so. The malady is universally shared by all proposed means of "correcting" the course of nature of which I am aware; and it seems hard to cure. That is because, in order to supply a credible theory that endorses as "right" actions that unbreak a broken state of nature, it must first supply a set of coherent and generally endorsable principles according to which the broken states can be reliably distinguished from unbroken states of nature. And this appears impossible to do.

Upon reaching this impasse, partisans of the corrective justice theory of human responsibility for nature commonly turn to "ecosystem service" talk. Basically, they say that broken nature is nature that doesn't work right. But a little prodding at "doesn't work right" reveals that this actually means "doesn't do what people want it to do" or more broadly, "people don't like it". In other words, it is speech that quickly devolves into expressions of disapprobation, still absent justifying reasons.[25] These expressions

---

[25] Of course, the theory of ecosystem services is most often wedded to economic theory, which appears to provide reasons. But aside from the fact that ecosystems that work "right" to increase the GDP are quite unlikely to resemble anything that Light or others would recognize as "the natu-ral world", economics is merely a formalization of what people prefer and the expression of those preferences in real or imagined market transactions.

translate the original rubric of "broken nature" into "disliked" and "fixing it" into "making it likeable". But they have no justificatory force in themselves and no more supporting justification than the originals.

As I have remarked, it seems that the choices involved in historical re-creations are extremely arbitrary. The same arbitrariness haunts corrective re-creations whose ends are couched in terms of non-historically characterized ends – an assertion that I support in the next subsection. Add to this arbitrariness the sometimes-desperate desire on the part of people to believe that they can redeem themselves. Circumstances such as these are notoriously vulnerable to both conscious and unconscious bias and manipulation. This warrants strong suspicion that the arbitrary choices that define the parameters according to which wholesale manipulations of environments are proposed and carried out might well be made for morally inappropriate reasons – for example, to serve the short-term desires and interests of those who make them.

I explore this suspicion in Sect. 8.1.7 (Moral corruption – a symptomology). Light (2000, 67) fuels the suspicion in summarizing one of his main theses – namely, that "Restoration… teaches us the actual consequences of our actions rather than allowing us to ignore them…" This statement shows a remarkable unawareness of what is relevant for bringing something under moral consideration. Merely knowing the consequences of actions, including those that are part of restoration therapy, is not the same as knowing whether those consequences count as morally relevant – and therefore, whether they are subject to moral approbation (or disapprobation), or whether bringing them about has anything to do with duties to, or appropriate relationships with, nature.

In sum, there is a remarkable absence of principled reasons to persuade us that restorative gestures towards nature have any redemptive or reconciliatory quality. This is layered on top of a similar and similarly remarkable absence of reasons that justify the ends of restorative actions, which are said to satisfy the need for, or at least gesture towards, correction. It seems sensible to at least tentatively suppose that such actions remain without justification because they *cannot* be justified as appropriate, let alone justified as obligatory. But this conclusion prompts another question with a potentially more damaging answer: The question is: Given that there is no good reason to think that the sorts of actions so far proposed to "unbreak" broken nature are appropriate, might these sorts of actions, in fact, be *in*appropriate? Might they even *diminish* the value of the natural world and therefore justifiably be viewed as destructive of nature? Should they therefore be subject to disapprobation? Nothing in Light's representation of redemptive reconciliation with nature militates against affirmative answers to any of these questions.

To review: Under consideration is a model of responsibility for nature, which presumes that there are norms for nature's structure and functioning that define an "intact" state whose sufficient disruption "breaks" nature. It presumes that, not only can nature be broken, but also that doing so commits offenses against nature, much like committing mayhem against a person would offend and violate her. This model further supposes that, in fact, people have acted in ways that push nature outside of these norms; we have, so to speak, lopped off nature's various arms and legs willy-nilly. This constitutes breaking nature; in breaking it, we have committed transgressions against it as well as transgressions against a proper human relationship to nature. This, the model finally supposes, dictates a responsibility for making amends to nature in a way that also mends our human relationship it; and this is properly done via gestures of restoration ("unbreaking" it by restoring those various severed appendages), which are regarded as acts of restitution.

I leave for the next subsection an examination of the questionable notion that there are norms for the intactness of states of nature, including whether there are norms according to which nature can be said to be broken. For the purpose of the current discussion, the salient feature of this narrative (of restoration as restitution) is that it stands on an analogy between people and nature to suppose that improperly relating to nature, like improperly relating to people, can violate and offend it; and that restitution is a means of making amends to nature as the offended party. Unfortunately, this analogy is devoid of morally relevant analogous elements. While restitution arguably has currency in interpersonal contexts, to uncritically transfer this practice to a relationship in which "the offended party" is the natural world is a leap of creative imagination, not a well-grounded step. It is creative fantasy because the natural world isn't capable of being offended. And it is fantastical because nature isn't the sort of thing that one can make amends to.

When it comes to serious moral judgments, fantasy should not be permitted to substitute for carefully cultivated theories and practice based on credible principles. This is not just because fantasies are unlikely to yield good practical judgments about how to act. It is also because they leave uncomfortably ample room for serious mischief. If there is no principled way to determine what is broken and what is not, then there is no principled way to determine what actions constitute "unbreaking" and restoration, and what actions are endorsable as gestures of restitution or reconciliation towards them. If these are matters of serious moral concern – as one must presume when talking about responsibilities – then there could be no greater moral hazard than to leave their definition to acts of creative imagination, no matter how well intentioned.

A second model of "responsibility for nature" does not start from a perceived need for redemption and reconciliation. Nor does it attempt to portray corrective action as necessarily a matter of "unbreaking" nature. Rather, it proceeds from a conception of responsibility based on the notion of a voided warranty. One spokesperson for this view is biologist Erle Ellis (see Sect. 7.2.3, Home is where you make it – in situ, ex situ, and ecosystem-engineered). He claims that "a lot of people interacting with a lot of land" have produced so much change in almost all the world's biomes that they can now really be understood only as anthromes. From this sound argument for how best to understand and categorize the world's various ecosystems in terms of past and ongoing transforming human activity, Ellis leaps to this moral pronouncement:

> We are responsible for the way that nature behaves now... and in the future. If we want to live in an environment that is desirable for all of us, it's up to all of us to make that happen.[26]

---

[26] http://www.youtube.com/watch%3Fv=sTUOHMkGa0Q. The quote is from the very end of the interview. See also the main text relating to Note 26 in Chap. 7.

In other words, because humankind has changed the natural world to the point where the original specification no longer applies (biomes have become, as he says, anthromes), people now ought to draw up their own specification for anthromes. And the primary design requirement is to satisfy human desires.

Far from being a fringe model of responsibility for nature, the voided warranty model is hard to avoid in conservation and restoration circles. Landscape biologist and geographer Eric Sanderson and his colleagues (2002, 902), whose work on the "human footprint" antedates that of Ellis and Ramankutty (2008), join many others in maintaining that

> The global extent of the human footprint suggests that humans are stewards of nature, whether we like it or not... Conservation organizations and biological scientists have demonstrated surprising solutions that allow people and wildlife to coexist, if people are willing to apply their natural capacity to modify the environment to enhance natural values...

... whatever might be meant by "natural values". For although Sanderson et al. prominently mention biodiversity in connection with them, they offer no real account of what they mean or entail. This is rather concerning in light of their suggestion that humankind "modify the environment" to realize them. But this gap in their account closely aligns with the salient attitude that people, and most particularly practicing conservationists, get to more or less arbitrarily define what they want "natural values" to be, which allows them to arbitrarily define what modifications of the environment might realize them. I shall say more about these untethered attitudes about ends in the next subsection.

I call Ellis' and aligned views the "voided warranty" model of responsibility for nature because the implicit logical leap from humankind's transformation of nature to sole responsibility for its future maintenance and enhancement is the logic of a warranty voided by contractually excluded tinkering. It is easy to see that this connective, "voided warranty" logic, just as surely as the just-examined "you break it you own it logic" (for the broken nature model), arrives at an owner-possession model of humankind's relationship to the natural world. However, while the two models end up at this same destination, there is at least one important difference. In the case of the broken nature model, I have suggested that there is an unfulfilled and possibly unfulfillable requirement to find endorsable general principles that distinguish between broken and unbroken states of nature. Ellis' voided warranty model of human responsibility for nature does not appear to require such principles and therefore seems to sidestep this issue. Nature itself need not be considered broken, only the warranty. But his model still needs something closely related – namely, some account of what the warranty is supposed to have covered and what, therefore, people are now obligated to maintain on their own and as they see fit.

Ellis describes this obligation in terms of responsibility for realizing a certain end – namely, "an environment that is desirable for all of us". He takes this to mean something roughly equivalent to "a state of the world that serves people well" or, one might infer, "a state of greater economic welfare". He seems to be suggesting something like this: Previously, the appropriate role of humans with respect to nature might have largely been one of passive beneficence, for while people had to labor hard in order to extract from nature its stuff and services in order to satisfy basic needs, these goods seemed to be more or less always there for the taking.[27] But human tinkering – whether or not one regards it to be deleterious and whether it was intentional or unwitting – has changed that. It thrusts the natural world squarely into the world of human social planning, design, engineering, and management.

This elaboration of the voided warranty model of human responsibility for nature makes it evident that it does not end up very far from the broken nature model. One need only interpret "breaking nature" in terms of human activities that are salient causal factors in transforming nature into a state that no longer serves people so well. This gloss removes all but three distinguishing characteristics of the "you break it you own it" link from causal to moral responsibility: This "broken nature" logic differs from the "voided warranty" logic insofar as it sometimes emphasizes ceremonial restitution, reconciliation, and redemption. It differs (as exemplified by Light's rendering) in its tendency to regard historical restoration as the most appropriate form of corrective action. And it differs insofar as it tends to demote the importance of the efficacy of restoration efforts. In contrast, the "voided warranty" model takes the voiding of the warranty as license for creative ecosystem design with few if any constraints, historical or otherwise.

But while the last two of these three remaining distinguishing tendencies typify the broken nature model, that model by itself does not strictly entail them. It therefore seems capable of shifting yet closer to the voided warranty model – almost coalescing with it – in a variant that regards as appropriate, acts of restitution, reconciliation, and redemption, which gesture towards the corrective action of "serving people well" by arranging and rearranging the natural world in whatever configurations are thought to best serve that end.

Perhaps the most noteworthy concordance between the two models lies in their convergence on a norm for how humans ought to relate to nature that is based on a proprietary interest. This is a norm that ensconces humans in a role tantamount to property owners with virtually unchecked discretion to do as they please with their property. Both theories are therefore subject to like concerns concerning the potential for abuse of the principles, if any can be found, that the human owners suppose to guide exercise of their responsibility. As I have observed, whatever dubious principles there are for restorative actions on based on the broken nature model's "you break it, you own it" logic seem to be alarmingly vulnerable to bias and manipulation. The "voided warranty" view that human desires are the ultimate arbiter of how to rearrange nature fares no better in this regard. There is little reason to believe that (in Ellis' words) "an environment that is desirable for all of us" jibes with what is valuable in or about nature. And certainly, proponents of these models of responsibility supply no such reason.

---

[27] Of course, this simplistic description skates right past the havoc that nature has regularly wreaked and continues to wreak on humankind via "natural disasters" such as flood, famine, pestilence, and disease.

This points to one more closely related concern: Even if the desires that define "an environment that is desirable for all of us" were deemed respectable according to some independent standard, this does not seem to point to anything that is special about nature. In this role, as an object of a desire – and even more specifically, as an object that is desired for its utility – it resembles legion others, for which the real value lies in the utility utilized for desire's satisfaction rather than in any other salient characteristic of the object. My bicycle and the computer on which I type also satisfy desires. And they do so in much the same way as Ellis and others suggest – as useful resources. But no one would say that they have natural value. It seems to me that a theory of natural value that in this way fails to account for why nature is valuable *as nature* fails a basic test of credibility. I shall return to this theme in the next subsection and again in Sect. 8.2.

In the remainder of the current section, I explore how the view of the relationship between people and the natural world, depicted in this subsection, combines with the view of nature as a diverse collection of bio-parts to arrive at an idiosyncratic interpretation of what "broken" means in terms of the natural world's bio-parts. This interpretation, in turn, leads directly to the exhortations – now characteristic of conservation and restoration science and echoed by many environmentalists and their organizations – to intervene in the natural world in order to remediate "incorrect" assemblages and their inventories of bio-parts. In examining such an environmental norm, it is good to keep in mind that it cannot be better judged than the notion of responsibility for nature on which it relies for its moral force. Nor can it be better judged than its deconstructed conception of nature as a warehouse of more or less valuable bio-parts that humans – as owners – are responsible for managing to achieve the "correct" arrangements and inventories.

Finally, I don't want to leave the owner-manager view of responsibility for nature without briefly recalling the "lesser evil" terms that lend it some initial plausibility: Those who subscribe to this view say that even though the human manipulation of stocks of bio-parts to achieve a "correct" state of nature might be distasteful or even repugnant, that is nevertheless what people ought to do in order to fulfill this responsibility. This is the lesser of two evils – for the alternative, according to this framing, is to entirely ignore the responsibility. And surely, it is better to meet a responsibility than to shirk it. I will show why this dichotomous choice does not represent the universe of alternatives. The dichotomy stands or falls with the concepts in terms of which it is couched; and in the end, they fall along with the normative force of the underlying conception of responsibility.

8.1.3 The Ends of the Biodiversity Project

It is by turns startling and dismaying that confusion and disagreement about ends pervades the biodiversity project, when ends are not just a matter of casual disregard. Most obvious are confusion and disagreement about what those ends actually are – that is, exactly which biodiverse state of the world should be aimed at. But it is a good bet that this first confusion is a consequence of underlying confusion and disagreement about what counts as a good reason for aiming at any one particular biodiverse state of the world rather than another, including just letting the current one alone. These dual and mutually supporting sorts of perplexity leave a general impression of singular impotence to muster good reasons for any bit of conservation or restoration work that is said to be done on behalf of biodiversity – or indeed, on behalf of any particular state of the natural world, whether characterized in terms of its biodiversity or not.

This book has encountered confusions and disagreements of this ilk at every turn – starting from how biodiversity is defined (Chaps. 3 and 4) and on to the various theories for why biodiversity is supposed to be a good (Chap. 6). For example, some (such as Daniel Faith (2007)) say that the most valuable state of biodiversity is the solution to what might be called the "traveling evolutionary biologist problem", which maximizes the combined phylogenetic traveling distances between a minimum number of leaves on the tree of life.[28] Others (such as Maclaurin and Sterelny (2008)) say that a state of biodiversity is more valuable the more it includes organisms that have great potential for further evolution. These two views directly conflict because there are some organisms (most famously, the venerable tuatara) on the twig-tips of evolutionary branches that have no apparent evolutionary path forward, yet their far-flung location in the tree of life requires that the traveling evolutionary biologist cover some great phylogenetic distance to reach them. And both these views conflict with yet others. Consider the view that the key to biodiversity value is some emergent property, such as great productivity (which might aptly be called "the kudzu theory of natural value"). Or the view that the best biodiverse state of a place is one inhabited by congeneric surrogates that resemble historical assemblages of now-extinct creatures. But of course, historical assemblages might well constitute notably unproductive ecosystems, while productive ecosystems might be achieved by means of completely novel assemblages. On the other hand, evolutionary potential cuts neither one way nor the other with regard to these two other considerations, and the venerable tuatara is likely to be left out of consideration by both.

The weekly science news often features sober pronouncements from some conservation biologist that we should manage for species X or manage for function F. Which might seem convincing until another biologist disagrees: No, she says, we should manage for species Y or for function G. In disputes such as these, it is hard to identify any legitimate grounds for adjudication. The preferences of one biologist over another – say one abhors exotic organisms while the other does not – do not, I believe, qualify as legitimate grounds.

The previous subsection suggested how a state-based view of natural value leads to shaky views of how and why people might be responsible for arranging and

---

[28] This is a reference to the antipodal (and NP-hard) traveling salesman problem, which requires finding the *least* distance route that enables the "salesman" to visit every vertex, whose pairwise distance with every other vertex in a graph is known.

rearranging the state of the natural world. In this subsection, I examine a parallel phenomenon: how the state-based view, coupled with its deconstructive interpretation in terms of biodiversity, confounds the biodiversity project's attempts to arrive at a credible set of ends, let alone ends supported by convincing reasons grounded in coherent principles. To put this in the abstract terms of Sect. 8.1.1, the biodiversity project sets out to create states of nature whose sliced and diced biodiverse parts are mapped by the "right" biodiversity evaluation function into a high-lying natural value topography for the world. Which leaves open the big questions of which is the "right" evaluation function, and what reasons suffice to finger one function rather than another as closer to the axiological truth? In other words: how, exactly, should the natural value topography maps be drawn; and what considerations can possibly justify drawing them one way rather than another?

Before proceeding with this discussion, I wish to interject a bit of metaethics to help frame it.[29] I have in mind the famous "open question" argument, which the early twentieth century ethicist G.E. Moore (1903, Ch. 1, §13, 15–17) pressed into service on behalf of his claim that *no* set of natural (or for that matter, supernatural) properties of the world can, in principle, circumscribe all and only the world's "goods".

The gist of Moore's argument can be understood by way of comparison to questions about matters of fact: Suppose that I claim that the large, long-nosed, striped animal before us is a zebra. You might legitimately question my claim. Perhaps you think that it could be a horse that has undergone a stripe-generating mutation, for example. Standing my ground, I say that no, it is really a zebra. To support this, I show that in every defining respect that distinguishes a zebra from a horse, the animal before us possesses the property that is zebra-defining and not horse-like. The creature's bone structure is unique to zebras; unlike a horse, it has a solid tail; I sequence its DNA and its genome proves to be that of a zebra and not a horse. Suppose that you claim to understand and agree with all these facts about the matter, saying, "All that is true, but is that beast really a zebra?" At this point, I am entitled to conclude that, contrary to your assertion, you simply do *not* understand what a zebra is, for there really are no remaining legitimate grounds on which a question about the matter can be raised.

Contrast the claim that a good biodiverse state of the world is one that maximizes the minimum traveling evolutionary biologist problem over the domain of the tree of life. You might understand and agree that if the tree of life that represents all extant organisms has a certain branching structure, then its minimum visitation distance is indeed quite large. You say, "Yes, that is true, but is this really a good biodiverse state of the world?" But in contrast to our little chat about zebras, it appears that I cannot question your understanding about the matter. You might completely understand "biodiversity" and "tree of life". You might be computationally and biologically expert in the traveling evolutionary biologist problem. And you, as

---

[29] I return to metaethics, though in a narrower context, in Sect. 8.1.5 (A metaethical gloss on biogeoengineering).

well as anyone, understand "good". But you might still quite legitimately regard as not settled the answer to the question of whether the biodiverse state under consideration is good. Moreover, it does not appear that this irresolution exists for lack of key information. It does not appear that I can settle the question by adding more details about bio-parts inventories, how they relate to some other historical timeframe, or whether or not they are jointly capable of manifesting some natural property such as high productivity.

Moore's open-question argument is itself open to question. A great deal of twentieth-century ink was spilled in swings and roundabouts in debating it. Whatever the merits of Moore's idiosyncratic formulation (of which I have just given a very loose recounting), I think that his argument fingers just the kind of difficulty that the biodiversity project wrestles with. There is a desperate quality to the wrestling match – a seemingly endless heaping on of ecological facts coupled with a seeming unawareness of a need to say why anyone should think any of these facts matter at all for what is good, let alone why one set of facts matter more than another. There is, in short, little in the debate that qualifies as good, let alone, compelling practical reasoning; this conveys the impression of an attitude that the difficulties of turning the corner from "what is" to "what is good" excuse the need to even make the effort.

This attitude might well help to explain the tendency to resort to arguments that pose choices in terms of two evils. I have already remarked on this phenomenon with regard to conceptions of responsibility for nature. And I leave most of my discussion of this sort of framework to Sect. 8.1.6 (Lesser evil arguments). What is most relevant to the current discussion is that this framing substantially lowers the bar for justification. The standard for adjudication then becomes the minimal one of tipping the balance – even just barely – towards one of the two poles of the supposedly dichotomous choice. So it might be thought that even if we are quite queasy about endorsing dubious principles marshaled in support of an end that involves keeping certain bio-parts (in preference to some others) in our bio-inventory, and even if we are also queasy about the sorts of actions that seem to be the only practical means to realize this goal, then that end and the requisite means for achieving it might nonetheless be justified. This would be true if, according to this framing of the question, the balance tips – howsoever slightly – towards that uncomfortable choice. But I would say that the queasiness – about the principles invoked in this kind of situation and the consequences of the actions – signals a need to investigate the root cause of the queasiness, which might be more evil than either of the choices that induce it.[30]

Sometimes, there is no pretense whatever of a principled defense of the ends of a conservation or restoration action. Andrew Light's suggestion for a kind of ritualistic, feel-good, community-building reconnection with nature through participation in

---

[30] This description stays at the most abstract level. The desire for something more concrete can be satisfied by skipping ahead to the examples, including the tale of San Clemente Island, in Sect. 8.2.7.2 (Implications for action, social policy, and conservation management).

restoration projects (discussed in the preceding subsection) goes hand-in-glove with a self-conscious eschewing of any attempt to propose, let alone justify, principles according to which one might judge whether or not it effects the right sort of connection. In lieu of hassling with principles, the suggestion is to adopt an "experimental approach". This suggestion is given a "philosophical" voice by a group of self-branded "practical pluralists". Among these thinkers is Andrew Light who combines practical pluralism with his brand of environmental pragmatism, discussed in the previous subsection. Light is up-front in endorsing an approach that rejects the need for any particular set of guiding principles, which he emphasizes, is key to forming a coalition of any and all – or at least, most – views. If they pass the most modest of qualifying criteria, any set of principles and any kind of reasoning from them are welcomed to the fold – so long as they happen to entail some agreed upon ends for nature and the environment. Light never characterizes these ends, but presumably they are ones that *he* endorses as worthy of pursuit. As he (Light 2003a, 237) puts it:

> As long as we are not contradictory about our ends, and as long as we are not advocating morally suspicious schemes of value, environmental ethicists should be making as many arguments as possible to appeal to as wide an audience as possible…

Part of Light's "modest qualifying criteria" (Light 2003a, 235, 243) is that one should not admit "obviously morally suspicious schemes of value". Elsewhere he says that one should shun dealings with fascists. Later writing (Light 2007b, 32–33) continues to firmly hold onto "not fascist" as the principal standard for admissible practical reasoning:

> One could also raise a similar objection that methodological pragmatism would appear to endorse the development of any argument for a given converged upon policy even, say, an eco-fascist argument just to make sure that the fascists are green too. But just as we have independent moral reasons to condemn lying and deceiving… we have independent moral reasons to reject fascism, racism, and other contemptible views. Those reasons are sufficient to reject working with such communities or making those arguments.

But what is "morally suspicious"? Light's answer – "not fascist", "not racist", and "not contemptible" – is hardly an acceptable, let alone a helpful characterization. He is otherwise mute on guidance for screening morally inappropriate from morally respectable ways of thinking.[31]

Light's only other constraint regarding admissible ends is his requirement that they be "not contradictory". But this qualification, too, is unhelpful when it comes to such restorations such as the Chicago Wilderness project, which he repeatedly holds up as a model of its kind. For one thing, as I've already mentioned, biologists

---

[31] I purposefully use the term "morally inappropriate", whose scope is far wider than the scope of "morally contemptible". For example, "morally inappropriate" includes reasoning that might appear benign, but that cannot sustain a conclusion of great moral or political consequence. Such reasoning is morally inappropriate because these reasons risk being exposed as inadequate for holding up the great weight of the conclusion. That, in turn, can severely compromise confidence in, and respect for what might be a worthy conclusion on other grounds. Economic reasoning – particularly reasoning from a principle of efficiency – to conclusions with great moral and political moment is often morally inappropriate in precisely this way.

often and emphatically disagree about what they should conserve for. I am not aware that any of these people are fascists, so their contradictory views cannot be dismissed on this account. More generally, Light's position on this arbitrarily dictates that the door be closed on any challenge to ends that he might endorse, even when that challenge is voiced by a non-fascist. I, for one, believe in ends that quite flatly contradict Light's, though I'm not a fascist and I believe that I'm as concerned as anyone (including Light) to ensure that the value of nature is widely recognized and taken under consideration.

So far as I can tell, Light is simply not interested in having a discussion about justifying ends with regard to the natural world, for that would necessarily require good reasoning from principles, and their vigorous defense as worthy of general endorsement. He believes that the ends can be agreed upon among "friends" (or at least, among non-fascists). Therefore, he focuses on the means.[32]

When it comes to the means, Light recommends a particular kind of rhetorical project wherein one persuades and recruits others by showing how their overall views and positions, even if befuddled or wrongheaded, converge on the "right" ends for the environment. It is hard to imagine what this project might look like because characterizing it immediately runs into the already-mentioned problem of identifying these ends. Though restoration often comes up in his discussion, Light never offers a general characterization of what is "right"; and his two constraints rule out few possible characterizations. One is therefore left with the unfortunate impression that "right" means "the ends that a person such as Light endorses". As he himself emphasizes, we should be out to persuade, not to evaluate – for critical evaluation thwarts cooperation. This explicit rejection of the core of practical reasoning, which involves a principled understanding of the values that attach to action and the values at which right action aims, coheres with Light's own characterization (Light 2003a, 241) of what he is concerned to do as "practical anthropology".[33]

---

[32] One might come to Light's defense on the grounds that his views constitute a public philosophy, which focuses (Light 2006, 177) on "ecological citizenship" in a "larger community... inclusive of the local natural environment". This philosophy, it might be said, can be held and judged apart from, and is therefore immune to, a criticism such as mine, which questions the worthiness of the community's restoration goals and points to a general failure to justify these goals. But if ecological citizenship entails pursuing restorations of the sort that Light describes, then I don't know of any legitimate reason why those who propose ecological citizenship as an ideal of citizenship should be held immune from the requirement of morally justifying these projects. Light himself seems to recognize this requirement because he attempts to supply those underpinnings in this thesis: Our individual characters are damaged by how we individually and collectively "harm" nature and our participation in restorations is a means of repairing our characters. Even if this thesis is thought to live within the realm of "ecological citizenship", merely classifying it in this way is not a legitimate protective shield. If an ecological citizen is a person who participates in restorations whose goal is unworthy or whose goal is pursued for unworthy reasons, then so much the worse for ecological citizenship.

[33] In (2007a, 9, 20–21), Light appeals to what he calls "sociological research" for evidence that, for example, working on a restoration left participants (quoting one study) "'feeling that they were doing the right thing.'" He takes this to be empirical evidence for a positive answer to his question, "Can ecological restorations be a source of... such normative ecological relationships?"

Light's neglect of ends supported by reasons grounded in well defended, generally endorsable principles leaves his view of right action completely rudderless when it comes to nature and the environment. This, in turn, leaves his position vulnerable to being usurped by those with whom Light and most other environmentalists might vociferously disagree. As I have now twice before remarked in this subsection, even among biologist friends, there seem to be irresolvable differences about what is right. What are representatives of the world's mining industry likely to regard as "right" for the environment? And if Light does not like their "right", on what grounds can he dispute it? So far as I know, these industrialists are not Nazis. They are just enterprising businesspersons.

Light (2003a, 244) suggests that "get[ting] the right sort of pluralism" with regard to "practical environmental ethics" is an "empirical question". So far as I can tell from Light's explication of this suggestion, he means that we can find the right kind of environmental practice by performing social experiments. This practical pluralist suggestion for an experimental approach combines with his environmental pragmatist suggestion that restoration be viewed as a redemptive ritual to imply that experimentation is the right tool for determining what kinds of "restoration" projects make people feel as though they have redeemed themselves. But a social experiment to find the best guilt-relieving ritual cannot demonstrate that such a ritual is a token of an *appropriate* end that people *ought* to strive to achieve. That, as I have already observed, requires some set of independent, principled reasons. Experiments that uncover feelings of redemption that might well rely on misguided beliefs are irrelevant as a means for uncovering a principled basis for evaluation.

I have attended in some detail to Light's self-consciously permissive attitude towards the requirement for credible principles to justify the ends of restoration, because I see it as a philosophical (and possibly sociological) counterpart of a similarly casual attitude towards ends on the part of many scientists. I hold both sets of views up for examination together in the interest of getting past any prejudice that might attach to the different disciplinary labels and thereby cast the same critical eye on both.

Scientists are not averse to making their own rhetorical move, which leverages science's honored social status in modern culture. Conservation and restoration projects are promoted as being based on "hard science"; the subtext is that science is an authority that mere laypersons cannot legitimately challenge. But as impressive as "hard science" might at first seem, it generally has little to do with justifying the *goals* of "scientifically grounded" efforts. Rather, it has to do with the means, which seek to tap an understanding of the workings of ecological systems in an effort to find a way to achieve goals that still want justification. And even in the service of means, the "hard science" routinely decomposes into fuzzy logic. To recall just a few of the many examples that this book documents: Biodiversity is claimed to be the means of keeping drinking water clean, when that end is actually accomplished by means of keeping sewage out of water sources, preventing cows from defecating in them, and stopping suburbanites from so liberally dousing their expansive lawns with toxic chemicals. Again, biodiversity is claimed to be the means of satisfying human biophilic tendencies. As I remarked in Sect. 6.6 (Biodiversity as progenitor of biophilia), even aside from how precarious is the credibility of this speculative end, what meager evidence there is for the means-end relationship shows that a potted plant in every cubicle will serve that end quite well.

Science also sometimes echoes Light's "experimental approach", especially under the rubric of "adaptive management". Volumes have been written on this topic. What one finds in them is a single-minded focus on means in the form of intricate elaborations of experimental methodologies.[34] These methodologies all boil down to a sequence of trial-and-error "experiments", which is supposed to compensate for epistemological limitations with regard to how ecosystems work. At bottom, this experimentation is a matter of figuring out "what works", with some appearance of sophistication lent to the process by analyses of results at multiple scales of space and time and with due attention to the peculiarities of the particular place on which the experimental trials are inflicted. The theory is that, after analyzing the results of a sufficient number of these experimental manipulations, some manipulative procedure will rise above the others as the one most likely to achieve the "right" result. But however efficacious this trial-and-error procedure is as a means to arrive at this evaluation of means to some end, it once again leaves untouched the question of why the end is thought to be "right".

On this critical question of ends, the theory of adaptive management, like Light's practical pluralism, is essentially mute; the theory is really about rules for using trial and error to achieve any imaginable end. This raises the same concerns that arise with respect to Light's philosophy – because this arrangement implicitly leaves it open to adaptive managers to "experimentally" determine those ends according to their own predilections, which are formed outside their "science" of adaptive management and held immune to critical scrutiny.

There is one notable exception to this neglect of ends. Oftentimes, the implicit context of these discussions makes it apparent that they are about the means of "managing a resource" in order to satisfy some *economic* end. I shall discuss the merits of the economic ends of the biodiversity project shortly. Bryan Norton (2005, 95) takes a different slant on the question of ends. According to him, "Adaptive management, like medicine, is a normative science." In saying this, he appears to be pushing towards the view that adaptive management is like managing a person's health. I shall shortly have something to say about this proposed end, too.

Unfortunately, what is supposed to be "right", by science's lights, fails even to meet Light's minimal requirement to be "not contradictory about our ends". The various and contradictory ends cited at the beginning of this subsection amply illustrate this. It is also a reasonable diagnosis of why Chap. 7's "inconvenient implications" are so discomfiting: In the cases considered there, the blind pursuit of an end at some point abruptly collides with realizing another end that seems more fundamental to nature's value. To cite one example from that discussion: The noble-seeming

---

[34] See Norton and Steinemann (2001) and Norton's later re-presentation in Norton (2005, 92–100) for an example of this fixation on methodology, which leaves aside the question of whether the methods serve any worthy end.

end of conserving for coral leads to a "Merry Maids" service involving vacuuming, sprig-setting, and recruiting grazing creatures, which are sprinkled over the reefs. In this way, "conserving for coral" becomes creating and managing an at-sea aquarium, complete with a complement of organisms that would not be there were it not for the aquarium managers and their idea of what a coral reef should look like in the age of warming and acidifying oceans. But from another perspective, transforming a coral reef – even one that is not its former self – into an at-sea aquarium is, if anything, an even worse nightmare incursion into the natural world.

Something similar is at work with the promotion of cows as nature's savior in a remarkable variety of contexts. This book has tripped over several: For Sahotra Sarkar, they are the key to creating and maintaining a wetland; for Stuart Weiss, they are saviors of the bay checkerspot butterfly; for Andrew Dobson, they are representative of the biodiversity that provides a shield against the scourge of malaria.

The domesticated cow might be the most surprising example of a creature whose introduction scientists urge for the ends of conserving and restoring nature, but it is far from the only one. Others are shipped in for sport; others for "biological control" of previously introduced but now unwanted organisms. Still other introductions are said to compensate in some way for the loss of another kind of organism. This latter kind of reasoning props up relatively modest proposals that focus on one creature; it reaches its grandest expression in proposals such as (Donlan et al. 2006) the one to rewild North America. Perhaps the fullest flowering of enthusiasm for a conservation project that transports the biodiverse hordes is The Mother of All Conservation Moves, which is supposed to be *en masse* conservation on a planet with a warming climate.

At the same time, introductions are reviled, with the revulsion expressed and encapsulated in the terms "alien" and "invasion", which contrast with the honorific title of "native". So in this quite important matter of remixing and reconstituting which bio-parts should go where, there are utter contradictions in proposed goals. Apparently, introducing organisms to new homes is detrimental to nature's value… unless this is part of a conservation plan. And no coherent principles are on hand to adjudicate whether the arrival on the scene of a new organism constitutes an alien invasion or conservation. If there is any reasoning behind proposals that are supposed to constitute conservation, it preponderantly has to do with whether the proposed means might be effective in achieving the proposed end.

Thus, proponents of The Mother of All Conservation Moves ask only: Is this conservation project likely to be an effective means helping the moved organisms by alleviating them of the need to deal with the dialed-up thermostat in their current environs? Mountains of bio-facts might be amassed in the interest of trying to answer this question. So long as attention does not stray from consideration of this particular means, this discussion can give the impression of sober "hard science". But in this process, there is a notable failure to consider other possible means, such as genetic modification of the organisms to help them thrive in warmer climes. I don't suggest this as an advocate of this alternative – even if it would, indeed be more efficient in realizing the end. Rather, I believe that it might be a first step to breaking away from the question of means to consider whether any means would court a worthy end.

In particular, the question that more urgently needs answering is: Is a world in which bio-parts are moved hither and yon as a means to maintain high bio-inventories a world that best realizes nature's value? Why, exactly, is it good – so far as the value of the natural world as nature (as opposed to the value of a well-groomed and well-stocked exhibit for the public's edification) is concerned – to have the kind of biodiversity associated with a wetland created by grazing cows rather than the bio-diversity of the grasslands that would exist in the cows' absence? What *natural* value is there in the mere persistence of an organism that used to exist as an integral part of the natural world but that now persists as vacuumed object? A cow-generated wetland and a vacuum-serviced coral reef might evidence an extraordinary ability on the part of some conservationists to manipulate ecosystems into something to their liking. But why should their preferences in this regard command general respect? Are their creations valuable as part of the natural world? The resounding answer from the world of conservation is… contradiction.

Section 8.1.1 (Biodiversity and the state of nature) counted four morally salient aspects of the state-of-nature portrayal of valuable nature. The third was the notable absence of people from the foreground of the value picture. The questions I have posed about human-deployed cows and human-wielded vacuums need only the slight adjustment of emphasizing "human" instead of "cows" and "vacuums" in order to see that, properly considered, they touch on precisely this omission from the biodiversity project: In pursuing projects that deploy cows and vacuums to create a "nature" that would otherwise not exist, are people acting in a way that is truly valuing of the natural world as nature? I return to this question and address it in Sect. 8.1.8.1 and at greater length in Sect. 8.2 (Natural value as uniquely valuable human relationships).

To ensure that I present a complete picture and to return to my previous remark about a notable exception to the neglect of ends, I must emphasize that *some* ends that the biodiversity project claims to serve are made quite explicit, comprehensible, non-arbitrary, and based on a "scientific" assessment of the effects of biodiversity. Those ends are all and only the ends of human welfare and more narrowly, economic welfare – as Erle Ellis says, the end of creating "an environment that is desirable for all of us". Chapter 6 took essentially all the major proposals along these lines under consideration, including, for example, the end of having various "services" at human disposal and the end of good human health. But on the evidence, biodiversity is not even a particularly convenient or efficient means to these ends. In fact, the unfortunate truth conveyed by Chap. 6 is that there is little or no reasonable basis for thinking that anything but what most would call the most impoverished state of biodiversity serves these ends at all, let alone in any remarkable or outstanding way.

But I wish to make a far stronger claim, one that does not depend on this fact. To make this claim, I need to introduce two concepts that weave themselves into the remainder of what I have to say about nature's value and that feature prominently in Sect. 8.2. Suppose for a moment that, contrary to fact, sound reasons supported the claim that some highly biodiverse state of the world promoted human welfare. Then my claim is: At most, these reasons could show that certain biodiverse properties are connected to welfare in a way that is at once so non-resilient and so generic that they cannot possibly account for the value of the natural world as nature.[35]

Part of what I mean by "not resilient" is related to what I called (in Sect. 2.1.1, Concepts and categories of value) "a certain precarious quality" of goods that are of merely instrumental value.[36] Even if biodiversity is now the lead player in supporting otherwise unrelated ends such as those that benefit the human economy, there is nothing about the nature of biodiversity – or even of nature more generally – that guarantees this status forever, or even tomorrow. In due time one should expect its preeminent role to be replaced – via either economic or technological substitution – by a superior up-and-coming ingredient for baking a yet richer welfare cake.[37] Moreover, insofar as goods ought to be provided to humankind in the greatest amount and with the greatest alacrity, it becomes a moral obligation to move to the greater efficiency of providing them with less biodiverse means.

In fact, the archetypal example of non-resilient valuation is an economic one, which subjects its commodified objects – including nature-as-natural-capital – to the contingencies and exigencies of value-in-exchange. The fickleness that arises in this context is not merely one about the means for satisfying human preferences. It also includes the fickleness of the ends at which the preferences are directed. Neoclassical economic theory holds human preferences and their satisfaction to be the stuff of ultimate and unquestionable value and essentially outside of its purview. Yet the aims and origins of preferences are surely open to question. In this regard, I have already cited Debra Satz's point (2010, 69, 70) that "our preferences… are sometimes formed by whim, confusion, tradition, peer pressure, and social context." Together with "whim", she also cites "misinformation or advertising" as contributing to the vagaries of preferences that can be "mistaken, fleeting, confused, maladaptive, conformist or inauthentic". To that list, one should add "vile". As she says (Satz 2010, 76), even the *market* role (putting aside the moral role) for preferences is open to question when, for example, "they are based on or express contempt for others".

The implication is that neither preferences nor the end of fulfilling them can, without a great deal of further inquiry, be presumed to be morally respectable. As Satz says, "they may not reflect what is really important for us"; and they might not be "worthy of satisfaction" for any number of other reasons. Nor can they be assumed to be stable, for they are subject to change without notice or good reason. The fickleness of preferences underlies (though it is not the only factor underlying) the fickleness of market values and an absence of any credible moral accounting for

---

[35] Section 8.2.1 (Natural value as appropriate fit) develops a generalized notion of "resiliency", which, following Philip Pettit, I also call "modal robustness" and "modal demandingness".

[36] Though, as I emphasized in that earlier subsection, an instrumental good can also be valuable in non-instrumental ways; and even if it is "merely" instrumentally valuable, this can be mighty valuable indeed if it is the only means for realizing an enormously important end.

[37] See Sect. 2.1.2.1 (Consequentialism) for an account of these notions of substitutability.

market fluctuations. These latter are generally beyond the control of most moral actors, whose inclinations to produce, trade, or consume are nevertheless sensitively keyed to them.[38]

This brings me to the second critical failing – that an accounting of biodiversity's value via human welfare can be, at best, a generic accounting. By "generic", I mean that nothing about the "biodiversity-ness", let alone biodiversity's connection to the natural world, really comes into play in biodiversity's role as an instrument of human welfare. The generic nature of this instrumental role means that it cannot truly capture anything *unique* about nature's value. I cannot emphasize enough how enormously compromising this deficiency is. For those of us who believe that the value of nature is distinctive and unique in the panoply of values, this deficiency alone should raise significant doubts that its economic accounting is a reasonable one or even has any central role. Furthermore, it should stir a further concern – that such a view might actually diminish nature's value by diminishing the credible basis for regarding it as distinctive in a way that holds it immune to ordinary market-based justifications for trading it away.

I use the word "accounting" advisedly, for the most generic imaginable kind of accounting for nature's value is the economic one, which "objectively" reduces all value to the calculus of satisfying human preferences without much concern for what they are preferences for. It is worth reprising some of what I said in Sect. 2.1.2.1 (Consequentialism) about this conception of the good, to place it in the current context. In economic terms, an inventory of the world's bio-parts is just another inventory of parts. And that collection of parts it is valuable precisely insofar as people are willing to express their preference for it by opening up their wallets to pay to maintain those parts.

In the course of examining the role of this increasingly dominant derivation of nature's value to justify biodiversity projects, this book has already brushed in sufficient strokes to paint a picture of what this evaluative program really buys: It buys bio-warehouses, whose shelving system (designed by ecosystem engineers) need not resemble anything one might imagine "in nature" or "as nature". The program justifies allocating the wherewithal (for ecosystem managers) to stock these bio-warehouses with some minimal assemblages of carefully selected bio-parts. These bio-parts are carefully managed at the lowest possible inventory levels so that they serve human preferences efficiently – that is, sufficiently well at sufficiently low cost to make people inclined to keep their wallets open to continue their subsidy. And by virtue of this subsidy, the bio-warehouses are kept open, too. Of course "sufficiently well" today might not be sufficiently well tomorrow, for economic competition demands ever more competitive efficiency. So one should fully expect that these bio-warehouses – already largely emptied of bio-parts that don't pay their

---

[38] For a jarring example, see Note 32 in Chap. 6 for how the natural capital investment in Costa Rican forests plummeted in value when a change in market conditions torpedoed the value of the coffee crop, which previously benefited from the peripheral forest's value in harboring insects that pollinated the coffee plants. For a further exposition of what this story means for natural value, see Sect. 8.2.3 ("Living from" nature, uniqueness, and modal robustness).

own way – will become progressively more barren. The picture is not pretty. It bears little resemblance to anything that, by most environmentalists' lights, could credibly be called "the natural world" or valued as such. What concerns me far more is that, as I argue in Sect. 8.2, holding up this way of relating to nature as the model of how to value it is perverse for how it undermines rather than promotes any truly valuing relationship. If this is the end of the biodiversity project, then anyone who thinks that nature's value is something quite different from this should battle against it.

It is hard to overestimate or to overstate the power of the grip on current thought of this vision of nature as most centrally valuable as the source of resources from which people live. Even virtue ethicists, for example, tend to glue into their theories recognition of the business of living from nature – in the guise of promoting such virtues as respect and benevolence *not* towards nature, but towards other *persons* by means of justly distributing nature's stuff. For example, virtue ethicist Ronald Sandler (2007, 53) finds some significant part of nature's value in its role as a resource token by means of which one person can express compassion and benevolence towards another:

> Since all people depend upon basic environmental goods for survival and health, a benevolent person will be concerned that they are available to as many people as possible.

It seems very strange to take Sandler's point to have much to do with *natural* value. That is partly because his conception gives us no reason to think that it has anything uniquely to do with how people value the natural world. On Sandler's account, one must presume that nature is no different from any (other) stuff within one's control that serves the needs and desires of people generally, and which benevolence requires be distributed to the needy and desirous according to some respectable principle of distributional justice. "That which is distributed to people out of benevolence" says something about benevolence and justice. But it does not say anything that is particularly about the natural world. Nor does it say anything that makes nature worthy of special consideration, let alone moral consideration. Section 8.2.6 (What appropriate fit is not) returns to virtue ethics and some of its deficiencies in accounting for natural value.

In sum, I believe that the biodiversity project, even aside from the failure of the facts to support its ends, is necessarily a double failure: It cannot account for the unique quality of natural value; nor can it account for natural value's resilience to decay in the face of substitutions that more adequately serve human welfare.

While I am finished with ends for the biodiversity project that are (as I characterized economic ends) explicit, comprehensible, and based on a "scientific" assessment of the effects of biodiversity, I feel obliged to mention some other ends – ones that, as I said, Bryan Norton gestures towards – that are commonly cited and therefore explicit, but that fail the two tests of comprehensibility and based on "scientific" assessment. Two candidates, familiar from previous appearances in this book, are the ends of "ecosystem health" and "ecosystem integrity". The word "ecosystem", which appears in both phrases, might lead one to think that they both refer to scientifically verifiable attributes of ecosystems. And the word "health" will bring to mind the role of doctors – medical persons of science – who are held in highest esteem as best qualified to use medical science in order to ascertain the health of a person and help her to maintain and regain it. This view suggests that in a similar way, ecological persons of science are the doctors of ecosystems who are best qualified to use ecosystem science to determine the state of an ecosystem's health, to identify symptoms of an ecosystem's disease, and to restore a sick ecosystem to health by administering the sometimes bitter-tasting medicine that is nevertheless good for ecosystem health.

But the analogy between medicine for people and medicine for ecosystems disintegrates under scrutiny. In Sect. 7.2.3 (Home is where you make it – in situ, ex situ, and ecosystem-engineered), I addressed the dubious claim that there is a duty to cure ecosystems of their ills on the basis of an analogy to caring for infirm persons; I shall say a few more words about it in Sect. 8.1.5. That part of the analogy crumbles before the observation that ecosystems, unlike persons, have no *interest* whatever in being healthy, no matter what "being healthy" might mean for ecosystems. As I also observed in Sect. 7.2.3, this disintegration of the moral case for healthy ecosystems makes it hard to avoid the suspicion that "ecosystem health" maintains an appearance – and only an appearance – of maintaining its moral grip by first bleeding into "ecosystems that keep people healthy" and from there into "ecosystems that keep the economy healthy". The latter squarely reconnects with nature's valorization as "natural capital" whose value consists in its ability to satisfy human desires… at least, for now. "The New Economy of Nature"[39] then becomes a division of the larger human economy; and the nature division is said to be in a healthy state when it's doing well economically by virtue of its continued significant contribution to economic welfare overall. In other words, this set of interpretive glosses lands right back into the ugly picture of nature as bio-warehouses full of bio-inventories of bio-parts and the salient failings of this picture to account for nature's value.

One might try to avoid these normative dead ends in either of two ways. The first attempts to take nature's economy out of the general human economy by insisting that it is not really an integral part of that domain in which any and every person's preferences, as expressed in market transactions, are taken into account. Rather, according to this line of retreat, only the experts get to cast their preference votes. But this route seems singularly unpromising. For who might these experts be, but those conservation and restoration biologists who contradict each other for lack of a coherent and consistent set of principles that might define "healthy" states of nature?

The second backpedaling move insists that the notions of "ecosystem health" and "ecosystem integrity" are each analyzable into a collection of natural properties that ecologists can measure – much as a doctor might press a stethoscope to your chest or take your blood pressure in order to assess the state of your health. I dispose of this approach in Sect. 8.1.5, where I place it in the context of a certain metaethical stance.

---

[39] This is a reference to the book (Daily and Ellison 2002), *The New Economy of Nature: The Quest to Make Conservation Profitable*. Its lead author, Gretchen Daily, is cofounder and codirector of the Natural Capital Project, whose philosophy of Natural Capitalism supplies the *principium agendi dirigens* for The Nature Conservancy and most other major conservation efforts today.

There I show that, insofar as health really can be reduced to a set of measurable properties of ecosystems, there appears to be absolutely no reason to regard these properties as ones that people ought to strive to foster.

Once again, the defects of all these approaches are obscured when the issues of ecosystem health and integrity are loaded into a lesser-evil framework. Would you rather have *un*healthy ecosystems than healthy ones? Do you, as a person of integrity, favor ecosystems with integrity, or do you want them to disintegrate? But questions framed in this way divert attention from the moral impotence of "health" and "integrity" with regard to ecosystems. And it is only by remaining unaware of this impotence that one might be inclined to condone the evil of turning an ecosystem into a garden of arbitrary restoration and conservation decisions – so long as "experts" maintain that they conduce to restoring or maintaining ecosystem health or integrity.

States of "ecosystem health" and states of "ecosystem integrity" are not the only preferred states that are explicitly cited as ends of the biodiversity project. States of nature can be, not just healthy, but pristine, Edenic, or beautiful. But these states are on as tenuous a conceptual footing as healthy states, so far as the coherence of their characterization and assessment are concerned. On the other hand, states resembling those of some preferred past point in history, at least on the surface, might at least seem to be scientifically determinable – whether that preferred historical state be a few decades ago, a few centuries ago, before the Columbian exchange, many millennia ago, before the influence of *H. sapiens*, or (why not?) the state that obtained at the end of the glorious Cambrian explosion. But these states fail the test of non-arbitrariness, for no generally endorsable principle appears capable of promoting one such state over another as a norm for the current state of the planet. So while some of these alternative proposed ends of the biodiversity project are less likely than ecosystem health to be undone by failing to connect with biodiversity or by devolving into economics, they founder for other, equally confounding reasons.

In sum, confusion about the ends of the biodiversity project – ends that are supposed to be grounded in "hard science" – is no less than the confusion about the ends characterized by the descriptive apparatus of retributive justice. Applying to the former Light's motto regarding the latter: So long as we are not contradictory about our ends, and so long as we are not advocating morally suspicious schemes of value, environmentalists should be making as many science- and economy-based arguments as possible to appeal to as wide an audience as possible. But valuing nature as (natural) capital asset seems to have much less to do with valuing anything unique or special associated with the natural world than with valuing... capital assets. Worse, there is no reason to think that the bio-warehouses of bio-parts, whose construction and maintenance might be justified on Natural Capitalist grounds, will be an assemblage of biotic parts in abiotic settings, which one would expect to regard as "in nature" or part of the natural world.

The raft of other suggestions regarding the ends of the biodiversity project is a raft of contradictions. The proposals fall into two classes. In the first class are those, such as a state of health or an Edenic state, that are incoherent and not at all scientifically characterizable. In the second class are proposals, such as aiming at resemblance to some historic state, that might be thought to admit scientific description. But no amount of scientific description compensates for the lack of credible, generally endorsable principles establishing that these states constitute norms worthy of striving to achieve. The stark absence of principles makes for a stark absence of any basis to debate the relative or independent merits of any of these proposals. Therefore, to achieve the only remaining and very minimal standard – of self-consistency in the set of ends – would require culling on an entirely arbitrary basis.

Yet, this state of anarchy regarding the ends of the biodiversity project mostly flies under the radar. I suspect that this is due to the rhetorical move that I noted at the outset of my discussion of the "hard science" of the biodiversity project: For the most part, discussion of the science simply avoids any debate regarding ultimate ends, and proceeds from the undefended assumption that biodiversity, under some description, is an end worth pursuing. Yet despite flimsy or nonexistent reasons for pursuing its ends, the biodiversity project pursues them via conservation and restoration means that manipulate and manage the natural world; and these operations have come to be viewed as authorized by principles of "hard science" – as if this gloss sufficed to lend it the normative authority that, in fact, it entirely lacks.

Some part of the biodiversity project's operations is commonly called "management regimes". But I need a term that extends to projects that are far more audacious than the word "management" suggests. I adopt the term "biogeoengineering" to capture the scope and grandness of current-day proposals and projects to make nature what (it is thought) it ought to be. It is to biogeoengineering, considered as the biodiversity project's principal tool of action, that I now turn.

8.1.4   *Biogeoengineering Value into Nature*

The conception of right action in terms of biogeoengineering, and the permission or even obligation to engage in it, is the fourth morally salient characteristic, from Sect. 8.1.1's list, of the biodiversity project's state-of-nature view of natural value's genesis. This element is endemic to the specifically scientific discourse about ecosystems generally and biodiversity in particular. It is the *modus operandus* of what its practitioners call "conservation", but which I prefer to call "nouveau conservation" or "neo-conservation" to suggest that this is a new form of "conservation" and that its connection with conserving anything in nature is tenuous, at best. Of all the elements that characterize the biodiversity project, the effect of this element on moral regard for nature might be the most debilitating. It is also the most intellectually and morally corrupting – a concern that I address at length in Sect. 8.1.7 (Moral corruption – a symptomology).

Biogeoengineering, as I have said, is the characteristic form of action that the biodiversity project supposes to be entailed by moral consideration of the natural world. It is the sort of action that frames the effort to "save the environment" or to "save nature". For advocates of the biodiversity project, biogeoengineering is more specifically regarded as the appropriate means of saving or "enhancing" biodiversity in order to realize a more valuable natural world.

The term "biogeoengineering" is apt for two reasons. First, it accurately conveys a prevalent and possibly prevailing view of what "saving the environment" and "saving nature" mean. On a popular elaboration of this meaning, economics provides the evaluative calculus that determines which bio-parts in which bio-inventories (which ecologists call "assemblages") pay their own way. Science then offers engineering techniques to assemble these inventories and inventory management tricks to keep the right bio-parts in stock. This is not the terminology that, for example, the three biggest conservation organizations use to describe the world's largest, best endowed, and most heavily promoted conservation efforts. But this is precisely how they structure their practice. Scientists and others who endorse this view and put it into practice persuade themselves and the public that their principal roles are those of design and management engineers. In every important relevant respect, these projects are engineering ones.

Unfortunately, these projects are extremely *bad* engineering projects according a criterion that should be applied before any of the engineering actually begins. The project must have clear and clearly justified ends; that is, it must be seen to meet requirements that are worthy of meeting. But as the previous subsection details, bio-geoengineering projects routinely lack clearly defined and well-justified requirements. Insofar as the ends of biogeoengineering are made clear, they lack justification. But mostly, the ends are simply enveloped in a cloud of confusion.

Of course, a complete assessment of an engineering project as an exercise in engineering must also look at its approach to finding the best means to solving a puzzle, even when that puzzle is entirely the wrong one to solve. But in the department of choosing and justifying an implementation, too, biogeoengineering tends to get abysmally low marks. This is especially true when the justification of ends puts aside the economic calculus, because most anything goes when the end of maximizing economic welfare is out of mind. This criticism is certainly justifiable when leveled at adaptive managers who give themselves permission to tinker on a trial-and-error basis with a natural system just to see how well they like what happens. The logic behind planned introductions of organisms to restock depleted inventories of "similar" organisms as is often even more obviously fuzzy. When it comes to debate over what is "similar" or similar enough, it is hard to avoid the impression of a complete absence of respectable principles according to which such judgments are nevertheless confidently asserted.

I discuss a couple of illustrative cases in Sect. 8.2.7.1. There, I examine the suggestion of the North American rewilders that modern lions are "close enough" to Holarctic lions to be justified as a means of realizing the end of a rewilded North America. One might be prone to dismiss this suggestion out of hand as madcap, until one realizes that the logic used in this case differs in no significant respect from the logic used, for example, to justify introducing a tortoise to the Mascarene Islands (in the Indian Ocean east of Madagascar) that is "close enough" to another, lately departed species of tortoise. And then one will realize that similar and similarly fuzzy logic for substituting one species for another is taken as *de rigueur* in any number of other, similar restoration projects. The authors (Hutton et al. 2007, 26) of a proposal to re-create a Lord Howe Island (in the Tasman Sea between Australia and

New Zealand) of some time past confidently specify a refined set of rules for the use of surrogates:

> We suggest a hierarchy of decisions on surrogate taxa based on the taxonomic relatedness of the candidates for reintroduction: the same species before subspecies before genera, with functional replacement being a further filter on candidates that are not the same species. In our opinion, taxa with functional equivalence but without taxonomic relatedness would not be acceptable candidates for reintroduction.

But the principles that justify this recommendation of how to achieve an historical reenactment remain as mysterious as those that justify the reenactment itself.

These conundrums about means, as I said, seem to crop up whenever economic ends are put aside. This observation helps to point up the extraordinary seductiveness of economic analysis: It appears capable of pulling off the dramatic coup of crystallizing ends in terms of economic wealth and health, while also making the means to achieve economic wealth seem to be a matter of small *normative* consequence, except insofar as they do effectively and efficiently realize it. The resulting normative picture is beguilingly simple: The value of nature is in the economic wealth that it bestows on humankind. The role of ecosystem engineers and managers is to design and manage the bio-creations that best realize this wealth or, as Ellis suggests, that satisfy our desires. Having charged these environmental guardians with direct responsibility for assembling inventories of felicitously interworking living and non-living parts that maximize the economic wealth from these sources, the rest of us can turn our attention elsewhere.

Which brings us to a second reason for the aptness of the term "biogeoengineering". It obviously builds on the term "geoengineering". In doing so, it encapsulates its inheritance of meaning and salient characteristics from that latter term, now commonly used to designate large-scale, indeed global, projects designed to intentionally manipulate the earth's climate. The inheritance works at two levels. At one level, it works in the obvious way that "biogeoengineering" inherits from "geoengineering" the supposition that the solution to improve or to repair some portion of the environment is primarily an engineering one. The conjunction of "bio" prepended to "geoengineering" perfectly reflects how biogeoengineering expands geoengineering's domain. The latter is "restricted" to the planet's atmosphere and oceans, while biogeoengineering claims for its domain the entirety of natural creation, including all living things in their abiotic geo-homes.

The semantic inheritance also works more subtly at a second level – the level of a narrow and problematic framing of the problem domain, which prejudicially excludes a variety of morally relevant considerations. Geoengineering proponents regard themselves as placing a pragmatic emphasis on evaluating various schemes to modify "components" of the planet so as to allow humans to continue unabated their profligate release of fossil carbon into the atmosphere without excessively ill effect, or so they claim. This framing of the problem gives welcome relief from any need to consider difficult questions about norms of human behavior, character, relationships with the natural world, and acts of valuing. It's not that the technology of geoengineering isn't fabulously difficult. But the difficulties are primarily technological. The question of how best to geoengineer the planet's fluid (atmospheric and oceanic) envelope so as to compensate for questionable human behavior is morally trivial. It remains trivial even when political difficulties involving international cooperation are also acknowledged; that merely suggests the additional, morally peripheral question of how best to get political agreement on deploying this technology. Geoengineering proposes to alleviate a felt need for society at large to confront and honestly assess the various behaviors and ways of thinking that are the problem's root causes. In other words, geoengineering is a technology of moral avoidance.

In precisely this spirit of avoiding moral issues, *bio*geoengineering projects propose to save biodiversity, nature, and the environment, to fix what humans have broken, and (again invoking Erle Ellis) to "create an environment that is desirable for all of us" while the rest of us get on with our life as usual. The avoidance occurs by way of several characteristic ways of thinking, increasingly entrenched in the professional identities of individuals and institutions – once again, in much the same way as this occurs with regard to geoengineering.[40] These characteristics are symptomatic of what is sometimes called "morally corrupting" influence on judgment of interests that have no legitimate bearing on the judgment. They most obviously present themselves as a seeming blindness to the real consequences, which appear to literally shred natural value rather than nurture and honor it. Worse, they badly confuse and distort the moral issues involved. That is because they do not just prejudice a single isolated issue or instance. Rather, the corruption infiltrates the very framing of basic questions about how people should relate to the natural world. The result is that the most important questions are not posed and the most important moral implications of not answering those questions are lost.

8.1.5   A Metaethical Gloss on Biogeoengineering

In Sect. 8.1.8.1 (Biogeoengineering as right action), I discuss the moral suitability of biogeoengineering insofar as it reflects an attitude towards nature expressed in certain characteristic activities and behaviors. In this subsection, I pick apart its moral suitability from above, so to speak. That is, I dissect it from a certain metaethical stance that strongly attaches to its normative stance.

I have already (in Sect. 8.1.3, The ends of the biodiversity project) made a brief excursion into metaethics, by way of G.E. Moore's "open question argument".[41] That excursion served to suggest one reason why the biodiversity project is so continually and so badly confused about its goals. After reprising those previous metaethical reflections, this subsection dives deeper into the topic in the interest of gaining an even more broad perspective on the biodiversity project's difficulties.

---

[40] This institutional dimension of moral corruption is just one of several, which are elaborated in Sect. 8.1.7 (Moral corruption – a symptomology).

[41] Section 2.1.1 (Concepts and categories of value) also veered into metaethical territory with its discussion of objective versus subjective value.

This book has been largely concerned with theories of natural value that attempt to anchor that value in some structured property characterized in terms of biodiversity. Details vary and are continually disputed, but the thrust of all of these biodiversity-based theories is that nature is valuable to the degree with which it exemplifies certain biodiverse states – that is, certain states of the world characterized in terms of their biodiversity. Or, to put it slightly differently, nature realizes natural goodness insofar as it exemplifies some set of scientifically ascertainable biodiversity properties.

Moral philosophers will recognize that this position incorporates a particular view, not just about the value of nature, but also about the nature of value itself – that is, a theory of metaethics. Broadly speaking, the metaethical position that seems to underlie biodiversity-based theories of natural value is one that asserts that there are facts about value (the value of nature or of anything else) and that these facts are essentially facts about how real, natural properties are instantiated, exemplified, and structured in the world. Here I use the term "natural" in the sense that contrasts with the sense of "supernatural", not in the sense that contrasts with "artificial" or "artifactual".

On this view, values in the world are quite literally in front of our noses. The task of finding them is a matter of being observant in the usual ways that people observe and find out about the natural (again, in the sense opposed to supernatural) world. We use our normal sensory capabilities, augmented by tools and scientific theory, in order to determine whether relevant properties are exemplified – and exemplified in the right structured relationships – so as to constitute a valuable state of things… or a not-so-valuable state.

This metaethical stance is known as "naturalism". It is one of two main branches of metaethical realism; philosophers cleverly anoint the other branch "nonnaturalism". Moore's "open question argument" is one possible crack in naturalism's armor. But as I shall point out after revisiting it here, there are more. In fact, this theory – that value is bound up in natural properties – has some quite formidable problems. In the minds of some (including Moore), their intractability suffices to justify abandoning naturalism and fleeing to metaethical nonnaturalism as the preferred theory.

Metaethical *non*naturalists refuse to give up on the idea that properties of *some sort* confer value. They also refuse to embrace supernatural properties. They therefore presume that there must be a realm of some other kind of property – *non*natural properties (what else?) – that does the value-conferring job. This move – of positing a universe of such properties as "goodness" that people can somehow come to know, except not by using the customary sensory faculties that enable us to detect the natural properties of things in the world – is the sort of jaw-dropping thesis that gives philosophy a bad name. Fortunately, it is not the only alternative to naturalism. But the willingness of Moore and others to embrace this extreme position evidences the seriousness of naturalism's problems. Yet, as I will describe, environmentalists are on the whole wedded to naturalism.

In the realm of the natural world and its value, those who hold biodiversity-based theories of natural value often seem to say, or at least seem to start out saying, that the value of nature is a matter of its instantiation of some more or less complex property of "being biodiverse", properly understood. The "proper" understanding of "biodiversity", in the main, makes natural value a matter of some combination of diversity of kinds in various categories, as set out in Chap. 3 (What biodiversity is). Candidates for the correct value-conferring recipe include (but are not restricted to): a sufficiently biodiverse mix of species, a sufficient representation of lineages, a measurably great total phylogenetic distance separating the totality of extant organisms, and an instantiation of a sufficiently diverse set of "natural functions".

This story about the source of natural value in the world has an obvious implication for guiding behavior with regard to the natural world. Insofar as people should aim to realize a more valuable natural world, our actions should be aimed at ensuring that the world exemplifies more and more strongly the natural value-conferring properties, which, according to biodiversity theories of natural value, are structured according to the various categories of biodiversity. Achieving this goal involves first, understanding the exact structure of the value-conferring biodiversity properties; and second, arriving at some viable means of constructing and maintaining a state of nature that exemplifies these properties in the required, structured way. This enterprise is recognizable as neither more nor less than the work of biogeoengineering.

Insofar as biodiversity theories of natural value build on metaethical naturalism, their systematic failure to provide a convincing account of natural value can be traced (at least in part) to difficulties that attach to naturalism generally. The great strength and attraction of naturalism (and realism generally) is that it views values as being incontestably "in" nature rather than being dependent on the whims and fancies of human psychology. Values inhere in natural entities just as solidly and just as tenaciously as the relevant, value-conferring natural properties.

But there are some very great downsides to this understanding of how value arises – in nature or in anything else. The first is the extreme difficulty of envisioning any convincing reason to think that value is conferred on some object merely by virtue of its exemplifying a set of natural properties, no matter how carefully chosen. I have already related G.E. Moore's classical challenge: that there is always a legitimate open question about whether any state of the world – no matter what natural properties it exemplifies – constitutes a "good". That same challenge is the crux of many of the conundrums that this book has shown to plague biodiversity theories of natural value: Take any reasonable definition of biodiversity in terms of the diversity of kinds in some combination of biotic and biota-encompassing categories. Use that definition to specify some particularly biodiverse condition. Then one can imagine some states of the world exemplifying that condition to be replete with natural value, while others are simply abhorrent. One can easily see this by way of an example in which biodiversity is simply defined as "species richness". Suppose that the current state of biodiversity is an exemplar of a particularly species-rich condition. It would continue to exemplify this biodiverse condition to exactly the same degree – that is, biodiversity would be completely preserved – were each extant, nonhuman species methodically replaced, one by one, by the invention of a genetic engineer. A central point of my suggestions in Sect. 5.3 (The moral force of biodiversity) for "enhancing" biodiversity along these lines was to lend visceral force to this conundrum.

Metaethical naturalism is vulnerable to yet another, even more confounding problem, which has to do with its role in motivating right action. It is legitimate to think that the ability to motivate action is not just *a* significant role, but also *the* most significant role of moral evaluation. Values that are supposed to be taken into account in guiding action are vacuous in the absence some reason to believe that acting according to them really matters. In fact, this is part of what it means for something to matter morally: It ought to be taken into account by moral agents in making morally significant choices about how to act. These are the kinds of choices that are closely connected with moral responsibilities and our moral character; they are the ones that are subject to moral approbation or disapprobation. None of this considerable moral weight burdens a decision made on the basis of arbitrary or frivolous valuations.

Metaethical realism is vexed by its insistence on building value out of some structured set of properties in complete isolation from moral agents. This makes it extremely difficult for a valuing subject to find a motivational connection to an object that exemplifies these properties. Why, exactly, should a person care that an object exemplify certain natural properties while its being an exemplar of others is (and should be) a matter of moral indifference? The number and diversity of species (for example) is a property of the world. But so is the pattern in which the hairs on the back of my neck grow. Likewise the total number of carbon atoms in the universe. Why should the first fact matter morally but not the second or third?

It is reasonable to suspect that it is this severed connection between valuing subject and object valued that renders so arcane and vacuous the conservationist wrangling over the precise values of the parameters definitive of biodiversity that humankind supposedly should seek to realize. Should it be total number of species? Or species that perform certain functions? Or species that expand the total phylogenetic distances traveled in and around the tree of life? Or species with the most "evolutionary potential"? Why, indeed, should any one of these matter more than any other – or more than those carbon atoms?

Abandoning a strict biodiversity rubric for characterizing value-conferring properties does not seem to help address this problem. A rubric centered on "functioning" seems equally impotent so far as motivating right action is concerned. Independent of its biodiversity, why should an ecosystem that exemplifies one way of working be held more valuable than another? An ecosystem comprising a nutrient-rich body of water functions rather differently from another (or its previous self) that is relatively nutrient-impoverished. Applying the term "eutrophication" to the former prejudices the discussion for implying an excess of nutrients, which in turn implies some norm for their abundance. Putting this bias aside, why should the relative abundance of nutrients determine whether an aquatic system (or for that matter, a terrestrial one) is a natural good or not?[42] One can attempt a retreat to yet other candidates for natural

---

[42] If one turns this comparison back to an issue about species diversity, then there is no doubt that the set of species in a eutrophied zone will differ from the set outside it. But as I observed in Sect. 6.3 (Biodiversity as service provider), the empirical jury is out about how, if at all, the *diversity* of species differs between the two zones.

goodness-conferring properties, including ones that are relatively complex or relational. Rarity, uniqueness, wildness all come to mind. But they all seem to share with biodiversity-based characterizations of natural goods an inability to make us think that they merit any more (or less) consideration in our evaluation and decision-making than any other of the myriad properties that entities in the natural world might exemplify.

This seemingly unbridgeable gap between natural properties and their pull on moral agents hooks directly into the biodiversity project's confusion about ends (Sect. 8.1.3, The ends of the biodiversity project). In fact, this gap is really just one manifestation of the confusion regarding which conservation ends are the "right" ones, based on biodiversity properties exemplified in the world. From this perspective, the extreme difficulty in justifying why people should strive to achieve one biodiverse state of affairs rather than another is that the properties that characterize these states of affairs are invented in a moral vacuum that abhors valuing subjects. There is no basis for their comparison because they are equally impotent so far as the moral force that they exert on valuing subjects is concerned.

At this point in the argument, something rather strange often happens, metaethically speaking. Some proponents of the biodiversity project actually take Moore's jaw-dropping tack, veering off the straight and narrow line of a naturalist account of natural value couched in terms of scientifically ascertainable biodiversity properties. Like Moore, they are loath to abandon metaethical realism. And so they head for metaethical nonnaturalism. If biodiversity and other natural properties cannot account for the value of nature, then (this line of thought goes) the source of nature's value must be some set of its *non*natural properties. In other words, they head for a *non*naturalist account of *natural* value.

Let me review what the move to nonnaturalism entails in order to clearly understand what is entailed by conservationists' move to it. I sketch two alternatives. The first one supposes that there is a general property of goodness, which is precisely the property that attaches to all good things, including good things in nature; and that this property is not detectable by the sensory apparatus of normal human beings, even augmented by the instruments and theories that aid and abet scientific investigation. But this is a rather difficult-to-defend version of the nonnaturalist story and not one that many, other than Moore, are willing to defend. That leads to a second alternative. It might seem more plausible to suppose that there are more *specific* value-conferring nonnatural properties. This, at least, is a metaethical gloss on the possible attraction of edging towards such rubrics as "ecosystem health" and "ecosystem integrity". These latter-day surrogates for the now out-of-favor "balance of nature" will by now be familiar to the reader for their frequent intrusion into the discussion as the biodiversity project's environmental norms of last resort.

Of course, insofar as these phrases are intended as shorthand for structured collections of *natural* properties – the sorts of properties that science can characterize and measure – we are back to naturalism. That would mean a return to the task of describing, for example, patterns of growth and decline of populations of prey, predators, and "innocent" onlookers to the predation mayhem; then identifying some such pattern as "good". But this undertaking is once again burdened with the task of proposing coherent reasons, based on credible and generally endorsable principles, that justify a norm according to which some population undulations are held to be good while others are held to be bad.

In fact, there seems to be little serious inclination on the part of those who resort to the rubrics of "ecosystem health", "ecosystem integrity", and other, related phrases to find their moral force in sober naturalistic analyses. The concepts that these rubrics attempt to capture do not seem to have much of anything to do with the detailed matters of fact concerning, for example, how successfully one creature is consuming another or how successfully the other is avoiding being consumed. Focusing as I have before on "health": one reason is that even as applied to a person, it is extremely difficult to reduce "a state of health" to some collection of that person's biologically ascertainable properties. Of course, some extreme values (for example, zero blood pressure) preclude continued survival for a human kind of creature. But within a very wide range, whether or not a person is in a healthy state is largely determined by what she has to say about how well she can pursue her most important interests.[43] But, as I have emphasized several times before, ecosystems have no interests to pursue.

This last observation points to what is most likely the paramount reason for a notable disinclination on the part of exponents of the biodiversity project to reduce ecosystem health or ecosystem integrity to natural ecosystem properties. The moral force of these terms springs from some aspect of goodness preloaded into the terms "health" and "integrity". And indeed, when it comes to their application to a person, that moral loading legitimately derives from the role of health and integrity in her capability to pursue her legitimate interests as a person. It is obvious that nothing that is heard through a stethoscope and no sphygmomanometer readout is equivalent to a human state of health from the fact that no one accords these natural properties the same moral attention that is accorded to good health. Moreover, the case for non-equivalence is clinched by centenarians whose irregular heartbeat and high blood pressure have not curtailed their well-being in any way; and by the circumstance that some of them spend their unusually large number of days oblivious of these particulars.[44]

Because the naturalistic reduction program looks singularly unattractive when it comes to human health, it looks advisable to also avoid it when it comes to ecosystem health. The reasons are similar; and in particular, it is hard to know why some attribute of an ecosystem, such as its containing certain fluctuating populations of certain organisms, is, under that description, worthy of moral consideration. But by avoiding that metaethical brick wall, the ecosystem health gambit runs straight into another – by way of the path that leaves "ecosystem health" unanalyzed, except

---

[43] See Jamieson (2002, 216–217) for a thoughtful elaboration of the varied and complex factors that bear on assessments of human health.

[44] In bringing long-lived persons into the discussion, I don't wish to imply that longevity – something else that's measurable – is equivalent to good health either. See Jamieson (2002, 217) on this point as well as the rest of his enlightening treatment of the subject of ecosystem health.

insofar as it supposes a heretofore undetected or neglected correspondence between nonhuman, biotic-and-abiotic combinations of entities to persons when it comes to their respective interests, health, integrity, and even autonomy. In this way, it seeks to preload moral value into ecosystem health by leveraging a general inclination to grant that moral value is preloaded into human health.

This anthropomorphizing move often has a deontological feel to it, wherein duties to ecosystems are said to arise, just as they do for persons, out of respect for their health, integrity, and autonomy. But deontologists are not alone in anthropomorphizing nonhuman nature. They have the company of some virtue ethicists – at least in trying to attribute to the natural world qualities that are typically thought to be the *bases* for the moral consideration of other *persons*. Virtue ethicists are often inclined to think that from these qualities in nature, such virtues as respect for *nature's* autonomy and beneficence towards *nature* arise.[45] But, as I have argued at several previous junctures, anthropomorphizing value into the natural world fails the test for minimal credibility.

The most basic reason that anthropomorphizing moves fail is the one cited in my earlier treatment of this topic in Sect. 7.2.3. To reprise it here: However one understands health for a person and possibly other individual creatures, it inevitably relies on a conception of that being's *well*-being and its interests in pursuing a characteristic way of going about its own life. On the other hand, ecosystems have no wellbeing, they have no interest in pursuing one kind of existence rather than another, and they have no life to go about. Nothing can thwart or constrain them because they have no goal whose achievement can be thwarted or whose striving towards can be constrained. Ecosystems are not the sorts of things that pursue any goal at all.

I have focused on health, but if anything, the application of the concept of "integrity" to ecosystems fairs worse than the application of "health" to them. Insofar as it applies to nonhuman creatures it is a near synonym for "health" – an ability to pursue their characteristic interests in their characteristic way. Insofar as it applies to people and is not used as a synonym for "health", it relies on a conception of a certain aspect of character according to which various beliefs and actions align. But ecosystems have no beliefs; nor do that take actions (which are not merely a matter of a state that changes in time). And so ecosystems do not even have any beliefs or actions to get into alignment.

There will remain some ecologists who claim that they know a healthy ecosystem when they see one; or one that is "vibrant" (see Note 38 in Chap. 6); or one whose integrity is intact. Perhaps they believe that a supporting analogy is irrelevant and unnecessary for these attributions. But if they are not making these epistemic claims on the basis of natural properties (which appears to be quite problematic), then they must be making them on the basis of nonnatural properties. It is a strange phenomenon, for a person of science to posit a realm of properties that are not subject

---

[45] I am here adopting the virtue ethics language that Ronald Sandler (2007, 40–41) expounds, as well as thinking of his particularly lucid account of how it applies to environmental virtues. I revisit Sandler's views on environmental values in Sect. 8.2.6 (What appropriate fit is not) by way of differentiating them from mine.

to sensory perception and not subject to scientific investigation (as opposed to scientific proclamation), but that some number of persons (albeit perhaps only properly trained scientists) can detect in some as yet unspecified way.

This subsection has suggested how the appeal of a realist metaethics with regard to the natural world leads the biodiversity project most directly to naturalism. It has shown how difficulties with naturalism provoke some advocates to abandon naturalism and heed the siren lure of nonnaturalism. And finally, it has observed how nonnaturalism posits a parallel universe of properties unavailable to normal experience – something that seems far more perplexing than anything that universe might help to explain.

These metaethical conundrums – associated with trying to account for the value in biodiversity or in nature itself – help to explain the tendency of the biodiversity project to retreat from the proposal that it should be valued as an end. By considering biodiversity as a "mere" means for what is hoped to be a less problematic and independently justified end, the need to find either a naturalistic or nonnaturalistic account of its value is avoided. And of course, casting about for what is hoped to be a less problematic end, the end that comes to preeminent attention is the end of economic welfare.[46] This is the metaethical gloss on why theories of biodiversity value so often go down an economic path – trying to make the case that biodiversity is an economic boon, a vital service provider, the key to human health, and so on.

8.1.6 Lesser Evil Arguments

I have previously remarked – in connection with both Responsibility for nature (Sect. 8.1.2) and The ends of the biodiversity project (Sect. 8.1.2) – on how a "lesser evil" framing of a responsibility for nature and a "lesser evil" framing of the biodiversity project's ends obscures or entirely pushes aside what might be the most relevant moral considerations. This subsection takes both a closer and more general look at lesser evil arguments, as they are deployed in the defense of biogeoengineering the natural world.

At its core, a lesser evil argument attempts to mash the salient complexities of a moral issue and its context into a dichotomous choice[47]: We can be responsible for nature; or not. We can atone for our sins against nature; or we live with the shame of our transgressions. Having broken nature, we can leave nature in its broken state; or fix it. We can nurse ecosystems to health; or let them suffer their terrible illnesses. We can create "an environment that is desirable for all of us"; or we can create one that we would pay to avoid.

Having pared the moral context down to a simple dichotomy, a lesser evil framing can freely admit that *both* poles of the dichotomy might well be morally problematic:

---

[46] In my judgment, economic welfare as an ultimate end is extremely problematic, for reasons that I hint at in Sect. 2.1.2.1 (Consequentialism) and elsewhere in this book. But in current society, it certainly enjoys the advantage of being largely immune to challenge.

[47] Bipolar choices between evils are obviously generalizable to multiple evil choices.

Being responsible for nature might entail assuming an uncomfortable, proprietary relationship with nature; redemption for perceived past transgressions against nature might require gestures of restoration that no identifiable principle can distinguish from the transgressions; fixing nature might require turning it into something that does not accord well with our feeling for what nature is or how people should relate to it; it is discomfiting that no scientist has yet revealed the sphygmomanometer whose readout shows an ecosystem to be unhealthy, and in any event, no ecosystem has so far complained of not being able to do what it could do, if only it were more healthy; and there is no evidence or reason to think that a "desirable environment" will have any close connection with a natural world worth valuing as nature. Yet, each of these choices is posed on one end of the dichotomy as less onerous than its alternative; it is better on balance. On these grounds, a lesser evil framing can make it seem urgent to condone and even embrace the morally problematic nature of the alternative that is less evil. Why, the reasoning goes, would one not do so when the only alternative is to choose a greater evil?

The moral hazard in a lesser-evil framing of a problem is that it pushes to the periphery serious consideration of why, exactly, the less evil alternative might still sacrifice something of moral importance. This clears the way to focus on the best means possible to secure that alternative, whereupon the prospect of biogeoengineering as the principal means presents itself.

The presentation of an argument in this form, which forces a choice between unpalatable alternatives, is in itself cause to be wary. It invites careful scrutiny because the principal suasive force flows from undefended suppositions about the structure and framing of a moral dilemma rather than from independent, well-considered grounds that actually justify regarding the lesser evil as the choice that, at least to some degree, reflects positive values and avoids negative ones. I have already remarked on one subversive effect: A lesser evil framing implicitly sets an extraordinarily low standard for favoring one pole of the dichotomy over the other: No matter how hideous, the lesser evil can be seen as justified merely as less objectionable than its even more hideous evil twin. No matter how slight the weight of a consideration, it is made to seem momentous if sufficient to just barely tip the balance one way or the other.

This framing's formal structure is also responsible for the second subversive effect mentioned above: It discourages scrutiny of the morally objectionable elements of the lesser evil and why they are objectionable. That is because it diverts attention elsewhere – to a highly restricted set of elements that compare, in only the simplest and most easily observable ways, the poles of the supposedly dichotomous choice. This connects with the first subversive effect: A supposedly decisive factor drawn from this truncated set might well be trivial or peripheral in comparison to factors that this evaluative framework leaves outside the frame.

A lesser evil argument is cause for wariness for other, nonformal reasons. For one thing, the dichotomy might be a false one – an illusion that is sustained by removing much of the relevant moral context that bears on the normative comparison. I suggested that this is the first of several reasons why the "assume responsibility for nature, or not" dichotomy misses the moral mark: The dichotomy presumes some understanding of "responsibility" that is based on one or another extraordinarily feeble analogy.

Moreover, not only does the dichotomous framing tend to distract attention from proper evaluation of the dichotomy's poles, but also the comparison's initial plausibility tends to derive from significantly undercharacterizing the evil that is supposedly the lesser of the two. A lesser evil argument concedes, up front, that the lesser evil might well be distasteful. But that concession should not disarm evaluative investigation. It should not seduce us into thinking ourselves relieved of the need to develop a more complete understanding and contextualization of just why the supposedly "lesser evil" has its own moral costs and just how grave those costs might be. Those costs might in fact be great, amounting to what environmental philosopher Stephen Gardiner (2010b, 300–302) calls a "marring evil". Or the nature of those costs might be repugnant in some other way that is not taken into account within the dichotomous framework. It can turn out that, even within that framework, the lesser evil is in some significant respects *not* lesser. Considerations drawn from outside the framework can even reveal that it is *antithetical* to the values that are at stake. In fact, I will suggest, for reasons mostly presented in this chapter's second section, that biogeoengineering is a case in point: It is antithetical to what is centrally valuable about the natural world.

Gardiner discusses lesser evil arguments in the context of the debate about geoengineering: He examines and criticizes an argument that purports to defend – as the lesser evil – one representative proposal to geoengineer the earth's climate system so as to avoid catastrophic global warming. He identifies a number of morally relevant elements that are omitted from this argument, by virtue of how it is framed. Finally, he shows how this undercharacterization can be understood as a consequence of the argument's vulnerability to moral corruption. In sum, Gardner (for example, in Gardiner 2010b) shows how framing an argument to geoengineer the planet's climate as a choice between two evils saliently narrows its evaluative focus. That narrowing is largely due to entrenched ways of thinking that shut off essential moral illumination under which geoengineering casts a harsh moral shadow.

I believe that something quite similar occurs in many arguments that are supposed to support *bio*geoengineering. At every level – from the smallest conservation effort on up to the most grandiose scheme – the choice is all too often posed as one between biogeoengineering a "better" nature; or not. It is said that we can choose to fulfill our responsibilities to nature; or not. We can choose to promote the health of ecosystems; or ignore their great ills. We can bolster the integrity of ecosystems; or let them disintegrate. Those for whom economics is the ultimate moral compass urge that we can choose to suck as much economic value from nature as possible; or foolishly ignore judiciously selected treasures, there for the taking. Even those somewhat wary of economics insist we can choose to account for nature's economy and thereby make nature's value more palpable; or leave economic value off the table and so leave nature that much less well defended.[48]

---

[48] My realization that much thinking about biodiversity and natural value is framed as a lesser evil problem owes enormously to Steve Gardiner's exposition of the climate change problem and how much dubious reasoning in that domain implicitly hinges on its framing as a lesser evil problem. See also Note 50 for more on Gardiner's influence on my thinking.

The posing of each of these dichotomies should set off an alarm that, in turn, should trigger a barrage of questions: Is the kind of responsibility that people have to nature really like their responsibility to repair a broken object or to take over the design of one that they've already vastly changed? What, really, is the norm according to which ecosystems are healthy or not; and what justifies supposing that people should play doctor with them? What reason is there to suppose that extracting the greatest economic benefit from nature would result in a natural world worth valuing as nature? That, at least, is a good beginning.

The persistence of these lesser evil dichotomies and the notable absence of questions that challenge them are signs of moral corruption, which I have previously mentioned in connection with the project of Biogeoengineering value into nature (Sect. 8.1.4). As Gardiner (2010b, 286) succinctly characterizes it, moral corruption is "the subversion of our moral discourse to our own ends." In other words, it is the human disinclination to properly weigh considerations that could militate against our own interests and goals. I will add to Gardiner's characterization that "our own ends" includes institutional interests and ends. For in the case of the biodiversity project and its biogeoengineering proposals, institutional interests – for example, an institution's interest in its self-perpetuation – do not always align with the values that the institution purports to promote.[49] There is one more important piece to this characterization of these sorts of misaligned interests: To affirm that they occur is not to impute evil intent to those individuals and institutions that fall prey to the moral malady of letting them derail good reasoning. Rather, it is the stuff of normal human psychology. Its recognition makes us routinely expect a judge to recuse herself from a case to which she has even the most indirect and tenuous personal connection. But if anything, the malady is all the more insidious for the preponderant tendency of people to be unaware of it.

Before leaving this general characterization of lesser evil arguments, I wish to mention one common and recurring form that they assume in the context of the biodiversity project – a form that I have not previously mentioned. Those familiar with the biodiversity project will be familiar with the claim that, by means of biogeoengineering, humankind can "buy time". The implicit subtext makes this claim a lesser evil argument, which runs along these lines: While biogeoengineering might itself be abhorrent (that is, itself an evil), at least it preserves *something* of the natural world until we can figure out some better approach to preserving its value. The trope that we are "at the end of nature" originates outside the biodiversity project. But it reinforces the subtext of an emergency, which justifies suspending customary norms to preserve whatever shreds and shards remain of nature.

---

[49] This is glaringly apparent, for example, in Goldman et al. (2008), which I cite elsewhere. These authors boast about attracting corporate sponsors to their ecosystem services approach by grotesquely distorting the rules of compliance. Their rules permit a corporation to do whatever it pleases to any place in exchange for a mere promise to reconstruct a facsimile of the original elsewhere. This "habitat banking" is money in the bank for corporations who profit from being seen as allied with environmental interests, while also profiting from being able to do business as usual. This phenomenon is further explored in the next subsection.

Unfortunately, the exhortation to "buy time" is an incomplete thought that lacks a convincing completion. First and foremost, as the premise concedes, there is something unsavory about biogeoengineering a new natural world that is a human design specified to meet human-defined requirements. One might think that it is evil quite simply by virtue of its sanctioning major transformations of the planet, which exemplifies the antithesis of nature-valuing behavior. Or one might think that it is evil by virtue of its sanctioning major transformations that serve no credible and generally principled end – if indeed, there is any overarching end at all. On this take, biogeoengineering is a kind of restless, aimless, undirected meddling. But on any of these takes, biogeoengineering should be regarded as an evil with a moral burden sufficiently onerous to warrant avoiding it under almost any circumstances.

The bare exhortation to "buy time" avoids addressing the question of what, exactly, the biogeoengineering effort is supposed to buy time *for*. This question gains some urgency from a legitimate concern about institutional momentum. Ignoring the question ignores the possibility that regularly responding to problems in a certain way can instill in people a strong disposition to continue operating in a like manner – even after the time has been "bought". Creating a paradigm and a culture of biogeoengineering, and investing resources, training, and infrastructure in that approach, could well turn out to be a self-perpetuating activity. Why should we trust that, having "bought time," the biogeoengineers would change tactics and cultivate a radically different approach that eschews biogeoengineering?

But even if biogeoengineering were morally neutral or, at least, not reprehensible, it is still important to answer the question of what such efforts are supposed to buy time for. Biogeoengineering proponents rarely try to answer that question. But so far as I can tell, what they do say hints at an end that, unsurprisingly, simply embraces continued biogeoengineering. The response boils down to buying time to ensure more choices for bio-parts with which people might wish to stock bio-warehouses "so as to serve people well". That end, as shown by Chap. 6's examination of ecosystem services, human health, and other ways of serving people well, is not defensible even on its own terms.

In the meantime, it seems that the very acts of slicing and dicing the natural world to realize a bio-inventory that efficiently hordes whatever stocks of bio-parts people most covet, is destroying anything that most people would recognize as the natural world. The fact that a bio-warehouse is stocked from bio-parts taken from nature does not make the bio-warehouse valuable as part of the natural world. Worst of all, the acts of evaluative and physical dissection reinforce just the kind of behavior that disvalues nature as nature. In other words, biogeoengineers urge that we "buy time" for an end that is hard to endorse as a legitimate way to value the natural world – or at least, as a legitimate way to value it as nature.

To summarize: The apparent indisputability of lesser evil arguments is largely due to their ability to obscure significant moral context. It can be extraordinarily difficult to recover that context, because morally corrupting influences often make it impossible to call it up. The symptoms and origins of this corrupting influence constitute the topic of the next subsection. By means of examining those symptoms, I propose to uncover and recover some of the moral illumination under which *bio*geoengineering does indeed cast a harsh moral shadow. At least as important, by exposing these corrupting influences and by deposing the false dichotomies that they prop up, this inquiry opens out the domain of possibilities for recovering some better sense of where natural value lies, and provides the freedom to explore some quite different paths to finding it. Section 8.2 in this chapter takes one such path.

8.1.7 Moral Corruption – A Symptomology

I now turn systematic attention to the symptoms of moral corruption that the biodiversity project presents.[50] The immediate impetus is, as I have said, to explore a legitimate suspicion that moral corruption is responsible for obscuring important moral context, in the absence of which, choices concerning the natural world's value can appear to be choices between two evils.

There is another, more broadly felt impetus for taking a look at moral corruption: Gross, systematic, and widely condoned failures of reasoning, such as those that I have recounted throughout this book for the biodiversity project, tend not to arise and persist in a vacuum. On the contrary, failures of this sort tend to be symptomatic of a shared framework – one that is rigidly and sometimes unknowingly adhered to by a community[51] in a way that imposes restrictions in concept, vocabulary, and even what constitutes a legitimate topic of conversation. This, in turn, can inhibit or preclude legitimate challenges – both from within the community and from outsiders who might be more inclined to push the stultifying weight of the framework off their chest. Under conditions such as these, a framework can persist indefinitely despite the fact that it serves its announced purpose quite poorly. It then persists because it serves the shared psychology of those who are accustomed to working within it, and because of how pertinaciously disinclined are its adherents to seriously consider its failings. There is more than a little evidence that the biodiversity project wraps itself around just such a framework.

The mere existence of morally corrupting influences does not in itself constitute an independent substantive argument that those influences have wreaked havoc on practical reasoning. However, the more one can locate the biodiversity project's entrenchment in its psychological attraction and inertia, the more one is relieved of

---

[50] This discussion of moral corruption, like my discussion of lesser evil arguments, owes greatly to Steve Gardiner's work in connection with global climate change, especially as it relates to arguments for geoengineering the earth's climate. See Gardiner (2010a), though his most complete discussion on this topic is in Gardiner (2011), Part E. Those familiar with Gardiner's work will notice that one of the more compelling sources of moral corruption in geoengineering is largely absent from biogeoengineering – namely the moral morass created by the temporal displacement of actors with respect to those affected. However, biogeoengineering shares with geoengineering sufficiently similar corrupting factors to justify suspicions of a similarly pervasive morally corrupting influence at work.

[51] The fabric of a community, in the sense I intend here, can be woven from scientific, institutional, political, social, and cultural threads.

the burden of supposing its entrenchment to be entirely rooted in compelling reasons, which are nowhere to be found. Perhaps more important, by grasping the psychology of the framing assumptions, one gains the capacity and freedom to step outside them.

For this discussion and for most of the remainder of this book, I put aside the question, much debated in environmental philosophy circles, of whether or not the sense of power expressed in the biogeoengineering approach to creating a more valuable nature is illusory or delusional; or as those who decry "faking nature" say, deceptive, hubristic, or reprehensibly domineering.[52] In this arena, such proposals as "The Mother of All Conservation Moves" and North American rewilding certainly lead with their chin. But I find this debate of lesser and even minor interest because it does not get to the core sentiments or core assumptions that motivate the debated proposals. These sentiments and assumptions could just as well motivate projects that are ill-conceived in ways that I am about to characterize and that are not so obviously open to such charges as expressing a vice of hubris or reprehensible domination.

So, on to presenting symptoms and sources of moral corruption connected with biogeoengineering. As will become plain, the various sources of corruption are not mutually exclusive, but rather often stand in mutually reinforcing or cause-effect relationships.

8.1.7.1 Blindness to Unwanted Implications

Biogeoengineering projects seem to have a siren lure. The idea that people might install or reinstate value in nature by dint of humanity's science-based engineering prowess seems to be irresistible. But it is hard not to suspect that this irresistible appeal is achieved and maintained by a blinkering effect, which places salient, unwanted implications out of sight.

The malady of blindness presents in several symptoms. One is the pervasiveness of basic fallacies of reasoning, which I have recounted throughout this book – most particularly Fallacies of accident (Sect. 2.2.3), which omit or simply ignore facts that disprove wanted implications or prove unwanted ones. To recall one of the most persistent, logic-defying examples from earlier in this section: Without assuming some profound blindness, it is hard to fathom citing biodiversity as the champion of New York City's clean water. Keeping it clean might have benefited biodiversity; but both that effect and the water's potability were effects of moving defecating livestock away from water sources and some modest amount of restraint in dumping toxic chemicals into it. Some leaps of logic are, if anything, more incomprehensible without presuming some blinding effect. How else can one account for eminent

---

[52] I do revisit this theme later in Sect. 8.2.6 (What appropriate fit is not), but only by way of differentiating my views about nature's value from those who harp on the vice of "faking nature" or of being domineering with respect to it.

ecologists and epidemiologists trumpeting the disease vector-diverting effect of cows as a triumph of *biodiversity's* health-inducing effects?[53]

A second symptom is that these fallacies of reasoning, like zombies, keep coming back. In 2010, a group of highly regarded experts in disease emergence and transmission reprised their 2006 argument (Keesing et al. 2006, examined in Sect. 6.5.2, Biodiversity as safeguard against infection) for the health benefits of biodiversity. The fallacious logic of the earlier work was transmitted intact:

> P1: Sometimes, the presence of an "unsuitable intermediate host… can reduce the probability of subsequent infection of humans" (Keesing et al. 2010, 648).
>
> P2: "a greater diversity of host species can sometimes increase pathogen transmission" (Keesing et al. 2010, 648).
>
> C: Therefore, "biodiversity loss frequently increases disease transmission" (Keesing et al. 2010, 647).

Insofar as the claim C is intended to generalize premise P1 and not merely repeat it, it is blind to the countervailing evidence of premise P2. Moreover, as I discuss in that earlier section, the argument is presented with no mention that reducing biodiversity – for example, draining "swamps" (recently redubbed "wetlands") to rid zoonotic disease areas of insect vectors or (in the northeastern United States) cutting down oaks to reduce the mast that supports Lyme disease reservoirs – is often an obvious means of reducing human infections.

A third symptom is the rampant confusion about the biodiversity project's ends, which was the focus of Sect. 8.1.3 (The ends of the biodiversity project). It is hard to imagine any more generous explanation than some unconscious blindness that accounts for the incoherencies, inconsistencies, and general eschewing of well-considered and endorsable principles observed in that arena.

The fourth and fifth symptoms are somewhat more difficult to discern, but they are nonetheless of even more concerning. The fourth symptom of blindness is biogeoengineering's preoccupation with its technological and methodological means compared to its gross inattention to ends. There is a desperate need for discussion of the question of whether and how a biogeoengineered world can possibly be a world in which much of the *natural* value of the natural world remains, let alone flourishes. But by dwelling on the nuts and bolts of dazzling science and dazzling biogeoengineering techniques and methods, the discussion of biogeoengineering walls itself off from such a discussion. There is no better exemplar of this than the methodological debates concerning adaptive management. It is plausible to suppose that this tendency to avoid discussing (and therefore trying to find credible principles that justify) ends is the doppelgänger of the third symptom, which is confusion about them. And indeed, they go hand in glove. Alongside the wrangling about whether or not, for example, biogeoengineering proposals are effective in "conserving for X",

---

[53] This example, from Sect. 6.5.2 (Biodiversity as safeguard against infection), is doubly confounding. As recounted there, the larger body of evidence suggests that the greater wealth of cow-owners and their consequently greater access to health care, not the mere presence of cows, provides the protective shield.

one finds wrangling about whether it might not really be Y – whose interests conflict and which requires conflicting means to serve – that should be served by biogeoengineering.[54] And at that point, as I describe in Sect. 8.1.3, the discussion drifts in a becalmed sea cut off from the fresh breeze of sound reasoning stiffened by discernible, credible, and defensible principles.

The fifth symptom also relates to the third. It is a seemingly unblinking and nearly limitless tolerance for, and failure to challenge or critically evaluate, the goals of experiments that are pursued, untethered from any visible or respectable set of justifying principles. I am talking about what must be the majority of conservation and restoration projects proposed and already under way: There are projects to create snow-globes of biodiversity such as those involved in vacuumed, sprigged, and urchin-planted coral reefs. There are those that stage fragmentary historical re-enactments (that is, "restorations") of previous, admittedly unrecoverable moments in the planet's history, such as the ones that seek to re-create historic river flows with the micro-engineered release of dammed water. There are those that usher in cows, promoting them to the heroic status of biodiversity saviors, which graze, trample, and defecate their way to making life easier for some select group of other organisms. And there are those that pursue ongoing sequences of "corrective" actions – righting previous ecosystem wrongs followed by the righting of what went wrong in the last round of wrong-righting. Such is the "Razorback Roundup", which sends razorback suckers to razorback kindergarten so as better to contend with the totally alien, dammed, and sport fish-seeded environs in which they now find themselves (Sect. 7.2.3).

The capricious quality of these projects cries out for serious examination of what, if any, natural value can be found in their goals. And (what might be the same thing) it cries out for examination of whether this way of relating to nature really embodies a truly valuing relationship. Instead, if such proposals provoke any debate at all, it mainly confines itself to details about the means: Might the Bay checkerspot butterfly be best served by cows, or might it be better served by goats, or by planting genetically engineered and better adapted variants of the dwarf plantain that it insists on eating, or by selective spraying that rids the dwarf plantain of its major competition, or by genetically tinkering with the butterflies to adopt a more sensible diet…?

While most everyone in the biodiversity project is embroiled in these kinds of debates, essentially no one is publicly asking – because they are framed out of the debate – the more systemic questions. For example, is the retention or enhancement of nature's value really the sort of puzzle that admits of engineering solutions at all? Would a world transformed into the imaginative creations or re-creations of biogeoengineers – even such a biogeoengineered world that serves humankind well – be a world full of *natural* value? These questions are not posed in public; therefore, no public discussion of them takes place.

---

[54] Section 8.2.7.2 (Implications for action, social policy, and conservation management) supplies graphic examples of this sort of wrangling.

This stubborn blindness to questions concerning what, really, is vital to "saving nature" is remarkable enough to justify suspicion that biogeoengineering partisans, even without their own awareness, might be wary of the distinct possibility that their projects, in which they have a great personal stake, would not come off looking well in a discussion of the norms according to which they operate. This suspicion is amplified by what I have to say in Sect. 8.2, which suggests that this wariness would be well founded. It is therefore all the more remarkable that the biogeoengineering projects, which seem to be the primary basis for neo-conservation, have established such a vise-like grip on the public and environmentalist imagination.

This claim – that in re-doing nature, neo-conservation is nature's undoing – will seem startling to some readers. So, although this is somewhat peripheral to the current topic, I will tip the cards that I play in Sect. 8.2. Entertain, if you will, a very different picture of how and why nature is valuable. Suppose that the value of nature as nature is not a generic kind of value in the sense that I characterized in Sect. 8.1.3. Suppose that nature's core value is not, for example, "value in trade" for something that one might regard as at least as desirable. Suppose, rather, that there is something quite distinctive about natural value, which captures something distinctive about the natural world. Suppose further that this value is not at all a matter of exemplifying certain properties that only exist in nature. As a consequence, suppose that scientists can spend careers characterizing the various properties of ecosystems and their relationships; and that biogeoengineers can spend careers figuring out how to poke, prod, dissect, and eviscerate "nature" into exemplifying certain of these properties and relationships; and yet, this would not yield a shred of natural value – because in fact, it is not a value that arises or can possibly arise from any identifiable or engineerable set of properties or relationships.

Suppose, instead, that the value of nature *as* nature is bound up in the relationship that persons can have to that part of the world that they do *not* poke, prod, dissect, and eviscerate into existence. Suppose that this relationship and the valuing interactions that sustain it are not promoted by, but rather destroyed by these property-changing and property-engendering projects. Suppose that this relationship and the valuing interactions that sustain it are not promoted, but rather destroyed, by these property-changing and property-engendering projects. Suppose that the relationship disintegrates to the extent that the natural world is viewed, not as anything uniquely valuable as nature, but as a human feat of engineering. And suppose that the primary achievement of these feats of engineering is to enable valuing transactions that are not the sorts that embody natural value, but instead at bottom are just another sample of the sorts of transactions that define human enterprises, embodied in the human economy and the marketplace. Suppose, that is, that they are feats of engineering to create "an environment that is desirable for all of us". All of this is to suppose that biogeoengineering is morally tainted in a quite profound way.

I will not just suppose but argue for all of this in Sect. 8.2. These defended suppositions underlie my claim that in re-doing nature, neo-conservation is nature's undoing.

8.1.7.2 Selective Attention, Distraction, and Delusion

Also nourishing the environment in which the biodiversity project flourishes is a group of corrupting elements, including selective attention, distraction, and delusion. These elements originate in the public domain. But they may well be related to the biodiversity project's blindness to unwanted implications, along the following causal lines: Techniques that are sexy, cool, "scientific", technologically challenging, and headline-grabbing rivet a science- and technology-worshipping public. It is easy to love a story about a difficult technological challenge and an unexpected, heroic, or astonishing solution; and it is easy to become completely absorbed in this kind of fascinating narrative. Blindness to unwanted implications fosters that love by fostering the illusion that the technological challenge is morally undemanding. Best of all, we ordinary citizens can and do appoint scientists and economists as our deputies to duly attend to these arcane matters – wrapped up in biodiversity indexes and obscure ecosystem relationships – that are beyond our ken. We feel that we can trust those high priests of technology to do their magic and afterwards regale us with tales of their quests, undertaken while we ordinary citizens redirect our attention to our everyday life.

Meanwhile, we have only the haziest and most narrowly contextualized idea of what we think we are deploying our biogeoengineer surrogates to do. We know that they invent technologies and methodologies to save nature or to save the environment. That is what we bid them to do and that is what they say they are doing. And we know that they proudly parade their inventions through popular and scholarly writings. This, we allow ourselves to believe, confirms their assiduous attention to our bidding.

A salient problem with this bargain arises from its delegation to technocrats a responsibility whose moral dimensions are ill understood before it is delegated. Lost in that shuffle is any moral concern that has to do with how we citizens structure our lives and societies with respect to nature. The situation has a salient hallmark of moral corruption: There is a general, shared interest in doing what is easiest – namely, shedding the burden of confronting potentially difficult moral questions that directly bear on our private and public life.

Perhaps worse than the "mere" dodging of these questions is the delusional self-righteousness with which it is commonly done. The delusion is that, despite gross inattentiveness to the moral crux of our relationship to the natural world, we are nevertheless doing our duty toward it. And indeed, we are doing so in the most technologically advanced way, and therefore (we can tell ourselves) in the best possible way. After all, we have appointed the best biogeoengineering experts to wrestle nature back into some ideal shape, whatever that might be. It is reassuring and thus easy to convince ourselves that we are doing everything that we reasonably can be expected to do. This, of course, reinforces a natural disinclination to wrestle with our own attitudes and our own individual and collective behavior as it relates to the natural world.

This last point – concerning a possibly huge misconception about what needs to be wrestled with – cannot be overemphasized. It is a variation on the theme of Sect. 8.1.1's third notable implication of the "state of nature" view of natural value.

This understanding, which underlies the business of biogeoengineering it into a valuable state, so tightly focuses on nature-as-object that it tends to lose sight of persons-as-valuing subjects. It loses sight of valuing persons, that is, except insofar as they engage in the transactions that define the market economy. Once we are caught in the gravity of "The New Economy of Nature" (the byword of the Natural Capital movement, ensconced in the title of the book by Gretchen Daily and Katherine Ellison (2002)), the pull towards objectifying nature only increases, and the prospects for nature's escape from this domain of market transactions correspondingly decreases. But I think (and will argue in Sect. 8.2) that the attitudes and behavior that people bring to the marketplace do not include the attitudes and behaviors that define the core of morally significant human relationships to nature. In fact those centrally valuing attitudes and behaviors, which embody expressions of moral regard for nature, are entirely excluded.

As I say, the attraction of "The New Economy of Nature" is strong. It is strong for how beguilingly clean, simple, and undemanding it makes (quite literally) the business of valuation. And it is strong for how effectively it frees our collective attention to wander elsewhere. It involves none of the messiness of hashing through the complexities of humankind's relationships to the natural world. It excuses us from anguished challenges to us personally and as a society. Instead, evaluation is capably handled by expert biodiversity accountants: Scientists produce detailed ledgers accounting for inventories of bio-parts in bio-warehouses. These ledgers provide the basis for computing a development plan ensuring that every bio-part taking up valuable shelf space in our bio-warehouse pays its own way. As proponents of this approach say, this is a matter of "getting the numbers right". The problem of developing and maintaining these inventories then becomes one of technically challenging but morally trivial biogeoengineering (perhaps combined with economic) theory and practice. Projects launched on these terms are morally uncontroversial exercises in hard-nosed technology development in support of "The New Economy of Nature". Wrapped in the cloak of scientific and economic language, this enterprise enjoys almost automatic cultural legitimacy. The salient point for this discussion is that the social approval that it enjoys is more easily given for our morally corrupting interest in avoiding discomfiting questions about the structure of our societies and how we live. It is far easier to let the biodiversity accountants and the biogeoengineers handle it all.

I wish to call attention to one final aspect of this sort of corrupting influence, which is its self-reinforcing and ultimately ironic quality. The situation resembles a fiduciary trust, but with a very odd twist. The terms of the trust are that the biogeoengineers, as fiduciaries, are supposed to act in the best interests of us ordinary citizens, the beneficiaries – but only in a certain, extraordinary sense that is alien to the customary concept of "trust". One must suppose that these fiduciaries cannot take up the interests of the natural world, for as I have explained (in Sect. 8.1.2, Responsibility for nature), nature cannot and therefore does not have interests. As a consequence, one must presume that these fiduciaries are supposed to represent *our* interests. One must presume that they do so by taking upon themselves the moral burden that we prefer not to shoulder directly. Having thus unburdened ourselves, we feel justified in pursuing paths of development, which systematically neglect to take into consideration what these activities entail for the natural world. That is, we do not just unburden ourselves, but we do so in a way that inevitably results in the perceived need for yet more biogeoengineering to ensure that nature assumes its "proper" shape and function. Thus we build the next dam; and that requires bio-geoengineering the bio-inventory of the denizens in its vicinity.

Here is the ironic twist: Back at the biodiversity warehouse, our biogeoengineering fiduciaries regard it as their fiduciary responsibility to perform their own type of meddling, manipulative, and transforming development of nature. Empowered by our directive, they hack nature to pieces – both conceptually and literally – and preserve or reconstruct what remains because that is what is required to fulfill their responsibility in a scientifically and economically respectable way. "Getting the numbers right" inevitably results in a highly selective inventory, one that is re-purposed to serve ends recognized within this framework. A non-zero inventory of any bio-part is maintained only with great trouble and at great cost. Therefore, each retained bio-part must earn this favor by virtue of its contribution – mostly with engineering help – to "an environment that is desirable for all of us".

In short, the end result of this form of moral corruption is that the natural world is assaulted from both sides – that is, by both parties who enter into the fiduciary trust.

8.1.7.3 Conflicts of Interest and Institutional Bias and Inertia

The credibility of a system of justice demands the recusal of a judge who has even the most tenuous personal interest relating to any party with even an equally tenuous interest in the trial's outcome. It is hard to know when or if a tenuous relationship to one of a trial's players will bias judgment despite the best, contrary resolve of the judge. But when the relationship is one of a clear and direct vested interest in the outcome, it is well known that even the most morally conscientious can fail to avoid bias.

Therefore, it is highly relevant that many of the key advocates of the biodiversity project – particularly, its core cadre of designers and implementers – have a clear and direct vested interest in the particular formulation of the problem of nature's deterioration in terms of (for example) changes in biodiversity, in their peculiar role as biogeoengineers to solve it, and in accepting whatever arguments they can muster to convince themselves and the general public of their projects' legitimacy. Very obviously, this description applies to conservation and restoration biologists whose work is based on "conserving biodiversity" or "restoring places to conserve biodiversity". It is also true of ecologists who, in the interest of demonstrating the good of biodiversity, devote a substantial portion of their professional life to trying to show how biodiversity correlates with the properties that the conservationists and restorations regard as justifying their work. It is otherwise hard to account for why so much ecological science is devoted to determining the correlation of biodiversity to "good" ecosystem "function" – productivity, resilience, health, integrity, human health, and whatever else might earn a conservation or restoration project's endorsement.

It is not an indictment on the character of scientists, but rather a reflection of their humanness, that many will reflexively recoil from a challenge to the assumptions underlying proposals that garner them grants to fund their work and on which they stake their professional careers and reputations. The conflict of personal interests is, for scientists, far more clear and far more directly confounding of their judgment than most conflicts of interests for which responsible judges routinely recuse themselves. If conservation is how you earn your living and feed your children, or if you live for the joy and excitement of inventing new and technologically exciting ways to restore, manage, or design ecosystems, then you have a very high stake in doing those things and in continuing to do them. And you probably have convinced yourself that they are the right and even noble things to do.

Recusing scientists as judges for what they do for a living does not in any way thin the pool of those qualified to make such judgments. Nothing in the details of ecological science and even more broadly, in the details of ecology or any other science better qualifies a person for making a value judgment about what our planet should look like or how people should behave with respect to the natural world and how people should regard it. The scientists on the Manhattan project were, pretty clearly, not experts in determining whether or not nuclear weapons were a good thing to develop. Their expertise in nuclear physics was entirely irrelevant in making this determination, while their eagerness to seize the opportunity to explore nuclear science cannot be discounted as a source of bias in their endorsement.[55]

The prejudicial effect of individual scientists' vested interests is all the more potent for being ensconced in academic and other institutions that self-identify as having an "environmental interest". These institutions sponsor the scientists who do the biodiversity, conservation, and restoration science. They thereby align their interests with the scientists' interests in pursing the biodiversity project, though it is likely that the forces of alignment originate from both sides. Moreover, each of these institutions has some substantial inertia of its own. This greatly amplifies the cumulative stake in, and the capacity for disinterested evaluation of the biodiversity project as well as in biogeoengineering being the right tool for the job. As a player embedded within this much larger institutional context, it takes an act of heroism for a scientist to question the basis on which her professional life depends.

Chief among institutional investors in the biodiversity project are a multitude of self-identified "conservation" organizations. This includes the three enormous, transnational groups, whose combined annual revenues are well in excess of $1 billion.[56] These organizations have latched onto the science of biodiversity-based or ecosystem services-based conservation, which includes all its ill-founded assumptions and terrible confusions about ends that result from trying to award natural value points on the basis of inventories of bio-parts. By now, they are enormously invested in

---

[55] Bill McKibben (2003, 182) makes this same point in his consideration of genetic engineering.

[56] The figure of "well in excess of $1 billion" is for the combined 2010 revenue of the The Nature Conservancy, the World Wildlife Fund, and Conservation International as reported in the annual reports for those organizations. TNC alone had revenues just shy of $1 billion in 2010.

conservation campaigns that will first and foremost promote the causes justified on these bases.[57]

This institutional investment comprises three tightly coupled components. It is an investment at once in identity, prestige, and power. Having defined their mission as promoting biodiversity or ecosystem services, and their means largely those of biogeoengineering, these coalesce into a core, institutional identity. By that, I mean that these institutions have come to see themselves quite essentially in terms of biodiversity and biogeoengineering, and cannot conceive of themselves otherwise. Furthermore, they have embedded that identity in public consciousness. As a consequence, their prestige is bound up in the portrayal of their activities as successful biodiversity projects pulled off with dazzling feats of biogeoengineering.[58] Finally, the cause of biodiversity has become their rhetorical lever of power – the switch that opens the sluice gates for revenues to flood their coffers.

By nature, an institution has an interest in perpetuating itself independent of any other interest it views itself as serving. This interest in self-perpetuation produces what I have already called "institutional inertia". Just like physical inertia, institutional inertia tends to increase with size. A billion dollars produces a billion dollars worth of inertia. The currency of inertia is spent in reflexively, rather than reflectively, resisting any challenge that an institution might perceive to subvert its identity, prestige, or power. It is also spent in the institution's efforts to bulk up its wealth, influence, and power for their own sake, and to do this by any means at its disposal – even when those means are harshly discordant with the institution's perception of itself and its public face. Institutions, in short, tend to be self-amplifying, often without regard for any other principle.

I am not making a novel point, let alone a startling one. Time and again, the political spotlight picks up some spectacularly self-amplifying program. A case in point was U.S. President Ronald Reagan's Strategic Defense Initiative. While opposed by some scientists and engineers, SDI was embraced by many other technologists (not just politicians) as a domain for the exercise of their imaginations and careers. And indeed, for a quarter of a century, many brilliant minds and fantastic sums of money disappeared below the event horizon to feed the rapacious appetite of that seemingly indestructible black hole of fantasies.

Institutional entrenchment and self-amplification are hard not to notice in the realm of conservation organizations. It is particularly noteworthy in the largest ones for whom, as I say, the terms of the biodiversity project have been fashioned into their rhetorical levers of power. It is unsurprising that this crushes any will to reconsider the real implications of their rhetoric, when it increasingly attracts big money

---

[57] At the time of this writing, TNC have let biodiversity slip into the background, while regrouping their rationales around the idea of ecosystem services, however those might cut for biodiversity. But this reformulation of their mission is, if anything, even more vulnerable to the corrupting influences that I recount here, because it cuts straight to the economics of nature without a possibly distracting stopover in biodiversity as a means to ecosystem service ends.

[58] One need only think of the TNC's coral reef vacuuming/sprigging/urchin-seeding project, which is undeniably a stunning biogeotechnological achievement.

from big corporations. For their part, those corporations grasp the rhetoric's power – whether or not the conservation organizations also "get it" – to drape their most flagrant intrusions into the natural world in a green mantle.

This phenomenon is best illustrated by how global conservation efforts have embraced the participation – and money – of the extractive industry, whose clear and unwavering interest is in ripping up any and every part of the planet to uncover its marketable buried treasures. For anyone not caught up in a judgment-corrupting conflict of interest, this is hard-to-ignore evidence that a self-amplifying cycle of interests has expanded to include ones that have little to do with the value of nature or the natural environment.

Mining is such a dramatic example because it makes the "biodiversity" in the biodiversity project irrelevant; it is hard to imagine any environmental or natural interest that it does not leave unscathed. It is difficult to wrap one's mind around the full scope of mining operations' intrusions into the natural world, including their routinely energy-consuming, toxin-producing, air-polluting, water-consuming, and water-polluting practices. One must take into account not just the actual extraction processes, which include such gross re-shapings of the surrounds as blasting the tops of mountains into adjacent river canyons and ripping out entire forests to get at minerals; but also infrastructure, such as roads, built to access the resources, and the effects of various forms of transport from the site of extraction – their construction and mishaps, including spills, in their use. It is sad but damning evidence for how badly critical judgment has been compromised that such organizations as The Nature Conservancy, with a distinguished cast of supporting scientists, boast about corporate liaisons with these interests rather than soberly reflecting on their *raison d'état*.[59] The all too plausible possibility for corporations' interest is that, while the rules of the biodiversity project place no substantive constraint on their behavior, association with its credo is tantamount to having the most powerful and prestigious conservation organizations bestow their imprimatur for doing business as usual.

Finally, one way that institutions serve their inertial interests is through competition. They compete for money and power. So the big conservation organizations also can be expected to actively discourage any view that might become the basis for competition.

In principle, it is possible for persons and institutions to rise above their own interests and biases. But it is next to impossible unless those interests are openly acknowledged and unless their possible corrupting influences are honestly considered.

8.1.8   A Parting Look at the Biodiversity Project

Section 8.1 (The disvalue of the biodiversity project) has been concerned to tease out the various and sometimes subtle respects in which the biodiversity project fails to promote or even subverts the value of the natural world. If one temporarily

---

[59] A previously cited paper (Note 27 in Chap. 6) by Goldman et al. (2008) is one example of this.

forgets its incapacity for identifying and justifying worthy ends, then one might be inclined to think the project's problem to be that today's biogeoengineers lack the requisite knowledge and expertise to pull it off – whatever "it" might be. There certainly are some famous cases of ecosystem engineering gone awry. In this regard, the exotic *Bufo marinus* (the infamous cane toad) brought to Australia in order to predate on the native *Dermolepida albohirtum* (cane beetle), which the recalcitrant amphibian stubbornly pushed off its plate in favor of a variety of varanid (monitor) lizards and land snakes, never fails to come to mind. At least that particular fiasco had an identifiable economic goal, even if the Anuran means to it was ill considered. The remarkably similar introduction of the exotic *Diorhabda elongata* (the tamarisk leaf beetle), which is supposed to predate on the now-naturalized *Tamarix* in the American southwest is representative of numerous similar cases where the proximate goal (in this case, nuke the *Tamarix*) is clear, but justifying principles (beyond certain biologists' feeling that the *Tamarix* is out of place in its new home) range from murky to opaque. The project to extirpate *Tamarix* by means of the tamarisk leaf beetle has so far not triggered any unforeseen event sufficiently disturbing to motivate this effort's exponents to extemporize countermeasures. But the arbitrariness of its goal is no different in kind from that of other, similar efforts where this becomes increasingly evident with each successive move in a sequence, which is supposed to "right" the "wrong" state of nature produced by its predecessor.

There is no better illustration of this than the tale of San Clemente Island, which I relate in some detail in Sect. 8.2.7.2 (Implications for action, social policy, and conservation management). On the surface, that story is a tragicomedy of macabre errors in which each in a succession of extirpation victims – first goats, then fox, then Golden Eagles, then cats, not to mention some number of plants – was supposed to be a means of fixing a previous choice of means to set things "right" with nature. But the story of San Clemente Island is at least as much a tragicomedy of a succession of uncertain ends, which continually makes it uncertain what, really, was "wrong", what was in need of fixing, and why. As in the case of Tamarix, there are proximate ends, such as: Remove the golden eagles to save the fox. Shoot the fox to save the shrike. And so on. But these exhortations provide no answer to the questions, "to what ultimate end?" and "on what credible principles?" The health of the ecosystem, its balance, and the unwelcomeness of aliens were cited. But these concepts are entirely incapable of identifying, let alone justifying a claim to the effect that the value of some envisioned state of San Clemente Island makes that state worthy of trying to create. Why, for example, is the natural world better, healthier, or in better balance for being in a state with fewer fox and more shrike – or should it be fewer golden eagles and more fox?

Unfortunately, San Clemente Island is only exceptional in the number of interleaving and overlapping elements that entered into its biogeoengineering. In other salient respects, it is an exemplar of biogeoengineering; insofar as this is so, the implications are devastating for biogeoengineering generally: Even if biogeoengineers had at their disposal exquisite manipulative capabilities that were informed by perfect and detailed knowledge of the effects of each manipulation, these capabilities would be completely impotent in making the natural world more valuable.

It seems that this conclusion – biogeoengineering's impotence as an enricher of natural value – is indeed generalizable, and for the reasons that I have set out to this point in Sect. 8.1. To review: Those reasons have directly to do with the key implications of the state model of nature, as set out in Sect. 8.1.1 (Biodiversity and the state of nature). First, it is hard to imagine any principled grounds for supposing, on the basis of its biodiversity or any other structured set of properties, that some particular state of nature is an end especially worth realizing (Sect. 8.1.3, The ends of the biodiversity project). Second, when the biodiversity project supposes that the natural value in a state of nature derives from deconstructed and reconstructed bio-parts that are supposed to constitute or engender some valuable property, it is hard to avoid noticing that these collections of bio-parts have no recognizable relationship to the natural world other than the fact that the bio-parts had some origin in it.

Third, in the biodiversity project's account of nature's value as a properly biodiverse state, the valuing scene is largely devoid of valuing subjects – except as coveting the largesse of nature, which delivers to them piles of goods and renders them services. There is an abiding sense that this anemic and fragile account of nature's value, which hinges on the vagaries of the markets for these goods and services, sells short some more solid, valuing relationships that people actually have with the natural world. But nor does it help the biodiversity theory of natural value to divorce itself from this market-based view of nature's value. Doing that strips away its most plausible explanation for why some deconstructed, but presumed-to-be-ideal, assemblage of bio-parts should have a morally motivating effect on people. And fourth, the biodiversity project proposes that the paradigmatic means of realizing suitably biodiverse states of the world is via biogeoengineering, which is therefore held up as the paragon of right action. But this model for acting on the natural world embodies a relationship to it that is terribly problematic as a *valuing* relationship.

Each of these themes has threaded its way through my discussion of the biodiversity project's disvalue. However, I have largely reserved comment on the fourth theme, while comments relating the second have largely been registered in the context of other considerations, such as how picturing nature as a bio-parts warehouse facilitates economic valuations. Sections 8.1.8.1 and 8.1.8.2 bring these two themes (in reverse order) back into individual focus.

8.1.8.1   Biogeoengineering as Right Action

I have the sense that biogeoengineers picture themselves as operating like sheepdogs of the natural world. In this picture, nature has strayed from its proper, natural path, and these persons of science are the panting agents that push, prod, and nudge it back on course. From a moral perspective, the significance of their vocation is entirely a matter of its effectiveness in attaining this salutary result. This paints a benignant face on biogeoengineering. While the enterprise might sometimes err, it is nevertheless nothing but a noble venture that aims at a better nature, which it might even sometimes achieve.

This, at least, is the pastoral picture that the biodiversity project itself paints of biogeoengineering. But it is not a properly panoptic view of biogeoengineering as behavior that is supposed to be emblematic of permitted or obligatory right action with respect to the natural world. Standing sufficiently far back, one can abstract from the intent, content, and consequences of individual projects and acts of bio-geoengineering. What remains is the form of action – the pushing, prodding, and nudging. This, I think, is highly questionable behavior; and I wish to scrutinize it with an eye towards making the question of right action the springboard for a fresh start at characterizing nature's value in Sect. 8.2. This is scrutiny, I need to empha-size, that cannot come from within the biodiversity project. It is yet another moral hazard of that project's framework that something so critical to its evaluation flies stealth-like, beyond its scope of detection.

There is a palpable drivenness to biogeoengineering as the principle instrument of implementing the biodiversity project. Part of this feeling comes from the sense of urgency conveyed by "lesser evil" tropes such as "buying time" and forestalling the "end of nature". It also comes from biogeoengineering's uncanny appropriation of the drivenness and unfettered quality of the development of our planet – in all respects, not just with respect to biodiversity – into (as Erle Ellis puts it) "an envi-ronment that is desirable for all of us". Ellis' rationale for engaging in biogeoengi-neering is an all-purpose rationale for development of all kinds. It is a rationale for meddling with everything that humans are capable of meddling with, or at least regard themselves as capable of meddling with. Seen in this light, biogeoengineer-ing appears as just another manifestation of a larger, intensely pursued, planet-wide development project.

In short, biogeoengineering is the principal tool of development in the domain of nature. I must emphasize that this characterization is not my invention, but rather (as I observed in Sect. 6.3, Biodiversity as service provider), part of the biodiversity project's self-portrayal. This quality reflects the view of those of the project's parti-sans who self-consciously ally it to the greater economic system by referring to it as "biodiversity development" (see Note 25 in Chap. 6). Insofar as this assimilation is achieved, it should be enormously worrisome. I shall argue in Sect. 8.2 that there is a well-justified sense in which the driven and unfettered development of the planet – and perhaps especially biodiversity development – is antithetical to any true valuing of the natural world as nature.

The view of development of biodiversity via biogeoengineering as of a piece with other, carefully engineered development raises another concern. This outlook thrusts this kind of development into a very particular and limiting evaluative con-text, in which it tends to be judged good or bad insofar as it meets requirements already presumed to have been vetted. The criteria invoked involve how well, how innovatively, in what respects, and to what degree the project produces an end prod-uct that hues to the designs of the engineers, thereby consummating their vision. At the same time, these criteria tend to exclude the vexed matter of evaluating the engi-neers' vision – that is, the ends the design is supposed to serve. More succinctly: Biogeoengineering invites evaluation primarily as a judgment about the quality of the engineering according to standards of engineering prowess. The context therefore excludes evaluation of the appropriateness of the engineering activity itself, considered as a principal, valuing expression of the relationship of people to nature.

As a consequence of the increasing dominance of biogeoengineering as the way that people – both individually and collectively – relate to nature, or delegate that relating to scientist-engineer surrogates, one should expect that more and more of the natural world will be evaluated largely on this basis. Insofar as the ends of the biogeoengineering designs are "an environment that is desirable for all of us", the evaluative context will also draw on economic analysis, which posits that the satisfaction of human desires is supremely valuable. When economic analysis is coupled to biogeoengineering – as it is in The Natural Capital Project and by the most powerful conservation organizations – then judgments about the value of nature become barely recognizable as being about nature. For they are structured as judgments about how effectively and efficiently biogeoengineers can manipulate the natural world in order to come out with assemblages of bio-parts that increase economic wealth.

Whatever judgment one might ultimately reach about this view of why nature is valuable – scientifically informed and based on "objective" economic analysis – it is important to keep in mind that science and economics have no special authority in telling us what ways of relating to the natural world are the ones that constitute valuing relationships. In fact, as I have argued in Sect. 8.1.7 (Moral corruption – a symptomology), moral corruption might seriously bias these sources for telling us how people ought to value the natural world.

Biogeoengineering plans devolve into manipulation of maintained inventories of bio-parts, which are its field-replaceable units – or "FRU's", as engineers are wont to call them. It is to that closely related topic that I now turn.

8.1.8.2   Not Seeing the Natural World for Its Bio-parts

The deconstruction of nature into its component bio-parts makes complete sense for the pursuit of ecological science. That discipline seeks to find the laws of the natural world that connect its various pieces, sliced and diced in strategic and tractable ways that permit these connections to be made. For this work, species, genes, phylogenetic distances, assemblages that function to maximize productivity – indeed, all constructs amenable to scientific investigation – qualify as epistemologically respectable building blocks of the biological world. This situation differs little from the situation in other sciences – for example, physics, which conjures up point masses as the physical building blocks of the world of Newtonian gravitation.

Science stops and trouble starts – or at, least, a qualitatively different kind of reasoning is required – when norms are set and evaluative lines are crossed. Valuing certain bio-parts (for example, species) above others, or assigning special value to carefully selected inventories of bio-parts for their natural description in terms of biologically useful categories, is not the stuff of ecology. Yet it is the stuff of modern-day conservation, restoration, and what is too casually called "environmental science".

Perhaps it is fortunate that when science masquerades as practical reasoning, the results are (or should be seen as) terribly disjointed. The situation is akin to valuing, not people, but an inventory of arms, legs, somatic cells, and various other body parts. We all agree that a collection of detached appendages and other body parts does not constitute the morally considerable entity called a "person", which we other persons relate to in morally good or bad ways. Distinguishing between appropriate and good ways of relating to a person from inappropriate and bad ways requires that a *person* be recognized as the entity valued in the valuing relationship. That person's particular inventory of body parts and the functions that they enable are, for the most part, morally irrelevant to her identity as an object of moral concern. Were that inventory, or the emergent properties that flow from it, diminished or eliminated because reduced by some number of unfortunate accidents or amputations, that would have little bearing on her value as a person. This circumstance in no way impinges on the legitimacy of scientific curiosity about human functions as they correlate to human body part inventories. That investigation crosses back across the evaluative line into safely scientific territory.

Let me quickly squelch any inclination to use this analogy of the relationship of a person to her body parts to as a cue to reprise the claim that biodiversity inventories are "proxies" for ecosystem health. I hope that my previous discussion of this unhelpful notion (among other places, in Sect. 7.2.3 (Home is where you make it – in situ, ex situ, and ecosystem-engineered), and more abstractly, in Sect. 8.1.3 (The ends of the biodiversity project) and in 8.1.5 (A metaethical gloss on biogeoengineering)) has adequately shown that whatever "ecosystem health" means, it is devoid of the moral force that attaches to a person's health. Its meaning is most clear when economic evaluation is nominated as the natural value evaluation function. In that case, "ecosystem health" devolves into something like the "the economic health that results from producing stuff and services that a lot of people desire". That, it seems, is a long, long way from any conception of "health" as it might apply to ecosystems. Yet even this economic rendering of "ecosystem health" does not cure the phrase's moral impotence, for a lot of people might like and often do like a lot of bad things – including a lot of things that are bad for their own personal health.

In any case, the notion of "proxy" is entirely inadequate for the job of transferring moral evaluation. An arm is not a proxy for a person, morally considered, even when the limb's severing clearly violates that person's interests. A resident species is even less an evaluative proxy for an ecosystem. It not only fails the basic "arm test" of evaluative substitution, but also, its extirpation cannot possibly violate the ecosystem's interests. The reason lies in the now-familiar refrain: An ecosystem does not have any interests to violate.

Nor would many more arms and legs a better person make. Yet the analogous sentiment with respect to bio-parts prevails in the biodiversity project. According to the project, when it comes to the role of biodiversity in how valuable the natural world is, the size of the collection of bio-parts – for example, the species richness, which is a measure of species diversity – really does matter. But this seems pretty obviously wrong-headed. In an oft-cited passage, the eminent political philosopher Ronald Dworkin (1993, 75) says:

> Few people believe the world would be worse if there had always been fewer species of birds, and few would think it important to engineer new bird species if that were possible.

What we believe important is not that there be any particular number of species but that a
species that now exists not be extinguished by us.

Of course, what Dworkin regards as important is not the diversity of species, but
the value of some particular species whose fate is in some way dependent on human
choices and behavior. The discussion of this thesis is eminently worthwhile, as witness its vigorous pursuit well before the invention of the neologism "biodiversity".
The thesis has to do with the value of existing species as species, whether and on
what grounds this value varies from organism to organism, and what all this implies
for human responsibility to try to extend the stay of some selected group of species
on the planet. It does *not* have to do with the value of species diversity or of biodiversity more generally.

Dworkin thinks that only a "few people... would think it important to engineer a
new bird species". He is correct to allow for some biogeoengineers – though probably many more than he is aware of – who are, in fact, quite enthusiastic about the
idea of creating new organisms to populate the planet, and to pursue this idea by a
number of means and for a number of purposes. But this observation stops short of
a more nearly complete truth, which is that the advocacy of most remaining biodiversity project advocates commits them to this position, even if they refuse to
endorse this implication of their advocacy. If, for example, the core value of nature
lies in biodiversity's ability to render ecosystem services, then consistency proscribes summarily rejecting the suggestion that perhaps the best way to achieve
this valuable service-rendering biodiversity is by designing it that way. Why should
this not include designing creatures that are engineered for superior robustness and
performance on the ecosystem job? No discernible principle of the biodiversity
project excepts these elements from the eligible set of design elements – which also
includes such elements as how fast the rivers run and at what volume, the regime
according to which fires are encouraged or discouraged, and the weeding out of
organisms that are believed to subvert service provision – that factor into how well
the ecosystem performs. Some scientists additionally believe that having more species in an ecosystem is tantamount to having more understudies capable of assuming the burden of an actor who might at some point fail to show up for a performance.[60]
Those who adopt this stance should, on similar grounds, espouse some additional
bolstering of the biodiverse minions.

Moreover, a figure-ground ambiguity in adding an organism to an environment
allows for the addition to be achieved either via the figure of the organism or via the
ground of its environment. Even biogeoengineers disinclined to define the figure of
a new species are often only too happy to accomplish much the same thing by redefining an existing species' environmental ground. Consider, for example, an animal

---

[60] This "prudential redundancy" argument goes hand-in-glove with precautionary arguments such
as those discussed in Sect. 6.3 (Biodiversity as service provider) – though, as I observe in that section, factual and epistemic requirements for application of such arguments to biodiversity are virtually never met.

that shows signs of succumbing to development projects, which obviously would not be *biodiversity* development projects. Suppose that these projects have drastically transformed the context in which the creature must struggle to make a life for itself. The creature's design does not take into account the consequently transformed requirements of living in its now-developed world. On the evidence of Sect. 7.2.3 (Home is where you make it – in situ, ex situ, and ecosystem-engineered), one need not be content with mere supposing. I have just described the urban peregrine falcons, which rely on nesting sites designed for them, whose fledglings require elevator rides back to their 33rd-story custom-made nest sites, whose ability to pass on their genes depends on a newfound ability to avoid regarding the glass walls of urban canyons as empty space into which they can fly, and whose hunting skills are commensurate with an environment in which their feathered prey have virtually no place to hide. And I have just described the razorback suckers, formerly river fish but now lake dwellers, which as larvae, are ferried to razorback kindergarten in order to compensate for their youthful inability to avoid becoming trout fodder.

For now, at least, biogeoengineers are mostly unenthusiastic about becoming bioengineers who create replacement creatures from whole cloth or via the patchwork methodologies that are the current state of the art – patching in a little DNA from here, a helpful gene from there. Rather, they favor engineering or re-engineering existing creatures' *surrounds*; and the new surrounds come with programs to assist the creature in navigating them. But in its human-designed home and with its human-assisted lifestyle, a creature inevitably behaves and functions in ways that significantly differ from those that typified the creature that formerly lived a very different life in a very different context. A new phenotype emerges; and eventually, a new genotype. Biogeoengineers therefore still create a new creature – by manipulating the ground for its survival rather than by directly manipulating its genome.

This story, which is central to the biogeoengineering narrative, reveals a second important respect in which Dworkin's succinct statement is incomplete with regard to valuing the natural world in general, and even with regard to any one particular, existing species: It matters, and it matters centrally, *how* people go about perpetuating a species, because some species-perpetuating behavior is at best questionable and perhaps even destructive of the kinds of human relationships that are genuinely valuing of nature.

Dworkin penned the just-quoted passage in 1993, just before the trickle of writing on the value of diversity became a torrent. It is unfortunate that the writers who unleashed that flood did not take Dworkin's sentiment to heart. A Martian wading through this literature might be forgiven for concluding that the most diehard earthly environmentalists yearn for a species-creating version of King Midas' touch. But that, I think, would be as much a curse for E.O. Wilson as was for Midas his ability to transform all that he touched into gold. I would go a step farther: The Midas biodiversity touch would remain accursed even if it gave us more services, more medicines, and more completely satisfied biophilic yearnings – insofar as the value of the natural world would become that must more inaccessible.

It is hard to account for the urge to defy Dworkin's common sense regarding the questionable value of humans laboring to produce new species. But I end my discussion of the biodiversity project's disvalue with one speculation, which traces back to the allure of John Locke's labor theory of value (Locke 1690, Chap. V "Of Property", Sect. 42):

> ... when any one hath computed, he will then see how much labour makes the far greatest part of the value of things we enjoy in this world: and the ground which produces the materials, is scarce to be reckoned in, as any, or at most, but a very small part of it; so little, that even amongst us, land that is left wholly to nature, that hath no improvement of pasturage, tillage, or planting, is called, as indeed it is, waste; and we shall find the benefit of it amount to little more than nothing.

Locke's view is that the value of things derives principally from the addition of human labor, and the required design, planning, and engineering (though Locke did not know that last word) that guide the labor. This view found its way into Adam Smith's cost-of-production theory, according to which the price of a good is, in the first instance, the labor "embodied" in it – a theory that David Ricardo, Karl Marx, and other classical political economists developed in their own distinctive ways.

Neoclassical economists spurned this labor theory of economic value, favoring an approach that values investment in mental states (desires or satisfied desires) through consumption of goods rather than physiological investment (of toil) in their production. However, I think it likely that Locke's basic idea still inhabits our consciousness; and perhaps it is impossible to resist. At the very least, it encourages us to think that anything, including the natural world, can profit from labor invested on behalf of good redesign. The core idea can be embraced without dragging along Locke's labor theory of economic value; it is even congenial to those wedded to the principles of neoclassical economics, which honor efficiency above all else.

The power of Locke's idea is that it elliptically folds into itself all four defining characteristics of the biodiversity project's state-based view of nature's value, which I described in Sect. 8.1.1 (Biodiversity and the state of nature). Couched in terms of those characteristics and compressed down to essentials, it is the idea that the primary valuing relationship of people to nature is one in which their labor creates "the far greatest part of the value of things we enjoy in the [natural] world"; and this labor more specifically takes the form of projects that evaluate each piece of nature, which by itself "hath no improvement" until by labor and design its place in the scheme of things is calculated so as to ensure its contribution to a "good" or "better" state of nature. This Lockean spirit inhabits the soul of the biodiversity project and most all current-day conservation efforts. Perhaps I should say "haunts their soul". For this poltergeist renders them startlingly impotent when struggling to provide a credible account for why anyone should think that their "improvements" yield a world with greater natural value. In the next section, I argue that this abiding spirit of the biodiversity project doesn't merely cause it to miss the bull's-eye of natural value by a small or even large margin. Rather, it destroys the entire target.

8.2 Natural Value as Uniquely Valuable Human Relationships

State-based views of natural value notably include biodiversity-based views. Their collective failures should be a cue for any alternative proposal to unburden itself by stepping outside a state-of-nature framework. It seems that there is little to lose and much to gain by abandoning what seem like fruitless attempts to view nature's value as, for example, deriving from its constitution as a collection of maximally biodiverse stuff, or as a warehouse of economically valuable bio-parts, or as a felicitous arrangement of parts that together perform certain "functions", or as an arrangement of stuff that is evocative of some previous state of the world, or as some state of health or vibrancy that cannot be specified except by way of broken analogies. In fact, I propose to take these multiple points of failure as guideposts for staking out an alternative proposal.

Taking additional cues from the four key elements that Sect. 8.1.1 identified in these kinds of frameworks, I propose to push past the notion that some states of the natural world are good or better than others by virtue of exemplifying properties characterized in terms of biodiversity or in terms of any other atom or atoms of a fracturing and fragmenting analysis. This means rejecting the methodology of determining the composition and shape of a good natural world, and only later, trying to figure out why people should hold such a state of nature in such high regard as to be part of the set of attitudes that guide action. It means bringing a primary focus onto what kinds of human attitudes and characteristic behaviors towards the natural world can be broadly defended as bringing people into an appropriately valuing relationship with it.

A theory of natural value that actually does this would be a seismic shift, indeed – requiring sufficient energy to decisively cut loose from the irresolvable evaluative conundrums that continually ensnare the biodiversity project. Such a shift would be a boon; but it would not suffice for a satisfactory theory of natural value. That would require the satisfaction of four other requirements, which biodiversity-based theories help to make conspicuous by their consistent failure to satisfy them:

First, one should require a theory that is guided by a few simple, clearly delimited, and general principles, which people might be able to endorse on clear and solid grounds. This would stand in dramatic contrast to the muddy and shifting ends that mire the biodiversity project in a normative morass.

Second, it should be a theory based on principles that elucidate rather than obscure why what I shall call the "core" value of the natural world is so unique that it is not possessed by anything else in the pantheon of valuable things. In other words, the theory should avoid pushing natural value into a value sea of faceless anonymity. This consideration alone makes economic valuations suspect for their evaluative transformation of the natural world into a stock of commodified parts that are (economically) indistinguishable from most other entities whose value is largely defined by the price they command in real or imagined trade. The core value of the natural world really is of an entirely different genre from that of mass-produced plastic toys. A theory of natural value must account for that.

Third, a viable theory of natural value should be grounded in principles that show why the value of nature is not subject to human whims and preferences. That is because whims and preferences might not reflect values to which people are truly committed; and even when they do reflect such values, they still might not be worthy of satisfaction. The principles of a viable theory of nature's value should suggest why this value should be a constantly present consideration for individual and collective decisions about how to arrange our human lives. This, once again, should make the marketplace an entirely unsuitable evaluative tool for nature. For when reduced to a collection of tradable commodities, there is no reason not to think that they might not all be traded away tomorrow. An acceptable theory of natural value must show how and why the value of the natural world has a kind of resiliency that is the antithesis of the reasoning that underlies market transactions.

And fourth, a viable account of the value of the natural world should adequately account for why anybody should care about it. The account should give us some reason to understand why people both individually and collectively should and would *want* to take nature's value under consideration in making the most important decisions about arranging their lives. A natural world constructed on such a principle as the "traveling evolutionary biologist problem", which seeks to prune the tree of life for maximum total phylogenetic traveling distance, is impotent in this regard.

All of this, I believe, can be achieved by a theory of natural value, which I call "appropriate fit".[61] It is a theory of nature's value, which most saliently differs from the dominant one in starting with people. It reflects broadly how people can best "fit" into the world in which they find themselves. More specifically, it reflects how people can best "fit" in with respect to some part of the world that is stringently regarded as "given". According to the theory of appropriate fit, it is by virtue of this latter kind of fitting in to some part of the world that that part of the world can be (and is) regarded as "the natural world". This conception stands on its head the customary formulation, which asks first and sometimes only how the entire world, without exception, can be pressed and formed into something that fits in with human needs and desires. Finally, the theory of appropriate fit reflects the kind of relationship – a valuing relationship – with nature that this "fitting in" entails.

The notion of "natural world" that I have in mind for consideration of persons' appropriate fit in the world is critical. It should be apparent by now that "natural world" is not and cannot be defined by means of a state description, including one that is couched in terms of biodiversity. It is not and cannot be characterized by any particular set of attributes, nor by a set of properties that obtain when certain qualities achieve specified levels. It is not and cannot be defined by how closely it approximates some ideal state. It does not help to supplement the characterization of a state with an account of the history that led to it. Nor can "the natural world" be defined by reference to good prudence – that is, in terms of a world whose state

---

[61] The notion of "fit" introduced here, including some of my vocabulary for explicating it, owes much to David Schmidtz's explanation – in Schmidtz (2008b) – of why altruism might be rational.

saliently includes the stuff that satisfies human desires and the stuff from which people can live, while perhaps not including the stuff that people can live without or even detest.[62]

Rather, appropriate fit requires thinking about the natural world as that part of the world to which people relate in a very particular way that is resilient and also unique to this particular relationship. It might initially seem that I prejudice the issue (a complaint that I level against many others) by pasting the normatively suggestive label "*appropriate* fit" on this type of relating. But in fact, this way of relating encapsulates the core of what makes the natural world natural. And I do not merely presume that nature, so defined, is valuable. I offer reasons for that proposition. I try to suggest why it is that people are not foolish to keep some portions of the world "natural" by relating to those portions in the resilient and unique way that I shall describe. I argue that they are not deluding themselves into thinking that this is a valuing relationship rather than some sort of capricious preference. I seek to justify the "appropriate" in the label "appropriate fit" by suggesting how the relationship underlying it is both enabling and constitutive of an appropriate fit of people in the world. What makes it appropriate – and valuable – is how richly it integrates human lives into the world at large. It should not be surprising to find that the attitudes and ways of behaving that embody natural value, as I understand it, are the antithesis of the attitudes and behavior that drive scientific biogeoengineering, or management, or marketing, or redemptive offerings, or more generally, molding nature into any chosen state – no matter what rationale is offered for that state's supposed evaluative preeminence.

It is to this notion of natural value as appropriate fit that I now turn. I expand on this introductory characterization and tease out some of its salient implications – including what it implies for conservation and restoration, as those activities are now commonly understood and practiced.

The reader should be forewarned: What I propose does not fall neatly into any single moral tradition, though it has some affinity to certain strains of thought in virtue ethics. Also, some of its most important implications turn currently accepted notions of conservation and environmentalism more generally on their head. But I think that these departures from traditional compartments of moral inquiry and of environmental "improvement" are a very good thing, because thinking within the confines of these old categories has been anything but fruitful.

8.2.1 Natural Value as Appropriate Fit

If one adopts the biodiversity project's conceptualization of nature primarily as a catalog of biota and biota-related entities, then one's concerns for nature are likely

---

[62] I use the locutions "natural world" and "nature" more or less interchangeably. However, I tend to favor "the natural world" to convey some sense of a location, a place or (somewhat more abstractly) a domain set apart from human endeavors and enterprise. I tend to use "nature" even more abstractly in connection with the attitudes and behaviors that are the basis for this setting apart.

to be expressed as concerns for the catalog's size, its contents, and the variety of goods it offers. If one also leans towards an economic evaluative overlay, then one will also be concerned to determine which items are good sellers, which ones sell poorly, and what levels of inventory each bio-part thereby merits. This extraordinarily prevalent view of nature and its value is also extraordinarily narrow and myopic.

A theory of appropriate fit radically shifts and broadens the perspective. It stands back to claim a vantage point with a more expansive view – sufficiently wide-angled to include in the foreground people capable of valuing nature, and at sufficient remove to take in, alongside those people, a natural world whose individual bio-parts blur into each other. This more panoramic view affords space in which it is possible to ask a key question that is not asked because not really ask-able within the biodiversity project. The question is: How can humans best fit into a world that includes that blur of nonhuman elements, but that is also a world largely and increasingly human-designed, human-developed, and human-managed? It is vital that, in considering this question, it is not presupposed, as popular conceptions of human responsibility for nature (examined in Sect. 8.1.2) suppose, that the world is a monolith, all of which is understood to be a domain for human-designed, human-developed, and human-managed projects. On this conception, "undesigned", "undeveloped", and "unmanaged" merely means "not yet". Rather, one must hold out the possibility of a domain in the world – some part of it – that can be conceived as outside the human-designed, human-developed, and human-managed in a sense that makes those attributes not contingent on shifting needs, desires, or conditions.

At best, as I have tried to convey, the biodiversity project is capable of essentially standing on its head this question about how humans can fit in with that which is not part of human enterprise. It asks: How can the natural world be subsumed into the world of human enterprise? How can it best be integrated into and made part of the whirlwind of development? What is the best way, for example, to pursue the development of biodiversity as part and parcel of an all-encompassing and literally, all-consuming general development project?

When regard for the natural world is framed by the answers to these kinds of questions, it becomes, at best, a kind of afterthought – a modest coda appended to, but ultimately part of, the grand symphony of human enterprise, which otherwise plays on with little or no consideration of nature as anything exceptional. Within this enterprise, nature is one instrument among many that together sound the theme of development… and then more development. That is what biodiversity development comes down to. To paraphrase Erle Ellis, it is a set of projects to make the environment what we desire, no matter what that desire might be. It includes projects that seek to achieve or more nearly approximate some ideal state of nature such as an historical benchmark, a tree of life pruned for "evolutionary potential", or a state of (ecosystem) health. And it includes projects that even more directly assimilate themselves into the world of economic development – where nature, as "natural capital", developed and maintained like any other form of capital, is fuel for the economic engine.

This orthodoxy presumes a world in which human enterprise has standing permission to expand into every domain. Some interstices – which might yet be squeezed out of existence, should the logic of development demand that – remain.

The salient concern about nature then shrinks down to a tiny, narrow, and mean extension of the greater enterprise, comprising a marginalized set of projects that poke, prod, and manipulate those interstices to achieve ends that please the designers and managers, but that (on the evidence of Sect. 8.1.3) have no reasonable or defensible justification. I am speaking very abstractly here. But this is essentially what conservation, restoration, and ecosystem management – as these activities are now understood – come down to. It is hard to find a conservation issue that escapes this orbit. For example: Having blockaded a river with multiple dams with fish-pureeing turbines, conservation is seen to be a matter of contriving to help and even transport salmonids up and downstream, so that people can continue to have tasty dinners or so that environmentalists can be satisfied that these creatures persist or subsist at certain GPS coordinates. Whatever the importance of these moves might be, I believe that it has little or nothing to do with the conservation of the natural world. That is because it has little or nothing to do with valuing nature as anything more than, in this case, stuff that can be developed into a foodstuff.

Consider, by contrast, a theory of natural value that starts by examining the attitudes and behavior of people and how these can be said to fit appropriately – or inappropriately – into the world at large. Consider starting without the presumption that the attitudes and behavior that are appropriate for pursuing human enterprise are everywhere and every time – in other words, universally – appropriate. Remarkably, merely taking this different starting point makes it possible to trace out a very different picture of nature's value.

The different picture emerges from trying to understand something about fitting humans and their activities into at least *some* portions of the world that are not continually or even *potentially* eyed as development opportunities. Its framing requires that some portions of the world be regarded as valuable just for what they "naturally" are and just for what they "naturally" do – no matter what led to their current state and no matter what we humans might desire their state to be. It emphatically requires that these portions of the world be regarded as outside of and independent of the human economy and without regard for any consequent benefit that people might enjoy or deprivations that they might suffer. The contrasting picture emerges from asking: How should we, as human animals, fit into a world, some portions of which do whatever they do without human contrivance, which are "off limits" as objects of human intervention or re-creation, and which are resiliently (or as I shall say, in a "modally robust" or "modally demanding" way) regarded as "given"?[63]

Asking this question is self-confronting. The fact that it is almost never asked is evidence that it is incredibly difficult to confront ourselves in this way. And the difficulty in answering the question is at least as great as the difficulty of asking it. It is certainly easier to ask biodiversity questions, to run biodiversity numbers, and to amass biodiversity analyses. It is easier to ask about ecosystem health and ecosystem

---

[63] I use "modal robustness", "modal demandingness", and "modal resilience" interchangeably. In my usage, which I shall shortly explicate, they generalize the notion of "resiliency", which I discussed in Sect. 8.1.3 (The ends of the biodiversity project). There, I identified the non-resiliency of the biodiversity project's values as one of its fundamental shortcomings.

integrity because no one would think that they stand against the health and integrity of anything. And it is easier yet to translate these various questions into economic ones, run the economic numbers, and amass the economic analyses, which show that some small number of the world's bio-parts are, after sufficient tinkering, worth keeping around for now. In fact, it is plausible to think that this relative ease might partly account for the otherwise unaccountable popularity of such approaches, which seem so grossly alien to a core of value built upon an entirely different and even antithetical way of regarding the natural world.

People are instinctively put off by confrontation; they are even more put off by discomfiting difficulties. But I take this reaction to be an encouraging sign that my question is well aimed. For once, we have a question whose asking and whose answering truly reflects the incredible difficulty of getting a handle on why, exactly, the natural world is valuable. One cannot avoid realizing that the answer rides on the very choppy and expansive waters of what it means to be human and to have appropriately human relationships to that which is not within the realm of human design, creation, or management. The answer rides a difficult-to-characterize and an even more difficult-to-justify notion of "achieving an appropriate fit in the world" for each of us individually and for all us as together in our various social collectives.

These difficulties notwithstanding, the value of the natural world is centrally founded precisely in attitudes and behaviors that enter into and are formative of relationships to nature that achieve this fit. The "stuff" of natural value is a sense of integrating ourselves into the world in a way that does not involve changing everything or even anything in our surrounds. It is a matter of changing ourselves. It is about finding a value in living with and accommodating a natural part of the world that operates according to rules not of our making, and whose operation neither demands nor could possibly be helped by human enforcement of its rules, because the natural world has no ends of its own, let alone an end that it has an interest in achieving. It is about finding a human place in a part of the world that is radically detached from human rules, human ends, and human interests; and where an anthropomorphizing projection of these vital human elements is worse than meaningless: It is at once dehumanizing and denaturizing in sense that, by conflating nature with persons, it effaces the distinguishing qualities of both – qualities that underlie the distinctive roles required for a valuing relationship. Truly valuing nature entails functioning well within given structures rather than (always) restructuring.[64] It is a realization that something of major significance in our human life derives from understanding its locus within something bigger that is outside ourselves and our individual and collective projects. In the case of the natural world, this is, must be, and *ought to be*, something that is and steadfastly remains outside human invention and enterprise. As well, it is and ought to be outside the rules, ends, and interests according to which human projects are pursued.

---

[64] I do not claim to have a formula that determines when restructuring is appropriate and when it is not. In fact, I suspect that operationalizing that balance is not possible. Of course, it might be possible to find and formulate some guiding principles and that might be an interesting project. But that is not my major concern, which is the far simpler one of pleading the case for at least thinking in these terms.

I said that natural value is *centrally* founded in a relationship with the natural world that provides an appropriate fit of persons in the world at large. I included that qualification to leave room for acknowledging that there are other valuing relationships that persons have to nature. However, I do wish to say that the kind of relating to that natural world that places nature outside of characteristic human enterprise constitutes what I have already informally called the "core" of natural value. That is, it both grounds and bounds other significant valuing relationships that people have to nature. Insofar as other forms of relating require the core for their grounding and bounding, they are derivative. I shall have more to say about derivative natural value later on.

The ethical keyword "ought" has just made its way into my discussion. In this introduction to appropriate fit, understanding the force of the "ought" involved in attitudes of valuing the natural world is crucial. These attitudes are, in philosopher Philip Pettit's words (Pettit 2008), "modally demanding"[65] or (Pettit 2011) "modally robust": In order for people to enjoy the core value natural world *at all*, they must resolutely place some portion of nature outside the domain of continual, casual, and even *possible* consideration as yet one more opportunity to launch a project to make it into an object of their desire. The supposed "requirement" for exceptions must be almost entirely waived. Again, I am speaking abstractly, though examples abound. Think about managing fires in U.S. National Parks, or managing dangerous predators in them for the safety of those near or in the parks, or managing river flows so that they do not flood structures in their flood plains. Such routinely sanctioned practices, which make exceptions to accommodate human projects, largely fail to meet the test of modal robustness.

Pettit elaborates the principle of modal demandingness in connection with the nature of (human) freedom. He drives home the truth that I do not truly enjoy the good of freedom if I merely happen to choose options that you (who are in a position to constrain me) are inclined to also favor, while rejecting options whose choice you would interfere with. For surely I am not free – even if I choose according to my preferences – when my ability to satisfy those preferences critically depends on their being adapted to yours.

Nor can I enjoy the good of freedom even if you are not inclined to interfere with my choice of *any* option – when your disinclination to interfere comes by virtue of my ingratiating myself to you. For then, my ability to exercise free choice is contingent on my continued ingratiation with you, which promotes and sustains your disposition to refrain from interfering. Thus, despite the fact that in this case, too, I might satisfy my choice preferences in the actual world, the dependence of this on your contingent beneficence excludes this as a case where I can truly be said to enjoy the good of freedom.

In sum, my access to the good of freedom requires that I be able to exercise my preferences, not just in the *actual* world in which they happen to match yours or in which you are disposed to beneficence with respect to my choices, but in the many

---

[65] In this paper, Pettit develops the notion of modal demandingness in the context of human freedom. Pettit (2011) provides a further elaboration.

possible worlds in which my choice preferences vary (and sometimes conflict with yours), as well as in the many possible worlds in which my degree of ingratiation with you and your disposition to interfere varies. Furthermore, the *likelihood* of realizing any of these worlds, characterized by variation along these two dimensions, is not a relevant consideration. Even the *least* likely of these possible worlds has as much bearing on my freedom as the most likely (and the actual) state of the world.[66]

I would suggest that enjoying the good of the core of natural value places quite similar modal demands on human attitudes and behavior with respect to nature. Like that person who is in a position to deprive me of my freedom, persons generally – as capable actors in and on the nonhuman world – are in a position to interfere, shape, mold, and constrain nature in ways that they find suitable. It is their modally stringent resolve to refrain from exercising these capabilities that realizes the value of the natural world.

However, I must be careful to recognize a salient difference between the good of human freedom and the good of natural value. In the simplest kind of relationship that defines the good of true freedom, each of the two relata is a person. In contrast, the relationship that embodies natural value places a person (or a collective of persons) at just one of its two poles. As a consequence of this difference, it would be an unfortunate mistake to interpret the choice to interfere and constrain as a matter of depriving nature of her freedom. That all-too-common move (especially, as Sect. 8.1.5 noted, in virtue ethics) requires anthropomorphizing the natural world and imbuing it with human-like interests – including an interest in acting freely to pursue its own interests as an independent agent. But as I remarked in that earlier discussion (mostly in connection with ecosystem health), there is no credible reason to believe that nature has any ends at all, let alone interests in achieving ends. Nature therefore cannot have the freedom, in the sense that relates to pursuing ends, which is the salient moral consideration for persons.

So what is saliently at stake with nature is *not* the freedom or autonomy of some nonhuman entity or collection of nonhuman entities. Rather, what is at stake is a good that can accrue to those persons who are in a position to interfere with the natural world – by making it part of a development, conservation, restoration, or management project – but who, in a modally robust way, choose not to do so. This good is the core of natural value. It is realized only because persons *are* in a position to develop, conserve, restore, and manage, but instead exercise restraint. In other words, they "let it be". Moreover, the restraint must be modally robust in the sense that it must not just be exercised in the actual world and just in cases where the cost benefit analysis or the biodiversity calculation shows the justification for a project to be weak on economic grounds, on biodiversity grounds, on the grounds of ecosystem health, or any other similar grounds. Rather, the disposition to not interfere

---

[66] This is a synopsis of the argument in Pettit (2008), which Pettit himself summarizes on pp. 217–218 of that paper.

must extend over many possible worlds – across the spectrum from likely to highly *un*likely – in which these calculations yield economic or biodiversity numbers (whatever they might mean) that vary from minimal (or negative) to very large. It is in this way that natural value is embodied in a way – a modally robust way – that mimics the value of human freedom.

If such a value for the natural world exists, then it is unequivocally unique, for it requires a relationship that is unique among value-forming relationships. Uniqueness is hard to argue for because it requires a proof of the nonexistence of something similar in relevant respects. I will not attempt this here, but instead challenge the reader to try to find some other entity, consideration for which is grounded in a "letting it be", but a letting be that is not coherently regarded to be in that entity's interest. That kind of "letting it be", I think, represents a unique kind of valuing with regard to the natural world.

The major lines of reasoning that dominate the biodiversity project founder centrally because they run completely walled off from any awareness of this sort of unique, modally demanding relationship that, I believe, people can and ought to have to some part of the world in which they live.[67] The very framework in which they operate leaves out this relationship, which I am suggesting constitutes the core of natural value. The path to bringing this relationship back into the frame requires serious consideration of what it means to be human in the world today; but that path, too, is completely cut off. Truly serious consideration of what it means to be human is not achieved by adding up the strengths of human desires (which is the main business of economics). It is not a matter of counting species, defining "ecosystem health" (if that is possible), or solving the traveling evolutionary biologist problem. Nor can the most sophisticated biodiversity index show what it means to be human. None of these exercises can show what it means for persons to find their appropriate fit in the world in which the counted species actually do exist or some particular biodiversity index value actually does obtain. This, then, is the deep reason why the core value of the natural world does not lie in its properties of biodiversity, in the identity of bio-parts kept in stock, or in anything else that one could imagine as world-state-characterizing ciphers that stand in for the natural world's "what is", and that therefore could, in principle, be biogeoengineered into existence.

The dispensability of the details of the "what is" for nature's core value does not, in any way, discount its value in helping us to understand something about the framework of the world in which people must find their appropriate fit. But a theory of natural value according to appropriate fit makes it clear that those details do not affect what makes persons' way of relating to nature a *valuing* way of relating.

---

[67] To this point in my reflection on appropriate fit, I have tried to characterize its essential characteristics in terms of a possible way of relating to the natural world, but I do not claim to have *demonstrated* that this relationship is one that we can and ought to have. In fact, it is fair to say that I never supply a demonstration of this proposition in anything approaching the sense of "a crisp syllogism from indisputable premises". However, I hope that it will appear more and more plausible with each incremental observation, that appropriate fit compellingly captures a critical aspect of how people can live well on the earth, as it now exists.

Suppose that the "what is" in the world as we know it were replaced wholesale by another collection of organisms, creatures, and peculiar environments as the result of the unfolding of another version of biogeohistory. Suppose that in that alternative world, the organisms and environments were not just altogether different but considerably less varied than those that actually came to be in our actual world. If by some miracle *H. sapiens* in more or less our current form happened to be among this strange entourage (strange, at least, to actual us in the actual world), then this would have little or no effect on the value of nature in that quite different and considerably less biodiverse world. That is because, for the people in that alternative world, the value of *their* (quite different) natural world would still be derived from their understanding of how, as people, they fit or *should* fit into that rather different world in which they happened to find themselves. Natural value would derive for them in precisely the way that it does for us – by way of an understanding of how a good human life is, in part, a matter of finding an appropriate way of fitting in with what is naturally "given".

In teasing out some of the salient implications of appropriate fit in Sect. 8.2.7.2 (Implications for action, social policy, and conservation management), I will go one step further. I will suggest that the strange, fantastical, counterfactual world of the just-related thought experiment can be replaced by places in our actual world that are equally strange and fantastical; and often by virtue of a history of human impingements. At least, they are equally strange in the sense of dramatically deviating from what the natural world "should" look like according to orthodox, state-based conceptions of nature.

8.2.2   *Difficulties with and Limitations of Appropriate Fit*

Section 8.2.4 continues my mainline elaboration of the notion of "appropriate fit" as the core basis for valuable human relationships with the natural world. I wish to place what I have to say there in the context of some hazards and limitations in this approach to natural value. I turn to these difficulties here and I do so by taking this perspective: Every approach will have its vulnerabilities; a theory of natural value according to appropriate fit is no exception. Assessing this theory should therefore partly be a matter of assessing its vulnerabilities relative to those of the most prominent alternatives, which attempt to locate value in the ends of the biodiversity project (Sect. 8.1.3, The ends of the biodiversity project).

8.2.2.1   Some Difficulties

1. **There is no science, calculus, or economics of appropriate fit.**

Appropriate fit does not admit of scientific measurement. It is not the sort of thing that one would think of putting up for bid on eBay. The appropriate fit of an individual person or persons generally in the world is simply not something that could possibly be part of any economy, let alone some "New Economy of Nature". Appropriate fit shuns the means by which scientists and economists seek to take the measure of the world. As a consequence, some of those who pursue these disciplines and others in a culture that yearns to see "the proof in the numbers" can be expected to reflexively shun this conception of natural value. This warrants some reflection on the place of science and economics in defining and validating norms.

First, regarding science: At its best, it provides a special epistemic window into the constitution and properties of the natural world. It structures our knowledge in ways that generally permit a more accurate and acute awareness of the circumstances of the world in which we live, and in which we make important choices for ourselves individually and collectively. If one is not a metaethical naturalist (see Sect. 8.1.5), then one does not believe that natural value arises from the exemplification of certain structured sets of scientifically ascertainable properties – including, for example, some more or less complexly structured property of being biodiverse. On this nonnaturalist view, science appears to have little to say about the value of the natural world, beyond providing some fine-grained context for morally inflected decisions, which attempts to take proper account of values otherwise determined.

Of course, nonnaturalists might eventually determine empirically that the set of all entities that have natural value happens to be coextensive with things that exemplify some interesting, scientifically assessable property or set of properties. They might discover that, as amazing it might seem, all entities that (on other grounds) have natural value have a certain, complex chemical signature; and no entity that (on other grounds) lacks natural value has this signature. But even if this were so, the scientific discovery of this fortuitous correlation would be of no help in fingering the valuable entities in the first place. Furthermore, even after discovering it, the role of science would be restricted to detecting the presence of this natural property. If no exceptions were ever encountered, its presence would constitute increasingly strong *prima facie* evidence that exemplifying entities were valuable. But on the continued supposition that this value-correlated property is not truly value-conferring, there would always be an open question about this. The final judgment about what was valuable in nature and what not would still always require appeal to considerations outside of science.

A metaethical naturalist might see a greater role for science in evaluation. But Moore's open question[68] would still lurk, as would the question of why anyone should care about the exemplification of some natural properties but not others. Another way to put this is: No scientific description – indeed no matters of fact in and of themselves – can ever be the sole basis for ascribing value. Even the metaethical

---

[68] Recall from Sect. 8.1.3 (The ends of the biodiversity project) Moore's suggestion that the ascription of value to some entity on the basis of its possessing some structured set of natural properties is always open to question. In contrast, one can legitimately settle questions of fact – for example about whether or not a certain striped, odd-toed ungulate is a zebra – by establishing that it possesses a certain set of properties including its distinctively patterned black-and-white coat and a single toe.

naturalist is obligated to find credible principles for discriminating between, on the one hand, natural properties that endow the entities that exemplify them with value, and on the other hand, those natural properties that do not have this value-conferring capacity. Reasons are required to convince us that these principles are worthy of guiding valuations, which in turn, guide morally weighty decisions. A recitation of numbers cannot serve this purpose; it only leads back to the question of what reason we have to believe that the numbers matter.

Furthermore, science has no special means or authority for answering the questions that concern the value of its objects of investigation. In fact, as Sect. 8.1.7.3 (Conflicts of interest and institutional bias and inertia) explains, the professional biases of biologists likely put them in a far worse position than most ordinary citizens for making unbiased value judgments about nature, despite their best efforts to avoid bias.[69]

In short, appropriate fit suffers little if at all for not being a "scientific" theory. Unfortunately, this salient observation is obscured by modern veneration of science, which mistakes the special epistemic status that science legitimately enjoys for a power – which it entirely lacks – of a universal and virtually indisputable validator. This erroneous view is really all that sustains the perception that a theory of natural value founded on, and emblemized by, (for example) scientifically assessable biodiversity has some special status as a "scientific theory of value". Biodiversity numbers – mash-ups of species counts and abundances, or phylogenetic distances, for example – might summarize something interesting about a natural system's characteristics. As scalar representations, they have special appeal for imposing a total ordering, which makes possible universal application of "more" and "less". But despite the seductive simplicity of this comparative language, it has nothing to do with value. The biodiversity numbers provide no greater insight into nature's value than the number of ecologists wondering about it.

Nor does appropriate fit suffer by not going by the economic numbers. The evaluative anchor of economics is "value in exchange". This captures the "what you can get for it" kind of value. But it fails to command moral attention on its own. I have previously touched on reasons for this – reasons connected with the potentially unrespectable provenance of many preferences that underlie this kind of value, as

---

[69] The misplaced regard for science as a privileged authority on value occurs in several of the most important moral debates of our time. Without exception, the call for scientific adjudication distracts from the real moral and political issues at stake. The political philosopher and constitutional scholar Ronald Dworkin (1993) brilliantly examines this phenomenon in the debate on abortion. He remarks that debate is derailed by futile discussions about when a fetus becomes a person – as if science were capable of defining the instant at which a human fetus suddenly has specifically human interests of its own or is suddenly imbued with the rights that are normally granted to competent adults. To do that, science would have to have some special insight into the question of what constitutes a person in the sense of a holder of such a right as the right to make her own choices about how to live her life, the right to some reasonable chance for a healthy life, and the right not to have that life terminated against her will. But this is a scientifically opaque question and, most likely, a distraction from the more central question of whether or in what respects cutting short a human life in the form of a fetus reflects on the value of human life more generally.

well as the failure to provide a principled basis for excluding preferences that are unworthy of being satisfied. It is possible to reframe and freshly illuminate this latter failing by observing that preferences are highly promiscuous in the compass of the interests that they serve. They ascribe values to such things as preferred colors and more generally, to what Ronald Dworkin would call "experiential interests".[70] The satisfaction of such interests undeniably provides pleasure. But a decision regarding the choice of a color in a purchasing decision or (to use one of Dworkin's examples) a sport to follow has little if any moral heft. Choosing to pay a premium for a paint job on a new car that gives me pleasure might be worth its cost to me. Spending time to watch a baseball game instead of scheming to earn more money might also be worth the cost. But decisions such as these do not rise to the level of a moral choice. Furthermore, even where its domain of valuation extends to an entity of high moral worth, that worth can be, and often is, largely or entirely independent of the entity's "what you can get for it" value. Respect for another person's freedom is not even marginally qualified by, and certainly not equivalent to, that person's value-in-exchange as a slave; nor is a person's value-in-labor as, say, a baseball player, relevant to consideration of her freedom.

One can understand how both science and economics can seem to satisfy a yearning for a science of numbers that, with the reassurance of their apparent precision, tell exactly how much the natural world is worth to us. But the comfort in numbers is a false comfort – one that relies on mistaking "number" for "value" and "certainty", and on mistaking "bigger" for "better". These gross errors evidence something in the modern mind, which struggles mightily against acknowledging that some value propositions are not amenable to scientific demonstration; and nor is it generally appropriate to consider moral tradeoffs as exercises in marketplace trades or development efficiency.[71]

In the end, not even armies of trained ecologists can ever tell us how to be human in a natural world. At best, scientists can help us understand, in greater detail than we otherwise could, the consequences of our actions and behaviors in terms of consequent states of the world. But as I have been insisting, states of the world are not, in and of themselves, revealing of nature's value, because the value of nature does not derive from its realizing some particular state rather than another.

Nor can armies of economists ever tell us the value of what humans value most. This seems so obvious; yet it is also alien to current-day orthodoxies of public and

---

[70] The phrase "experiential interest" is from Dworkin (1993, 201ff.), who sets them apart from "critical interests".

[71] My point is in no regard compromised by the absence of real markets for some (environmental) goods. As I have previously remarked, when real markets, with real transactions involving a good do not exist, economic evaluation readily resorts to fictional markets. These fictions *imagine* transactions via such indirect evaluation methods as hedonic evaluation and declarations of willingness to pay (WTP) and willingness to accept (WTA) in contingent evaluation surveys. My critical point is indifferent to this routine substitution of fantasy for reality in economic evaluations. Nor is my point qualified by economics' imaginative creation of hybrid real-and-imaginary markets in which the real markets are "corrected for externalities".

even private valuation, which strong currents of modern thought sweep into prominence. In this context, a little reflection on friendship can help. The value of a friend is caught up in the value of a certain kind of relating between two persons, which does not essentially depend on the stuff that a friend might offer us or on the services that she might render. A friendship does not become less valuable when, for example, one's friend is ill and less capable of delivering such benefits. To think of a "friend" in terms that vary with the efficiency with which she provides benefits is to enter into *some* relationship or other, but not one that bears much resemblance to friendship. Nor can the marketplace help in her evaluation. The value of a friend is not measured by how high eBay bidding for her (or her services) would go.[72]

Of course, the value of the natural world is quite different from the value of a friend. So the way that one enters into and serves valuing relationships of these two very different types of entity also differs. But the example of friendship and its value opens the door to a clearer view of how valuing something need not be a matter of "the numbers" – scientific or economic. And more strikingly, the example alerts us to the possibility that, for some significant valuing relationships, an eager and close consideration of the rendering of stuff and services to the valuing subject can be largely irrelevant, and sometimes even antithetical to the relationship's value.

2. Appropriate fit can be argued from the other side.

Those who find little of value in nature *except* in its stuff and services can easily usurp the concept of "fit" and bend it around their own views. Much the same can be said about those who regard nature's stuff and services as the gold standard of its value. This is precisely why the philosophy of such corporate extractors of natural wealth as Rio Tinto are in surprisingly close accord with the philosophy of Natural Capitalism, which receives expression in such organizations as The Nature Conservancy.

The question of "how we humans appropriately fit" into the natural world of the planet we inhabit can be tossed up in the air, spun off the fingertips like a cut fastball, and batted over the left field fence by the other team. Most obviously, one should anticipate claims to the effect that people simply fit into the world at large as designers, creators, manipulators, extractors, and managers of all their surrounds. Those who believe this would say that people most appropriately fit into the world essentially as beavers on steroids; that they are at their best precisely when they do *not* take their surrounds as given; that it is intrinsically and emblematically human for people to view their natural surrounds primarily as a given (and perhaps God-given), ready-to-hand hammer, just waiting to be picked up to bang in a world full of unsunken nails; or that people are at their finest when a world of inconvenience and woes sparks their re-imagining and transformation of it into increasing degrees of suitability for an easier life.

---

[72] This is a freely rendered gloss on what Aristotle (350 B.C.E. b, VIII.4) has to say about friendship in his *Nichomachean Ethics*. I shall return to his conception of friendship shortly.

This challenge, I believe, cuts to the heart of the question of where natural value truly lies. Like all deep moral challenges, this one demands consideration of how people best live in the world in which they find themselves. It confronts my contention that the best way of being human requires holding apart some part of the world as uniquely and resiliently valuable by virtue of its peculiar role in circumscribing human enterprise and thereby shaping the human condition. An environmentalist intuition – that something *is* worth holding apart – is justified, or not, insofar as it can be grounded and sustained by a convincing account of the "appropriate fit of people in the natural world" in which a modally robust "letting it be" is central. This challenge to it requires that one argue, as I do, that this is a far more convincing account of a valuing relationship to nature than regarding it as beaver bait.

In stark contrast, an economic test of nature's value barely skims the superficies. Stripping from the natural world all that is distinctive about how people relate to it, "The New Economy of Nature" re-imagines the natural world as an inventory of bio-parts, which represents a capital investment much like any other. The test is then whether and to what degree one can prune and structure this inventory to earn its place – by literally paying its way – within the domain of human enterprise. The tragedy of this test's extraordinary popularity is that, by embracing an understanding of humanity as a race of steroidal beavers, it already concedes the anti-environmentalist answer to the true test.

Of course, I don't concede this anti-environmentalist rendering of a theory of appropriate fit. I push back by sketching three immediate responses to its challenge. First, while the steroidal beaver conception of humanness might seem initially plausible, modest reflection shows it to be, at the very least, far too narrow. It is true that great deal of meaning in a human life comes from pursuing our very human kinds of projects. But it is a gross error to think that "pursuing meaningful projects" means "endless and frenetic bending of everything in the world around us". Even when the goal is "an environment that is [most] desirable for all of us", it is hard to conceive of a plausible defense for achieving this by means of developing every last square meter of turf and appropriating every last cubic meter of water on the planet.[73]

I would submit that the difficulty of defending such a position is that human projects derive some part of their special meaning as an expression of human striving and creativity insofar as they are set off from that which is *not* such a project. If everywhere we looked, we saw ourselves reflected back as designers, creators, and

---

[73] This observation has a venerable lineage. Despite the ultimate failure of pre-biodiversity-era environmental advocates to make "wilderness" the cornerstone of nature's value, many eloquent pro-wilderness arguments hit on a similar point. Wallace Stegner did so in his 1960 letter (http://wilderness.org/content/wilderness-letter) advocating for what became the 1964 Wilderness Act. In it, he famously noted that human beings need not just clean air and clean streams, not just certain resources drawn from nature, and not even just recreational opportunities, but also a "geography of hope". That, he said, will be carved up and lost with the carving up of the world to serve our more physical and pragmatic needs and desires.

managers; if we could not look at anything or at any place without seeing in it our own desires and ambitions; if the most visible connection of our projects to nature were their palpable balancing on a pedestal cobbled together from parts evidently derived from nature; then many of us would say that something enormously valuable would be absent from our world. I would say that this reaction to a thoroughly human-contrived world is deep-seated and tied to a regard for our own projects as distinctly human in a way that is lost when they inundate us. Insofar as this is so, it contradicts the portrayal of humankind as purely or even dominantly a race of steroidal beavers. For if that portrayal were true, it seems that we would always find unalloyed pleasure and satisfaction in creating and occupying such a world.

Perhaps we humans are, in part, beavers on steroids. But if that is so, we are many things besides that. We simply are not at our best when, without abatement, with feverish intensity, and with essentially single-minded focus, we try to make the world "right" or convenient or comfortable for ourselves.[74] We are not at our best when we always and without hesitation transform our surrounds to make them "ours". We are not at our best, for example, when we build our largest cities in the middle of torrid and waterless deserts, consequently ensconce ourselves as water managers on a continental scale, and then build and manage ecosystems according to the "ecosystem qualities" that, given all this, serve those cities well. Just so, this is not what most would call a valuing of nature as nature.

Insofar as this behavior is characteristic of humanity, nature becomes one domain among others for the exercise of a restlessness that is oblivious to valuable relationships not served well by that restlessness. Our relationships to "nature" are reduced to ones that involve constantly tearing it apart and patching it back together. But this structuring by restructuring of that to which we relate seems incapable of sustaining any uniquely valuable or resilient relationship that persons might have to the natural world. This is unsurprising because, as I have already suggested, it is antithetical to a valuing that regards some part of the world as "given".

The environmental philosopher Baird Callicott (1995, 439) lends his voice to a frequently voiced, self-righteously approving variant of the view of humanity as a race of steroidal beavers. According to him, humans, like beavers and all other creatures are

> … part of nature [and so] we have a rightful place and role in nature no less than any other creature… What we do in and to nature – the transformations that we impose upon the environment – are in principle no better or no worse than what elephants, or whales, or redwoods, may do in and to nature.

This declaration veers towards the language of rights and compresses much confusion into a small space. Chief among the confusions is the proposition that a human right to transform the environment is secured by the notion that humans are "no better… or worse" in doing this than non-moral creatures and organisms. Of course,

---

[74] Of course, trying to make the world "right", convenient, or comfortable for ourselves does not guarantee success. A number of writers, taking a broad view of what is "right" or comfortable, have argued that these efforts regularly or even systematically fall short of their goals. My argument does not depend on accurately assessing this track record and so I do not pursue this point.

humans are animals and one organism among millions. But this does not in any way militate against the possibility that, as the peculiarly moral creatures that we humans are, we are capable of valuing relationships with the natural world that do not and cannot obtain for other creatures.

A second response to the challenge – that appropriate fit can just as well point to humans as bipedal castorids – is to point out that all theories of natural value, not just a theory of natural value as appropriate fit, are vulnerable to co-option. As I have argued, scientific and economic "proofs" that the natural world is valuable not only are vulnerable to usurpation; it is a done deal. There is simply no other reasonable interpretation of the fact that such an enterprise as Rio Tinto can do business as usual with the imprimatur of The Nature Conservancy's scientific methods of conservation. As a consequence, I regard it as a very good thing that, as this book has shown in its detailed analysis of such views, they do not simply fail, or fail simply on their own terms, but rather their failures are multiple, dramatic, and cascading. To be clear: This failure is a very good thing, because these views fail as badly for those who have no special regard for nature as for those who evidently do.

The third response expands on the observation just made. The approach of appropriate fit avoids a basic misstep that underlies most of the failed attempts to argue for biodiversity's value as the cardinal emblem of nature's value. This misstep consists in adopting a blinkered view according to which demonstrating the value of nature must be a matter of demonstrating how nature can be bent to fit in with human endeavors and preferences. That viewpoint has produced one unconvincing argument after another. So it should be a welcome advance to finally have an account of this systematic failure's root cause as lying in a systematic failure to properly frame the valuing relationship. This, I should think, is ample justification to abandon these collective failures in reasoning in favor of trying to understand how we people fit into the world as given to us, rather than the other way around.

3. **The obvious, supporting analogies are not apt.**

One might think to turn to the social or political domain for an analogy that might help to anchor a notion of "fit" that characterizes valuable human relationships to the natural world. The fittingness of one person's attitudes and behavior towards another – as in the example of the relationship between two friends mentioned above – might come to mind. Or one might look towards a model that draws inspiration from the fit of citizens in the polis along the lines of Plato's model city. This latter model has achieved latter-day currency – resurrected in the guise of neoclassical economics' idea of "fit" as that which most efficiently serves the market. On some accounts, this is held to include the fit of citizens, whose contribution to society is largely understood and (of course) measured in such economic terms as how efficiently they do their job.

I would avoid the economic notion of fit, for all the reasons that have to do with its inability to reliably capture anything of moral significance. That leaves the social and political notions of fit – or at least their implications for norms and value. However, these notions critically hinge on reciprocal or mutually beneficial kinds of relationships. Other people are as much moral agents as oneself. They are just as much capable of giving or denying consent to actions that impinge on them; and just

as much capable of acting towards you and me as we are towards them in morally
meaningful ways. And the polis is a collective arrangement requiring certain contributions from all in order to achieve the collective satisfaction of the needs and
desires of those who belong and contribute to it.

In this critical respect, these analogies (between human/nature relationships, on the
one hand and human/human or human/polis relationships, on the other) fail to illuminate the fitting relationships of persons to *nature*, which do *not* incorporate social reci-
procity, recognition of interests, and mutual benefit of the sorts that are essential to
norms of social and political interaction and related norms of fit. One cannot strike a
bargain with nature, or lie to it. One cannot be selfish with respect to it by taking and
not giving back, or be unfair to it in any way. That is because there is no role for people
in the natural world that is jointly agreed upon by us and by it. Nature simply does not
enter into relationships with persons in the requisite coequal or cooperative way for
any of these value-laden terms to have their customary normative sense with respect
to those relationships. And nature has no interest in being treated fairly or according to
any of other norms (such as those that respect consent given or denied) that customarily
apply to cooperative human behavior.

Nor can one do ill to nature – considered an anthropomorphized moral patient –
for example, by failing to keep it in good health, or by failing to respect its autonomy or integrity. For, as I have argued at several previous junctures, nature simply
does not have an interest in maintaining its health, making autonomous choices, or
leading an existence in which its beliefs and actions (whatever those might consist
in for nature) are well integrated in the morally relevant sense. Therefore, if any of
the above-mentioned interactions with nature make any sense at all, they do not
have the sought-after moral implications. Whatever initial plausibility they might
have relies on erroneously supposing otherwise – that the metaphor implied by such
words as "promise", "fairness", "health", "integrity", and "autonomy" automatically extends to the natural world in the morally relevant sense.[75]

In sum, the anchor points for obvious analogs to the relationship entailed by appropriate fit do not hold, let alone hold fast, that valuing relationship's moral weight.
Consequently, the move to appropriate fit requires, not facile analogies, but rather a
fundamental shift in thinking. It requires starting with ourselves as human animals,
and taking a long and long-viewed look at the importance that attaches to locating
such an animal as ourselves and the projects that are important for us to undertake
within some "given" framework, which is not entirely of our own design or making.

8.2.2.2 Limitations

My proposal for a theory of natural value according to appropriate fit is an attempt
to identify an axiological core. That is, it seeks to find a central consideration that

---

[75] There nevertheless is a broader sense in which fitting into social structures is an instance of
a more general notion of fitting into a "given" environment, including the natural environment.
I return to this kinship in Sect. 8.2.4's further elaboration of "appropriate fit".

explains why the natural world has some core value, which seems to many of us at once extraordinarily important and unlike any other. The theory locates this axio-logical core in the value of a particular relationship that people can have with nature. The relationship in question appears to be a crucially defining one – not just for individuals, but also for societies, cultures, and humanity as a whole. I would say that entering into this type of relationship in the modally demanding way that it requires sharpens and deepens the contours of most everything else that makes a human life worth living. Even if the kind of relationship that I have in mind does not create new ends to live for, it elevates the worthiness of existing ends. If we feel as though we are living a good life, it is important, as Ronald Dworkin (1993, 206) notes, "that we *find* it good" [italics in the original]. Valuing nature according to appropriate fit is one basis on which we might find the good of our life's projects better than they otherwise would be.

This is saying quite a lot. But this is also *all* that my proposal pretends to be. It is *not* many other things, including a lot of things that many an environmentalist yearns to find in an environmental ethic:

First, I do not mean to imply that there is one best way for a person to live. Nor do I wish to defend the supposition that the kind of human relationship to nature that I think engenders its core value is the only relationship that it is possible for people to have with the natural world. That we even have a theory of humankind as a race of steroidal beavers proves otherwise, independent of how plausible it might be as a complete account of humanness vis-a-vis nature. I do not even wish to imply that what I regard to be the core nature-valuing relationship is the only legitimate one. The strength and persistence of the appeal of "man the indomitable conquering hero" makes it highly likely that this vision is rooted in something important about being human.

The theory of natural value according to appropriate fit supplies no knockdown argument for "saving nature", let alone one for saving it at all costs. But I believe that the theory merits a preeminent place and a constant presence in our thinking about our many and frequent actions that impinge on nature. At the very least, it should make us aware that every time a field is cleared, a road is built, a mountain-side planted with wind turbines, a desert scraped clean and covered with solar reflecting panels, a river dammed, or a fire regime implemented, we have decided and acted in a way that torques our human relationship to the natural world; and this is morally significant.

Second, it should be understood that, like all moral considerations, consideration of appropriate fit with respect the natural world both potentially and actually con-flicts with other legitimate values. However, I would hasten to add that when the inevitable conflicts with economic development arise, economic valuation of "alter-native development paths" is not a proper tool for settling the score. That means of comparison presumes that the relevant values can be characterized in terms of the economic benefits of development. Making such comparisons is therefore a pre-scription for willful disregard of natural value conceived as appropriate fit.

Third, and related to the previous disclaimer, I do not deny that many conditions in the modern world conspire against achieving something like an appropriate fit into the world. Appropriate fit requires some significant restraint with regard to viewing the world as resource. Yet, very saliently, one cannot disregard some strong obligation to meet the needs of a still-increasing population of 7 billion humans for living at least a minimally decent life. However, difficulties in achieving appropriate fit do not militate against appropriate fit as a matter of serious moral concern. Rather, the importance of appropriate fit adds weight to the case for alleviating whatever difficulties stand in its way. Therefore, consideration of appropriate fit bolsters the case for addressing such a concern as the ballooning population of humans on the planet.

Fourth, I offer no operational interpretation of appropriate fit – no prescription, no timeless formula, no universally applicable rule for when the natural world is best "engaged" predominantly in terms of appropriate fit. Section 8.2.7.1's discussion of historical perceptions of nature's value suggests that the importance of appropriate fit is highly contextual; it has only recently become worthy of moral consideration. For only in the last 200 years or so have conditions stemming from technologically leveraged, ubiquitous development coalesced to frame a picture of the natural world that focuses the moral eye on how the appropriate fit of humans with respect to nature is preeminent in shaping the human condition.

Even within the context of current life on earth, I do not claim or even wish to compete with conservation biologists by way of offering my own recipes for choosing places to cordon off. Nor do I claim or even wish to possess a magical protractor capable of drawing a crisp line of demarcation between those portions of the world that are or ought to be part of nature and those that are or ought to be part of human enterprise. I harbor a suspicion that these kinds of bounding exercise might not even make sense, given the complexity of human entanglement with the world at large. Or if some sense could be made of them, I wonder whether an implementation based on credible principles could ever be devised. But I also believe, as the unlikely examples of "letting it be" presented in Sect. 8.2.7.2 (Implications for action, social policy, and conservation) suggest, that *these* kinds of bounding exercises might not even be particularly important or even morally relevant. The important boundaries, I believe, are not physical ones, but rather those defined by a modally demanding attitude of "letting it be". This realization helps to keep my proposal focused on two central aims. The first is to make the best case that I can for the proposition that this latter kind of boundary should be an ever-present and abiding moral concern in evaluating the worthiness and scope of human projects. The second is to underscore the urgency of holding that core concern in mind by spotlighting how it is routinely cast aside, even by those who view themselves as friends of the natural world.[76]

A theory of natural value according to appropriate fit cannot provide moral "proof" that a "letting it be" of nature trumps all other considerations and that further development of the planet ought to be banished. It is useless and foolish to pretend that building a road to the next big mine is, in moral terms, akin to murder. Appropriate fit has no pretensions along these lines.

---

[76] I am grateful to participants in the 2011 Northeastern University Workshop on Applied Ethics for pressing me to clarify this limitation of my work.

On the other hand, I believe that consideration of appropriate fit provides strong reasons for rejecting – as antagonistic to nature's core value – the increasingly prevalent practice of conservation as a misleadingly labeled enterprise to improve-by-developing-and-managing nature. By ignoring appropriate fit, conservation (so understood) pretends to champion a value that its practice actually destroys. This topic is further pursued in Sect. 8.2.7.2.

Fifth, it would be foolish to deny that the human relationship that grounds the core of natural value is the only relationship connecting people to the natural world in a positive or morally salient way. Most salient among other relationships is that which has recently wreaked so much havoc in thinking about biodiversity in particular and nature more generally. This is the relationship (discussed in the next subsection) that hooks into the fact that people must, in a very basic way, make a living from nature. People have yet other normatively salient relationships to nature, which I discuss in Sect. 8.2.5 (Natural value outside its core). However, I believe that other relationships generally, and the "living from" relationship in particular, do not in any fundamental way ground nature's moral standing as nature.[77]

8.2.3 *"Living from" Nature, Uniqueness, and Modal Robustness*

I wish to visit one more topic before getting to More on appropriate fit in the next subsection. I should say "revisit", for this is a topic that I addressed in Sect. 8.1.3 on The ends of the biodiversity project. It also received attention, though couched in different language, in the immediately preceding subsection's discussion of the first "difficulty" with appropriate fit. The topic concerns the business of "living from" nature.

Nothing that I say in this book should be interpreted as denying that nature *is* a resource and that it *does* have value as such. People ultimately live *from* nature. The current-day orthodoxies of the biodiversity project and conservation generally build on this unexceptional fact. However, by expressing this basic fact in the currency of economics' evaluative language, the natural world is exchanged for a considerable superstructure, which ultimately makes out of nature something unrecognizable as its actual self.

In the language of economics: People, being animals, must have some amount of "natural capital" in the bank to provide needed food, water, and shelter. Moreover, nature *should* be used and (in part) regarded as capital whose value is value-in-exchange for the service of human needs and desires. From this limited view, a

---

[77] The language of "responsibility for nature" can mislead one into thinking otherwise – until one realizes that this language lacks moral credibility (Sect. 8.1.2, Responsibility for nature). The language of virtues might also mislead, insofar as the role of nature in environmental virtues is that of supplying generic resource tokens, subject to the principles of distributional justice (Sect. 8.1.3, The ends of the biodiversity project).

chunk of the natural world is evidently valuable as a commodifiable and tradable means to living at least a minimally decent life, and maybe even a pretty darn good one. Moreover, as I remarked in Sect. 2.1.3 (Note 27 in Chap. 2), it seems that someone committed to valuing nature as a capital investment is not necessarily committed to regarding this as nature's only value. On this account, nature-as-capital-investment is just one environmentalist card to play as the need arises. However, while the value of "living from" nature should be acknowledged, so must its limits; so, too, the corrosive effects of pushing this value into preeminence, as well as the consequences of assimilating nature's value into the domain of the greater human economy. Several salient concerns stem from these reservations.

The first and most easily expressed concern is with the panglossian gloss on nature-as-natural-capital whereby it is regarded solely as a kind of trump card that can be played, when needed, to win a close game of nature valuing.[78] Two distinct problems arise from this reckoning. The first is that it builds into itself a tacit pre-supposition of nature's high standing before consideration of it as capital (but which nature-as-capital only bolsters). This would seem to require some independently defensible and defended theory of why nature is valuable aside from the proposition that it is a capital asset. But in the main, it is hard to find anything but vague gestures at what those other values might be. They are, it seems, the Unspecified "moral reasons" (Sect. 6.1), which, unexplained, remain undefended.[79]

The second problem with this reckoning is that appears to suppose that playing the natural capital card always increases the cachet of the natural value hand. But this supposition appears to be false. On the evidence of Chap. 6, it seems likely that nature, at least nature as represented by its biodiversity, is generally a poor to terrible capital investment. Those who push nature-as-capital into preeminent consideration must then answer the question: On what principled basis can this circumstance be withheld from determining that on the whole, nature is not worth keeping as it is? The suggestion that economic evaluation must be circumscribed in such a way as to not develop nature out of existence lacks any credible principle for determining where the containing perimeter lies. It will not do to say that economic evaluation should be proscribed only when the consistent working out of Natural Capitalist principles prescribes a wholesale gutting, rearranging, and reassembling of "nature" into a Frankenstein monster of available bio-parts, merely because it is disagreeable to the champions of those principles. Consistency requires Natural Capitalist to condone – and perhaps obligates them to pursue – projects that transform ecosystem

---

[78] Recall the example of the forest periphery around Costa Rican coffee farms described in Note 32 in Chap. 6.

[79] The prototypical response of those who are economically inclined is to subsume these "unspecified moral reasons" into the economy. They are preferences, which like all preferences, can be expressed in market transactions. Since economic theory is agnostic about why people value things as they do, the requirement for an explanation is removed at a stroke. A reductionist program of evaluation built on a premise that excuses it from supplying principled justification for its rules should be regarded with enormous suspicion. I have previously suggested several reasons for why this suspicion of economic valuations is warranted.

after ecosystem into a richer warehouse of usable stuff and to engineer ecosystem after ecosystem into a more efficient service provider. This train leaves the station without usable brakes.

Second in the list of general concerns is that, while it does seem clear that nature provides stuff that people need, and that it can be harnessed to serve human needs and desires, it is much less clear that nature should be regarded principally *or even at all* as that which has value-in-exchange for providing any, arbitrary benefit that a person is willing to pay for in a market transaction.

The reason for this gets back to the fact that not all benefits are worthy of being conferred. Insofar as a person benefits from the satisfaction of an unworthy, "experiential"[80], or whimsically formed preference, her benefit does not merit moral consideration. For example, there is a difference, not just in market value, but in kind, between the value of being able to drink drinkable water and that of acquiring an iPhone – even though neoclassical economic theory has enormous difficulty in distinguishing them. From its viewpoint, they are both "benefits" whose "consumption" contributes to "human welfare", in the economic co-option of that latter phrase. To observe this is to once again (as in Sect. 8.2.2.1) take note of the promiscuousness of this form of valuation. This promiscuity seriously blurs the distinction between considerations that demand serious attention as bearing on living a decent human life, and those that fall below that threshold and therefore merit at most casual attention or are so frivolous that a coin flip is the most appropriate means of making a decision about them. As a consequence, the claim to moral seriousness of any good placed inside the container of market valuations is seriously diluted to the point where even its respectability is open to question. In the marketplace, my access to water that will not poison me is just as well measured by iPhones as by anything else that, in principle, one might trade for it.

Furthermore, benefits understood as fulfillment of preferences are subject to question regarding their moral credentials insofar as those preferences are questionable. This is the point that I made in Sect. 8.1.3, with the help of Debra Satz's observations regarding preferences as the driving force behind markets. Most saliently, not all preferences are worthy of satisfaction, despite the fact that their satisfaction might well increase utility.

Third (in the list of concerns), from the economic view of "living from nature", the natural world and its various "components" blur into faceless commodities. They become things whose main evaluative identity is embedded in their exchangeability for another (human-made or natural) commodity in a market transaction. Thus masked, nature-as-commodity is barely distinguishable from any other commodity. Once out of the barn and freely roaming the great commodity range, it is hard to corral nature inside any special evaluative category. This, in turn, promotes a narrow and ungenerous view of how people relate to the natural world – as the stuff of transactions that drive human enterprise and development. This placement of the

---

[80] See Note 70 and the accompanying main text for an explanation of this term, which is Ronald Dworkin's.

natural world squarely within that of human development is plainly antithetical to the "letting it be", which is at the core of natural value according to a theory of appropriate fit.

Still, it is true of many valuable entities that they admit multiple kinds of value through multiple kinds of relationship that persons can have to them. It is possible for these different kinds of value to be at odds; and it is plausible to think that this happens with some considerable frequency. Values that conflict are the rule rather the exception in the human condition. However, in many cases, significant limits and qualifications to this plurality of values apply. Natural value is one case in point.

Most saliently, the value of some things is marginalized and sometimes even foregone entirely by excessively regarding them as things of utility or preference.[81] Aristotle's famous exposition of the nature of true friendship and other forms of relating that are, at best, imperfect friendships, hinges on just this observation. He remarks that if one values a "friend" mainly or solely on the basis of the pleasure one derives from her, or on the basis of her utility in serving one's own ends, then this is at best an imperfect friendship. A friendship that is largely framed in these terms dissolves with the dissolution of the advantages that one derives from such a relationship (Aristotle 350 B.C.E. b, VIII.4). In such a case, using Philip Pettit's vocabulary, the good of a true friend is not realized because the terms of the relationship are not modally robust. Yet, one can still value a friend for the benefits that she confers. And one might insist that this quality of a friend's value – that it admits value from "pleasure or utility" (in Aristotle's words) – closely resembles that of nature's value to people who must "live from" it.

However, this appealingly pluralistic and tolerant gloss is also too hastily drawn. Some ways of relating to and valuing a friend *are* entirely antithetical to realizing the core value of friendship. Suppose that I were to have in the back of my mind that

---

[81] Satz (2010, 80) considers, and rejects, Michael Walzer's similar view (though more narrowly concerning markets) that "Markets can *change* and *degrade* the meaning of a good." [italics in the original] Satz (2010, 82–84) ultimately rejects Walzer's position. Her strongest reason is that a market does not require any consensus on what makes a good good: A market allows different participants in a transaction to understand the good of something traded in entirely different ways. She concludes from this that the mere fact that certain goods – such as human organs – are traded does not account for any discomfort in the market for such goods. *A fortiori*, nor does it cause their goodness to be degraded.

However, I believe that Satz's conclusion does not follow from her premise. Suppose that one party to a potential trade of some thing justifiably regards its salient good to be its value as friendship, health, political equality, or abstractly, as X. And suppose there are strong grounds to think that thing's good *as X is* corroded by making it an object of trade in a market. Then it seems to me that even if the other party to a potential trade has a different conception of the thing's value – say, its good as Y – then this does not undermine the claim that the thing's good *as X* would be subverted by its trade. Of course, there might be a restricted set of such market-corrodible values. My claim depends only on the value of *nature* being one such value. Though, insofar as this corrodible quality mainly hinges on a value's modal demandingness and values other than natural value have a modally demanding requirement for their realization, nature's value is unlikely to be the only one to have this quality.

should I ever come to covet some thing to an extraordinary degree, and should I have no other means of acquiring it, then my "friend" would be valuable – perhaps as a highly marketable slave – for her value-in-exchange for that coveted object. In that case and even if I never came to covet some thing so strongly, this would incontrovertibly show that, whatever my relationship is to this person, it is not one of true friendship. In this way, a friend's value might be thought to differ from that of the natural world, whose value as nature survives its actual participation in market transactions. But this is a matter of degree rather than kind. It is highly doubtful that nature's core value survives an unrelenting and essentially unconstrained demand that its "components" pay their own way. Nor does it survive continual defense of them on this basis.

This line of thought leads to a fourth consideration. Of course, not all valuing of a friend for benefits conferred by the friendship is altogether antithetical to the relationship being a true friendship. The more balanced view is the one suggested by Aristotle's notion of "imperfect friendship". There is something secondary about the value of a friend considered *insofar* as it derives from considering her as a person who confers benefits – that is, insofar as the friendship is "for the sake of utility". But more than that, there is something about this way of valuing that corrodes and subverts the core value that attaches to the friendship. Just so, there is something secondary, but also subversive, about the value of nature to the degree that its valuation focuses on conferring utilitarian benefits. This is not to deny that "conferring benefits" contributes a utilitarian kind of value. Rather, it is to say both that it is *not* centrally constitutive of the core value of the natural world, and that it corrodes and subverts-by-displacing the different relationship that defines this core.

To this point, I have suggested that the corrosiveness of a focus on "living from" results from its displacement of the core evaluative basis for nature's value. But a fifth consideration harkens back to the first and ties together the immediately preceding three to reveal how "living from" does not merely corrode by *displacement*. It also has an acid quality that penetrates more deeply by dissolving natural value's two most vital qualities – namely its uniqueness and modal robustness. These latter two qualities, which in Sect. 8.2.1 I identified as forming the nuclear core of nature's value as nature, are intertwined in a way that I can now describe by continuing my reflections about friends and friendship.

A key observation is that a friend's conferring of utilitarian benefits does not identify anything particularly unique or modally robust about valuing her for those benefits. So far as uniqueness is concerned, it is not difficult to think of other persons and "mere things" that are not and could not possibly be friends, that could nevertheless confer benefits of a similar, generic utilitarian nature. My faithful servant, paid sufficiently well, will do just as well as my friend in this regard. But even my servant is not safe from possible displacement by robots, which could largely supplant and even surpass her in performing the duties that are of utility to me. This last observation weaves in the fact that, when I consider her as a thing that serves my desires, my valuing of the servant is not modally robust. There are many possible worlds – and in this case, worlds likely to obtain – in which her value as a servant is severely diminished or even eliminated. This observation makes it easy to see that the same lack of modal robustness corrupts a relationship with a "friend" considered valuable for the services that can just as well be performed by a servant. It is precisely the requisite modal robustness that Aristotle's "imperfect friendship" sorely lacks.

When it comes to valuing the natural world in a way in which "living from" is preeminent, the corrosive effects on the uniqueness of nature's value and its modally robust quality parallel the patterns with regard to friendship. The moment that one begins to view nature as a set of component parts and to view these parts as commodities of utilitarian value, one eats away at an ability to see nature as something uniquely valuable. A tree is a source for wood that can make a chair and valuable on that account. But in this value, it must compete on an equal footing with synthetic resin.

These days, an emphasis on "living from" often takes the extreme and extremely acidic form expressed in the language of neoclassical economics. Almost the instant nature becomes "natural capital", anything unique about its value is dissolved. At a stroke, the natural world becomes an anonymous cipher that is absorbed into the larger cipher of capital assets as these enter into the computation of economic welfare along various development paths. Along such a development path, "natural parts", in the sense of "once existing in the natural world", might for some time persist. But they would persist outside of any context that could be regarded as "the natural world". They would no longer be "components of nature" so much as they would be components of economic development – selected bits and pieces, which work as convenient cogs in this entirely different context of human enterprise's machinery.

But if anything, the profound failure of "living from" to capture nature's core value is eclipsed by its utter incapacity to capture natural value's modal demandingness. This point is best made by way of examples. Some of the best and most representative ones come from those who subscribe to the extreme, economically packaged form of "nature as natural capital". Consider the value of an insect that efficiently pollinates and therefore increases the value of a commercially valuable crop. If the value of the forest that houses the insect depends on the value of the insect's contribution to the value of the crop, and if the value of the crop is vulnerable to the vagaries of market conditions, then the value in this way attributed to the insect and to the forest utterly fail the test of modal robustness. This is the story of the Costa Rican coffee farms, which is one of the seminal stories of the Natural Capitalism movement. When the pollinated coffee crop's value fell precipitously, farmers naturally switched crops – to unpollinated pineapple, whence "The New Economy of Nature" no longer had a place for the insects or their forest home.[82]

Or again: If the global economy dictates that the value of minerals underlying a forest vastly exceeds that of the forest itself, then by the precepts of Natural Capitalism, this literally removes any ground for the forest's existence unless some other place for it in "nature's economy" can be found. This is precisely the story that justifies the Rio Tinto's decision to "develop" a coastal Malagasy rain forest by cutting

---

[82] I also make reference to this situation in Sect. 6.3 (Biodiversity as service provider) and in Note 32 in Chap. 6. See also Note 38 in Sect. 8.1.3 (The ends of the biodiversity project).

it down in order to extract the ilmenite beneath it. By the precepts of Natural Capitalism, the true natural value lay in the market for ilmenite; the forest merely obstructed the realization of that value. Of course, the situation would be reversed had the wood in those now-uprooted trees been found to make especially good materials to build boats or guitars. Think of it: Is the value of the natural world really contingent on a decision by a Costa Rican farmer to switch crops? Does it hinge on the vision of Rio Tinto's CEO who cannot see the Malagasy forest for the ilmenite?

This kind of valuing, which depends on consideration of nature on a case-by-case basis – and most saliently, economic state by economic state – is antithetical to what I believe is the common understanding of natural value as being resistant to collapse from such contingencies. I take this common understanding to be at once our best understanding of the natural world's value: It is unrecognizable for what it uniquely is, when constructed on such shifting sands – let alone sands that shift with market conditions and corporate strategies. Nor is it held hostage to the vagaries of fashion; nor to any other force in the broader human economy that makes it more or less valuable as "natural capital" or "ecosystem service". Nature's value does not shift with shifts in the ecological face of the planet – including its biodiversity. Nor does it shift with the shifting epistemology based in the scientifically understood details of these matters.

Rather, nature's value seems to be modally robust and this robustness extends over a wide spectrum of likely and unlikely contingencies. On our best understanding of natural value, its core is something quite solid, and that solidity requires a solid foundation. This is one of the major attractions of the theory of appropriate fit, which proposes that the value of the natural world is sunk into the bedrock of humanness-in-the-world.

This is such a central point that it is worth reprising in the context of modal demandingness. The "humanness-in-the-world" to which I refer is specifically, human modes of relating to the world. Some modes of relating apply to the undifferentiated world in its entirety; but for some part of the world are reserved a special and modally robust form of relating, which confers on that part, not just its designation as the natural world, but its (natural) value as such. The special qualities of this latter relationship require looking beyond the relationship that humans have to things that constitute capital or that are objects of trade in market transactions. It requires looking beyond the somewhat broader class of relationships that humans have to things from which are derived "pleasure or utility".

People are capable of entering into a unique and uniquely valuable relationship with the natural world because people are not the sorts of animals whose needs stop at mere subsistence, nor at the continual satisfaction of desires, nor even at the accumulation of an outrageous amount of stuff. People are peculiar, indeed unique, in having a consciousness of their place – as living on the planet at this time of its history. I believe that this awareness gives rise to another uniquely human and very profound need – namely, to locate the human place in things as they are "given" to us by the natural world – that is, as they have emerged from the 4.5 billion year history of the planet on which they find themselves.

8.2.4 More on Appropriate Fit

I now return to the main topic and thesis for this part of the book. It is easy to confuse a theory of natural value according to appropriate fit with a number of other approaches that have long been a part of the discussion of natural value. Section 8.2.6 (What appropriate fit is not) is devoted to clarifying how these other approaches differ. However, to avoid the most likely confusions in further elaborating appropriate fit here, I would like to perfunctorily proclaim it distinct from two other approaches: First, appropriate fit is not a plea to save an inventory of natural stuff that is current humanity's bequest to its descendents. Nor is it a theory of intrinsic value, of the sort described in Sect. 2.1.1, whereby some kind of stuff "out there" – namely the stuff of the natural world – generates a moral force field that commands the consideration of moral agents in the vicinity. Those proclamations made, I proceed to dig deeper into appropriate fit:

For people to achieve an "appropriate fit" in the world at large is, in part, for them to integrate themselves into the world as "*given*".[83] As previously noted, this principle is most often seen to apply to how people integrate themselves into the social structures that frame their social environment. Good relationships to other persons, such as friendship, cannot thrive when our means of relating is, without exception, first, and foremost, making others bend to our desires. A person for whom this is always the first inclination and who rarely or only as a last resort considers how her own attitudes and behavior might profit from modification is generally considered to be a sociopath. Rarely does the life of such a person go well.[84] There are also strong grounds for thinking that this is not even a *rational* attitude for a person to have.[85] The prospects for a society filled with such people or populated with citizens who collectively make decisions on this basis are equally bleak.

The theory of natural value according to appropriate fit proposes that a principle of fit does not just apply to social environments. It extends to how people individually and collectively integrate themselves into natural environments – although the way in which an appropriate fit is achieved in this domain radically differs from what works in the social domain. In the final analysis, I believe that *core* value of the natural world hinges on this single, slender, but surprisingly strong silken thread of a uniquely human way of relating to the natural world that takes it as a given structure or framework for living a human life on this planet.

To embrace such a principle is not to categorically or unconditionally reject the idea that it is sometimes rational to raise questions about the structure of the

---

[83] Compare Schmidtz (2008b, 72).

[84] I cannot here dive into all that "going well" means, except to suggest that it means much more than the professional and financial success that even a sociopath might enjoy for some significant stretch of her life.

[85] Schmidtz (2008b) makes an enormously convincing case for the rationality of being altruistic. If one understands "rational" as "reasoning well about practical decisions", then this can be seen as equivalent to moral endorsement. See also Note 61 and related main text.

environs in which one finds oneself. Nor is it a categorical rejection of the notion that we sometimes do well to move for change in that structure. Such exceptions occur for social as well as for environmental structures. But insofar as people are creatures who need to belong to, or, more generally, need to understand their place with respect to something larger than themselves, they do not do well to always, or customarily, or casually, or reflexively presume that those enveloping structures should be transformed so as to satisfy desires that do not seriously take into account the value of finding their place in this larger context.[86] Nor can they be expected to do well if they regard choices that impinge on, and possibly compromise, these enveloping structures as routinely subject to, and contingent upon careful, case-by-case computation. A fabric that is regarded as disposable and regularly *is* discarded for another based on unforeseen and unforeseeable contingencies is not something against which or within which people can reliably plan a life, or seek to understand and carve out their meaningful place in the world.

We all are thrust into both social and natural frameworks that are not of our own design, not of our own creation, and not of our own choosing. However, this circumstance should not be regarded as an invitation to continually dismember, reconstruct, or hack away at them. It should not automatically be seen as a reason to discount and discard old structures, create new structures (including new ones that mimic no longer but previously existing ones), or more generally, to always deform, re-form, or substantially re-create and manage the surrounds in which we find ourselves. That is emphatically true when the primary grounds for these transformational activities are that the surrounds, as given, are not of our own creation, or that they are not what we would have chosen to create had we been in a position to do so, or simply because we feel that they do not suit us. Certainly, these are not good grounds for failing to at least consider how we might form a life within and with respect to our social and natural frameworks as they are. People who reflexively move to change and manage their surrounds, social or natural, to their liking (no matter for economic development and gain or even for reasons such as what currently passes for "conserving biodiversity", which have a nobler ring) and who have little or no inclination to explore how they might establish various relationships to those surrounds as given, have a narrower view of themselves because they summarily surrender even the possibility of entering into such relationships. To forfeit such relationships, and even the very possibility of them, is to forfeit much that can define, refine, and give meaning to a human life.

Part of "functioning well" with respect to these "given" structures is a matter of habitually seeking the most satisfactory ways to exercise our human capacities and to meet our human needs within with respect to these structures, as we encounter them. It is a matter of seriously considering how we might form a good life by folding

---

[86] I think that this description applies to the decisions of individuals, but it is more obvious in our collective decisions, for example, to divert water in grandiose schemes, build roadways through parks that were intended as refuges from roadways, and leveling redwood forests for crops such as wine grapes, which serve no essential human need.

it round what is "given" and thereby taking advantage of the availability and ability of what is given to frame our life. It is an inclination to view the contours of the encountered world as by and large defining the possible paths through it, rather than as the unfortunate starting point for altering the contours when they appear to hinder our way, or do not immediately appear to satisfy our preferences, or do not meet with our approval. Functioning well comes from finding a satisfactory place for ourselves as a part of, or with respect to the given whole – our integration with it. It comes from finding some large part of our life's meaning from how we choose to thus integrate ourselves, individually and collectively, into the world.

I have loaded a lot into the word "given" and the role it plays in providing structure and meaning in a human life. The concept is familiar to most any parent who is responsible for framing the life of a child. This is largely a matter of giving the child a rich "structure", which serves as a framework and which the child can take as a reliable context within which she can, with a parent's guidance, develop her human capacities. You and I are not children. Yet even we adults cannot flourish without some structure – something "given" that helps us to explore and eventually identify where we fit in the scheme of things.

As I said at the outset of this subsection, appropriate fit within a social fabric is achieved in a rather different way from the way that appropriate fit with respect to nature is achieved – despite these being instances of some more general principle. Social structures define roles that persons can assume within an encompassing structure designed and maintained to accomplish goals for us all that we cannot possibly accomplish as individuals. In contrast, nature is not designed and maintained in the same way as a social or political structure such as on might encounter in the polis. Unlike the polis, it is not something of human design – not part of the "stuff" of culture and civilization.

As a consequence, nature's claim on us does not play out in the same way as the claim of a social fabric in which each of us finds a way to thread our way into warp or woof. Rather, nature's value is bound up in a peculiar relationship with people who self-consciously set it aside and steadfastly do *not* allow their life and life's projects to thread their way into that warp and woof. Valuable nature is that which is literally and metaphorically outside the walls of the city, considered as the embodiment of civilization. It is the boundary in opposition to which civilization and its development are defined. It is that which sets off by contrast those enterprises that are identifiably part of human culture. At its core, it is that which is *not* emblematic of human desires and striving and designs to achieve them. That is the reason why appropriate fit's meaning for human behavior with respect to nature is a "letting it be" – a modally resilient refraining from breaking down nature's bounding function, a steadfast pushing aside of any impulse to make it part and parcel of the human economy, human enterprise, or part of human culture more generally.

To embrace some principle of appropriate fit, no matter the specific domain of concern about fit, does not require that we automatically remove worthy goals from the set of goals from which we choose – on the grounds that they impinge on our surrounds – social, natural, or of any other kind. A too-rigid self-censorship of goals would be a tragic surrender of human freedom. Or it would be the achieving of the false kind of freedom that comes from always taking care to choose an option that is unconstrained.[87] Rather, the force of the principle of appropriate fit, as it applies to the natural world, is that no end is worthy of pursuit unless it, along with the available means to achieve that end, is respectful of the given surrounds in a modally robust way.

It is important to keep in mind that, according to appropriate fit, being "respectful of the given surrounds" is, underneath, being respectful of a uniquely human relationship to those surrounds, which endows those surrounds with their unique, natural value. According to appropriate fit, "respect for nature" is a matter of entering into this valuing relationship. It is emphatically *not* a respect for nature-turned-person, whose health, autonomy, and integrity command respect so that it might pursue its rightful business. There is also a sense in which the respect involved in a valuing relationship with nature is self-respect. This is not self- aggrandizing anthropocentrism, but rather respect for a range of valuing relationships with nature that are especially valuable for *specifically human* creatures to enter into.[88]

"*Specifically* human" has some significant force beyond "identifiably human" because it serves to distinguish in yet another way the characteristic relationships that define the core of natural value from those – shared by other, nonhuman organisms – that do not. Most saliently, relationships that are part of our living *from* nature are not at all unique to a human kind of creature. A view of the natural world's value as centered on "natural capital" or "ecosystem service" is entirely defined by relationships with their natural surrounds that humans share with other creatures.

"Not unique to humans" likely *dis*qualifies these economically-oriented conceptions as grounding a plausible theory of natural value. However, "unique to humans" is not, by itself, qualifying. The view of the natural world as that which achieves a certain state of nature, whose characterization and realization can be the goal of scientific endeavors, might pass the unique-to-humans test. It is likely that humans, alone among all creatures, are capable of forming such a conception, which relies on a sense of history and some capacity for systematic visualization. But stripped of its scientific pretension to know what a state of nature looks like, and stripped of its moral pretension to know what it *should* look like, it is at bottom another human enterprise to make nature into what humans wish it to be – albeit an enterprise that is not, on the basis of that description, hooked into the market economy. Even a non-economic assimilation of nature into the world of relationships that frame human enterprise should give major pause.

In direct and stark contrast, the theory of natural value according to appropriate fit commands a broader perspective from a perch where the domain of human enterprise, culture, civilization, and "development" (including economic development)

---

[87] I have in mind Philip Pettit's (2008) analysis of freedom, which I leveraged in my initial presentation of appropriate fit in Sect. 8.2.1.

[88] See Sect. 8.2.6 (What appropriate fit is not) for more discussion on why a theory of natural value according to appropriate fit is not anthropocentric in any objectionable way.

does not engulf the entire field of view. This domain no longer looks to be all-encompassing, and furthermore, that is so precisely because it is circumscribed and bounded by another domain, which is the domain of the natural world. The two domains are characterized by sets of relationships that give them their respective characteristic values. But the attitudes and behavior implied by the different values in these two domains are diametrically opposed. Moreover, the distinct relationships that make the natural world the natural world and that give it its value *as* the natural world cannot be realized except when some significant expanse of nature is, first of all, placed outside the boundaries of culture as a matter of contingent fact. And yet more essentially, this expanse must be, as a habitual way of thinking, *regarded* as exterior to the domain of human endeavors of *any* kind. This last condition is the condition of modal demandingness.

One might be tempted to regard appropriate fit in the natural world as humankind's proper "integration" with it. Indeed, environmentalist literature is replete with paeans to some melding of humans with the natural world. But this conception contradicts and undermines the fundamental basis for a valuing relationship to nature. In the important sense in which that core valuing relationship is an attempt (however imperfectly realized) to draw a bright line between human culture and what lies outside cultural enterprise, that relationship is the *opposite* of integration.

This might at first seem counterintuitive, because it requires that people suppress a strong impulse to be central actors and interactors in their important relationships. However, with the exercise of due care, the struggle to honor a bright line between human enterprise and the natural world can still be understood as integration. The trick is to understand it, not as integration with the *natural* world, but rather as integration with the larger world, which encompasses both human culture and nature. The integration of the two requires recognition of the distinctive qualities of each of these two realms. It consequently requires acknowledging the value of the boundaries that separate them – boundaries that set off and sustain those distinctive qualities, in which reside distinctive values for people.

This core role of the natural world – of bounding human endeavors – is necessarily oil in the waters of any system of evaluation that attempts to assimilate nature's value into the sort of evaluative scheme that is appropriate for those endeavors. This is notably true of such an evaluative system as Natural Capitalism's, which self-consciously tries to locate natural value squarely within the bounds of human enterprise. The contrast with the proposition of appropriate fit could not be more dramatic. According to the latter, the core of natural value is not and cannot be represented in the common currency that greases the channels of actual and contingently evaluated market transactions. It is not and cannot possibly be counted as the value of capital or services. Rather, the value of the natural world can survive only in a culture of recognition of its vital function in bounding this enterprise. Moreover, by bounding the scrum of human-defined development projects, the natural world affords that latter domain a recognizable evaluative shape – a distinctive and honorable profile – that it would lack, were both domains collapsed into a unitary, amorphous continuum. This view of nature's value yields a double dividend. For by virtue of being set outside the domain of development, nature is seen to occupy its own fertile domain with its own distinctive qualities, which are the germs of a separate set of relationships that flow from people "letting it be". I shall shortly show how these relationships, and their value for people, are embodied in the symbols, metaphors, and myths that lend meaning to human lives.

The intuitions of those who have not allowed economic theory to confound them might have anticipated this result, which shows why nature-as-instrument-of-economic-development is antithetical to nature's core value. Perhaps less expected is the result that the relationship that underlies the core of natural value cannot be sustained, and in fact is placed under assault, by the development projects of the conservationists and restorationists of our day who extend the range of human enterprise to manipulating, molding, and reforming the natural world into something that is supposed to better approximate a more ideal state of nature... or the economy.

These projects flail away at the natural world in an attempt to "improve" or "repair" it. When the definition of "improvement" or "repair" merely doubles back to the economic business of "more natural capital" or "better ecosystem services", then the matter is made clear, but in a way that clearly does not escape the natural-value-crushing gravity of nature-as-instrument-of-economic-development. And as I have argued, outside the economic orbit, it is hard to find grounds with any normatively respectable standing according to which people ought to vigorously engage in conservation and restoration projects as they are now typically defined. This distressing situation regarding ends defined in terms of some state of nature is already sufficiently exposed in Sect. 8.1.3 (The ends of the biodiversity project).

And yet, this enormous lacuna in practical reasoning is peripheral, at least in one important respect. For even if there were a credible defense of some norm for nature improvement or "nature development", a more fundamental objection would still remain. The objection is that the development and management of "better nature", including biodiversity development, is at least as subversive of natural value as the development of the natural world into cities, farms, and all the infrastructure in and in-between them. Both kinds of development vigorously violate the terms of the relationship that grounds the core of natural value. This is an obvious consequence of the relationship that underlies natural value – a relationship that requires that people let the natural world be what it will be, on its own terms, without human meddling, and precisely not as the grand vision and goal of a development project *of any kind*.

It is the loss of all grip on this core value of nature that, I believe, most environmentalists most viscerally feel when confronting the prospect of a world in which the sort of relationship that sustains it is essentially impossible. This is a world in which virtually no place remains that is not, in some way, a reflection or a part of the striving, achievement, and management of human enterprise; and (in violation of modal demandingness) a world in which each and every square meter of the planet not already caught up in this development enterprise is continually viewed as held in reserve as subject to consideration for also being swept up in it. Regarded on these terms, every part of the world comes to be valued largely on terms dictated their human planners and designers, who in turn, are valued according to their acumen in achieving their questionable goals. It is a world in which everywhere we look, we see our human faces reflected back. It is a world in which literally and metaphorically, there is nothing outside the walls of our cities, for our cities and the vast expanses of farmland and other infrastructure that support them have grown to engulf almost the entire planet in actuality; and the planet is entirely engulfed in most of the many possible and even quite probable worlds.

The kind of good that we forego in such a world is hard or perhaps impossible to further analyze by means of traditional philosophical tools. But as I suggested above, it can be pursued further via symbols and myths, which have been and continue to be emblematic of such goods. It is not surprising that the best guidance for this pursuit comes from outside the traditional philosophical, scientific, and economic sources. I start with Dante scholar Robert Pogue Harrison. By way of the metaphor of *Forests, The Shadow of Civilization*, Harrison vividly explores the crux of the idea that I have been propounding – that some uniquely valuable human relationships stem from our ways of viewing ourselves as acting with respect (and regard) to that which exists "outside of ourselves" (Harrison 1992, 201). By this, Harrison means: outside our actual endeavors and even outside our *imagined* efforts to humanize the world. Harrison weaves a tapestry in which the natural world bounds and in that way defines those human endeavors. His narrative also shows how this same opposition fuels an antithetical human urge to break through nature's bounding confines and thereby destroy the relationship.

The narrative is both new and as old as the first cities. Harrison recounts how Gilgamesh, the legendary and real king of the ancient Sumerian city Uruk, was symbolically the "builder of the walls of Uruk" at around 2500 B.C.E. As Harrison (1992, 15) says, "Walls protect, divide, distinguish; above all they *abstract*." The actual and metaphorical significance of Gilgamesh's achievement in raising Uruk's walls lay in their role of dividing off the natural world to make possible the emporium of culture within the city. Yet, as the tale goes, Gilgamesh cannot abide this division. Assuming a very modern-seeming stance, he laments that, unlike the forests, "Man the widest, cannot cover the earth". He seeks to assuage his frustration by venturing out on the land to slay the forest demon Huwawa – the symbolic deforestation of the Cedar Mountain (Harrison 1992, 13–18).

Harrison claims that the very word "forest" came into common usage during the rise of the Frankish empire starting in the sixth century. According to him, it originated as a legal term, denoting (here drawing on Harrison's vocabulary) forested land that was "afforested" – that is, declared off limits to cultivation, exploitation, and encroachment. Afforested areas were reserved for the monarch's pleasure in the hunt – a symbolic enactment of the king's dominion over nature, regarded as a domain quite separate from that of domestic culture. It was a sanctuary, whose regard as such precluded allowing it to be assimilated or constrained by "the voracious world of social humanity" (Harrison 1992, 74).[89]

---

[89] This is part of Harrison's discussion of the royal forest (Harrison 1992, 69–75).

In his Elizabethan era *Treatise of the Forest Laws*, John Manwood (1717, 37) writes about this separation of domains in terms of walls, whether or not they exist in the material world:

> Forests must be meered and bounded with *unremoveable Marks, Meers and Boundaries*, either known by matter or Record, or by Prescription; for though a Forest doth lie open and not enclosed with Hedge, Ditch, Pale or Wall, yet in the Eye of the Law it hath as strong an Enclosure by These Marks, Meers and Boundaries as if it was enclosed with a Wall. [italics in the original]

Harrison (1992, 120) captures the broader significance of this and other parts of Manwood's description in this way:

> … a forest was essentially an asylum from the human world, a natural sanctuary where wildlife could dwell securely in the king's protection. It had nothing to do with the public interest, nothing to do with usefulness. On the contrary, forests marked the limits of human exploitation of the wilderness.

But this view of the forest as bounding human industry did not persist. It yielded to the modern-day forest of Natural Capitalism. Many mistake that ideology for a great and novel advance in thinking about how people should regard nature. In fact, it is nothing of the sort, but rather the latest incarnation of German *Forstwissenschaft*. This was the "forest science" in which "algebra, geometry, stereometry, and xylometry came together" to compute tree-felling schedules from wood volumes and growth rates (Harrison 1992, 122). And in truth, *Forstwissenschaft* was no more original than was Natural Capitalism. Both are but lightly embellished variations on the Enlightenment theme of "forest" understood in a way that obliterates the boundaries between culture and nature. In fact, it is hard to find any significant difference between the view of Natural Capitalists – of nature as a pile of stuff and service-rendering arrangements – and those of Charles-Georges Le Roy.[90]

Le Roy, a game warden of Versailles, was a contributor to the *Encyclopédie*. That groundbreaking expression of Enlightenment philosophy, edited by the philosopher Denis Diderot and the mathematician Jean-Baptiste le Rond d'Alembert, had its first volume published in 1751. Le Roy's entry for the word "*forêt*" depicts the forest as a very specific inventory of trees of certain ages, which serves "*l'utilité*

---

[90] The Natural Capitalists themselves have created a remarkable mythology of the origins of their doctrine, according to which Plato was the first of their kind. This curious and ironic wrapping of Natural Capitalist axiology in the cloak of ancient wisdom appears in multiple books and papers on the subject – always in strikingly similar language and never (so far as I can tell) with a citation of anything that Plato actually wrote. My best guess is that this modern myth arises from the *Critias* (http://classics.mit.edu/Plato/critias.html), in which the eponymous character lambastes the inhabitants of Atlantis (mythical Attica) for the progressive corruption of their character. Critias offers an account, which I believe the Natural Capitalists would want to shun at all costs: Early on, these people "[thought] lightly of the possession of gold and other property". But this situation deteriorated and eventually reversed. Having lost track of such virtues as friendship, the citizens of Atlantis destroyed their land because they came to value it primarily as a means to the end of ever-increasing wealth.

publique". This vision is the more starkly and startlingly modern for how this man – a *game* warden by trade – "scientifically" abstracts the forest from wildlife. As a modern ecologist or conservation biologist might say, the fauna constitute merely another "component" of biodiversity, which is of little or no interest on this view of *la forêt*, except insofar as it plays a supporting role in maintaining wood inventories or provides some independent stuff or service of its own. Le Roy's enlightened view of the forest is precisely the view of Natural Capitalism – the view of it as a warehouse, which stores or ought to be managed to store an economically advantageous and taxable inventory of wood. This is understood to be a quantity that efficiently satisfies the needs for fuel, construction, and manufacture (Harrison 1992, 114–124).

The enlightened forest collapses into the enlightened human economy. It is (Harrison 1992, 121) no longer something separate and apart "from the havoc of humanity going about the business of looking after its 'interests'." Natural Capitalism changes no essential element in this picture – of the forest as a pile of usable stuff; it merely reformulates this enlightenment portrayal in the languages of modern ecology and neoclassical economics.

What is missing from the enlightened forest is precisely a representation of anything that lies outside the walls of the city – that is, something other than (Harrison 1992, 74) "the voracious world of social humanity".[91] The enlightened forest cannot serve as that which specifically, emphatically, and essentially is not, and ought not to be, caught up in exchange, nor even the *idea* of exchange of goods. Thoroughly integrated into the domain of human vocation and industry, it is just another means to just another end of just another human project. An assimilated part of human culture, it is no longer a defining boundary. Chiseled away, no *alto-rilievo* remains, rather only a plain background in which not only the forest, but the potentially special and valuable qualities of human culture, are lost. The domain of the city, in which people exercise their capabilities to design and create, becomes more ordinary, less meaningful for being ubiquitous. Harrison (1992, 201) puts it in this way:

> We dwell not in nature but in relation to nature. We do not inhabit the earth but inhabit our excess of the earth. We dwell not in the forest but in the exteriority with regard to its closure. We do not subsist as much as transcend. To be human means to be always and already outside of the forest's inclusion, so to speak, insofar as the forest remains an index of our exclusion.

Later, he (Harrison 1992, 247) reiterates this idea in somewhat different terms, which hook tightly into modern-day plaints about the "loss of nature":

> The global problem of deforestation provokes unlikely reactions of concern these days... because in the depths of cultural memory forests remain the correlate of human transcendence. We call it the loss of nature, or the loss of wildlife habitat, or the loss of biodiversity,

---

[91] I hope that it is clear that I'm following Harrison in using the notion of "the walls of the city" metaphorically. In my usage, the wall is not an actual structure, nor even a political boundary. Rather, it stands in for the modally demanding attitude of "letting it be", which excludes drawing a place into the business of supporting our life in the cities or anywhere else.

but underlying the ecological concern is perhaps a much deeper apprehension about the disappearance of boundaries, without which the human abode loses its grounding.

One need not drag Heidegger into Harrison's Heidegger-tinged language to get at a tremendously important yet simple truth in his words: People cannot fully appreciate the deeper significance of what they design, create, and manage, except by reference to what they do not design, create, and manage. People cannot fully understand their proper place in the world nor the distinctive importance of their achievements – that is, their appropriate fit in the world – by making or even regarding the *entire* world as an unrestricted *arena* for the projects of human culture. Nor can this be understood (à la mode of modern conservation and restoration science) by regarding the entire world (including any part that might up to that point was excluded) to *be* the project itself – not just the grand arena for other projects. This understanding can only be achieved by placing human projects within the broader context of a world that reserves room for a natural world in the sense of not itself being a project or the arena for one.

The urge to turn nature into a project, as I have observed, finds expression in modern conservation, restoration, nature management, and what some conservation biologists forthrightly call "the gardenification of nature". Sticking to the concept of "forest", Harrison's exposition again affords a metaphorical understanding of the significance of these activities – as the incarnation of Dante's *selva antica*. As he puts it (Harrison 1992, 85–86), the *selva antica* "is merely a denatured *selva oscura*":

> Here there are no more lions, no more leopards, no more she-wolves. Thanks to the purgatorial process, this forest has ceased to be a wilderness and has become a municipal park under the jurisdiction of the City of God. In Christianity's vision of redemption, the entire earth and all of its nature become precisely such a park, or artificial garden.

One could be forgiven for mistaking this description for a description of even many places that we have come to regard as "natural" and "wilderness". For example, it applies to such a place as Yellowstone National Park, which was redeemed by driving out its wolves (and then re-introducing them) and by managing its fires.

In the next subsection, I return to parks and gardens and human relationships to them, which I believe *are* in certain ways expressive of natural value. But human relationships to such entities do not sustain the *core* of natural value. Rather, the natural value of parks and gardens is derivative and dependent on the core value of the natural world. That core value does not and cannot come from nature denatured, nature redeemed, or nature re-created or reborn any more than it comes from nature as development opportunity. That is because this is not the kind of "nature" that admits of the unique kind of relationship that humans can have with a part of the world that is *not* appropriated as a means to achieving some human project (economic value), and *not* appropriated as an end of a human project (the value of nature redeemed, "improved", or in "better balance").

Nor can the core of natural value come from a natural world that, just for now, is not appropriated – not, that is, when that state of affairs obtains only for the moment and only by virtue of the contingencies and vagaries of the current human economy and scientific fashion. If our decisions to meddle with and intricately manage our natural surrounds are routinely subjected to intricate computation – economic or scientific – and if they can therefore go now one way but tomorrow quite the other, then the natural world cannot possibly serve as a framing structure. A boundary that is contingent on and itself framed by principles that are poised to "justify" a boundary violation at any time is no longer a boundary because it is not regarded as such. This is a world whose boundary signs can be yanked out with casual disregard. In such a world, we rationally presume – and our attitudes reflect – the proposition that our projects will continually transform the world according to continually transformed market conditions and preferences that are shaped by forces that have little or nothing to do with our fit in the world. So while it might seem that no single act of transforming a single part of the natural world can be subject to moral criticism, that is not so. Not when such an act is the natural fruit of attitudes that countenance any and all such acts on any and all parts of nature – provided only that the winds of human economies and preferences (including the predilections of scientists) are blowing in the "right" direction.

As the world is more and more transformed and presumed subject to transformation in this way, as it is more and more a thing of human design and management and regarded as such, and as it is more and more evaluated and modified just to achieve those design and management goals, there is less and less of it recognized and recognizable as something outside of, setting off, and framing those characteristically human endeavors. Insofar as the world is a world of more or less successful development projects, it is a narrower world that admits a narrower set of relationships, which excludes the one that is definitive of natural value. That relationship requires a modally demanding "letting it be" of at least some portion of world, which by virtue of this, is valuable as nature.

In the next subsection, I will explore facets of natural value that lie outside its core. Before moving ahead to that, I wish to very quickly dispatch one view of "letting it be" that should not occupy us on its merits, but which nonetheless seems to have "legs". This view summons the charge that what I have characterized as a modally robust "letting go" of something that we therefore take as "given" is itself an imposition of human design and management. On similar grounds, I can take part credit for designing Vincent van Gogh's *Starry Night over the Rhone*. For, in a recent museum visit, I showed exemplary restraint by not whipping out my plein-air oils to add a few stars of my own and to reposition and diminish the brightness of Magrez (the root of the tail and faintest star) in Ursa Major, which van Gogh so carelessly misrepresented. Or with equal justification, I can claim to have left my mark in the design, management, and intactness of my critic's cranium. For I have not applied my baseball bat to the task of reshaping it to express my displeasure at her misunderstanding of what it means to design, create, and manage something.

In both these cases, my behavioral choices are real in the sense that I have the capability to leave my mark (or at least launch a serious attempt to do so), were I to put my mind to it. But it is absurd to say that, by actually choosing *not* to design or to interfere, I am designing or interfering in some covert or particularly subtle way.

Still, one might think that an elaborated variation on my non-interference stories might succeed in demonstrating interference where the unelaborated versions failed: Suppose that were I constantly calculating my chances of getting away with "enhancing" van Gogh's painting with a paintbrush, or constantly calculating my chances of getting away with reshaping my critic's pate (and thereby, her patterns of thought) with a baseball bat. Suppose I were merely biding my time for the best opportunity. Then at least in some important sense, I could be said to be exerting a managing influence in both these cases. On these accounts, the mere condition of being constantly poised to exercise my will to repaint and reshape the world around me suffices to properly attribute to me the role of designer and manager.

Thinking back to Philip Pettit's explication of true freedom, these variations can be seen to make a valid point. However, they do so precisely by violating the condition of modal robustness, which is a core requirement for valuing nature, according to appropriate fit. For I claim that natural value cannot survive a constant plotting, consideration, and reconsideration of whether or not to intrude into an as-yet unpenetrated part of the world – even if, as a matter of contingent fact, the intrusion never occurs. The variations on my original stories therefore fail to include a crucial element of the analogy that they try to press; this suffices to scuttle them.

Beyond that, the criticism that steadfast refusal to engage in an enterprise is itself an enterprise becomes uninteresting semantic carping about my use of the word "enterprise". But even on that score, I would lobby against abandoning my usage: The critic who favors folding into the meaning of "enterprise" all human choices, attitudes, and dispositions risks trivializing that meaning by making it too inclusive – running together salient differences about different choices, attitudes, and dispositions, which people might adopt.

In this subsection, I have abstractly characterized the kind of relationship that occupies what I believe to be the core of natural value. I have sought a means to do this that includes both traditional analytic philosophy and recurring metaphors, which endue with insight and meaning the way that people live their life. My characterization has major implications for what valuable nature "looks like" and doesn't "look like". I have already gestured at them here and they are further fleshed out in Sect. 8.2.7 (Some implications). Some of those implications follow from one final observation I'd like to make here: The sort of relationship that I've been at pains to describe as governing the value of the natural world is a sort of relationship that excludes basing that value on a state description, no matter how scientifically informed.

The next subsection expands on the core of natural value by fleshing out some derivative kinds of value that depend on this core.

8.2.5   *Natural Value Outside Its Core*

There is more to the value of nature than its core value, which has so far concerned me. Here, I suggest how natural value outside the core might arise as deriving from that core. "Derivative" does not mean "unimportant". However, I would say that all the derivative values that I describe are dependent on a solid core in the specific sense that they cannot survive in its absence. In addition, these other ways of (as I shall say) embodying natural value also embody values other than natural value. While some of these are entirely concordant with the core of natural value, they can sometimes overwhelm whatever natural value they might also embody, and even severely compromise it. As a consequence of all these factors, it is fair to say that it would be destructive of natural value generally to neglect the core by way of focusing entirely on derivative ways of realizing natural value.

What I have said about the core of natural value and its underlying, modally demanding relationship of "letting it be" might make it appear exclusive of other, familiar kinds of relationships that people have to nature – relationships such as those that are bound up in the creating, maintaining, and inhabiting of gardens. One might also think of public parks and landscape art, and the relationships that people have to them. After all, these are all human creations, well ensconced within the domain of human culture.[92]

However, as I am about to try to show, at their best and when sufficiently constrained by vigilant awareness of the need to moderate possible contravening effects on nature's core value, the human behaviors and attitudes connected with at least some kinds of these entities in at least some contexts embody valuable relationships. Moreover, these relationships as well as their value are legitimately characterized as constituting some important part of nature's value by virtue of supporting and reinforcing natural value's core.

How can this seeming contradiction be accounted for? There are several answers to this question; I shall focus on one answer that felicitously ties together the seemingly diverse kinds of entities mentioned just above.

Let's first think about gardens. On an initial, cursory assessment, one might say with Rudolf Borchardt (2006, 30) that, "The garden speaks of human modes of order, where man is master, subduer, and transformer."[93] But that is not the only "mode of order" that a garden can embody. For as Borchardt notes, the plants and flowers that populate a garden are part of a "prehuman", pre-cultural history, which even under the gardener's cultivating hands, grow and change in ways outside of human control.

There are many kinds of gardens and correspondingly many kinds of human attitudes and behaviors exemplified both in their creation and in how they are experienced by people who later view, inhabit, and pass through them. From this variety of attitudes and behaviors, gardens can come to have a variety of different meanings. In *A Philosophy of Gardens*, David Cooper (2006, Chapter 4, "The Meanings of Gardens") assembles an impressive menagerie of meanings, including

---

[92] I do not claim that gardens, parks, and landscape art are the only candidates for derivative natural value. But they are salient ones, which serve my limited purpose of suggesting an approach to this important topic, rather than propounding a comprehensive theory of it.

[93] This passage also finds its way into Harrison's work (Harrison 2008, 47).

mereological (part-to-whole), depictive, allusive, expressive, symptomatic, and associative meanings.[94] I wish to focus on yet another kind of meaning that certainly not all gardens have, but that a garden might have and that some actually do. When present, I believe it gives a garden a kind of natural value that is both derivative on, and reinforcing of, the core value of the natural world.

I will call the meaning that I have in mind "embodiment", to which I assign a "technical" meaning that should be read into the word (instead of its ordinary meaning) as I use it in the remainder of this discussion.[95] I believe that a garden's natural value derives from the core value of the natural world by (in my technical sense) embodying the latter in at least three different ways. Embodiment, as I have said, is a certain kind of meaning. In my usage, its meaning is expressive in any or all of three different and fairly specific ways.[96] The first way requires that the garden both possess certain properties and also refer to them[97]; the second, that those who create, maintain, or experience the garden stand in a certain relationship to it by virtue of its possessing and referring to these properties; that the properties in question are ones that we perceive to be present and prominent in *un*-gardened nature; and finally that the relevant relationship to the garden is one that requires of persons attitudes and dispositions that embody some part of the core of the kind of relating to *un*-gardened nature that give un-gardened nature its core value. Let me elaborate.

The first way in which a garden can embody the natural world involves the second of Borchardt's two "modes of order" – an acknowledgment and glad acceptance of the garden's fate as being something that is not entirely bounded by the gardener's preferences, her botanical acumen, or the creativity and artistry of her designs and means of striving to realize them. The gardener cares for the garden in a way that would be inappropriate for the natural world at large, for this care does not constitute, nor does it pretend to be, the "letting it be" outside the walls of culture, which realizes nature's core value. Rather, it is a caring-combined-with-acceptance, where the acceptance *is* a matter of letting go and recognizing that, after all the caring, the garden – both given that caring and despite it – will do what it "will".[98]

---

[94] My discussion, though taking a different course from Cooper's, is indebted to Cooper's for its inspiration.

[95] I appropriate this term from Cooper, but as explained in the next two notes, I use it in a slightly different way.

[96] This differs from Cooper's notion of embodiment, which he characterizes as expressive in a more general way.

[97] This is what Cooper (2006, 125–130) calls "exemplification", which is his riff on a Nelson Goodman's "metaphorical exemplification". To arrive at what I call "embodiment", I add to Cooper's notion two additional layers. The first has to do with the relationships that those who create and those who experience a garden have to it. This layer is inspired by Robert Pogue Harrison's deep exploration of the various manifestations of "garden" in Harrison (2008). The second layer concerns the specifically non-gardened-nature-referring properties that are exemplified.

[98] Here, as elsewhere in this reflection, I use scare quotes for the application of words to nonhuman entities that, when applied to persons, are words of volition. Of course, I do not use them with this connotation in their application to the nonhuman entities in my discussion.

The gardener makes a bet with the soil and with the seeds, and she tries to stack the odds. But she cannot take on the plants' own labor to gain a roothold in the soil from which to push skyward. As Karel Čapek notes, sometimes in the course of *The Gardener's Year*, forces outside the gardener's control conspire against this. The rain does not come; it comes in an over-soaking mildew-inducing deluge; the snails have a feast; the birds fly elsewhere and take with them their fertilizing guano; a frost destroys, weeds invade, or buds come too soon with an early-season warm spell. On the other hand, sometimes forces conspire to give unplanned "gifts", such as when, in January (Čapek 2002, 14), "without the gardener having suspected or having done anything, crocuses and snowdrops have pricked through the soil."[99] But always, there is a sense that whatever ultimate form it takes, whatever arrangement nature ultimately "chooses", whatever flowers flower and whatever fruit emerges is a "gift" more than the gardener's creation. As Harrison (2008, 56) says:

> … however much art may play a role in their design, gardens have a natural life of their own which exists independently of their formal determinations. Not matter, not idea, but life is the phenomenon that finds articulation in gardens.

This leads into the second way in which a garden might embody the natural world. Though gardens are placed "within the walls of our cities" – in the sense of being objects of design and cultivation, and therefore, culture – they nonetheless constitute space set aside. That space is a world in itself, a world that is physically and formally bounded – just as the natural world, properly considered, is truly outside the walls of our cities. Even despite the porosity of a garden's boundary – for some gardens are purposely allowed to "look out" on their surrounds – and even despite the ministrations of the gardener, it can be a world in which nature ultimately is permitted to "take over" and determine what emerges – what flourishes, what withers, and therefore, what, exactly, the soil yields. Such a garden both constitutes and refers to a welcome expression of the working of the natural world even though its existence owes to human culture. It is a haven from, and a counterweight to, the bustle of human economies in which gardens morph into agricultural enterprises whose value, in contrast to the garden's natural value, is measured in quantity of food produced and the efficiency with which that is accomplished.[100]

A third way in which a garden can embody the natural world is by emphasizing some one or more salient aspects of nature, which thereby become more evident to those who experience the garden.[101] The means, somewhat ironically, are "devices that impinge on the senses" – isolating, perhaps exaggerating, even to some extent

---

[99] Also quoted in Harrison (2008, 28). Čapek's book is largely a reflective devotion to a kind of caring about a nature that, for all the gardener's ministrations, still largely defines itself.

[100] I have been careful to say "*can* embody" and "*might* embody" because, as Harrison observes, certain gardens fail to achieve this potential. Among others, Harrison (2008, 109–113) fingers the oppressively ostentatious artifice of Versailles and (Harrison 2008, 167) the "depthless… flower beds that deck the entrances of corporate high-rise buildings", which are replaced overnight to maintain the illusion of "the 'perpetual spring' of Eden".

[101] This notion trades on one that Cooper suggests, in Cooper (2006, 141).

distorting what the ungardened natural world might offer up. I have in mind mostly effects that will be visual, though they could operate on other senses as well. They will be designed to suggest what a more practiced eye might pick out in the natural world without the aid of the gardener's conceits. These cues can be as simple as the placement of plants in positions that allow their form to be viewed to good advantage. Or they might consist in isolating groups of plants, that nature often "naturally" groups together. In such a way, a garden can (to appropriate Robert Harrison's turn of phrase for a different purpose) reanimate nature by making it more palpable, more present to human sensibilities as something separate from, beyond, and defining-by-bounding human culture.

Some might single out this last kind of embodiment as associated with certain public parks – particularly municipal parks and botanical gardens. But unfortunately, while many of these clearly refer to some isolated aspects of the natural world, they often do so via "devices", so extreme and so intrusive, that they are more emphatically a pronouncement about their designer and themselves as a cultural artifact. A park that falls under this description takes on the affect of a museum or a warehouse with a cleverly displayed inventory. I do not claim to know where the dividing line is, much less how, exactly, to define it. In some cases, I believe that it has something to do with size and the grandeur of the effort, which at some point pushes it into the domain of cultural pretension. But I think that overstepping the line has more to do with the "natural" elements in the park predominantly taking on the role of means to some goal of its creator, which is entirely disconnected from the core value of the natural world as bounding culture and setting it off in relief. This is particularly true when that goal falls squarely and obviously within the domain of culture and could possibly be served by other means. A park that trumpets this goal in any of a variety of ways is less an embodiment of the natural world, in the sense that I have described, and more a type of didactic lecturing, which only re-creates someone's problematic idea of what others ought to see in, and as, nature. In this way, it mimics the sort of problematic conservationist or restorationist ideal, which never connects to a clear or well-principled norm.

Such a place as the New York Botanical Gardens meets this description (Rothstein 2008):

> This garden in the Bronx, for example, began in 1891, when the New York State Legislature carved out 250 acres of the city's undeveloped land for "the collection and culture of plants, flowers, shrubs and trees." It was meant to advance "botanical science," but would also provide "for the entertainment, recreation, and instruction of the people." It was to be "a public botanic garden of the highest class," in which visitors were as essential as the collection.

Some parts of it comprise a kind of "scientific stagecraft", as Edward Rothstein puts it – little ecological niche vignettes staged in greenhouse dioramas. This is the modern-day version of yesterday's more straightforward presentation of "collections", presented in what amounted to curio cabinets. Today's collections are still largely mere collections, though ones with some visually enabled ecological pedantry. And (as noted by Rothstein (2008)) they tend to be placed in landscapes of "unremitting human effort".

Look at what it took to make this landscape so seemingly natural, the way bridges had to be raised across the Bronx River, or how roads were created so visitors could see the sights. Horses carted immense mounds of fresh hay off mown meadows or were harnessed to carve drainage canals in the unspoiled landscape.

The New York Botanical Gardens is an effort squarely in the tradition of Frederick Law Olmsted, who might be considered one of the first and perhaps still one of the most brilliant of landscape artists. By way of describing how Niagara was "invented", Ginger Strand (2008, 145) tells us that

Olmsted's best parks [particularly New York's Central Park and Niagara Reservation] are so naturalistic they are often perceived as nature itself, slices of preserved wilderness, neatened up and made more user-friendly. In fact, they are massive engineering projects, carefully designed, constructed and maintained. Central Park feels like a nicely tended piece of New York's precolonial environment, but it's really New York's largest art object, a monument build from the ground up by 4,000 laborers who blasted out rocks, built watercourses, installed drains and moved 5 million cubic yards of soil to create nature-seeming rolling hills, meadows and woods where there was only flat swampland before.

Olmsted's parks are highly artificed works of design expertise. He considered himself an artist; plants and trees were his paints.

Much of what Strand says about Central Park is strikingly apposite as a description of current-day restoration projects.

Strand's main topic is Niagara Falls, and she goes on to describe how Olmsted designed Niagara Reservation to "improve" nature. She (Strand 2008, 154) presents this excerpt from the proposal that Olmsted and Calvert Vaux drafted for this project:

We are far from thinking that all that is required to accomplish the designed end is to "let Nature alone". Inconsistencies, discordancies, disunities and consequent weaknesses of natural scenery may result, even at Niagara, from natural causes.

As Strand (2008, 154) recounts, "These natural causes included landslides, rock-falls, erosion, even ugly trees." Part of Olmsted's genius was his ability to more or less completely obscure his "improving" touches. Strand (2008, 157) instinctively recognizes how this effort runs counter to natural value:

Olmsted and his companions had inaugurated a new era at Niagara: The era of fake nature, an artificial wilderness designed to hide all evidence of design. Their park, while beautiful, solidified an opposition between "natural" and "man-made" that misrepresents our relationship to nature, obscuring the very real, increasingly critical role we play in the ecosystems of which we are a part.

As Strand says, such a park "misrepresents our relationship to nature". I would add to this the observation that the human role in ecosystems is not just the role of inadvertent influence. With increasing insistence and vigor, conservationists and restorationists have assumed Olmsted's mantle in designing and managing ecosystems, and even (or especially) in obscuring evidence of this. And like Olmsted, we tend to join these biogeoengineers in mistakenly insisting that our creations are more "natural" than the natural world.[102]

---

[102] The business of "faking nature" is not central to the current discussion; nor is it central to the notion of appropriate fit. I take up this topic in Sect. 8.2.6 (What appropriate fit is not).

But to return to parks, which are molded by the hands of a designer such as Olmsted: they are means to certain ends. For Olmsted, the ends were art and the edification of the public by exposing them to "beauty", according to eighteenth century ideals of the picturesque and pastoral. Our twenty-first-century sensibilities readily detect this dedication to cultural ideals of a bygone day and are not so likely to be reflexively sympathetic to them. But to view nature as a means to such ends has little to do with viewing and relating to it in the way that I have suggested lies at the core of nature's value. Whatever kind of value an entity such as one of Olmsted's parks might have, it is clearly not natural value – because it clearly does not embody the natural world in the required sense.

As I have already suggested in passing – by classifying Olmsted as the archetypal landscape artist – that modern landscape art might be read in much the same way as parks in Olmsted's tradition. Think of Michael Heizer's *Double Negative*, Nancy Holt's *Sun Tunnels*, Christo's *Running Fence*, or Robert Smithson's *Spiral Jetty*. These works are alike in their use of the landscape as a means or a prop to serve some imaginative purpose firmly embedded in human culture. However, they stray from Olmsted's aesthetics in the absence of pretense to create nature from whole cloth. And I would say that the imaginative purpose of some, but not all, landscape art resembles that of the kind of garden that I have described, which reanimates nature, as I said, by making it more palpable, more present to human sensibilities as something separate from, beyond, and defining-by-bounding human culture.

Of course, landscape art reflects natural value by means rather different from gardens'. Gardens are most often places well within the walls of the city, where parts of the natural world are brought in to realize the gardener's design. Landscape art commonly reverses this arrangement, bringing the artist out to the natural world, which then provides the setting for an expressive purpose realized through plainly human-made means. Perhaps this is most evident in a work such as Michael Heizer's *Double Negative*, in which the artist's "contribution" was to take something away – by excavating two trenches following the natural, eastern edge of Nevada's Mormon Mesa, and aligned across the negative space of a gap formed by a rise in the mesa.

To the extent that this work focuses attention on the inimitable contours of sandstone mesa and canyon formations in the United States' desert southwest, it is fair to say that it embodies natural value. But such a grand project – 457 m of trench that is 15.2 m deep and 9.1 m wide and the result of removing 218,000 tonnes of rock – also embodies the antithesis of natural value by its ostentatious intrusion and violation of the boundary between culture and the natural world. And insofar as the artistic purpose does *not* have to do with referring to the qualities of the desert southwest, Heizer's idea joins Olmsted's parks in violating natural value by relegating the natural world to the role of an available, grandiose, and outsize dioramic setting to serve the end of embodying a rather modest metaphor.

So I would say that public parks and landscape art are more problematic than at least certain kinds of garden insofar as their embodiment of the natural world's value is overwhelmed and contradicted by other values that attach to them. But it should be emphasized that neither can gardens embody natural value if there is not a truly natural world outside of them. The reason harkens back to the requirement that embodiment not only have certain properties, but that it refer to the existence of those properties outside itself. For natural value, that requires that some part of the world outside of gardens be reserved – that is, for which we exhibit a stubborn (modally resilient) unwillingness to impose any structure and a stubborn unwillingness to shape and bend it around our desire for stuff, services, health, convenience, beauty, or anything else. If a garden can no longer effectively refer to something outside the walls of the city because everything is within those walls, then it becomes self-referential and thereby loses its ability to embody natural value.

8.2.6 *What Appropriate Fit Is Not*

According to a theory of natural value according to appropriate fit, natural value at its core is a matter of people finding their appropriate fit with respect to the world. This entails a willingness to let some part of the world frame their life as "given", which is expressed via a relationship that consists in a modally stringent "letting it be". This "letting it be" requires steadfastly resisting the temptation to create, re-create, manipulate, or manage. It is a resolve to avoid transforming this part of the world – the natural world – into yet one more project among many other human endeavors. The core value of the natural world lies precisely in its unique ability to serve as a counterweight to culture – to be the "negative space" against which the positive values of cultural endeavors can be best understood and appreciated for what they are not.

My suggestion for identifying the core of natural value is a major departure from most all other suggestions of which I am aware – including from philosophers, scientists, economists, and environmentalists generally. I have touched on its most salient differences. But as a means to further fleshing out what appropriate fit *is*, I wish to more systematically catalog some things that appropriate fit is *not*. I believe that this exercise yields an additional major benefit: By distancing a theory of appropriate fit from other theories with which it might be conflated, the need to defend positions that it neither is nor entails should vanish.

"Anthropocentrism" is the 4 × 4-letter dirty word of environmentalism. But appropriate fit is **not anthropocentric in any obviously objectionable way**; it certainly is not anthropocentric in any way that justifies reflexive rejection by environmentalists. As Sect. 2.1 (An environmental philosopher's conception of value) remarked, the term "anthropocentric" has come to be used in a way that promiscuously tars any theory of nature's value that connects it with humans. But insofar as anthropocentrism is confined to its core assertion that everything that is valuable is human, it does not accurately describe appropriate fit.

Of course, appropriate fit certainly is anthropocentric in the sense that it brings human valuers into the value equation from the very start. But the vital role of people as initiators of valuing transactions is not a legitimate basis for censure. As I shall explain, to arrive at this jaundiced judgment requires a mashing together of several different respects in which a theory of natural value might be said to be anthropocentric.

Insofar as a theory of appropriate fit regards natural value to be built on valuing relationships between people and nature, it regards value as, in this sense, anthropo-*genic*. It thus takes exception to any theory that seeks to find in nature an intrinsic value propped up by a realist metaethics in which human valuers primarily function as circumstantial discoverers of value that would exist independent of their fortu-itous discovery: It abandons the sort of view characterized in Sect. 2.1.1, according to which, by dint of certain of its properties, nature emanates a moral force field – whether or not it ever impinges on valuing subjects – and nature's value is therefore "objectively" present in the world. In contrast, appropriate fit proceeds from a con-sideration of what it is to be a human being – a being who is immersed in both cul-ture and in a world larger than the world of cultural endeavor. It finds the value of nature through valuing behavior, which entails people choosing, in a modally demanding way, to keep some part of the world outside of, and therefore bounding, their various endeavors, projects and enterprises, which "live" within the realm of the human culture and economy.

Therefore, this is the sense in which appropriate fit regards the value of nature to be anthropo*genic*: People might be said to generate value by virtue of forming a special relationship with some part of the world outside of human enterprise. However, the value generated flatly contradicts anthropocentrism's core tenet that everything that is valuable is human. I suggested in Sect. 8.2.4 that, at the risk of misunderstanding, one could emphasize one pole of the relationship or the other. One could say that natural value is a matter of the respect for nature entailed by the sort of valuing relationship that I have been concerned to characterize. One could equally well say that natural value is a matter of self-respect in the sense that it takes seriously some important aspect of our humanness in the world in which we live. But this latter way of speaking is as much half the appropriate fit story as the former. It is not a basis for supposing that appropriate fit covertly slides into valuing something that is, in the end, entirely human. In fact, the theory of natural value according to appropriate fit is a direct protest against that view. Most saliently, this theory characterizes value in the natural world as deriving from "letting it be", with complete disregard for achieving any particular human end within the realm of human endeavor.

It once again helps to contrast this axiological picture with Natural Capitalism's antithetical one, according to which a valuing relationship with nature involves taking command of, and management responsibility for, its bio-parts so that they might best be selected and arranged to serve the human economy. Add that "the human economy" is ultimately defined in terms of transactions that satisfy human preferences – no matter whether (Satz 2010, 69) "formed by whim, confusion, tradition, peer pressure, and social context" and no matter whether reflecting values worthy of being satisfied. Then Natural Capitalism can be seen to endorse valuing nature in a way that ultimately circles back to valuing human desires or their satisfaction; this view, not appropriate fit's, is vulnerable to the change of circling back to valuing only that which is human. That is, it circles back to the core of anthropocentrism.

Appropriate fit is **not a naïve, scientifically uninformed, or Muir-esque pantheistic worship of nature**. From Muir (1911, 195–196), we have thrillingly graphic depictions of supernatural spirit suffusing all creatures:

> A few minutes ago every tree was excited, bowing to the roaring storm, waving, swirling, tossing their branches in glorious enthusiasm like worship. But though to the outer ear these trees are now silent, their songs never cease. Every hidden cell is throbbing with music and life, every fiber thrilling like harp strings, while incense is ever flowing from the balsam bells and leaves. No wonder the hills and groves were God's first temples, and the more they are cut down and hewn into cathedrals and churches, the farther off and dimmer seems the Lord himself.[103]

This wonderfully imaginative passage, among many similar ones, shows how Muir thinks that, or at least writes as if, natural objects are spiritually animated. According to him, nature possesses a vitality that derives from a supernatural spirit in all things natural and nonhuman. It generates a force field that, Muir exhorts, should bend the *human* spirit to worshipful attention. Insofar as this description also applies to intrinsic value force fields, Muir's pantheism resembles intrinsic value theories. The major difference is a metaethical nicety: Muir embraces supernatural properties to be value-conferring, whereas most current-day intrinsic value theorists nominate natural properties to play that role. But this difference is irrelevant to the shared way in which they stand in fundamental opposition to a theory of appropriate fit. That latter theory views the value of nature as arising, not from moral force fields that emanate from objects that exemplify some particular structured property, but by virtue of people properly judging their relationship to the natural world. John Muir might have felt the pull of the kind of relationship to nature that, the theory of appropriate fit proposes, is the basis of nature's value. But if so, his writings do not articulate it as a source, let alone the primary or original source of nature's value.

Appropriate fit is **not a romantic yearning for a lost, unspoiled Eden** – whatever time or place might be assigned to such an historic or mythic, and typically static, state of ideally valuable nature.[104] Edenic theories of natural value assess the value of nature by virtue of how closely it approximates this ideal state, which tends to be a static picture of the "natural world" *sans* traces of humanity, or *sans* the influence of selected civilizations, or *sans* some human influences but not others. I believe that this conception of the natural ideal founders before it is even launched, for being unable to satisfy two prerequisites – that Eden can actually be characterized, and that there are some practical means of achieving it. But even if it were to survive those rocky shoals, it is sunk (as I argue in Sect. 8.1.2) by its inability to justify any chosen ideal in a principled way.

I am therefore concerned to distance a theory of appropriate fit from anything resembling this picture of natural value, which Bill McKibben famously projects in

---

[103] This is the entry for July 24, in Muir's recounting of his 1869 experience in California's Sierra Nevada mountains.

[104] That appropriate fit incorporates no Edenic vision is vividly illustrated in Sect. 8.2.7.2 (Implications for action, social policy, and conservation).

The End of Nature.[105] In that jeremiad, McKibben (1989, 58) both seizes upon and is seized by the implications that humankind is generating a new kind of climate:

> We have changed the atmosphere and thus we are changing the weather. By changing the weather, we make every spot on earth man-made and artificial.

Moreover (McKibben 1989, 84),

> By domesticating the earth, even though we've done it badly, we've domesticated all that live on it.

But what most upsets McKibben (1989, 73) is that, according to him, humankind has upset the natural order of things – a harmony whose permanence is bound up in his idea of nature:

> The chief lesson is that the world displays a lovely order, an order comforting in its intricacy. And the most appealing part of this harmony, perhaps, is its permanence – the sense that we are part of something with roots stretching back nearly forever, and branches reaching forward just as far.

Nor does McKibben hesitate to position himself within the theological tradition of Aquinas and Leibniz (discussed in Sect. 2.3.2), in which this natural order is directly connected to God. In McKibben's exegesis of the Biblical story of Job (McKibben 1989, 75–76) the Creator expresses His great pride in His creative accomplishment even insofar as it exists apart from people. According to McKibben, Job shows how He expresses His displeasure at the self-aggrandizing imposition of human presence and will on His creation.[106] McKibben quotes Henry David Thoreau's awed reflection, while (Thoreau 1864, 70) "sloping down some miles toward the Penobscot" River (Thoreau 1864, 71), that "here not even the surface had been scarred by man… It was a specimen of what God saw fit to make this world". Channeling Thoreau, McKibben (1989, 72) echoes: "The earth is a museum of divine intent."

I hope that it is clear how McKibben's vision of nature as the Museum of Eden – whether "of divine intent" or not – is fundamentally antithetical to appropriate fit's view of the natural world. It is true that not everything that he says coheres with this vision; sometimes he even says something that might initially appear to fit comfortably within appropriate fit's ambit. When he laments (McKibben 1989, 210) that what used to be nature now "bears the permanent stamp of man", this might recall the value, expressed by appropriate fit, of some portion of this world that is set apart

---

[105] In later works, McKibben manages to move beyond *The End of Nature*'s unremittingly despondent view that humans have thoroughly and for all time defiled the natural world. In *Hope, Human and Wild*, he is prepared to grant (McKibben 1995, 11) that, however much humans have spoiled nature, we are capable of redeeming it – and ourselves – by pursuing "a vision of recovery, of renewal, of resurgence".

[106] This theology also grounds McKibben's (1989, 84) account of human vices connected with nature. On that account, God's pride in His Creation rightfully supplants man's pride and pretensions to domination with respect to it. The theme of virtues and vices connected with nature's value – in more secular form – shortly comes up for discussion and contrast with appropriate fit.

from the projects and enterprise of mankind. But this tentative connection with appropriate fit is loosened by McKibben's claim – which coheres with his static conception of an ideal state of nature – that the stamp of man is everywhere and everywhere permanent. Everywhere includes "the remaining forest primeval" because by virtue of humanity's tinkering with the climate, the drops of rain that fall there are manmade. And the connection with appropriate fit is severed entirely when one realizes that by McKibben's theological lights, man has fallen and taken nature – conceived as the Eden of its pre-human state – with him.

A theory of appropriate fit has no place for a static, state-based view of an Edenic ideal, or indeed, of any other kind of ideal. It pushes this conception of "the natural" to the side along with all other state-based conceptions. Instead, it focuses on how people should act and behave in relation to the natural world in whatsoever state we might find it in now.

Closely connected to the previous point, appropriate fit is **not a resurrection of old state-based ideas of wilderness,** even though Robert Pogue Harrison and Ginger Strand include that notion in passages that I cite in the previous two subsections. Traditionally, the wilderness idea is tacitly based on ideas that hew close to McKibben's – by incorporating some historical (mis)conception of an ideal state of nature, which human meddling and intrusion necessarily taint. Appropriate fit parts ways with these wilderness theories of natural value in its very different basis, which does not require defending an indefensible conception of a norm-imposing ideal – Edenic, pre-Columbian, pre-industrial age, or any other. Even more dramatically, it parts ways with wilderness theories with regard to *what* kind of place it regards as part of the natural world. This is a crucial point; I expand on it later in the next subsection.

Appropriate fit is **not biophilia in disguise**. Biophilia, as Sect. 6.6 (Biodiversity as progenitor of biophilia) explains, is a speculation in evolutionary psychology. Even its most insistent champions acknowledge its speculative status. While consistent with evolutionary theory, it simply has no credible evidence. Moreover, even if such a thing exists, the argument for its *value* falls into the trap of a genetic fallacy. The argument assumes that, since this alleged, characteristic mind-set of humans once might have served people well, it is still a good thing; therefore there is a moral imperative to ensure that it continue to be satisfied. Finally, even one were to ignore these dual fatal defects – that biophilia has no defensible basis in fact or norm – it appears that the force of any biophilic norm would be extraordinarily anemic. At most, it would justify practices such as placing potted plants in work cubicles and decorative greenery around suburban housing tracts.

In stark contrast, appropriate fit has no need for, and does not make use of, speculation in evolutionary psychology or in any other domain. It focuses quite self-consciously on the business of living a good human life in the world that humans find themselves in today. It therefore avoids all of biophilia theory's fatal flaws. And its implications (the topic of the next subsection) – that people ought to regard the natural world as something to be held separate and apart from their characteristically human projects – are far more serious and far-reaching than biophilia's concern about decorating work cubicles and the like.

Appropriate fit is **not a norm for making a bequest of today's goods to future generations,** according to some theory of intergenerationally just distribution of natural goods. It is not a theory that attempts to justify an obligation on the part of current humankind to hold in reserve – in order to hand over to future humankind – some state of nature, or some pile of stuff, or (in today's oft-preferred economic parlance) a supply of natural capital. In fact, appropriate fit is not a forward-looking theory of natural value in any sense suggested by the concept of a bequest.

These differences are due, in the first instance, to the fact that appropriate fit does not recognize any grounds for thinking that there is a moral stake in the obtaining of *any* particular state of the world whose natural properties can be preserved for future persons. Nor does it find any reason – based on nature's value – to preserve some particular pile or amount of stuff or of natural capital. *A fortiori,* it does not find any reason to suppose that our current generation should strive to realize some particular state of nature or pile of natural capital as representing a fair or adequate bequest to future generations.

When it comes to appropriate fit's protest against designing, re-creating, restoring, and managing the natural world, this is **not a protest against "faking nature", or the vice of "hubristic domination of nature", or a violation of some "right" that the natural world is supposed to hold, which enables it to pursue its will**. Railing against the human perpetrators who have wrecked the "museum of divine intent", Bill McKibben (1989, 84) pounds on the theme of hubris with a characteristic flourish:

> It is a brutish, cloddish power, not a creative one. We sit astride the world like some military dictator, some smelly Papa Doc – we are able to wreak violence with great efficiency and to destroy all that is good and worthwhile, but not to exercise power to any real end.

Similar in spirit, but perhaps more carefully considered and more compelling, is Robert Elliot's line of reasoning. According to Elliot (1982), nature can (and should) be regarded as an original piece of art. Consequently, restoration can (and should) be regarded as a kind of deception. The deceit involved is the deceit of fabricating a replica, which, like forged art, cannot have anything like the original's value – even if some or most observers cannot distinguish it from the original.

Nor does a theory of appropriate fit bear any kinship to Eric Katz's "visceral reaction" (Katz 1992, 234) that "The fundamental error [of restoration and management projects] is… domination, the denial of freedom and autonomy" (Katz 1992, 240). In other words, Katz views restoration and management through the lens of the violation of certain rights, which the natural world is supposed to hold in order to ensure its exercise of autonomy.

Dale Jamieson (2008, 165) has his own version of a story that Elliot and others tell, and which is intended to help make the case for natural value being a matter of something like artistic authenticity:

> Imagine… we are out hiking and see a landscape filled with amazing mud-covered large termite mounds. You are beside yourself with admiration. But when I tell you that these are

fake termite mounds, put up by the local chamber of commerce to amuse people who aren't interested in bush-walking, your face falls.[107]

Such a story relies for its impact on our understandably grumpy reaction to being taken for a fool.

This is not the basis on which appropriate fit makes its judgment. Nor does appropriate fit buy into Jamieson's different take that our grumpy reaction derives from the fact that the beheld landscape is the "product of human influence". We would not be grumpy in this way (in fact, we might be delightedly astounded), had we been introduced to the place as a showcase for Christo's landscape art. But also, in the absence of intentional deception, appropriate fit does not join Jamieson (or Elliot) in denying natural value to places that are quite obviously and in the extreme a "product of human influence". According to appropriate fit, our justified reaction to some such places again can be the rather different one of delighted astonishment. That reaction – to nature's bravura improvisation in such a place as Chernobyl's Zone of Alienation – is an example, which I discuss in Sect. 8.2.7.2 (Implications for action, social policy, and conservation).

In some cases, appropriate fit might arrive at conclusions similar to those of Elliot and Katz. But if it does, it will do so on entirely different, and (I believe) more plausible and defensible grounds. Its justification does not depend on a tenuous analogy to manmade art. According to a theory of appropriate fit, the problem with nature restored or nature managed is not that it is a sort of a mocking deception, something deserving of outrage.[108] Rather, nature restored or nature managed is not nature valued in the way that is required by an appropriately valuing relationship, whereby it is left to be what it will be.

Moreover, a theory of appropriate fit is not propped up by, and in fact eschews a questionable application of, the language of human rights and entitlement to the natural world. As a consequence, it does not encumber itself with the onerous task of giving good reasons for why nature (or ecosystems) should be thought to have *human* or human-like rights. The insuperable difficulties of pulling off this task can be seen in the brief remark of Katz's cited above. This commonly presented rationale for the claim that nature has morally relevant rights merely appeals to the related claim that rights-violating behavior such as "domination" should be subject to censure. But this second and undefended claim is not a legitimate defense of the first one. That is because it merely (though correctly) asserts that rights-violating behavior is subject to censure, while leaving open the question of whether there are legitimate grounds for according rights to nature. We can all agree that censure is an

---

[107] Jamieson draws a different, and I believe, questionable conclusion from his story. It is conveyed by his next sentence, which draws on his shaky principle of "human influence" as the destroyer of natural value:

What you had thought was natural, you now see as the product of human influence.

See Sect. 6.12 (Biodiversity as the natural order).

[108] However, something like that reaction might be warranted when it comes to one of Frederick Olmsted's parks, as discussed in the immediately preceding subsection.

appropriate reaction to a person who violates the rights of another *person*, including those rights that relate to her autonomous pursuit of a life as she sees fit. But that is because there are independent grounds, based on that other person's interests in the context of her human life, for believing that rights-holding norms actually do apply to her. In contrast, there appear to be no independent grounds for thinking that the natural world is entitled to rights in this morally interesting sense.

I have visited at several previous junctures the topic of anthropomorphizing attempts to attribute value to nature on the basis of its integrity, health, and autonomy. I remarked specifically about ecosystem health in Sect. 7.2.3 and reprised those remarks in Sect. 8.1.5 where I also added a few regarding ecosystem integrity. Here I wish to expand my brief treatment of autonomy in Sect. 6.12, while also indicating how integrity is entwined with it.[109]

Autonomy undoubtedly commands moral respect when embodied in a person. But one must look to the reasons for why this is so in order to determine whether or not those same or equivalent reasons apply with equal, equivalent, or at least some analogous and substantial moral force to nature.

On any reasonable account, the human right to autonomy is rooted in the perception that every person, or at least every competent adult person, has a right to choose (within bounds defined by factors such as similar rights of others) how to live her life. She has a right to make important decisions – for example, to choose which goals and projects to pursue. In part, this right resides in the presumption that each person has interests of her own and the concomitant presumption that she herself is generally in the best position to make choices that serve those interests and thereby serve her own welfare. This is what Ronald Dworkin (1993, 223) calls an "evidentiary account" of autonomy.

However, the evidentially account of autonomy seems incomplete. Even when a decision clearly does not serve its author well (something that she herself might sometimes realize and acknowledge) – that is, even when she decides imprudently, in a way does not serve her genuine interests, or even in a way that is self-destructive – we do not revoke her right to autonomy because the basis for that right does not entirely reside in the right for a person to best serve her own interests. Completely accounting for her right to autonomy requires another consideration, which Dworkin (1993, 224) characterizes as the protection of a person's "capacity to express [her] own character – values, commitments, convictions, and critical as well as experiential interests – in the life [she] leads." We respect the exercise of this capacity whether or not we judge, or even the person exercising the capacity judges, her bundle of values, commitments, convictions, and interests to be entirely worthy or even coherent. Dworkin calls this the "integrity view" of autonomy, where the integrity in question is the integral expression of a person in these various ways. It lets us (Dworkin

---

[109] For much the same kind of discussion regarding the concept of "ecosystem health", see Jamieson (2002), Chapter 15 on "Ecosystem Health: Some Preventive Medicine", 213–224. Jamieson, however, is disinclined to apply a similar analysis to the notion of ecosystem autonomy. My treatment of autonomy largely follows that of Dworkin (1993, 222–229).

1993, 224) see the right to autonomy as a means of protecting each person's "general capacity to lead their lives out of a distinctive sense of their own character, a sense of what is important to and for them."

On either the evidentiary view or the integrity view of the source of autonomy, or on a view that combines these sources, the autonomous subject must have interests. According to the evidentiary view, the subject must be capable of a self-conscious exercise of a capacity to choose goals as well as the means of trying to achieve them. On the integrity view, the subject must additionally hold, and therefore be capable of holding, a bundle of values, commitments, and convictions. Except on a view that endues the natural world with a vigorous kind of animism, it appears that nature does not meet even the weaker of these requirements. That is because, as I have said in those previous discussions, nature cannot be said to pursue its own interests. As for the integrity view of autonomy, it is hard to defend the view that nature (or an ecosystem) pursues a life according to held values, commitments, and convictions – except perhaps in the derivative and morally irrelevant sense that it might not be under complete human control. These dual failures likely stem from a common root: Nature is simply not the kind of thing that has a life at all.[110] As a thing that is not the subject of a life, it has no interest in, let alone a sense of importance about, how its "life" goes – even putting aside the problem of ascribing to it a sense of its own commitments to it. And it certainly has no capacity to make choices in a way that constitutes self-conscious expression of its character.

One attempt to salvage a respectable and morally weighty conception of autonomy for nature is to fall back on a more inclusive conception of "interests" – for example, one that could be said to apply to any individual and nonhuman organism. This would be something like Paul Taylor's *telos* or Holmes Rolston's defense of a life (see Sect. 2.1.2.2, Deontology). However, Taylor's conception still requires a subject of life with a purpose; and Rolston's still requires a subject that defends a characteristic way of being. Neither conception survives its stretch to the natural world, which is neither a subject of life, nor something that has any characteristic way of being, let alone defends it.

So it appears that there is no reasonable basis for supposing that the natural world has a right to autonomy in the morally relevant and compelling sense that applies to persons or even nonhuman organisms. This realization makes clear that charges of "paternalistic", "dominating", "hubristic", or "disrespectful" behavior with respect to nature cannot be grounded in violations of autonomy (or integrity, or health) – for nature does not have autonomy (or integrity, or health) in anything like the sense required to put normative teeth into their bite. As a consequence, character-assassinating

---

[110] Supporters of the Gaia Hypothesis in some of its forms will dissent from this view. More generally, so will those who subscribe to a purely autopoietic definition of "living", which they believe, qualifies the planet as a living organism. I believe that these views ultimately fail roughly because the principles underlying them show that far too many things – not just the earth, but candle flames, for example – live. Unfortunately, it is dubious whether any of the customary moral precepts, such as those stemming from a living thing's autonomy, apply to all entities that are "alive", according to this distended conception.

charges such of these devolve into dramatic, but undefended and apparently indefensible, expressions of disapproval. In the end, disapprobation perched atop anthropomorphizing projections carries no moral weight.

In contrast, a theory of appropriate fit does not entail or endorse any anthropomorphizing norm relating to nature. It has no need to suppose that the value of nature is bound up in an obligation to respect nature's supposed interests, to let it make self-conscious choices about how it lives or to defend its "life", whatever that might be. It stakes out a much less dramatic position, but also a more credible and defensible one: It suggests that the foundation of the natural world's value derives from the value of a human relationship to it that is founded in the stable attitude of people to let it be what it will be. It has no need to suppose that this relationship is grounded in nature's rights to be healthy, autonomous, or to maintain its integrity.

Sect. 8.2.3 ("Living from" nature, uniqueness, and modal robustness) in combination with many other things that I've said along the way to this point should make it clear that appropriate fit is **not a theory of the instrumental value of nature**. In particular, it is not a theory of how valuable nature is because of benefits that it confers on people. Most emphatically, it is not a view of natural value through the lens of "The New Economy of Nature". Consequently, it **has nothing to do with "existence value"**.

I remarked in Sect. 2.1.2.1 (Consequentialism) on a widespread misunderstanding about existence value as a kind in the taxonomy of economic values; in Sect. 6.9.1 (Option value and conservation) I touched on its relationship to option value as another of those economic value kinds. The first of these earlier subsections distinguished existence value (which is sometimes attributed to biodiversity) from intrinsic value, which is often mentioned in the same breath as existence value, as if the two were equivalent. The second remarked on the similarity of existence value to option value as, using economic parlance, a "non-use" – and therefore non-rivalrous and non-excludable – kind of economic value. The meaning of this is based on the economic meaning of "consuming" some thing, which is "to satisfy certain preferences with regard to" it. Given that, the non-rivalrous and non-excludable qualities that existence value shares with option value mean that, when you and I consume some thing for these kinds of value, we can satisfy our preferences with regard to it without either of us using it up and without either of us interfering with the other's consumption (at about the same time) of that self-same thing.

In other words, existence value is a categorical nicety, characterizing a certain class of values that sometimes drive consumption (in the economic sense), including the consumption of nature. Insofar as some piece of the natural world is held to have existence value, it is part and parcel of "The New Economy of Nature" conceived, Natural Capitalist style, as a division of the broader human economy. As such, its value is conceived as its value in exchange[111] for potentially any other thing, according to their relative ability to satisfy preferences.

---

[111] I take "value in exchange" in the broad sense that includes neoclassical economics' conceptual extension of the real marketplace to include contingent evaluation, as well as "trades" inferred from hedonic or other proxies. See also Note 71 regarding this broad conception of the marketplace.

The story told by a theory of natural value according to appropriate fit could hardly be more different from the one told by means of existence value. According to appropriate fit, the core value of nature is entirely disconnected from the shifting sands of human desires and vagaries of the market. It cannot be conceived centrally as the value of what might be gotten in exchange for it. It is not something that people stand in a relationship to as consumer to thing consumed. It is not something that even can or ought to be "consumed". Most of all, natural value-as-existence value cannot account for the uniqueness and resilience that I, along with most environmentalists, perceive it to have. In contrast, that is precisely what appropriate fit does – by virtue of the unique and modally demanding kind of relationship that underlies natural value according to that theory.

Appropriate fit is **not a construct whose credibility relies on pushing rival theories down a slippery slope.** That is, the case for appropriate fit does not hinge on leveraging difficulties in drawing the line between genuine conservation or restoration and human development. It has no need for an argument along these lines:

P: It is impossible to draw a bright line between the "conservation" of nature and nature-destroying intrusion.
C: Therefore, there is no valid distinction between the two.

The theory of natural value according to appropriate fit does, indeed, arrive at a negative assessment of conservation, as it is now widely conceived. But the preceding argument has no role in this. Rather, appropriate fit finds current-day (neo-)conservation anathema simply because it insists on the central importance of a basic relationship, whose terms neo-conservation projects routinely violate. Worse, actual neo-conservation practice violates the terms of a truly valuing relationship to the natural world, not merely as a consequence of repeated but circumstantial oversights, but by virtue of what "conservation" is fundamentally (mis)understood to be. Section 8.2.7.2 (Implications for action, social policy, and conservation) further explores this topic.

Appropriate fit is **not a theory of nature as having intrinsic value** – at least not in three of the four senses of "intrinsic value" characterized in Sect. 2.1.1 (Concepts and categories of value). The first sense (on that previous accounting) hinges on the fragile dichotomy of instrumental versus non-instrumental values, whereby those latter are considered as ultimate. This sense has little or no relevance to the core of natural value, considered as arising from a unique, morally stringent kind of relating to the natural world. Appropriate fit regards the natural world neither as an instrument to some other good – any more than one would regard a friend as such – nor as a non-instrumentally valued end. Insofar as any ultimate end for the natural world will be conceived as a state characterized by some structured set of natural properties, appropriate fit will reject it. For, as I have previously emphasized, appropriate fit finds value, not in achieving certain states of nature regarded as ultimate ends, but rather in a certain way of *relating* to a part of the world not of human design or under human management. In truth, appropriate fit resides outside the realm of the false dichotomy of instrumental versus non-instrumental. It is perhaps best characterized as a form of human relating that is a constituent good in pursuing a human life on this planet.

Turning to the third sense of intrinsic value: Appropriate fit self-consciously takes as its starting point a position antithetical to one that views intrinsic value as "objective value" in the sense of "existing independent of actual or even possible human valuers". A theory of appropriate fit has no place for the notion that some entities or states of the world emanate moral force fields, quite independent of actually existing humans and even the concept of a valuing human. The notion that it is the job of humans, as moral agents, to align their behavior along the field gradients is entirely alien to it. Quite to the contrary, a theory of natural value according to appropriate fit *starts* with human valuers and the unique (and I would say, valuable) kind of relationship that persons, individually and collectively, can have with respect to the natural world.

With regard to the fourth definition of "intrinsic value" in Sect. 2.1.1, appropriate fit clearly does not attempt to define natural value in terms of some recognizable, structured set of properties that inhere in nature.

That leaves the second sense of "intrinsic value" – as "worthy of admission into the domain of moral consideration". Of course, whether or not the natural world is intrinsically valuable in that sense is the cardinal question. My answer, based on this chapter's account of nature's value according to a theory of appropriate fit, is that it does rise to this high level.

At first blush, appropriate fit might seem to be closely aligned to virtue ethics because it trades on certain attitudes and characteristic behaviors of persons, which, I have argued, are conducive to lending meaning to characteristically human projects. But a theory of appropriate fit **does not comfortably fit under the rubric of environmental virtue ethics, as those theories are commonly propounded**.[112] Accounts of environmental virtue ethics saliently promote such supposed virtues as benevolence towards nature, and denounce such vices as hubris in domination over nature. They denounce, too, other vices that are said to abrogate nature's rights or to diminish the welfare of other *persons* by way of depriving them of nature's benefits.

These central themes within environmental virtue ethics fall into many of the same traps that ensnare other approaches. Most prominently, the virtues and vices just mentioned tend to be underwritten by anthropomorphizing language, which tacitly imputes person-like characteristics to the nonhuman world – for example, health, autonomy, and integrity. These characteristics then become an illegitimate basis for leveraging, on behalf of the natural world, their legitimate moral implications for the beneficent, non-dominating treatment of other persons. I have already shown, including earlier in this subsection in my discussion of nature's autonomy, that the required lever bends or melts away entirely under the heat of scrutiny.

---

[112] What follows is obviously not an exhaustive or detailed examination of environmental virtue ethics. The modest dimensions of my discussion do not even allow a comprehensive treatment of the views of any single proponent. However, I hope that it does serve my purpose of suggesting both how far my views range away from some prominent environmental virtue ethics themes, and why.

The sole item remaining on the brief list of virtue ethics stratagems suggested above is a principle of beneficence towards the environment or the natural world. Even standing by itself, this is no small thing. It is a broad and sweeping gesture because benevolence is an archetypal virtue whose compass is usually taken to include more specific and important virtues such as respect, compassion care, and love. A principle of beneficence purports to demonstrate how the virtue of benevolence is appropriately expressed in beneficent acts towards an "other" whose care or whose good and goods are entrusted to us – no matter whether via explicit request or consent of the beneficiary, or implicitly via recognized social and moral norms or rules. But in contexts of central moral concern, "beneficent acts towards an 'other'" means "acting in that other's best interests". In its application to nature, a principle of beneficence founders yet again on the stubborn truth that nature itself has absolutely no interests in the morally relevant sense. One can plausibly claim that other persons benefit from some action that impinges on the environment. But that does not answer the question of what would justify saying that *nature itself* benefits from such an action. That justification is required to uphold a specifically environmental virtue of benevolence towards nature, in contrast to a more generic virtue of benevolence towards other *persons*.

Moreover, even if (contrary to fact) nature had interests of its own in the morally relevant sense, and even if (also contrary to fact) there were some clear sense in which nature itself benefits from certain treatment, beneficence would still have little meaning with respect to nature if nature were not the sort of thing that, at least in principle, could properly receive or acknowledge the benefits received from its benevolent trustees. A normally competent person is capable of this receipt and acknowledgment; she is capable of understanding herself as the beneficiary of a benevolent trustee. This is what makes her *capable* of explicit request or consent for a trustee to assume a beneficent responsibility, even in a circumstance in which consent – explicit or inferred via well-accepted convention – is not required to set up the trust. But nature has no such capability. Without it, there is a corresponding absence of the customary moral force that attaches to moral discourse concerning a trustee's responsibilities and a beneficiary's beneficial treatment.

These problems – with trying to connect a virtue of benevolence to nature as an appropriate recipient of beneficent behavior – make this conception highly problematic if it is taken literally. It is possible that the relevant locutions connected with "beneficence towards nature" are meant to be metaphorical expressions of approbation. In this case, we should expect some reason that justifies this normative use of the metaphor. But so far as I can tell, virtue ethicists have little to offer in the way of justification that does not lean heavily on consequentialist or deontological theory.

Of course, as I remarked in commenting on Ronald Sandler's environmental virtue ethics at the end of Sect. 8.1, *persons* benefit (or not) according to how other persons act on shared environments. This especially applies to environments that are, or are regarded as, *resources* for us all. Insofar as some persons (the beneficiaries) rely on other parties (the trustees) who are responsible for acting on behalf of the beneficiaries' interest with regard to some portion of these resources, the parties who take up this responsibility – the trustees – can legitimately be said to be beneficent (or not) insofar as they seriously take into account the interests of the beneficiaries. But first and foremost, this is a matter of interpersonal beneficence. The happenstance that the "goods" placed in the trustees' hands are the stuff and services offered up by the natural world is mostly peripheral to the moral picture painted in terms of the various virtues of benevolence. The goods in question could just as well been a pile of hammers or a pile of money. Therefore, it is hard to endorse Sandler's positioning of these virtues (Sandler 2007, 52–55) as preeminent *environmental* virtues.

Moreover, Sandler's and similar virtue ethics views can be seen to suffer a special strain of the general malady of emphasizing moral concern for nature as prominently related to human needs and desires to "live from" it. Insofar as this emphasis infects virtue ethics views, they are subject to the entire suite of objections that generally apply to it. With regard to Sandler's exposition of virtues of benevolence, for example, there is the clear danger that the trustees can fulfill their fiduciary responsibility and act according to these virtues by hacking the natural world to some minimum number of pieces and parts whose rearrangement still yields the requisite "living from" benefits. So it seems that, insofar as this kind of benevolence occupies the core of environmental virtue ethics, it gets no further than a rather straightforward kind of consequentialism – and perhaps more specifically, a Natural Capitalist or ecosystem services viewpoint. There is little or no advantage to serving up this failed axiology, which centers on the appropriation and distribution of environmental resources, in the language of virtues rather than in the language of human welfare. It suffers from the same fatal defects, however it is couched.

A virtue ethicist might insist that, no, the environmental virtues of benevolence really are a matter of being beneficent directly towards the natural world, as opposed to indirectly (via conferring benefits on other people), or merely as a manner of speaking, about beneficent acts towards other persons. The natural world is, in the language of virtue ethics, "an appropriate basis" for beneficent acts.[113] But it is difficult to interpret this objection as anything more than a restatement of the proposition that nature is to be viewed as a collection of goods, which must be properly distributed to human beneficiaries, and whose environmental nature is incidental to the moral point. The natural world is a basis for a virtuous response because it is one drop in an ocean of goods whose fair distribution best serves *human* interests.[114]

To this point in my discussion, I have focused on an archetypal environmental *virtue*. Of course, on the other side of the virtue-vice divide lies environmental vice. In one version of the dark side, Jeremy Bendik-Keymer (2010, 79) expresses the view that the thoughtlessness of human behavior, which impinges on the diversity of life, "destroys the order in which we live". It thereby creates (Bendik-Keymer 2010, 76) "a gutted, looted museum of the history of Earth". That description, reminiscent of McKibben's accusation (McKibben 1989, 72) of humanity defiling the "museum of divine intent", is anything but normatively neutral. It suggests what

---

[113] This specific language for framing virtues follows Sandler (2007, 40–41).

[114] I take this to be the thrust of the section on "Human Flourishing" in Sandler (2007, 52–55).

Bendik-Keymer also says explicitly – that the sort of behavior that contributes to this planetary condition is, to use his word, not "decent". According to him (Bendik-Keymer 2010, 75), "… decent people do not destroy thoughtlessly" and certainly decent people do not gut and loot a library. Bendik-Keymer prefers to avoid the language of harm. But it is hard to avoid thinking that those actions and behavior, which preoccupy him for their indecency, do so precisely because he believes that they greatly harm people; and that they harm people, in part, by depriving them of previously available libraries with their accustomed resources and services.

Of course, merely calling the natural world a library does not automatically make its value proposition that of a library of books. Nor does merely labeling people "indecent" thereby make them indecent. Unfortunately, as discussed in Sect. 6.8 (Biodiversity as font of knowledge), none of the myriad repetitions of the trope of the natural world as a library of genes and species – including Bendik-Keymer's imaginatively embellished version of it (Bendik-Keymer 2011, 15) – tell us why the genes, species, and other biotic "books" on the library's shelves are so saliently important, why they are more important than, staying with this metaphor, high circulation volume. Or why are they are more important than the evolutionary processes that run the library's circulation department? Or why might there not be greater decency in pursuing goals that wholly revamp and reduce the library's holdings in order to better promote other salient human ends?

For Bendik-Keymer, the central consideration bearing on what he calls (Bendik-Keymer 2011, 15) "the library of life" is (Bendik-Keymer 2011, 13) an "*extended respect* for a life being alive" [italics in the original]. By this he means "extend[ing] to other living beings something like the consideration we give each other as humans". I think it is fair to understand this proposition as a re-presentation of Paul Taylor's deontologically oriented position, famously elaborated in his *Respect for Nature* (discussed in Sect. 2.1.2.2). Bendik-Keymer (2010, 72) does follow Taylor, who anchors this respect to the teleological center of an individual organism's life. But Bendik-Keymer adds little to Taylor's argument, except a layer of virtue ethics coupled with a moral psychology, according to which the lack of respect is a vice born of thoughtlessness. So the heavy burden for providing an account of why this is a vice – and not the virtue of ignoring what is not important – remains on Taylor's shoulders; nothing that Bendik-Keymer adds offloads any part of it.

Bendik-Keymer (2010, 72ff.) also falls in line with Taylor's concern to promote moral concern up the scale of increasing collective inclusiveness. He moves from "a life being alive" to species, or "forms of life", then quickly on up to the "diversity of life" or "biodiversity". It appears that this moral escalator is powered by consideration of how impinging on the planet's biodiversity also impinges on the ability of humans to flourish. According to him (Bendik-Keymer 2011, 8), the possibility of a major extinction of species amounts to "a disintegrating of… nature *as we have known it*" [italics in the original]. This, says Bendik-Keymer, is something that runs the risk of humans "deracinating" themselves (Bendik-Keymer 2011, 24), losing track of "sense of what it is to be human" (Bendik-Keymer 2011, 14), and "[creating] an alien world with respect to our natural history" (Bendik-Keymer 2011, 17). These are strong claims. I think that they require more than the trope of the natural world as a library (Bendik-Keymer 2011, 15) to sustain. They also require more than the rhetorical use of such normatively loaded words as "deracinating" and "alien".

Insofar as Bendik-Keymer ventures beyond the library trope, he mostly leans (Bendik-Keymer 2010, 75) heavily on the line of reasoning propounded by deontologist Holmes Rolston and discussed in Sect. 2.1.2.2 alongside Taylor's views. It is sometimes enormously helpful to pose the archetypal question of virtue ethics: "What kind of people (decent or indecent) are we not to respect the diversity of life?" But at least this sample of virtue ethics suggests that, if this question has a plausible answer and cogent defense, these likely come from work done outside of the virtue ethics tradition.[115]

As I observed awhile back, the tendency of Sandler and other virtue ethicists to reinterpret consequentialist or specifically welfarist views about the good of the natural world in terms of virtues seems singularly unpromising. But ushering deontological considerations or intrinsic value theory under virtue ethics' skirts seems no more promising. It is true that nothing prevents a theory of virtue from supposing that biodiversity or the natural world generally is composed of the kind of stuff that makes it an appropriate basis for beneficent responses or for respectful ones. Virtue ethics can alternatively or additionally suppose that the natural world exemplifies certain structured properties that generate moral force fields, which act on morally sensitive agents to induce in them benevolent attitudes and disrupt disrespectful ones towards the natural world. It can suppose, to head off a possible problem with that view, that the world does not exemplify properties that radiate forces nullifying those that induce those environmental virtues. But the credibility of these suppositions requires a credible argument for why we should accept them. It calls for something like Paul Taylor's account of the "inherent worth"[116] of individual organisms, which derives from their having a good of their own. For his part, Taylor argues that this kind of good is not just good for the individuals whose good it is, but a good independent of any valuing subject and yet worthy of any valuing subject's respect. A theory of the virtues could assimilate this line of argument, but it would be an uncomfortable accretion. Installing it as a crucial piece of foundational structure in a theory that is supposed to be about the virtues should cast doubt on whether the virtues are doing any real, foundational work.

---

[115] In another line of discussion, Bendik-Keymer (2011, 20) agonizes over the question of how long it takes to "recover" from an extinction event. This question, insofar as it concerns a major change in the world's biodiverse state, is derivative in the sense that it is morally relevant only if such a change is morally significant in the first place. Moreover, it is hard to avoid the impression that it relies on the state-based view of what is right in the world, which is built into the word "recover": To recover, it seems, is to recover to a previous state. For that to be the morally right thing at which to aim, that previous state must have some special standing; and that standing must be established via independent considerations. For more on this topic, see Sect. 7.3 (Biodiversity value in human timeframes).

[116] Taylor (1989, 75) uses the term "inherent worth" for the intrinsic value – in the sense of "independent of valuing subjects" – of an individual that can be said to have a good of its own.

This unsympathetic assessment of environmental virtue ethics applies quite specifically to what virtue ethics has to say about the core of natural value, which derives from the significance to humankind of the natural world as a place apart from the humanized world of human enterprise and culture. The assessment does *not* extend to other areas also addressed by environmental virtue ethics. For example and saliently, some cadre of nonhuman organisms will always exist "within the walls of the city". A theory of appropriate fit is not obviously well equipped to comment on what moral consideration might be due individual creatures and organisms that make their life within human-made environments – right down to the potted plants, whose place in every cubicle biophilia labors to secure.

I do not intend my critique to challenge the merit of using the language of virtue and vice to characterize and evaluate, in a qualified way, certain ways of acting and behaving with respect to nature or the environment. In the end, there might be good grounds for the claim that much meddling with nature in the name of "conservation" or of "restoration" is hubristic insofar as it grossly overestimates the ability of humans to control outcomes. But (here is the qualification) this is a judgment about possibly misjudged *means* to the end of shaping and controlling nature. This, I have argued, sidetracks the more central concern about the appropriateness of that end. A theory of appropriate fit is well suited to call such an end into question. It is not at all clear that the sorts of virtue ethics considerations that I have considered are so well placed.

There might be a good case for characterizing as a vice an attitude towards nature that regards it solely or primarily as a domain for conquest and control. A theory of appropriate fit might find itself in sympathy with such a position, but only as a minor corollary. A tendency to regard any and every domain of human interaction as a challenge for control is almost certainly a vice. But just as many theories of natural value fail to find anything distinctive about nature that makes it valuable as nature, this vice also does not identify anything that is distinctively or especially vicious with respect to the domain of nature. That is, while it is a vice, it is not a distinctively environmental one. In contrast, a theory of appropriate fit seeks to delineate those characteristic relationships of persons to the natural world that define, distinguish, and are unique about the value of nature.

In sum, while one might be able to couch a theory of appropriate fit in the language of virtue, there is no particular advantage to this. And because appropriate fit finds its basis for moral consideration outside the customary haunts of environmental virtue ethics, there is a distinct advantage in disentangling itself from the rubric of virtues and vices.

Finally, in my list of what appropriate fit is not, a theory of natural value according to appropriate fit should be held apart from a varied collection of approaches that parade under the banner of "respect for nature". That phrase inevitably calls to mind Paul Taylor's eponymous, groundbreaking book on environmental ethics. I have just described how a certain strain of virtue ethics thought has an affinity for it. The phrase certainly sits comfortably alongside the project of virtue ethics to justify a virtue of benevolence. And when restraint is viewed as a primary manifestation of respect for the natural world, "respect for nature" sits equally well with appropriate fit's admonition to "let it be". But **despite occupying philosophical territory adjacent to Taylor's respect for nature, the road to appropriate fit traverses rather different philosophical terrain on its way to that neighborhood.** One need only recall Taylor's biocentric ethics (described in Sect. 2.1.2.2), which is firmly rooted in individual organisms – nonhuman and human alike – regarded as "teleological centers of life". For Taylor, it is the *telos* of individuals that ultimately is worthy of moral respect. Respect for individuals grounds his respect for nature. In contrast, appropriate fit is firmly rooted in the good of a human willingness to keep the natural world outside of, and so circumscribe, the domain of human enterprise; and to do this by adopting a modally demanding attitude of restraint towards nature. The *telos* of individual organisms does not enter into this founding consideration.

Other philosophers have co-opted Taylor's phrase "respect for nature" despite sharing little if any of the bedrock from which Taylor sculpts his biocentric ethics. I have already mentioned Jeremy Bendik-Keymer. Dale Jamieson's version of "respect for nature" (Jamieson 2010, §6, "Respect for Nature") is one that stands out for clearly and succinctly covering many of the plausible bases. He (2010, 441) first wrestles with the term "domination", which ecologist Peter Vitousek and his colleagues (1997) use to characterize the substantial influence of people on the biosphere globally.

Insofar as "to dominate" means "to influence by a commanding exercise of control", its use in this context sits a bit askew – for no one could plausibly suppose that people are really masters calling the shots, no matter how dramatically their activities have changed the workings and course of the natural world. This is not the irrelevant semantic quibble it might first appear to be, because it bears on Jamieson's move (Jamieson 2010, 442) to imbue the "domination of nature" with moral significance by way of connecting it with nature's autonomy. Even in the human domain, from which the terms "domination" and "autonomy" surely derive their normative leverage, we view a circumstance in which one group of persons commands another as their slaves rather differently from a circumstance in which a group of persons have constructed cities in which others then live. The first circumstance exemplifies domination that should be subject to universal moral censure. The second circumstance exemplifies a powerful influence – for to live a life in urban surrounds (many of whose effects on that life could never have been anticipated by those who contribute in multitudinous ways to the accretive growth that eventually creates a city) is a quite different thing from living a life in the country. One might even go so far as to call this influence a "dominating influence". But few would insist that this sort of dominating influence be subject to the moral censure that most would insist is appropriately directed at the practice of slavery.

It appears that the "domination" of nature, so-called, bears on nature's "autonomy", so-called, in much the same way as the city case, not the slavery case. As a consequence, when it comes to nature, domination appears to be of the benign variety and something not worth getting morally worked up about. This consideration alone suffices to undermine Jamieson's contention (Jamieson 2010, 442) that "if there is a duty of respect for nature, then human domination violates that duty."

But the misjudged connection of domination to autonomy is not the only basis for calling into question Jamieson's "respect for nature". In addition to that flawed connective link, the plausibility of Jamieson's case rests on the supposition that there is *some* normatively weight-bearing meaning of "autonomy" that can legitimately be attributed to nature. This, too, is highly questionable. I can happily grant his definition of "autonomy" (Jamieson 2010, 442) as "being self-caused". Of course, in the case of nature, the "self" is neither a person nor any other organism pursuing a *telos*. It is hard to see what that self could be other than a "system" – understood, for example, as a scientist might understand a gas-containing vessel in studying thermodynamics. Of course, that self could also be an eco-system, which is an imaginary vessel containing biotic and abiotic "molecules" that collide and interact with each other in various, complex ways. In both cases, "self-caused" most likely means "absent human causal interference". But also in both cases, there is no easily identified "self" that is "respected" by not interfering. In fact, it appears that the autonomy of these systems is an autonomy in which the notions of "violation" and "respect" simply do not apply.[117] A theory of appropriate fit does not fall prey to this conundrum because it does not rest on any dubious conception of autonomy or domination as those notions might be supposed to apply to nature.

With both legs cut out from under this conception of "respect for nature", it cannot go far. Jamieson does not place his entire bet on the domination-autonomy nexus, and turns to some other candidate bases for the intuitive appeal of "respect for nature". He (2010, 443) starts with prudential concerns, which does not help his cause. This book has shown such concerns, in myriad manifestations, to be ill-founded – at least in their application to "respect for biodiversity". That aside, there is an uncomfortable strain in claiming that I respect something because I really want to avoid the cost of harming myself. That sounds a lot more like respect for my own welfare. As I have before remarked, this self- and human-serving respect could well entail an obligation to destroy the natural world, were its harmful effects determined to outweigh its beneficial ones. Fortunately, a theory of appropriate fit is not vulnerable to this outcome because it eschews prudential considerations.

Jamieson offers two other interesting gestures. He (2010, 443) suggests that "respecting nature… provides a background condition for our lives having meaning". But the meaning he has in mind is that of specific places in specific, distinctive states that "define" these places. Whatever attraction this kind of "respect for nature" has, it finds its basis in identifiable states of the world. A state-based view of what engenders value in nature, this is the antithesis of the view offered by appropriate fit.

Jamieson's final suggestion (2010, 444) is that respect for nature might derive from "a concern for psychological integrity and wholeness", which requires respect

---

[117] This observation shows up a relative strength of Taylor's conception of "respect for nature". At least he builds it atop the more plausible proposition that other organisms are selves in the relevant sense in which "self-caused" for those organisms *does* (according to this proposition) command moral respect.

for "others". According to Jamieson, in failing to recognize other selves, people succumb to self-damaging narcissism. Unfortunately, his interpretation of this psychology once again projects nature as an other "self" in a sense that is applied to other persons. This also represents a significant departure from appropriate fit, which has no need to posit that nature is a "self" worth respecting as such.

8.2.7   Some Implications

My exposition of a theory of natural value based on appropriate fit concludes by sketching some salient implications. This is neither a comprehensive nor a detailed working out of the theory. Rather, it is a limited discussion of several topics, selected to suggest this theory's reach and power to provide fresh insight and to avoid the seemingly irreconcilable conflicts and paradoxes that entangle the more familiar approaches to grounding the value of nature.

A theory of appropriate fit should be welcome for its ability to more clearly and with less incongruity re-imagine the value of the natural world. Other implications will be controversial for challenging the operational goals of many neo-conservation organizations and for challenging the alleged basis of these goals in the widely espoused but (I believe) indefensible norms of today's theory and practice of conservation and restoration biology. Yet other implications – for its constraints on individual behavior, and social and political policy – will be disconcerting to those who see no greater good than development. But environmentalists generally and those who value nature might find all these implications particularly refreshing and appealing.

8.2.7.1   Implications for Natural Value Generally and Biodiversity in Particular

1. Native versus "alien" species in a particular locale

At several previous junctures, I have had occasion to raise questions about whether the classification of certain organisms in certain places as "alien" has any legitimate place in an evaluation of nature in terms of its biodiversity. I wish to refocus on just how tangled this notion is, before seeing how a theory of natural value according to appropriate fit glides through this thicket.

Much reasoning concerning aliens is categorically fallacious, with fallacies of accident (described in Sect. 2.2.3, Fallacies of accident) a leading cause of logical failure. Joining the logical frailty of indictments against "invasive aliens" is a raft of inconvenient facts. The vast majority of exotics (the more nearly normatively neutral term that I prefer) are grossly maladapted for their new surrounds, fail to come to grips with it, and slip into oblivion – a failed experiment in translocation.

Most of the remainder settle in without displacing any "natives". Much like human "aliens", they *contribute* to the locale's local and regional diversity and even ecosystem function. Despite the preponderance of studies that try to implicate "alien invasions" with the "degradation of biodiversity", a number of recent studies portray a rather different picture. To cite a few: The ecologist Karsten Reise and his colleagues (2006, 77) say that, "Although aliens accelerate change in European coastal biota, we found no evidence that they generally impair biodiversity or ecosystem functioning." Nicolai Aladin and Igor Plotnikov (2004) recount how wave after wave of invasion has made the Caspian Sea one of the most biodiverse places on the planet.

On terra firma, and specifically on California serpentine soils, Susan Harrison and colleagues (2006, 695) obtained results that

> … indicated similarities and differences in the conditions favoring exotic, native, endemic, and rare species. Our results suggest that, in spite of some localized impacts, exotic species are not exerting a detectable overall effect on the community richness of the unique native flora of Californian serpentine.

And Dove Sax with Steven Gaines (2008, 11490)

> show that the number of naturalized plant species has increased linearly over time on many individual islands. Further, the mean ratio of naturalized to native plant species across islands has changed steadily for nearly two centuries. These patterns suggest that many more species will become naturalized on islands in the future.

In short, despite the impression left by a small number of sensational and ubiquitously and insistently cited cases, it is rare for an "alien" (other than *H. sapiens* and that bipedal species' companion organisms) to have a diminishing effect on biodiversity.

Oftentimes, without any organism ("native" or "alien") needing to make any adaptive adjustment, an exotic presence aids and abets existing *natives*. Ecologist Laura Rodriguez (2006, 927) surveys a number of studies that document cases in point and that provide

> … evidence for several mechanisms that exemplify *how* exotic species can facilitate native species. These mechanisms include habitat modification, trophic subsidy, pollination, competitive release, and predatory release. [italics in the original]

Her list is not exhaustive. For example, Ariel Lugo (1997, 9) demonstrates how the "reestablishment of tree species richness on degraded sites with arrested succession could be facilitated through plantings of tree monocultures". In a separate study, Lugo (2004, 265) describes how "Invasive alien tree species in Puerto Rico often form monospecific stands on deforested lands that were previously used for agriculture" but that "most native pioneer plants are incapable of colonizing". Richard Caldow and colleagues (2007) describe how an "alien" Manila clam provides wintertime sustenance and so reduces the winter mortality of Eurasian oystercatchers. Francisco Padilla and Francisco Pugnaire (2006) discuss a "nurse effect" whereby plants are introduced to attract pollinators and shield other plants from grazers and the sun, as well as providing "hydraulic lift".[118]

Other times, the natives adapt – behaviorally, or genetically, or both – to the presence of their new neighbors, as do those recent arrivals to the established residents they encounter, often for the first time in the history of all species concerned. Genetic selection is almost never brought into the discussion of these events. But this is a grievous omission. The general principle – of environmental pressure being the single most potent lever for promoting adaptation and speciation – is well known. So by inducing new pressures on "natives" and "aliens" alike, recently arrived species have significant potential for adding, not just to local, regional diversity, but to global diversity as well. Despite the fact that few scientists have been inclined to investigate this biodiversity-enhancing force, there is increasing evidence that this potential is often realized. I have already remarked on this in Sect. 6.7 (Biodiversity as value generator) and in Sect. 7.3 (Biodiversity value in human timeframes).

The further undoing of the proposition that "aliens" subvert biodiversity has to do with the fact that many habitats are now so humanized or have undergone and continue to undergo such great and rapid alteration that erstwhile "natives", which did not evolve in anything like those same habitats in which they now find themselves, become as exotic as the "aliens" hailing from afar. It is not surprising that "natives" often find themselves ill equipped to deal with a now-alien terrain covered with structures, roadways, parking lots, agricultural crops, and suburban landscaping.

Even not-so-humanized habitats that have recently changed at a pace that is rapid by geohistorical standards have become inhospitable to organisms without respect to native/alien boundaries. At the same time, if any "aliens" do manage to find a toehold or roothold in a transformed landscape, then there are simply more "aliens". In such a circumstance, it is an error to mistake for a causal relationship the correlation of more "aliens" with fewer "natives", which have succumbed to the unneighborly habits of their new bipedal neighbors in their erstwhile home. An "alien" can make life harder on a native, just as it can make life better. However, in most cases in which an exotic is fingered as "the" cause of an extirpation, there is little hope of disentangling this possible explanation from the typically far greater effect on the unfortunate native of changes in hydrology, biogeochemistry, physical disturbances, and many

---

[118] A recent trend in ecology is to resort to the term "invasive species", which, by Executive Order 13112 of the United States Department of Agriculture (www.invasivespeciesinfo.gov/docs/council/ isacdef.pdf) means "an alien species whose introduction does or is likely to cause economic or environmental harm or harm to human health." If "environmental harm" is taken to include "unwanted change in biodiversity" (as it often seems to be), then the need for empirical evidence for the claim that invasive aliens are likely to harm biodiversity is defined away. That claim amounts to the proposition: exotics that are likely to cause unwanted changes in biodiversity are likely to cause unwanted changes in biodiversity. Of course, this move evades the interesting empirical question of the extent to which exotics contribute to, or detract from, biodiversity.

other factors that are not the doing of any "alien invader". In a study already cited, Aladin and Plotnikov cite such factors impinging on natives as alteration of river flows by hydroelectric plants; their discharges of toxins; the shredding of migrating fish by turbines; the concentration of toxic substances behind dams; over-fishing; the lowering of sea levels for a number of reasons, including reduced flows from dams and diversions; the construction of a dyke intended to prevent draining into a bay, which only succeeded in draining the bay; and pollution from industrial hydrocarbons, phenol, surfactants, chloral-organic pesticides, and heavy metals such as lead, cadmium, copper, and zinc. In such a case, it is easy to post a "Most Wanted" bill with a picture of the "alien outlaw". But it is hard not to suspect that this is a relatively minor diversion from the main narrative.

Overall, the more "aliens", the more widespread, the more biodiversity. There are exceptions, and the conditions under which they occur are a matter of ongoing investigation. But if one supposes biodiversity to be a good in itself, then for the most part, the more "aliens", the better. Of course, ecosystems with a complement of new denizens might work in a different way from their historical predecessors. They might sometimes work in ways that are detrimental to the human economy. But that does not cut against the fact that they are more biodiverse and not merely with respect to their species. If an ecosystem's newly naturalized citizens sufficiently alter how it functions, then the resulting ecosystem is likely to be sufficiently novel to contribute to ecosystem diversity. Some persons might prefer the ecosystem of old. But that is a matter of preferences (which might be reflected in the human economy), not a matter of diversity.

Biotic xenophobes also sound the alarm that "alien invasions" cause "biotic homogenization". Zoologist Frank Rahel (2002, 291) defines this latter term as "the increased similarity of biotas over time caused by the *replacement* of native species with nonindigenous species, usually as a result of introductions by humans." [italics added] "Homogenization" amounts to a certain kind of state change; this definition in terms of a specific cause should raise eyebrows. Surely, homogenization might have any number of causes – including, for example, the poisoning of environments. This would give an adaptive advantage to toxin-tolerant organisms, which might have salient similarities worldwide. The causal bias of the definition comports with the tenor of the discussion that surrounds it to convey the distinct impression that Rahel is not so much upset about homogenization as with aliens insinuating themselves into places where Rahel thinks they do not belong.

I favor a definition of "biotic homogenization" that lops out reference to any specific cause, to yield: "a change that results in fewer differences in the species composition between different ecosystems".[119] In other words, the supposition that the biotic world is undergoing homogenization is a claim that, when it comes to the category of ecosystems, there is a decline in the diversity of the kind of species composition (here using "category" and "kind" in the ontological sense suggested by Fig. 3.1).

---

[119] This is what some ecologists would call a reduction of "beta diversity".

The advantage of my alien-blind definition of "biotic homogenization" over Rahel's is that it does not restrict attention to the relatively rare phenomenon of aliens supplanting natives. For example, homogenization often occurs by the addition of species, which increases the diversity of biota in compared ecosystems. Rahel (2002) subverts his own "replacement" definition of homogenization with his report that the number of fish species common to at least two states in the United States increased – mostly due to the introduction of game fish – while extirpations of natives were uncommon. Dov Sax and his colleagues (2007, 467) observe that "in Hawaii freshwater fish richness has increased by 800% with the introduction of 40 exotic species and the loss of none of the five native species". They make a similar observation about "vascular plants, which generally show a pattern of few native species lost and many exotic species gained, invasions over the past few hundred years have led to a doubling of richness on many islands (e.g. from 2,000 to 4,000 species in New Zealand) and on the order of a 20% increase in many mainland regions, such as individual states in the USA."

One remaining maneuver for salvaging the thesis that "aliens" are bad for biodiversity is to find some credible basis for distinguishing between good biodiversity and bad biodiversity. Bad biodiversity, of course, would be the kind that results from an influx of "aliens". I have followed this and similar dead-end paths through several previous discussions, which examined various perfectionist views of nature. In the current proposition, which aims to exclude "aliens" from the good biodiversity club, the normative distinction seems to ride on one or both of two specific misinterpretations that accrues of evolutionary theory: On one misinterpretation, evolution is a perfecting or optimizing process. It makes a native more perfect than any alien, which has not enjoyed the perfecting opportunities that only accrue from evolving "in place". But it is hard to make any sense of this. If "perfection" is defined as *best* suited to the environs", the proposition is disproved by the fact that adaptation is just the process whereby an organism has to be just *good enough* to survive. There is no reason in theory, and also none in fact, why a recent arrival might not be better in this regard.

Nor can the normative distinction be maintained on the basis of a (different) misinterpretation of evolution as a process that produces organisms in their "appropriate place". As the famous evolutionary biologist Stephen Jay Gould (1998, 7) says, "… organisms (and their areas of habitation) are products of a history laced with chaos, contingency, and genuine randomness…" The history of creatures, how they assemble themselves into collections and subsequently disassemble themselves, and the places that are the loci for this constant process of species remix cannot serve as a credible norm.

I have so far been taking for granted that there is some real and valid distinction between "native" and "alien". But there is considerable doubt about whether there is any principled way to maintain it. Most visible "native" species were at one time "alien", given a sufficiently long timeframe, which is often not particularly long. Many "native" species stumbled, flew, or were carried into the environments in which currently respiring or previous generations of humans found them. This is true not just true of animals, but of plants. Charles Darwin already realized that these sessile organisms are, in fact, quite capable of arranging for means to get around.[120] So where is the cutoff for being a native? Ten years ago? One hundred? A millennium? Many thousands of years? What could possibly justify one choice over another?

Even by the lights of those who believe in the distinction, "native" flows into "alien" and "alien" to "native" according to fluid rules that seem to serve unstated preferences or other ends, all of which are left unjustified and appear to be unjustifiable. Consider, for example, the proposal of some scientists to introduce captive-bred tortoises of the species *Aldabrachelys gigantea* in order to re-create an aspect of the Indian Ocean's Mascarene Islands ecosystem that owed to the presence of the now-departed and different species of *Cylindraspis* tortoises. Proponents of this "rewilding" scheme acknowledge that *A. gigantea* has never before made its home in the Mascarenes. So they are concerned to explain (without explaining why this needs explanation) how being a member of the same, albeit non-congeneric, taxonomic family earns *A. gigantea* the status of non-"alien" or even honorary "native". It seems that these particular reptilian giants are given a green card on the grounds that they are "ecosystem proxies", which serve to re-create lost "ecosystem functions". In other words, while not "native" by birthright, they are "native" by behavior. Or close enough, by some unstated criterion. According to its proponents (Griffiths et al. 2010), this is good enough justify facilitating their invasion to re-create a lost "ecological niche".

The tortoise rewilders admit to committing the ecological sin of introducing "aliens" and try to defend it with their proxy argument. All the while, they vociferously protest comparisons with more ambitious rewilding schemes. But it is hard to take their protests to heart, for their proxy argument is cut from the same cloth as arguments marshaled in support of such grandiose schemes (Donlan et al. 2006) as the "Pleistocene Rewilding" of North America (which came up for discussion in Sect. 7.2.3, Home is where you make it – in situ, ex situ, and ecosystem-engineered). In that grander scheme, too, "alien" species are proxied in by virtue of their functional resemblance and (taxonomic) family ties to previous inhabitants. Proponents are not reluctant to realize the full potential of such "flexibility". The Pleistocene Rewilders (Donlan et al. 2006, 668) are keen to "repatriate" tortoises, too – for their burrow-digging function.[121] But that is just the start of a program that envisions, once again roaming the "wilds" of North America, equids, camelids, cheetahs, proboscideans, and Holarctic lions. The latter get their green card, for example, on the grounds that *Pathera leo atrox*, the late Pleistocene's American

---

[120] Darwin was enormously concerned to account for plant diasporas that would explain observed distributions. He found several convincing *modi operandi*. He determined that seeds could float across great expanses of salt water without damaging their ability to germinate. They could also ride on log rafts, hitchhike in clumps of mud adhering to the feet of wide-ranging fowl, as well as within the guts of those feathery means of transport. See, for example, Darwin (1859, 386–387). Gould (1998, 7) makes a similar observation about Darwin's concern.

[121] Essentially all Testudinidae meet the functional requirement for burrowing: Female tortoises burrow to build the nests in which they lay their eggs. But of course any number of other terrestrial creatures live a burrowing life, too – including rabbits, moles, and a variety of rodents. The proxy argument does not obviously exclude any of these adept burrowers.

lion, is considered to be a subspecies of the same species as modern subspecies of *Panthera leo*, though one quarter again the latter's size.

The North American rewilders' first stated goal is merely a more majestic version of the tortoise rewilders' – that is, to restore "ecological function". But they do not produce any compelling reason for why we should allow that certain similarities in function (or anything else) count, others don't, and why it would be good for the ecological function of modern North America to mimic, in just those respects that are said to count, that of its Pleistocene past. Of course, a big part of the problem in determining why some particular similarity counts as a good is that the proposal gives no clear specification even of what counts as similar or similar enough. On the evidence of what is proposed, one must suppose an extraordinarily liberal criterion. But then it is hard to understand why this doesn't also allow any number of unmentioned substitutions, such as one that mimics burrowing tortoises with burrowing rabbits. Even aside from these obstacles to good mimicry, enormous changes in climate, hydrology, fire regimes, biogeochemistry, and a multitude of other factors independently guarantee that something rather different from Pleistocene North America – in fact, something entirely unique – would be created.

Perhaps realizing that they offer no legitimate reason for pursuing their project, rewilders resort to the justificatory option of desperation: That is, they resort to economics. They float the idea of the possible economic value of camels (as sources of meat, milk and fiber). But it is unclear why this could not be pursued with greater success by establishing camel farms and hiring the best marketers to convince the North American public that camel meat is the new "white meat". They also talk about tourists' enjoyment of their "wild experience". This is not unexpected, for the introduction of "aliens" has been and continues to be justified as a means of enjoyment – as targets for hunters and fishers, and to decorate the scenery. So it seems that in the end, this scheme totters on a shaky economic foundation. One cannot help thinking that the proponents just think that it would be really cool; and of course, economic reckoning has no problem factoring in preferences based on this feeling.

The "native"/"alien" distinction is yet further undermined by the circumstance I mentioned at the start of this discussion and brought up in connection with "Assisted migrations" and such (Sect. 7.2.1). The ecosystems in which all organisms live on the planet have changed and continue to change – right underneath their feet, wings, fins, and roots. It must be acknowledged that, even in places on the planet that have not been subject to direct and obvious human intrusion, more and more of the long-time residents are now "aliens" in the EPA sense of "species that have become able to survive outside the habitats where the evolved or spread naturally".[122] Sometimes, these long-time residents (call them "natives", if you insist) find themselves in the same position as most "aliens" – in a now-hostile environment in which survival is uncertain. Which motivates proposals to groom their environments back into a state more amenable to their survival. Or it leads to proposals to purposefully make such creatures aliens twice over by providing human-devised transport to some other place never before visited by their species.

---

[122] See Note 10 in Chap. 7 and the associated main text.

The entire discussion is rife with selective and arbitrary definitions of normatively loaded categories, conflicting sentiments, and misinterpretations of the evolutionary process as a perfecting one, an optimizing one, or a thing of the past that abruptly ceased operation some time ago. This approach is a mess and it is best avoided. That is precisely what a theory of the value of nature according to appropriate fit permits us to do.

Appropriate fit counsels us to shun turning creatures of any kind – "natives", "aliens", aliens twice over, or labeled in any norm-laden way – into a science project. Most forcefully and most particularly, it counsels this with regard to those few creatures and organisms that have so far managed to survive outside the walls of our cities. It counsels us to reject turning their habitats into re-creations of the past, or terrariums, or aquariums, which biologists or conservation organizations aver are friendlier to their way of life. This counsel is based on the insight that *no matter what the outcome of these kinds of projects*, they constitute an interaction with the natural world that is antithetical to the spirit of the relationship that defines the core of natural value. That relationship is possible only by steadfastly withholding some part of our world from science projects or human projects of any kind. It is that part of the world – left outside the walls of our cities – which is valued as the natural world.

Of course, taking this stance and refusing to intervene or intrude for any reason – for example, by refusing to weed out "aliens" – entails that we must countenance loss insofar as this is part of what it means for the world to move away from the currently familiar state. This might seem a tradeoff difficult to accept – unless one realizes that neither does intervention prevent loss. The fact is that places and their resident creatures change despite as well as on account of the best of human efforts. But even if our most skillful re-creations and management efforts to maintain some-thing with the outward appearance and even scientifically measured function of some familiar habitat succeed in these terms, our relationship to that place radically changes by virtue of that effort. In fact, it has changed in precisely the way that entails the loss of the core of natural value.

The real choice is not between a lossless stasis – which neo-conservation and restoration projects continually and furiously try to maintain – and loss-incurring change. Rather, the choice is whether or not to embrace a world where something significant remains outside the domain in which human endeavors prevail; a world in which there remains something to juxtapose and give context to human projects; a world in which something remains to set off in relief the creativity of human cul-ture; a world in which not everything is evaluated and judged according to the suc-cess or the ingenuity with which a human project is realized; and a world whose perspective is not entirely flattened in these respects.

2. The uncomfortable fit of natural value in economic valuation

In Sect. 8.2.3 ("Living from" nature, uniqueness, and modal robustness), I described several salient respects in which pushing the economic value of nature into preeminence ultimately subverts the value of anything that can be credibly valued as nature. I argued that nature-as-natural-capital is a Frankenstein creature assembled from bio-parts in ways unconstrained by respectable moral principles, but rather dictated by the vagaries of markets; and that this "nature" is a nature that both displaces and corrodes nature's core value.

There is no need to recite those arguments in order to emphasize here how appropriate fit skates completely clear of these concerns. Appropriate fit entirely eschews identifying as the core of nature's value its role as a re-machinable and replaceable cog in the machinery of human enterprise – a role that is supposed to help create "an environment that is desirable for all of us". Instead, it stakes the core of nature's value on its unique role in crystallizing the value of that human enterprise by delimiting its domain, and by doing so in a modally robust way that markets, by their fickle nature, can never achieve.

What appropriate fit has to say about the value of nature should not be misinterpreted as a general denial that entities generally admit of being valuable in multiple ways. For some entities, economic value is one among several. Obvious examples abound. Certainly, a work of art can have not just aesthetic, but also economic, value. Moreover, an artwork's possession of the latter does not necessarily undermine its possession of the former. However, according to appropriate fit, the core value of nature differs from art's value in an enormously significant respect. Unlike the value of art, the value of the natural world is tightly linked to, and dependent on, its separation from human culture generally and from the human economy in particular. Because that is the principle source of this core value, it *is* subverted by putting the natural world into economic play, where it assumes the role of a direct, undistinguished, and undifferentiated competitor against all other goods for their role in enhancing economic welfare.

Different attitudes and behaviors are appropriate in the valuing of different things; these differences give rise to different ways of reasoning about these different ways of valuing. In the end, I believe that but a small niche of valuing has to do with, or can be usefully described as, considering the worth of something in comparison to something else for which it might be exchanged. And this corner of the evaluative world – the economic corner – for the most part does not rise to the level of moral consideration.

Economists commonly offer two reasons for thinking otherwise; both strike me as terribly weak. One reason is couched in terms of what is supposed to be a rhetorical question: How can values, which often conflict, possibly be compared and traded of with respect to each other unless there is the possibility of exchange (which markets facilitate)?[123] But this question has a strong and obvious answer. People have access to a rich moral discourse, which they routinely use to jointly reason and sort through the various values that come into play in complex situations. This discourse attempts to weigh, for example, whether and what kind of consent is required from those affected by some action (even one beneficial to them); whether the consent (including the consent to enter into a market transaction) might be coerced or given from a position

---

[123] The nobel laureate economist Kenneth Arrow posed this challenge in conversation about my position.

of ignorance or relative powerlessness; whether some way of acting, which promotes some general good nevertheless diminishes the ability of some to exercise basic human capacities; and so on. It is rare that this discourse resembles anything connected with paying for a commodity or service, or indeed, entering into any market transaction where the underlying values held by the transacting parties are mostly irrelevant and routinely held close to the chest. What is really invisible in the market is not Adam Smith's "invisible hand", but rather a genuinely moral discourse.[124]

The second reason on offer from economists has to do with the fact that compensation is often considered legitimate recourse for the unjust deprivation of a good. But while compensation might be the best that retributive justice can offer, it would be a mistake to think that the good consists in, or is a consequence of, the compensation it might command in a market transaction. More generally, as Debra Satz (2010, 78–79) says,

> … not all goods mean the same thing as money… I can think that there are goods that people have a claim to – life, public health, and public safety – without thinking that these people have *the same claim* on the cash equivalent that might (or might not) be used to buy these goods. [italics in the original]

Most obviously, the value of a human life, *why* this is something precious, and the place of a life's value in practical reasoning about how that life might best be lived and how the living of that life must be respected by others, is not well illuminated by considering what market transaction might tempt its owner (or worse, another person), to extinguish or cripple it, or even to incur the risk of doing so. Many of the things that people most dearly value follow suit: rights, such as the human right to autonomy; important relationships, such as those that we have with our family and genuine (not "imperfect") friends; our most important choices in life connected to a meaningful vocation and an intimate partner; generally living a fulfilling life and choosing the best way to live it; and even the best way (and if we are fortunate) the best time to die. Economic welfare has, at best, a peripheral role in properly considering any of these things. Most well-reasoned decisions about some of the most valuable things in life and society self-consciously and categorically set aside consideration of economic welfare.

I would say that the natural world is one of these "most valuable things". One of the salient strengths of a theory of natural value according to appropriate fit is that, in stark contrast to economic theory, it provides a compelling basis for understanding why this is so. Appropriate fit suggests how and why the core of natural value is unique in the pantheon of values. It shows how and why this requires very specific ways of valuing the natural world for very specific reasons, based on a very specific understanding of how the fit of people in the world at large is best when they relate to some part of it – the natural world – as given. With a theory of natural value

---

[124] See Debra Satz's enlightening discussion of this feature of neoclassical economics in terms of an "envy-free" division of resources. In this division (Satz 2010, 68), "Differences in people's resource bundles… reflect only their different preferences, attitudes, and life ambitions." And none of these need be revealed in the set of transactions that effects this division.

according to appropriate fit, in contrast with economic theories of natural value, there is no conflating the natural world for a warehouse or a service industry. The uniqueness of the core value of nature is satisfactorily accounted for.

3. **The conundrum of more and less biodiversity**

A third topic revisits the role of science in valuation, and specifically, the conundrum of how more and less biodiversity can be related to greater and diminished natural value.

On a theory of appropriate fit, the more and less of anything in the natural world is simply not relevant to nature's core value. What kinds and how many kinds of creatures arise and flourish; what kinds and how many kinds of collections assemble and disassemble; how many ecosystems stay intact and how many change into something unrecognizable as their former selves – none of these interesting facts about the dynamics and content of the world bear on what is central to their value as nature. Rather, that value derives from a relationship in which people let these "goings on" go on their own. By the very nature of what makes nature valuable, its value cannot possibly be positively affected by aiming at some set of species or ecosystems according to some scientific design specification. There is no scientific measure of whether natural value has increased; and more importantly, there is no *possibility* of a project to increase it – no matter what the readout on such a thing as a biodiversity-meter.

The vexing, biodiversity-enhancing projects listed in Sect. 5.3 (The moral force of biodiversity) are well justified if one believes that the good of biodiversity is served by greater biodiversity, which, in turn, is paramount in the good of nature. But the theory of appropriate fit gives us good reason to avoid every one of those suggestions. It removes all need to agonize about whether or not we should create genetically modified creatures to increase biodiversity. It shows that there is no good reason to cleverly induce adaptive pressures as a means of turbo-charging natural selection and to more quickly realize "evolutionary potential" for the end of greater biodiversity. According appropriate fit, these are all destructive of natural value. And no paradox attaches to this conclusion because, as I have said, the readout on the biodiversity-meter has nothing directly to do with the value of the natural world.

This is not to deny some moral import – even great moral import – in significant changes to the world's biodiversity. However, the significance of these changes has nothing to do with the amount of natural value in the world as a function of its changing state of biodiversity. Rather, the changes are significant as a very visible sign that we humans have rent the barrier – the walls of the city – between human culture and nature; and that we have done so utterly, and that our behavior has become so detached from recognizing any remnant of these walls and their importance, that we both self-consciously and inadvertently continue to plow right through them on a routine basis. The changes are significant as evidence that humankind has increasingly lost touch with any semblance of a valuing relationship to nature because few parts remain that are regarded as immune to the logic of economic development – putting down more roads, extracting more "free" stuff from the earth, and building more and more structures and infrastructure to serve humanity in an increasingly humanized world.

On a theory of appropriate fit, the loss of natural value is squarely centered, not on the loss of biodiversity as such, but rather on the loss of sense of the barriers that keep our valuing relationship with nature intact. According to appropriate fit, the loss of valuable nature is the loss of the recognition that there is a great value to make the space for some "projects" in the world – those within the realm of nature – to be ones that are not human projects; and that the role of humans is to zealously guard and fence these nonhuman "projects" off from human enterprise. Science has no special ability to help restore this recognition in which the fence posts must be sunk. It can only hold up to our faces the revealing mirror, which reflects, to those already capable of recognizing it, that the fences are in terrible disrepair.

4. **Valuing abiotic as well as biotic components of nature**

Theories that find natural value in the sentience of individual creatures, or in the good of an individual life of any organism, or even in the good of collectives such as species, struggle to connect this conception with "the land".[125] According to these "non-holistic" doctrines, the good of "the land" is primarily an instrumental good – as a means to the good of the organisms (and their kinds) that live on (or in) it and are well served by it. The places in which creatures and organisms generally happen to live are thus relegated to the status of a prop for acting out their life.

This leads to conundrums such as this one: When relatively meager numbers and kinds of organisms are at stake, regard for "the land" that they live on should be correspondingly meager. Such views therefore have difficulty justifying the place in our world of such places as the Grand Canyon, the world's great deserts, and other "extreme" or, for any reason, sparsely populated regions. More generally, nothing in the requirement to ensure the flourishing of even a large number of organisms and their kinds proscribes any and all human intrusions, so long as these organisms "do not care". To put this graphically: Theories of this ilk have, on principle, nothing bad to say about a McDonald's in the Grand Canyon, atop any mountain peak, or any other place where its construction and use allow the local nonhuman inhabitants to go about their business of making a life for themselves.

This is not the end of trouble for this collection of doctrines, which focus on the good of organisms and their kinds. From these doctrines also arise a universe of apparently irresolvable conflicts that pit the interests of one organism against the interests of the next. Almost inevitably, some changes to "the land" – whether human-induced or not – will favor some organisms and disfavor others. I recount some graphic illustrations of this in Sect. 8.2.7.2.

Placing ecosystems in the valuing limelight seems to push these conflicts to the side, but at the expense of a plethora of another set of equally impenetrable dilemmas. Why do ecosystems merit moral consideration? If they have interests, whence do these interests derive? Furthermore, the problem of adjudicating conflicting interests persists. Is one ecosystem wronged by another's impingement? Is it murdered by its successor?[126]

---

[125] This is a straightforward reference to Aldo Leopold's "land ethic" (Leopold 1949, 223).

[126] Dale Jamieson (2008, 151) suggests these and other, related questions that arise when ecosystems are taken to be primary objects of moral concern.

A theory of appropriate fit easily bypasses this entire raft of conundrums. Quite simply, its conception of natural value has no investment in putting up arbitrary value fences between the biotic and abiotic parts of the natural world. The natural world is valuable as natural for its being set apart and regarded as set apart from uniquely human endeavors. This view places no particular emphasis on one "component" – biotic versus abiotic, individual organisms or species versus "the land" – over another.

Nor does appropriate fit get entangled in the problematic business of defining the boundaries and identities of ecosystems, or in the more problematic business of trying to justify the value of particular ecosystem configurations. I have previously shown how this latter business aids and abets anthropomorphizing theories, which attempt to translate human autonomy, health, and integrity into like-named ecosystem qualities, while everything that bears on moral consideration gets lost in the translation. Appropriate fit does not sink into this morass, because it has nothing whatever to say about what constitutes a healthy ecosystem or what its integrity consists in; and it has nothing to say about this because these things don't matter, so far as nature's core value is concerned. Nor does appropriate fit have any need to offer an indefensible theory according to which nature is an autonomous agent who would be harmed by curtailing its freedom.

What matters, according to a theory of natural value according to appropriate fit, is the relationship that removes people from the job of determining what an ecosystem should look like, how it should function, and what particular entourage of organisms should there abide.

5. Historical changes in the perceived value of nature

A theory of natural value according to appropriate fit might at first seem to covertly rely on some questionable conception of "naturalness". This supposition brings along with it two related and persistent intruders that crash any party celebrating nature's value on this basis.

First is the charge of social constructivism. How can the deep and abiding value that one might suppose nature to have depend on a socially constructed conception of what nature is – something that has varied radically in time and over cultures and might be expected to continue to do so? Indeed, how can any respectable theory of natural value depend on some socially and culturally evolving conception of the world without devolving into some kind of radical relativism? Such value, it seems, would not be respectable, because arbitrary in its dependence on a seemingly chance and fickle history of attitudes regarding the place of nature in the world.

Insofar as this charge arises from supposing there to be no "natural world" apart from a human idea of it, it builds on a metaphysics of extreme philosophical idealism. Most of us who are not Berkeleyan idealists insist that things exist outside and independent of our concepts or ideas of or about them.[127] Over the course of the

---

[127] George Berkeley, Bishop of Cloyne, was the prominent early eighteenth century philosopher who defended the position that nothing whatever exists independent of minds and their ideas. Thus, his famous motto: "*esse est percipi (aut percipere)*" – to be is to be perceived (or to perceive).

history of human thought, people have generally acquired more and more powerful ideas of and about things that existed well before those ideas were formed. Gravity existed before Newton's idea of it.[128]

A second ground for dismissal is that, even if "naturalness" – as an approximation to some ideal state of nature (which has indeed changed from time to time and with the winds of fashion) – were found guilty as charged of social constructivism, a theory of appropriate fit would not thereby be implicated. A theory of appropriate fit gets off scot-free because it distances itself as much as possible from this state-based conception of nature.

Still, there is a sense in which appropriate fit can be said to be a social construction – in much the same way that friendship is. Both revolve around a particular relationship, which requires of people certain attitudes and behaviors with respect to certain elements in their world. In the case of friendship, these elements are some select group of persons; in the case of appropriate fit, the elements are some select part of the surrounding world. Each of these two kinds of relationship require of a person who enters into it some conception of that relationship and her role in it. I suppose that this is another way of saying that acts of valuing are transactions that people enter into with some significant understanding of their meaning. If this, in turn, means that friendship along with other forms of human valuing are social constructions, then so be it. But it is hard to see why this is a reason to reject friendship – or appropriate fit – as an idealist fantasy.

Related to the charge of social constructivism is what moral philosopher Bernard Williams (1995, 240) calls the "paradox" of preserving nature, which he equates with "using our power to preserve a sense of what is not in our power."[129] This charge of "paradox" also cannot be sustained. If there is any paradox in this, it comes only from slides in the interpretation of "power". Using a sequence of subscripts to cue the sequence of distinct meanings: On the one hand, people have the $power_1$ or capacity to adopt and develop moral attitudes and characteristic ways of behaving. These include the modally resilient "letting it be", which, on the theory of appropriate fit, is the core of a valuing relationship to the natural world. We are capable of developing our nature-valuing character by means of our capacity to choose how to act and behave in a way that embodies a modally robust walling off some part of the world from the development of human culture. On the other hand, we have the $power_2$ – now everywhere on display – to act and behave in ways that dissolve this barrier. That is, we have an ample capacity to absorb or annex the entire world – including any remnant of the natural world – into the domain of human enterprise. Because we have the first two powers or capabilities and because they are diametrically opposed, there is a choice. And people are creatures capable of making choices. So on the third hand, we have the $power_3$ to choose between these alternatives. And on the fourth hand, nature has some different kind of $power_4$

---

[128] Jamieson (2008, 164–165) similarly insists on distinguishing between concepts and facts that are conceptualized.

[129] The posing of this "paradox" is part of Williams' answer to the titular question of his essay: "Must a concern for the environment be centred on human beings?"

to do whatever it will, with or without human design and management. But of course, it does not choose how to act or behave. It just does what it "will" and exercises this power$_4$ whether or not people attempt to exercise restraint. Whatever paradox there is in Williams' proposition hinges on mistakenly thinking that the exercise of power$_2$ and the exercise of power$_3$ are incompatible with the "exercise" of power$_4$. To bludgeon this point home: The proposition is as mistaken as the view (brought up at the end of Sect. 8.2.4) that I can be said to design and manage the intactness of my critic's cranium by virtue of my steadfast resolve to refrain from crowning her with my bat; and it is mistaken for essentially the same reason.

It is hard to see how any paradox arises from people exercising their uniquely human abilities to make moral choices while the natural world does what it does. Perhaps Williams and like-minded thinkers feel that those who believe that valuing nature requires a restraint with respect to nature are somehow fooling themselves into thinking that they are actually powerless to break loose from this self-imposed restraint. Speaking for myself, I do not have any such illusion. Thinking back to the good of friendship, I also do not think that there is any paradox in restraining myself from bullying my "friends" into serving my interests. Yet, some part of the description of this relationship also falls under Williams' description of "using [my] power to preserve a sense of what is not in [my] power". The good of a friend cannot be realized by a slave or an automaton with respect to whom I use my power to ensure that she bend her behavior to accommodate my every wish. Instead, I use my power to choose to refrain from acting like a bully, for that is required to enjoy the friendship of a person (and my accurate sense of that person as a person) who is not in my power.

A theory of natural value according to appropriate fit also is compatible with natural value's temporal relativism and, in fact, handily explains it. I have in mind the oft-observed phenomenon that prevailing views about the value of nature change over time. In discussions about nature and its value, most famously in the context of discussions concerning wilderness, it is often noted that in the not-so-distant past, nature and wilderness were reviled. The Puritan Michael Wigglesworth was a preacher whose preaching was so reviled that he sought refuge in the writing of diaries and poetry. These works provide good evidence for why his sermons were so detested. Wigglesworth nevertheless has enjoyed a remarkable revival in the environmental literature for citations of his jeremiad on "God's Controversy with New England", which remained unpublished in his lifetime. This Puritan lament (Wigglesworth 1871) paints a decidedly jaundiced picture of the natural surrounds of New England "in the Time of the Great Drought, Anno 1662 / By a Lover of New-England's Prosperity":

> Beyond the great Atlantick flood
>   There is a region vast,
> A country where no English foot
>   In former ages past:
> A waste and howling wilderness,
>   Where none inhabited
> But hellish fiends, and brutish men
>   That Devils worshiped.

This verse might have well reflected general sentiment in seventeenth century New England. But human-held values of all kinds can, do, and should change with changing circumstances and context. Sometimes the change is a matter of the veil being lifted from some value that, in retrospect, should have been evident all along. Obvious examples abound in the realm of human rights – the expansion of basic rights and privileges to the property-less underclass and to women, and the general realization that slavery is an abomination. Other times, as in the present case of nature, values change as the result of changing circumstances, which bear on what, in those circumstances, is required to live a worthwhile human life. In Wigglesworth's New England, the natural world was everywhere and "cities" were relatively isolated enclaves of desperate human self-protection. Now, the situation is reversed, and this reversal makes it possible to recognize the value that would be lost were the isolated enclaves of the natural world to be entirely absorbed into our cities and integrated into the domain of human enterprise.

6. **Nature inside and people outside the walls of cities**

Before taking leave of this brief survey of the implications of appropriate fit for natural value generally, I wish to say something about appropriate attitudes and behavior towards vestiges of the natural world that remain "*within* the city walls". The importance of this topic derives from two diametrically opposed consequences of the expansion of the domain of human enterprise with the concomitant displacement of more and more of the domain that is not caught up in some human project. On the one hand, this trend *pro*motes the importance of gardens, city parks, potted plants in cubicles, and grass lawns instead of cement between houses – for there are more and more human-made environments in which these vestiges of nature might make a difference, however small. On the other hand, it *de*motes the independent importance of any principle, which aspires to identify what is naturally valuable within the city walls. As I argued in Sect. 8.2.5 (Natural value outside its core), such a principle must summon its force from a discernible connection to the *core* of natural value. That connection becomes more and more tenuous as the very possibility of its core value-defining relationship is more and more swallowed up by the expansion of "the city".

Whatever the ultimate resolution of these opposing forces, I believe that there is still good reason for saying that, for now, we *should* make an effort to invite and welcome other creatures and plants to live in and among us human creatures in our unambiguously human-made, made-for-human environments. Moreover, I do not see any basis for thinking it impermissible for us humans to wander outside the confines of these environments of our own making – so long as we strictly maintain our "intentional" distance as project designers and managers and so long as the conditions of maintaining that distance suits us. I believe that it can be argued (though I will not fill in the details here) that the reverse of this principle also applies: We can and ought to regard our human-made domain as welcoming of any and all organisms – insofar as it suits *them* as well as us.[130] By the same token, we should

---

[130] In these surrounds, it would be foolish to insist that people *must* welcome, for example, bedbugs or disease-carrying rats, mosquitoes, viruses, bacteria, and mould.

accept that this cadre of companions will be the organisms best suited for an "eco-system" that was not designed for any creature but one. Insofar as that is the case, the particular organisms that reside in our midst will be yet another reflection of human enterprise and development. Therefore, complaints about the particular creatures that do show up – a perceived overabundance of rats, corvids, gulls, and cockroaches, which happily make a life alongside ours – ring hollow.[131]

Nevertheless, I also see little harm in devising special projects that make special accommodation for certain creatures to live in our midst. But we should honestly acknowledge that this is an afterthought and that, as an afterthought, it carries no great moral burden. We should not make ourselves look foolish by contriving fancy ecosystem-based justifications for any such effort. The rationale is simple and light-weight: The "ecosystem" in question is already part of the human-designed world and there seems to be no harm in modestly modifying that design to suit its design-ers' fancy or the fancy of those human occupants for whom it was designed. So it is okay to give elevator rides to city-dwelling peregrine eyasses whose reckless first attempts at flight land them on city streets and who would otherwise perish for their inability to return to their human-designed skyscraper-perched nesting site. But pro-viding elevator services for peregrine falcons should not be mistaken for anything that could be justified on the grounds of "enhancing natural value" or recovering the "natural ecosystem" that the city long ago displaced.

In their skyscraper habitat, peregrines are our invited, "alien" guests. Just how alien they are is obvious from their tendency to self-destruct at falcon-speed – crash-ing into the glass walls of the strange, steel-girdered, glass-walled canyons. As with many "aliens", new selective pressures – elevator rides that reduce the importance of capable upward flight at initial fledging; fatal crashes that disfavor the continued transfer of genes that make the birds perceive a pane of glass as open flying space – could eventually yield a new kind of creature. But this increment to biodiversity would have little to do with natural value. It would have as little to do with natural value as a falcon made city-wise via genetic engineering.

The city remains the city and human relationships to the city as a place remain relationships to a human creation designed with human goals in mind. This circum-stance cannot be changed by decorating the city with birds of prey, trees, or other remembrances of a truly natural world, which requires an entirely different sort of relationship. Within the concrete canyons, the meaning of these bio-parts is absorbed into the meaning of the city's overall design.

8.2.7.2   Implications for Action, Social Policy, and Conservation Management

According to a theory of natural value according to appropriate fit, there can be no state description for what nature should look like. There is, for example, no historical reference, no particular state of biodiversity, no state of "wilderness", no particular

---

[131] See Note 4 in Sect. 7.2.1 ("Assisted migrations" and such) for a relevant story about California gulls.

structured set of properties. I realize that shunning these graphic depictions might leave an itch to know something more than I've already conveyed in describing the core valuing relationship of people to the natural world. My sketch of appropriate fit would be incomplete without scratching that itch. I offer welcome relief in the form of illustrations, which more graphically suggest appropriate fit's vision of a valuable natural world. These illustrations, in turn, would not be complete were they not set off against some more graphic illustration of the (literally) aimless and frenetic development of "nature" that is the hallmark of neo-conservation.

I take a path cleared by Alan Weisman's remarkable thought experiment, *The World Without Us* (Weisman 2007). But before setting out, I must earnestly emphasize that I do not take Weisman's thought experiment to suggest the perverse thesis that the best of worlds is one in which his experiment is actually performed. Nor do I endorse Weisman's own suggestion, expressed in the prelude (Weisman 2007, 4) and elsewhere in his book, that the arc of transformation in a world absent people could be characterized as a welcome process of attaining a post-human Eden. This runs counter to appropriate fit's central theme that natural value is an essentially human-valuing relationship. And it runs counter to appropriate fit's eschewing of any idealized "state of nature", including an Edenic version.

Vividly and with impressive, evidence-supported detail, Weisman imagines what would happen to the world if, without cataclysm or even minor disruption, all of humanity were suddenly spirited away, as if in a rapture. I am not concerned with one major focus of the book, which is on the titillating details of how almost all human-made infrastructure, no matter how apparently indestructible, would inexorably crumble by dint of natural forces and processes. Rather, I am mostly interested in Weisman's word painting of the kind of place that is left to its own devices, and I am interested in the ability of such a place over time to define or redefine itself as a natural place with relation to the "left behind" people. Given sufficient time, such a place reclaims its *potential* for natural value. However, on a theory of appropriate fit, that potential is not realized if the place is merely held in reserve for yet another human-designed project. Nor is the potential realized if the place is coerced into a restorationist's idea of what it should look like. However, the place's potential for being valuable as nature *is* gradually realized as the human practice of restraint gradually becomes more and more robust and more and more secure – eventually reaching the point where it is no longer possible to regard that place saliently as a human creation, nor the remains of a human creation, nor even a future site for a human creation.

The most enlightening parts of Weisman's book are his depictions of a few places on the planet where his mostly-imagined rapture experiment has actually been performed – at least in approximation. Without the intervention of a rapture, at their own pace, and by their own ways of working, these places have actually been allowed to express and transform themselves, not according to how people think they ought to look or to function, or according to rules about the diversity of kinds and numbers of organisms that people think ought to live there, but rather according to their own kind of non-teleological creativity. These are probably not the kinds of places that the concept of Eden would bring to mind; nor ones suggested by "wilderness". Rather, they are places – some with a quite checkered history – that have been withdrawn from the domain of human enterprise. For the most part, the motivation for their withdrawal has little to do with consideration of their natural value. Despite that, the human relationships that have developed with these places as a consequence of their withdrawal are precisely the kinds of valuing relationships that are the core of natural value. Yet, as I shall relate, this developed and developing reserve of places with genuine natural value is threatened, mainly by neo-conservation managers and restorationists who cannot wait to turn them into their ideal of nature.

On the list of such places is Białowieża Puszcza, the Polish wood. This is a remnant of a place once governed according to *John Manwood's Treatise of the Forest Laws.*[132] For centuries, it was reserved for the nobility's hunt; now it is threatened by modern ecology managers who vow to manage its primeval quality – via such practices as culling – into perpetuity (Weisman 2007, 9–14). Perhaps more initially odd-seeming is the Korean Demilitarized Zone. This 4-km-wide, 260-km-long swath of land separates the two, still-officially-warring Koreas. Its removal from the world of human enterprise is a matter of armistice. The two sides still eyeball each other and angrily remonstrate across this divide. But meanwhile, it is a land left to itself – some of it mountainous, some of it bottomlands formerly sown with rice and now, as the result of the Korean War, with land mines. There has been no restoration, no conservation. Creatures that might well have no other home in the region make it their home. Weisman (2007, 185) stresses the beneficent effect on some of its most visible denizens:

> Asiatic black bears, Eurasian lynx, musk deer, Chinese water deer, yellow-throated marten, an endangered mountain goat known as the goral, and the nearly vanished Amur leopard cling to what may only be temporary life support – a slender fraction of the necessary range for the genetically healthy population of their land.

This might be a happy circumstance. But what appropriate fit helps us to understand is the even more fundamental importance of such a place as one that is set off beyond the walls of our cities. This importance is in no way diminished by the fact that no restorationists had a chance to obliterate vestiges of previous human activity. Nor is it diminished by the fact that it was not added to an optimal set of biodiverse places, according to some conservation biologist's heuristic for solving the NP-complete set cover problem. That is because, according to appropriate fit, solving a problem of any kind – no matter how hard, how amusing, or how career-enhancing – captures nothing about the basis for relationships that ground how humankind can best fit into the world.

Two more examples italicize the cardinal point. They are related by loosely sharing a heritage rooted in a signal horror of the twentieth and twenty-first centuries. The first, not included in Weisman's book, is Hanford Ranch – part of the United States' Department of Energy's Hanford Nuclear Reservation in Washington state.

---

[132] See Sect. 8.2.4 (More on appropriate fit) for an explanation of the significance of Manwood's treatise.

This is the site at which most of the plutonium for the Manhattan Project was produced. Plutonium production there ceased a quarter of a century ago, but the Reservation continued and still continues to serve as a dumping ground for military radioactive waste, because no one has a better idea for what to do with it. The Ranch is 790 km^2, carved out in the year 2000 from the buffer zone for the entire 1,450 km^2 of land in the Reservation. Untouched and unmanaged for more than 60 years, it is a place that, as David Wolman (2010, 15) says, "actually has a chance to thrive on its own." He is onto something that points towards appropriate fit when he suggests (Wolman 2010, 18) that human relationships to this place are the key to its value, rather than its approximation of some "pristine" ideal state:

> All places are more complicated than pithy descriptors like *pristine* or *poisoned*, as are our relationships to them. [italics in the original print version]

Finally, back in Weisman's book, there is the Zone of Alienation, the 30 km-radius area evacuated around the infamous Chernobyl Nuclear Power Plant. This is the still-radiation-oozing, festering wound that is the legacy of Chernobyl's explosion and fire. It is hardly anyone's idea of Eden or wilderness. But that 1986 disaster pushed this place fairly completely outside of the walls of the city, outside the domain of human enterprise. And so, again, without the "benefit" of restoration or conservation projects, and without some management regime that serves no clearly defined ends, except, perhaps economic ones, the Zone of Alienation became a place beyond the reach of these and other human projects – a place capable of supporting relationships at the core of natural value. Weisman (2007, 215) again focuses on the flourishing of creatures:

> … skylarks perch on [the] hot steel arms [of cleanup machinery], singing. Just north of the ruined reactor, pines that have re-sprouted [from the pine forest that died within days of the blast] branch in elongated, irregular runs, with needles of various lengths. Still, they're alive and green. Beyond them, forests that survived had filled with radioactive roe deer and wild boars. Then moose arrived, and lynx and wolves followed.

This is clearly a natural world, as I understand that designation. That it springs from a baseline established by one of humankind's worst blunders does not diminish its natural value.

Unfortunately, just as at Białowieża Puszcza, ecosystem managers are ever eager to get into the act. Already, they have (re)introduced into Chernobyl bison from Belovezhskaya Pushcha, Belarus' counterpart of Poland's Białowieża Puszcza. This, according to them, has an ecological justification: It pushes the place towards a lost and irretrievable centuries-removed past. This "justification" is an exemplar of the hollowness of conservation rationales. It is impossible to understand how this goal can be considered rational even on its own terms: The place is utterly transformed, totally unlike the land that the bison once roamed. If nothing else, the nuclear blast has ensured this. So even if there were good reason to think that transforming Chernobyl into its former self would be a good thing, herding bison in from elsewhere cannot possibly help to realize such a metamorphosis. If there were anything to the category of "alien" creature, it would apply to these bison, which are surely aliens on their former stomping grounds.

And in fact, even if humans had the capability to wrestle Chernobyl into a state that more closely resembled something in its past, there is no reason to think that that would be a good thing for Chernobyl's natural value. One could say with much better assurance that Chernobyl would then have value as a setting for a remarkable feat of engineering. The successful execution of such a project would be admirable. But the focus for that admiration would be on the human capacity to make socially approved landscapes. On an appropriate fit theory of natural value, this would have little or nothing to do with the value of the place as part of the natural world.

This is the cue for the second part of my task for this subsection, which is to more graphically illustrate the aimless and frenetic development of "nature" that is the hallmark of neo-conservation. One can expose the multiple confusions about goals – confusions that have to do with achieving certain states of biodiversity, ecosystem health, a balance of nature (Sect. 8.1.3 and 8.1.5), or nature's autonomy (Sect. 8.2.6). And one can show (as summarized by item (2) in Sect. 8.2.7.1) why, to the degree that nature's value is seen to be bound up in its ability to compete in the market-place, the more it is sapped of the unique, modally demanding value at its core. But it is quite another thing to hold up to view the grotesque consequences that can result from acting in the more or less complete absence of consistent, coherent, or generally endorsable principles. These consequences are best illustrated by the phe-nomenon of what I call "pinball conservation". In this quite serious game, the bum-pers nevertheless tend to have labels reminiscent of arcade pinball, instead of well-reasoned rationales. They therefore deflect the course of conservation projects off in arbitrary directions; sometimes, the results are wild and wildly discomfiting.

Here is a small sample of the bumpers that are installed in conservation pinball and their possible effect on the ball:

- The "alien" bumper. This bumper can make the ball career unpredictably in either of two diametrically opposed directions: "Alien" creatures are introduced (for example, the Tamarisk beetle) on the grounds that "alien" organisms (for example, *Tamarix* spp.) are undesirable.
- The "native" bumper. The effect of this one can depend on how the conservation machine is tilted: "Native" species are defended against "aliens", but an excep-tion is made when the "natives" fail and an "alien" is perceived to be "function-ally equivalent", thus justifying its introduction.
- The "balance" bumper. More up-to-date versions of conservation pinball tend to "balance" with the labels "ecosystem health" or "ecosystem integrity": In the vicinity of two "native" bumpers, a "balance" bumper can cause the ball to ping-pong wildly between the two.
- The "prey" and "predator" bumpers. The effect of these can depend on the prox-imity of "alien" bumpers and on tilting by recreational interests: Populations of prey species (for example razorbacks) are nurtured so that they survive the attacks of predators (trout), whose populations are also nurtured, because although "aliens", they themselves are the favorite prey of sports fishers.

Natural history in North America is especially replete with stories whose plot lurches from "prey" to "predator" and back again. *Puma concolor* (mountain

lion) and *Canis lupus* (wolf) tend to find themselves on the predator side, while
ungulates of various kinds – principally, *Odocoileus hemionus* (mule deer) and
Odocoileus virginianus (white-tailed deer), but also *Ovis canadensis* (bighorn
sheep), *Cervus elaphus* (elk), *Alces alces* (moose), and *Antilocapra americana*
(pronghorn) – on the prey side. Predators are welcome when prey are thought to
be too numerous and responsible for changes that are characterized as "ecological damage" (that is, subverting the balance of nature). But they are reviled when
thought to deprive human hunters of their prizes. A gaudy but not atypical display of this sentiment is in Alaska, whose wildlife management system systematically guns down, ensnares, and poisons both bear and wolf that might kill
moose… before people do (Ross 2011). But perhaps predators are reviled even
more when thought to have committed crimes of "depredation" against the human
guardians of domesticated animals.

   P. concolor seems particularly adept at confounding the quest for balance in
nature.[133] It is seen as tipping that balance against various populations of bighorn
sheep – *Ovis Canadensis canadensis* in southern Alberta (Ross et al. 1997), *O. c.
sierrae* in California's Sierra Nevada, and *O. c. nelsoni* (desert bighorn) in New
Mexico (Rominger 2007). But sentiments are not entirely on the side of the victim-
ized sheep. Bighorns compete with domesticated sheep for forage. Worse, they have
been the vulnerable targets of domestic sheep-transmitted pathogens. As a consequence, sentiments have also run high for culling the *wild* animals who have had the
misfortune of being exposed by their domesticated brethren, for fear that they might
return the favor. If there is any guiding principle behind this kind of conservation, it
is that the balance of nature demands that nothing in the natural world get in the way
of the human economy.

   The mother of all neo-conservation parables is not a parable but the actual history of San Clemente Island off the coast of southern California. The complete
story has convoluted, interleaving, and cascading plots involving many human
and nonhuman players. The cast of nonhumans includes feral goats, feral pigs,
San Clemente Island fox, Bald Eagle, Golden Eagle, San Clemente Loggerhead
Shrike, and more. The plot is hopelessly tangled; what follows is just one track
through the thicket.

   Sheep farming was established on San Clemente Island in 1868. Somewhat later,
goats were brought to the island – most likely from Mexico – as part of a lively
contraband trade. The United States Navy took over the island in 1934. In 1955, it
decided, with the approval of the California Department of Fish and Game, that it
would be a fine thing to have pigs and deer to hunt, and so brought in these animals.
One does not need a Ph.D. in ecology to predict the effect that of all these grazers
and rooters-around had on the island's vegetation.

   So in the 1980s, on the grounds that these "aliens" were eradicating "native"
plants, the Navy took to eradicating the once-welcome goats and pigs – via traps,

---

[133] Terry Glavin's chapter on "The Ghost of the Woods", in Glavin (2006), places the life and times
of mountain lions in North America into the broader perspective of a brief natural history of North
American megafauna from Pleistocene times.

helicopter-shooting forays, and, for the goats, radio-collared "Judas" goats who unwittingly led the rifle-wielding brigades to their caprine brethren. The rationale? The demise of the native plants was more than enough evidence that these animals were upsetting the balance of nature and placing ecosystem health in jeopardy.

To this point, conservation management was cast in terms of managing a contest of "natives" versus "aliens". But the presence of both goats and pigs helped to set off another contest between *Lanius ludovicianus mearnsi* (the San Clemente Loggerhead Shrike) and *Urocyon littoralis clementae* (San Clemente Island fox) – both of which were regarded as "natives". The Loggerhead Shrikes suffered from the island's goat-thinned cover. The previously thick tangle had protected them against their mainly raptor predators and had provided them with favored nesting sites. In the transformed landscape, the bird most likely also fared poorly locating and capturing its own prey – primarily various large insects, but also small lizards such as skinks, small rodents such as mice, and small birds.

Meanwhile, the fox was having its own problems, which largely arose from the confluence of a seemingly unrelated story – the DDT poisoning of the Bald Eagles that used to frequent the Island before the 1970s. Bald Eagles do not customarily prey on fox. But Golden Eagles took advantage of the Bald Eagle's decline to fill their cousins' talons.[134] And they found the San Clemente Island fox much to their liking. At just one-fourth their size even as adults, the fox are easy pickin's for the Goldens. And much to the Golden Eagles' delight, San Clemente also had succulent feral piglets on its bountiful menu (Roemer et al. 2001).

The fate of the Loggerhead Shrike and the fox then collided. Under pressure for different reasons, both creatures retreated to the same, limited, steep-walled canyon habitat, which was the least-changed and most congenial to their lifestyle. Diminished in choices of both habitat and menu, and finding lots of Loggerheads in its neighborhood of last resort, the fox became the Shrike's primary predator. Nature was out of balance again. For awhile, the conservation solution was to shoot the *native* fox – the conservationists thus assuming the predatory role for which the Golden Eagles were removed. But that upset the presumption that "native" is good and so that practice was abandoned in favor of shooting non-native rather than native predators: Feral cats (*Felis catus*) are easier to dislike than native foxes; they subsequently found themselves in the crosshairs: Shooting them is not a violation of privileged nativeness.

Finally, tiring of endless rounds of managing one species against another, the Department of the Interior decided on what they called an "ecological approach" – yet another rubric (along with "ecosystem health" and "ecosystem integrity") for "balance of nature". It embarked on a program to create a new and improved San Clemente Island that gestures towards an imaginary San Clemente of some unspecified time past – the perfect restorationist-tinged ending for a fairy tale. According to biologists Gary Roemer and Robert Wayne (2003, 1257), the Department's

---

[134] Golden Eagle populations did not suffer from DDT exposure as Bald Eagle populations did, because their primarily-mammalian prey did not tend to concentrate the pesticide in their tissues the way the Bald Eagle's fish prey did.

actions include the livetrapping and translocation of Golden Eagles, the main source of fox mortality; the reintroduction of the extirpated Bald Eagle (*Haliaeetus leucocephalus*), to restore this component of the fauna and as a potential hindrance to future colonization of the islands by Golden Eagles; the eradication of feral pigs (*Sus scrofa*), to prevent further loss of native flora and to remove the food supply Golden Eagles require; the removal of exotic plants, such as fennel (*Foeniculum vulgare*), to increase the distribution of native plants; and the captive propagation, release, and monitoring of island foxes...[135]

How does the frenzy of San Clemente conservation pinball look from the viewpoint of appropriate fit? Unsurprisingly, appropriate fit's view is a jaundiced one, but this is not rooted in a judgment concerning the prudence or imprudence of letting the project careen from bumper to bumper. Appropriate fit does not frown on neo-conservation and restoration for "pragmatic" reasons. Its warning does not run along the lines of "leave it alone or it will get worse". A view based on a negative judgment about prudence would be subject to revision – if, first of all, "better" and "worse" were (counter to actual fact) well defined and the norms soundly justified; and second, if we became substantially more proficient at achieving what is "better". In fact, "better" and "worse" appear to be a hopelessly jumbled amalgam of balance, nativeness, and a lamented lost past – all vague, all normatively loaded, and all lacking any normative justification.

In contrast, a theory of natural value according to appropriate fit takes a stance outside the dichotomies of prudent/imprudent and better/worse. It stakes out a position that is not subject to revision based on evidence that tips the balance towards what the ecological whim of the moment labels "prudent" or "right". Rather, it calls for a total cessation of pinball logic with its pinball effects. And it does this for two principled reasons, neither of which is empirically contingent on the predictive power of ecology, or on human proficiency, or on capability in any undertaking:

First, neither conservation nor restoration, understood according to current orthodoxies, serves any environmentally justifiable, coherently describable, or generally endorsable end. The first seven chapters of this book showed how one line of reasoning that purports to show that biodiversity is an end emblematic of natural value and worthy of conservation fails. The current chapter has expanded and generalized this examination of conservation beyond biodiversity by laying bare its framing assumptions. These it has found to be not just weak, but mostly indefensible, and even antithetical to natural value, understood in terms of appropriate fit. The disastrous results of such a project as San Clemente Island are but a symptom of this pathology, which infects neo-conservation efforts independent of their results, including when they are not so obviously discomfiting.

This last observation is paramount, because it says that there is no fixing conservation practice, as it is now widely understood. Whatever one thinks about the actual history and outcome of conservation management for San Clemente Island, it could

---

[135] For a description of how the U.S. Department of the Interior views this effort to create a new island ecosystem reminiscent of the old, see U.S. Fish and Wildlife Service (2009).

not have gone "better" – if the real end is to honor the natural value of the natural world. Even if there were (contrary to fact) some identifiable and endorsable end for re-inventing San Clemente's ecology and even if there were some reasonable hope for designing and engineering some means to achieve it, in undertaking this project, people would enter into a relationship with the natural world that is antithetical to the relationship that embodies the core value of nature.

8.2.7.3 Final Words: Appropriate Fit in the Shadow of the Prevailing Alternative

I wish to acknowledge that, in the limited way in which I have presented and developed it, a theory of natural value according to appropriate fit remains a mostly abstract ideal – albeit one that suggests a radical departure from conventional ways of viewing human relationships with the natural world. This theory is not likely to be widely shared at the moment, partly because this book is its first widely disseminated presentation. Resistance is likely to be strong and widespread due to conflicts of interest and institutional momentum. Our human nature is such that it will be so much easier to keep our evaluative scope narrowly but comfortably confined to bumper labels that substitute for rationales. It will be far easier to "go by the numbers" that science and economics are tooled to generate. It will be easiest to follow along this familiar path – even though the labels and numbers reveal nothing about what really matters.

But the facts that my theory is not yet widely shared (because not yet widely known) and that the road to its acceptance and adoption is likely to be difficult are not *reasons* for spurning it. At the very least, appropriate fit provides a plausible ideal. It is an ideal that, more clearly than any other I am aware of, identifies the special nature of natural value – by virtue of riding on a unique way of relating to the natural world. I would say that it is good to have in mind an ideal – even at an early point at which prospects for its widespread adoption are uncertain. The only way to press towards an ideal is to keep it firmly in mind. And it is good to have a guiding principle that even a single individual can use to determine how, in living her own life, she can best establish a valuing relationship to the natural world. In this way, she might better integrate herself into the world at large.

Of course, there are other options. Among them is the rationalizing option, which co-opts the notion of appropriate fit so as to rationalize our collective moral failure. The rationalizer says that people are at their best as steroidal beavers. Since this characteristic might well be genetically imprinted, one could think that we would do well to embrace it as our defining identity. This option forthrightly denies the existence of any special value attaching to nature other than in offering our castorid selves a challenging arena for human enterprise.

There is the pragmatic option, which is sustained by a low expectation of realizing a world in which natural value plays anything but a token role – an afterthought – in individual and collective practical reasoning. It recommends that those of us who subscribe to the view that there is unique and compelling value in nature are nonetheless well advised to avoid wasting our energy trying to live by it, let alone convince anyone else of it. We all might as well maximize our benefits by joining the party that repurposes the natural world to satisfy any and all other interests, for any and all other reasons, or it seems, for no particularly good reason at all. We might as well, as they say, just "get the numbers right".

And there's the bio-technologist option, exemplified by biogeoengineering, which clings to the precept that if we sufficiently well understand how natural systems work and how to manipulate them, then we can pick up the tattered bio-parts and build out of them something that works more or less the way some designer thinks that they are supposed to work. We should strive to revamp the world's ecosystems so that they have greater biodiversity, and so that they are better balanced, healthier, and more vibrant.

I have argued as well as I can that we should heed the poet who tries to disabuse us of these fancies:

> Sweet is the lore which Nature brings;
> Our meddling intellect
> Mis-shapes the beauteous forms of things: —
> We murder to dissect.

Giving leave to our predilection for meddling, we murder nature by dissecting it into its component parts. These we take to be tokens of arrangement and rearrangement into configurations that suit our narrowly conceived interests or our narrowly conceived visions for what nature should properly be. But this meddling so grossly misconceives or so totally ignores the basis of a valuing relationship with nature that, inevitably, it mis-shapes the beauteous forms of things. Those forms are unrecognizable as bio-cogs in the machinery of the human economy. They are equally unrecognizable as the scientific pronouncements, handed down from the heights of a free-floating Laputa by learned technologists who use quadrant and compass to measure the proper dimensions of nature.[136]

We can also choose to reject these easy options. And this, I believe, is what we should do. We should refuse to reduce nature's value into something quite astonishingly disconnected from nature. We can and should join the poet in remonstrating:

> Enough of Science and of Art,
> Close up those barren leaves;
> Come forth, and bring with you a heart
> That watches and receives.

---

[136] This, of course, is a reference to the flying island and the technological machinations of its inhabitants (down to the tailor who employs quadrant and compass to measure Gulliver's clothes) who disdain practical reason – in Jonathan Swift's 1726 *Gulliver's Travels into Several Remote Nations of the World* (http://www.gutenberg.org/ebooks/829).

Acknowledgments

First, I wish to express great gratitude to (in temporal order of my contact) Carlo Ricotta, Dov Sax, Félix Forest, Jan van Dam, Richard Warwick, Tom Dudley, Maud Huynen, Bill Snyder, and Jon Erlandson for responding so graciously and promptly to pleas for copies of their papers. My pleas came in the first stretch of writing, when their materials were, for all practical purposes, otherwise inaccessible to me. Special thanks are due to Tom for permitting me to cite correspondence in response to some questions that I posed to him. Thanks, too, to Philip Pettit, for sharing with me a pre-publication draft of the 2011 paper cited in this book.

I owe much to discussion of my work with my philosophy colleagues, whose infectious enthusiasm for this project greatly stimulated my own enthusiasm and resolve to turn it into a book. Much of this help came from those who were exposed to the first incarnation of this work at the 2009 ISEE/IAEP Joint Conference in Allenspark, Colorado. Stimulating conversations with many people pushed the project forward. Special thanks are due to Neil Manson, for his suggestion to reflect on the value of diversity in general; to Darren Domsky, for pointing me towards the realization that informal fallacies of many kinds pervade discussions of biodiversity; again to Darren for providing inspiration for considering ways to "enhance" biodiversity; to Alan Habib for pointing my nose at the issue of diversity in the context of social justice; to Jennifer Welchman for some good discussion about that; and most especially to Baylor Johnson, for his encouragement, and for his presentation of my work at the conference. Despite the unusually great length and complexity of the material, he somehow managed to present it so cleanly and clearly that he succeeded in stimulating an enormous amount of beneficial discussion and enthusiasm that spilled out into the entire remainder of the conference and, I hope, into the manuscript.

Since then, a number of expert friends and colleagues have generously taken the time to share their insights on various parts of the manuscript. Dan Haybron shared his expertise on the biophilia hypothesis and helped me understand some of its subtleties as well as some of the subtleties of his own thinking about it. Dale Jamieson graciously responded to my request for comment on my challenge to his

D.S. Maier, *What's So Good About Biodiversity?: A Call for Better Reasoning About Nature's Value*, DOI 10.1007/978-94-007-3991-8, © Springer Science+Business Media B.V. 2012

distinction between "affecting" and "being a product of". Jeremy Bendik-Keymer was enormously generous with the time he devoted to a spirited discussion with me regarding several pieces of the manuscript that touched on aspects of his work. Andrew Light, too, shared perspectives on his work, which helped greatly to broaden my own perspectives on it. Derek Guerney and Larry Goulder vetted and offered their perspective as economists on the option value material. Jeffrey Lockwood was painstakingly thorough in examining the chapter on theories of biodiversity value, which greatly benefited from his remarks on it. Nicole Hassoun did as well as anyone could to suggest how to scour the chapter on Preliminaries clean of unnecessary distractions. Vas Stanescu's insightful commentary on the final chapter was invaluable in helping me see where clarifications and qualifications were in order. Several anonymous reviewers offered criticisms that prompted clarifying revisions in the first seven chapters, which focus on critical review of dominant views, and that spurred me to write the last chapter, which offers an alternative way to find value in the natural world.

I also wish to express great gratitude for the encouragement, comments, and valuable materials supplied by several personal acquaintances. Among those persons, to Pat Musick is due very special thanks for her yeowoman-like assistance in editing the first draft of the manuscript for the ISEE/IAEP Joint Conference, yet more help with the first draft submitted for publication, and an enthusiasm for this project that helped to make even me – a professional skeptic – believe in it. I am also indebted to Linda Winter, whose encouraging and nurturing presence had outsize influence in reinvigorating my belief in this project and in anchoring my resolve to finally bring it to completion. Last but not least, I cannot fail to mention Squeaky, whose uniqueness – taxonomically speaking and in many other respects – and whose refusal to be "managed" has constantly reminded me of what natural value is all about.

Two wonderful, highly interdisciplinary fora at Stanford University – the Environmental Norms, Institutions, and Policy Workshop coordinated by Debra Satz and the Environmental Humanities Project coordinated by Ursula Heise – helped enormously to expand the breadth of my thinking on the issues covered in this book and beyond. Their workshops exposed me to material that was as varied as the training of the enormously talented and collegial scholars who participated in them. Those biologists, climate scientists, environmental engineers, economists, political scientists, sociologists, anthropologists, information scientists, historians, classics scholars, literary critics, and even a few token philosophers are a testament to the possibility of harvesting rich fruit from intellectual fertilization that crosses disciplinary boundaries. I particularly benefited from the spirited discussion, which Carol Boggs played a key role in initiating, of parts of my book in the Environmental Norms, Institutions, and Policy Workshop. More recently, the 2011 Northeastern Workshop on Applied Ethics provided a lively forum for discussion of parts of the book's last chapter.

To Jack Lyon go many thanks for his heroic efforts in helping me coax the book's index out of his DEXter software.

I also cannot fail to mention Sarah Rabkin, my extraordinary, keenly perceptive editor, who helped me prepare the final version of this manuscript. She uncovered multiple places and multiple ways to say it better.

To my biologist son Paul go heartfelt thanks for a characteristic act of generosity, which planted a seed from which this book eventually grew. Seeing that I was desperate to find reading material for a shared vacation, he graciously permitted me to appropriate a book that he had purchased for himself. That book launched my renewed and excited interest in the biology-based topics that eventually led to the writing of this book.

I dedicate this book to my dad, Howard Maier. He had within him many books. Those of us who knew him are still reading and growing wiser from them.

Glossary of Scientific, Computational, Economic, and Philosophical Terms

The discussion of biodiversity inevitably brings into play the enormous body of scientific research bearing on the topic. I have not spared the non-scientist from exposure to this science, as scientists present it, complete with their technical vocabulary. I hope that this list of terms includes most that might be unfamiliar to the non-scientist. Scientific terms, particularly terms from the field of biology, are in the majority. I indicate this disciplinary association by *omitting* a discipline-identifying tag for them, except when the biological sub-discipline provides useful information (for example, "biological taxonomy").

In discussing algorithms connected with the conservation of biodiversity and the quantification of biodiversity, I use some terms from the field of computation theory, which are also included here. These are tagged as "computation".

Some technical vocabulary from economics is used in the introductory section that characterizes various philosophical approaches to environmental ethics. These bear the tag "economics".

Some terms familiar to philosophers but possibly unfamiliar to others are included. These are tagged "philosophy" and sometimes "philosophy/psychology" when shared with the field of human psychology.

abiotic Not part of, conceptually distinct from, or not the direct result of living organisms. Abiotic factors include such chemical and physical influences as temperature, pH, and relative availability of reactive nitrogen. Contrasted with "biotic".

affective (philosophy/psychology) Relating to feelings that are expressive reactions to something perceived – for example approval, pleasure, fulfillment, happiness, and the like. See also "conative".

agamogenesis Asexual reproduction. More precisely, reproduction that does not involve the fusion of specialized, gametic (male and female) germ cells. This is the primary or only means of reproduction for prokaryotes and many eukaryotes such as both unicellular and multicellular protists. Some fungi and some plants also reproduce via agamogenesis.

D.S. Maier, *What's So Good About Biodiversity?: A Call for Better Reasoning About Nature's Value*, DOI 10.1007/978-94-007-3991-8, © Springer Science+Business Media B.V. 2012

algae A diverse set of eukaryotic, single-celled to multicellular, and mostly autotrophic organisms whose largest and most complex forms are the "seaweeds".

allele, allelic A variant of a gene consisting of a specific sequence of DNA that occurs at the gene's locus. An organism's genotype is the collection of all its allelic variations, taken together.

allopatric speciation The emergence of a new species as the result of diverging adaptations of physically disjoint populations of an organism. Both genetic drift and differing adaptive pressures can contribute to the differentiation of the populations.

anoxia, anoxic Something approaching a total absence of dissolved oxygen in an aquatic environment. Anoxic conditions can occur as the result of a hyperabundance of organic matter that is then oxidized by decomposing bacteria. See also, "hypoxia".

anthropocentrism, anthropocentric (philosophy) Any of a variety of ethical stances that in some way make humans – as opposed to other organisms or natural systems – the nexus of value. Contrasted with biocentrism, among other views.

archaea, archaean One of the three domains of living things. Archaeans are unicellular organisms. On the one hand, the composition of their cell membrane sets archaeans apart from organisms in the domain of the prokaryotic bacteria. On the other hand, their lack of internal cellular structure sets them apart from organisms in the domain Eukaryota (such as humans). Archaeans of various kinds are famous as extremophiles, which live in extremely acid, alkaline, saline, hot, or oxygen-depleted (anoxic) conditions; though not all extremophiles are archaeans.

arthropod An organism in the phylum Arthropoda of jointed-legged invertebrates with exoskeletons. Arthropoda encompasses the vast majority of species, except, perhaps for bacteria. On land, the representatives of this phylum are mainly the insects and arachnids (which include spiders, scorpions, ticks, and mites). In water, they are mainly the crustaceans.

autotroph An organism that produces organic matter from carbon dioxide (inorganic carbon) and possibly other inorganic material, rather than by consuming other organisms or the organic material that they produce. An autotrophic lifestyle is most commonly supported by photosynthesis, though some autotrophs use chemosynthesis. Contrasted with "heterotroph".

axiology (philosophy) The general study of value, which includes the study of what things are good, how good they are, and why they are good.

benthic Associated with the bottom of a body of water, particularly its sediments.

benthos The complement of organisms that live in benthic environments.

biocentrism, biocentric (philosophy) The ethical stance that not all that is valuable or worth moral consideration is human or is considerable because it is good for humans. Taken to contrast with anthropocentrism.

biome An ecologically characterized type of habitat that takes into account such characteristics as terrestrial versus aquatic; marine versus freshwater; climate conditions; physical terrain; biogeochemistry; and the typical mix, adaptations, and interactions of plants, animals, and soil organisms that are present. The notion of "anthropogenic biome" (or "anthrome") extends these conditions to include the environmental influence of ongoing human activities.

biota, biotic Part of or the product of one or more living organisms. Biotic factors are ones that result from their direct influence. Contrasted with "abiotic".

biotope A relatively small region of relatively uniform environmental conditions that defines the extent of a biological community or relatively uniform assemblage of organisms (as opposed to a single species or population). The word "relatively" is flexibly interpreted in accord with the size of a community. A biotope is usually used to designate a particular place rather than an environmental kind.

Cambrian period The first geological period of the Paleozoic era, which started approximately 545 mya. From a biological point of view, the Cambrian is most notable for the "Cambrian explosion" of life forms, which produced many of the still-extant phyla.

category mistake (philosophy) An error of reasoning, which involves representing an entity as if it belongs to a category of things to which, in fact, it does not belong. A common form of this mistake is the ascription of a property to a thing that, by its nature, cannot possibly have that property.

chemoautotroph An autotroph that chemosynthesizes its organic matter via oxidation-reduction reactions. A chemoautotrophic kind of autotrophic lifestyle is less common than a photoautotrophic (photosynthesizing) lifestyle.

chronospecies A species that evolves into something so different in form or function from what it originally was, that according to some definition of "species" (other than the one for "chronospecies"), individuals at a later time are not members of the same species as individuals at an earlier time. The biological species concept can serve as the basis for distinguishing a new chronospecies when the organism is thought to have evolved so as to render it incapable of interbreeding with its ancestors, even supposing ancestral individuals still existed. See also "species".

cladistics A discipline for the systematic classification of organisms based on shared evolutionary ancestry. Sometimes called "phylogenetic systematics". Contrasted with other taxonomic systems in the Linnaean tradition that are rank-based and that place greater emphasis on morphology and other phenotypical characteristics.

cladogram A diagram that represents evolutionary relationships of ancestry. A cladogram of all life is called "the tree of life".

class (in biological taxonomy) In the biological hierarchy of categories of life forms, the major category just below phylum and just above order.

clathrate A compound in which one substance forms a lattice cage around a second substance. Methane hydrates are clathrates in which water molecules encage methane gas, which effects a naturally occurring, chemical sequestration of this greenhouse gas. There is increasing evidence that this type of clathrate is very abundant. Sufficient warming of the earth could destabilize deposits of it and release the methane into the troposphere.

coextinction The loss of one species consequent on the loss of another on account of a dependent relationship between the two. The archetypal example is the loss of a parasitoid as the result of the loss of its one and only host.

collective (philosophy) An entity that is composed of multiple other entities. In the simplest case, the constituent entities are objects in the ontological sense of "individual" (see Fig. 3.1). Among biologically significant collectives, some, such as species, are homogeneous – composed of multiple individuals of the same kind. Other biologically significant collectives, such as communities or ecosystems, are heterogeneous – composed of many different kinds of biotic and abiotic objects.

competent, competence (as for disease reservoirs) The relative ability of a host for an agent of infection (pathogen or parasite) to maintain a viable population of that infectious agent. In the context of diseases that are transmitted not directly from host to host, but via an intermediate vector, the competence of a host is largely a function of how likely it is to be infected by an infected vector, and conversely, how likely a vector is to be infected by an infected host. The term "competence" also applies to vectors, which, independently of hosts, vary in their ability to carry an infection from host to host.

composition, fallacy (philosophy) See "fallacy of composition".

computation theory, computation-theoretical (computation) As used in this book, the theory that classifies algorithms by their inherent difficulty or "complexity" and that classifies the difficulty of problems according to the complexity of the algorithms required to solve them.

conation, conative (philosophy/psychology) Relating to the mental faculties of desire and purpose – for example, preferences, wants, desires, and urges.

consequentialism, consequentialist (philosophy) One of the three major traditional classes of moral theory, according to which the rightness of an action is judged based on the action's consequences. The consequentialist starting point for moral theory differs from that for deontological theories and virtue ethics.

consumer surplus (economics) the value of a resource or service to consumers over and above its cost to them. A surplus represents a net benefit that takes costs into account. See also "option value".

convergent evolution The phenomenon whereby the distinct evolutionary lineages of two phylogenetically distant organism converge on the same or a similar trait.

crustacean A very large subphylum of mostly aquatic arthropods, including such animals as lobsters, crayfish, crabs, shrimp, krill, and barnacles. This set of creatures can be thought of as the aquatic insects.

cyanobacteria The photosynthesizing bacteria in the phylum Cyanobacteria. Also known as "blue-green algae".

Delgado button (philosophy) The activating control of a "stimoreceiver", which José Manuel Rodriguez Delgado devised to electrostimulate the brain in order to elicit various affective and perceptual responses. In philosophical thought experiments, used to conjure up the possibility of sensations and perceptions that are not "real" but that the experiencing subject does not or cannot distinguish from "real" ones.

deontology (philosophy) One of the three major classes of moral theory, which focuses on rules that are supposed to guide the actions of moral agents, often as they relate to entities of intrinsic value; and the actions' qualities of being permissible, impermissible, or (in particular) obligatory (that is, duties). This orientation differs from that of consequentialist and virtue ethics theories.

detritivore A heterotrophic organism that obtains its nutrients from non-living, particulate organic matter, known as "detritus".

disruptive selection A type of divergent evolution that can lead to speciation. In it, the heterogeneity of the environment of a place encourages a population of an organism to develop disjoint clusters of trait values that are adaptive and therefore favored, separated by intermediate values that are maladaptive and therefore suppressed in the population.

division, fallacy (philosophy) See "fallacy of division".

domain (in biological taxonomy) In the biological hierarchy of categories of life forms, the highest, most inclusive category. There are three domains – Bacteria, Archaea, and Eukaryota (the latter including you, me, and all other organisms with more complex cell structures).

dominance index An index of species diversity, such as Simpson's index, that more heavily weights the most populous species while relatively discounting species with lower abundances.

ecosystem A biotic community as it exists in its abiotic environment. The concept is very elastic. As for habitats, there is no clear or consistently established classification system or criteria to determine where one ecosystem leaves off and the next begins.

ecosystem service A property of an ecosystem that either directly or indirectly benefits humans and their endeavors. The word "service" in his phrase is customarily given an economic interpretation according to which it is an intangible good that people are willing to pay for.

egalitarianism, egalitarian (philosophy) The political doctrine that accords equal political, economic, and social status to all people. More broadly, a doctrine that accords equal status to any set of entities.

Elvis species (Elvis taxon) A species that goes extinct, then apparently reappears in the paleontological record. The phenomenon results from the convergent evolution of a later species whose morphology closely resembles the extinct one, despite the absence of a close phylogenetic relationship between the two. This concept applies, not just to species, but also to higher taxa. The phenomenon of Elvis taxa is among the best current evidence that biologists' sense of humor is *not* extinct.

endemism, endemic The property of an organism of being unique to a particular place or, sometimes, a particular kind of habitat. Endemism is a type of geographical rarity.

entropy (computational/information theory) The uncertainty associated with a random variable that represents information. More specifically, the Shannon-Wiener entropy, which provides a way to estimate the average minimum number of bits needed to encode a string of symbols, based on their frequency distribution. The term was self-consciously adapted from thermodynamics because of the resemblance of Shannon's information entropy equation to Gibbs' equation for thermodynamics.

epiphenomenon, epiphenomenal (philosophy) An event or state that arises from physical causes but itself has no causal efficacy. The term originally arose with reference to a theory of mental states.

eudaimonia, eudaimonistic (philosophy) One of the two terms (the other is "euzen") that Aristotle uses to characterize the ultimate good. It is commonly translated as "flourishing". A eudemonistic virtue ethics is one that views human flourishing as the justification for all virtues. See also "noneudaimonistic".

eukaryote, eukaryotic An organism (such as you or me) in the domain Eukaryota, which contains all organisms composed of cells with membrane-encased, complex internal structures, including a cell nucleus. Eukaryotes are distinguished from the prokaryotic bacteria and archaea. They comprise the kingdoms Animalia, Plantae, Fungi, and Protista.

eutrophic, eutrophication Nutrient rich. Eutrophication is the increased biotic productivity that results from an influx of nutrients.

expected value (economics) The sum, over all possible outcomes, of their likelihood-weighted values.

extinction "The End" for an entire species; that is, the death of the last remaining individual of a species.

extirpation "The End" for an identifiable population of organisms of one species in one place. Extirpation is local extinction.

fallacy fallacy A fallacy committed by inferring that a conclusion is false on the grounds that an argument for it is fallacious.

fallacy of composition A fallacy committed by asserting that something true of each or some of the parts of a whole is also true of the whole.

fallacy of division A fallacy committed by asserting that something true of a whole is also true of some or each of its constituent parts. This sometimes, though not always, results in a category mistake, as when one reasons that because water is wet, so are its constituent H_2O molecules.

family (biological taxonomy) In the biological hierarchy of categories of life forms, the major category just above genus and just below order.

fauna Animal life. That is, eukaryotic, heterotrophic, and mostly motile organisms (though some are sessile) in the kingdom Animalia, which comprises multiple phyla.

flora Plant life. That is, eukaryotic organisms in the kingdom Plantae, comprising various phyla, which encompass seed plants (dominated by angiosperms, the flowering plants), ferns and their allies, bryophytes (mosses, hornworts, and liverworts), and green algae. The latter two groups together make up the non-vascular plants.

frugivore An animal with a special fondness for fruit as a food.

fungibility, fungible The property of a good that makes it a commodity – namely that individual units are regarded as interchangeable.

generalist An organism that is relatively nonselective in its choice of food, habitat, or other resources.

genetic fallacy A line of reasoning that either endorses or discredits a proposition based on its origin or history, which is not normatively relevant to the current context for its evaluation.

genome The genetic makeup of an individual organism, or sometimes, the entire genetic complement of all the individuals in a species. More precisely, the complete DNA sequence, which includes both genes and non-coding segments, of each of an individual's chromosomes.

genotype The genetic makeup of an individual organism, or sometimes, cell or species. "Genotype" differs in usage from "genome" with respect to its focus on alleles.

genus (biological taxonomy) In the biological hierarchy of categories of life forms, the major category just above species and just below family.

habitat The sum of environmental conditions – both biotic and abiotic – in which an organism or group of organisms live. The concept is very elastic. As for ecosystems, there is no clear or consistently established classification system or criteria to determine where one habitat leaves off and the next begins.

halocline A relatively sharply defined vertical gradient of salinity in a body of water. Water above the halocline is typically considerably less fresh (more salty) than water below it, as the result of the greater effect of evaporation towards the surface.

haploid Having one copy of the genes in a genotype. Many organisms are haploid. Many others have both haploid and diploid (two-copy) states or generations.

hedonic evaluation (economics) One of several indirect methods that economists use to evaluate goods that have no actual market. A hedonic evaluation posits that the price of some good traded in a real market reflects the price of another, non-market good, which therefore makes the market-traded good a legitimate proxy for the non-market good. For example, the travel cost to a park is used to put a price tag on the park; the reduced price of houses in a polluted area is used to arrive at the cost of pollution; and the increase in wages for performing risky work is used to put a price on the life of the workers. (The travel cost method is sometimes, though not always, classified as "hedonic pricing".)

herbivore An animal with a special fondness for eating vegetative matter.

heterotroph An organism that produces its own organic matter by feeding on the organic matter that other organisms produce. Contrasted with "autotroph".

heuristic (computational) An algorithm used to solve a problem, but for which there is not a formal guarantee that it will produce a correct result; or more commonly, an algorithm that is not guaranteed to produce an optimal solution; and sometimes, an algorithm not even guaranteed to avoid a worst-possible solution. Heuristics are used to solve "hard problems" – that is, problems that, because of their complexity, require unreasonably large resources to compute provably correct or optimal solutions.

histogram A graphical representation of the number or frequency of cases that fall into each of a number of pre-determined categories.

homoplasy The similarity in some trait of two different, phylogenetically unrelated organisms that results from their convergent evolution.

hypoxia, hypoxic A deficit of dissolved oxygen in an aquatic environment. Hypoxic conditions can occur as the result of a large abundance of organic matter that is then oxidized by decomposing bacteria. See also, "anoxia".

index (as in "index of biodiversity") A single number (scalar value) that is used to quantify some property as a function of the quantity and value associated with each of many items or categories that enter into that property. An index of biodiversity purports to be a scalar representation of the "amount of biodiversity".

information entropy (computation) See "entropy".

instance, instantiate (philosophy) In the ontological sense used in this book, an object (particular individual) that satisfies an "is a" relationship to a substantial universal (kind); or a kind that satisfies an "is a" relationship to a category of (either substantial or non-substantial) universals. You and I are objects and instances of, among other substantial universals, *H. Sapiens*. However, not all instances are objects. For example, the species and substantial universal *H. sapiens* is itself an instance of the category species. See also "substantial universal" and "non-substantial universal".

kingdom (biological taxonomy) In the biological hierarchy of categories of life forms, the major category just below domain and just above phylum. The three-domain model defines six kingdoms. Animalia, Plantae, Fungi, and Protista jointly make up the eukaryotic domain. Bacteria and Archaea together used to be regarded as the single category of prokaryotes. However, their ribosomal RNA and cell structures set them apart from each other as well as from Eukaryota (and because of that some scientists would prefer to dispense with the category of prokaryotes altogether). The kingdoms of Archaea and Bacteria are coterminous with their eponymous domains.

lineage See "phylogeny".

Linnaean taxonomy A hierarchical, rank-based classification system for organisms, largely based on morphological and other phenotypical similarities, rather than evolutionary lineage. Contrasted with "cladistics".

macrofauna Animals visible to the naked eye. The term is typically applied to animals that inhabit soil or benthic environments and that have a dimension greater than 0.5 mm.

mast Usually, the fruit of shrubs and trees – such as oak, chestnut, and beech – that is a major food source for a variety of animals in their habitats.

megafauna Big animals. The most common definition of "big" is "weighing more than 45 kg. (100 lb.)"

metaethics, metaethical (philosophy) The study of the nature of axiological and moral claims – what they mean, whether or not they are about matters of fact, how they can be known, and what, if any, modes of reasoning about and justifying them are legitimate. In other words, metaethics is supposed to elucidate what normative ethics does when it addresses questions of good and bad, right and wrong. See "normative ethics".

metazoan Any member of the kingdom Animalia.

methanogen An organism that respires using methanogenesis – a form of anaerobic respiration (respiration without oxygen) that produces methane as a byproduct. The other major form of anaerobic respiration is fermentation.

microbe An organism too small for a human to see unaided. This category comprises a very diverse group, which includes bacteria and archaea, as well as a variety of eukaryotic protists, fungi, green algae, and zooplankton.

mollusk An animal in the phylum Mollusca. The most species-rich class of mollusks is the gastropods (snails, slugs, and whelks). Mollusks also include the bivalves and cephalopods (principally octopuses, squid, and cuttlefish).

monophyletic Roughly, a group of organisms at some taxonomic level (not necessarily species) that share a common ancestor.

moral agent (philosophy) A being who is capable of making moral judgments and who is justifiably held responsible for her actions.

moral patient (philosophy) A being to whom moral agents have obligations. More generally, a moral patient is a being to whom a moral agent owes moral consideration.

morphology, morphological The forms, structures, and shapes of the internal and external parts of an organism and the organism as a whole.

morphospace A formal representation of the morphologies that a group of creatures can instantiate. It characterizes variations in form within a group of creatures by means of multiple, distinct parameters. The value of each parameter is represented as one dimension in an abstract space. Each creature in the group occupies one point in this space. The subspace of points that have never been occupied represents (*prima facie*) potential space for evolutionary exploration.

mutualism, mutualist A relationship between two species that benefits both.

naturalism (philosophy) The metaethical stance according to which goodness consists in some structured instantiation and exemplification of the ordinary properties of real objects in the world.

nematode A member of the phylum Nematoda, comprising the unsegmented roundworms. This is one of the most species-rich and populous phyla. The vast majority of nematodes are entirely free-living, but a still significant percentage (up to around 20%, by some estimates) is parasitic.

net present value (economics) The net economic value over the duration of an investment or, more generally, any stream of benefits and costs. The economic computation of net present value discounts future benefits and costs on the (controversial) theory that this makes them comparable to current benefits and costs.

noneudaimonistic (philosophy) In the context of theories of virtue ethics, not directly relating to or affecting human flourishing. On some accounts of virtue ethics, some virtues are justified by reference to noneudaimonistic ends. Contrasted with "eudaimonistic".

non-substantial universal (philosophy) A kind that characterizes or qualifies a substantial universal and that is exemplified by that substantial universal's instances (objects). Non-substantial universals are often referred to as "properties". See also "instance" and "substantial universal". A non-substantial universal can itself be an instance of a category of non-substantial universals, but does not itself have instances.

normal good (economics) A good for which, at a constant price, a consumer will have greater demand when her income increases and less demand when her income decreases.

normative ethics (philosophy) The systematic study of the nature and categories of goodness, which goods pass a "demarcation test" whereby they rise to the level of moral consideration, and what these considerations imply for right action. Normative ethics is distinguished from metaethics, which tries to make sense of normative ethics' moral claims. See "metaethics".

NO$_x$ gas Any of a set of oxides of nitrogen, most notably nitric oxide (NO), nitrogen dioxide (NO$_2$), and nitrous oxide (N$_2$O). Unlike dinitrogen gas (N$_2$), all of these compounds are very reactive gases and make up a substantial portion of the world's reactive nitrogen.

NP-complete (computation) A class of problems that are very hard. There is no known algorithm for their solution whose running time is (merely) a polynomial function of the size of the problem. As a consequence, computing a solution takes tremendous computing resources – either lots of time or lots of storage space or both.

object (philosophy) A particular, concrete, individual entity existing in space and time.

ontology, ontological (philosophy) The study of the categories of being ("what there is") and their interrelationships.

option value (economics) The premium – over and above the expected net consumer surplus – that people are willing to pay to retain the option of consuming it sometime in the future rather than immediately. Oftentimes, the term is mistakenly applied to the expected net consumer surplus posited for future use. See also "consumer surplus".

order (biological taxonomy) In the biological hierarchy of categories of life forms, the major category just below class and just above family.

order of magnitude/Order() An estimation rounded to the nearest power of ten, which is an integer contained within the parentheses.

Ordovician period The second geological period in the Paleozoic Era, starting around 488 mya and following the Cambrian period. The Ordovician both started and ended (around 444 mya) with a major extinction event. The period following it, the Silurian, was relatively more stable, though it saw a major land invasion, which included the first vascular plants.

parasite An organism that lives in intimate association with an individual of another species – its host. The parasite benefits from this association while the host is harmed. By some estimates, the parasitic way of life is the most common one on the planet.

parasitoid A parasite whose association with its host is usually fatal to the host. Parasitoid associations are typically very specialized. Most but not all parasitoids are insects in the order Hymenoptera, which includes the parasitoid wasps.

Pascal's wager (philosophy) An argument, made by the seventeenth century philosopher and mathematician Blaise Pascal (in §233 of his *Pensées*), that it is rational to bet for rather than against the proposition that God exists. The argument is based on two premises. The first is that reason cannot determine whether God exists or not. The second is that there is everything to gain and nothing to lose in betting that God does exist, while if God does exist and you bet against this proposition, then you suffer "error and misery".

pathogen A microparasite that causes disease in its host.

phenology The study of the timing of periodic or seasonal phenomena connected with organisms, with particular attention to abiotic conditions such as weather and climate. For example, the timing of a plant's budding and flowering.

phenotype The physical expression of an organism's genotype in the environment in which it lives. This includes, but is not restricted to, the organism's morphology.

phylogeny The evolutionary relatedness of different kinds of organisms. The greater the degree to which two organisms share a common evolutionary lineage (for example, the fewer steps removed they are from a common ancestor), the more closely they are related. The basis for cladistics.

phylum (in biological taxonomy) In the biological hierarchy of categories of life forms, the major category just below domain and kingdom. Phyla are the categories that best correspond to an intuitive notion of a major, distinctive design for living.

phytoplankton The class of plankton – free-floating, aquatic organisms – that are autotrophic photosynthesizers. Phytoplankton span the prokaryotic/eukaryotic divide, and include such diverse groups of creatures as diatoms, cyanobacteria, photosynthesizing dinoflagellates, and coccolithophores. See also "zooplankton".

polyphagous Of animals: Feeding on more than one kind of food.

polyploid, polyploid speciation Containing more than two homologous sets of chromosomes (2n or diploidy). In the simplest case, polyploid speciation occurs when tetraploid (4n) organisms, capable of interbreeding with each other, are formed from diploid gametes. Such gametes might derive from mitosis (rather than meiosis). Or they can derive from somatic cells. However, attempts of a tetraploid individual to breed with an individual of an ancestral diploid species (a backcross) form triploid (3n) offspring. These are usually sterile because they cannot form gametes with a balanced complement of chromosomes.

Precautionary Principle (philosophy) Not a single principle, but rather a family of principles, that stake a claim to a moral or political responsibility to consider future harms, despite uncertainty about their occurrence or cause; and if the harms are sufficiently great, to take actions in order to either mitigate or avoid them. A Precautionary Principle commonly is invoked in circumstances in which the uncertainty results from an incomplete scientific account of how the projected harms are brought about. A Safe Minimum Standard is a form of Precautionary Principle.

predator A heterotroph that kills the organisms that it eats.

primary production The production of organic matter (known as "biomass") by autotrophs, primarily green (photosynthesizing) plants.

productivity The ability to produce organic matter – said of an individual organism, population, community, ecosystem, or sometimes the entire biosphere.

prokaryote, prokaryotic An organism in either the domain Bacteria or the domain Archaea whose cell lacks a membrane-encased nucleus. Prokaryotes are distinguished from eukaryotes in this and some other respects.

protista, protist A kingdom within the domain Eukaryota that comprises a grab-bag of disparate phyla that have little in common except for being either unicellular or having undifferentiated multicellular structures.

protozoan A category of eukaryotic microorganisms – typically a unicellular, heterotrophic protist such as an amoeba or ciliate.

quadratic entropy (or quadratic diversity) A measure of species diversity constructed from the likelihood of the pairwise encounter of individuals, possibly weighted by some measure of dissimilarity (e.g. phylogenetic distance) between the pair. Such a measure is quadratic because it is constructed from the (weighted) sum of products of the species' relative abundances.

quasi-option value (economics) the value of the future information that is made available through the preservation of a resource.

rational drug design The invention of new drugs based on knowledge of a target molecule present in a medical patient – particularly, knowledge that indicates possible ways in which the target molecule's bioactivity can be modified for the patient's therapeutic benefit.

reactive nitrogen The element nitrogen, not in its most abundant and very stable gaseous form of dinitrogen (N_2), but "fixed" or incorporated into ammonia (NH_3) or into oxidized forms, such as the NO_x gases. Only a very few organisms can utilize (and themselves fix) dinitrogen. Most organisms can only utilize and require some form of reactive nitrogen.

reservoir (epidemiology) A longterm host of a pathogen or parasite that sustains the presence of the infectious agent in a disease system. Reservoir hosts often are negligibly affected by their infection or altogether asymptomatic. However, the pathogen or parasite populations that flourish in them can then infect and cause disease in other hosts, which would not alone sustain those populations.

ring species A chain of physically neighboring populations of the "same" species of organism, such that individuals of two adjacent population links in the chain can interbreed, but the end links cannot. This phenomenon makes the relation "is the same species as" intransitive. This, in turn, jeopardizes the consistency of the biological definition of species, for which interbreeding (or the capability of interbreeding) is a requirement: On the one hand, if one maintains that individuals in each successive population link belong to the same species, then one must admit that some individuals of this species (the ones in the end link populations) cannot interbreed. On the other hand, designating each population a separate species disregards the fact that almost all adjacent populations satisfy the interbreeding requirement for being the same species.

riparian Relating to an area at the margin of a stream or river.

risk neutrality, risk aversion (economics) The absence of any preference between two options that produce the same expected value, even when one option is riskier than the other (neutrality). Risk aversion is the preference for the less risky option, even though its expected value is equal to or even less than the alternative. Risk aversion is distinct from loss aversion – the preference to retain an asset even despite an expected net benefit in giving it up. This is sometimes called "the endowment effect".

ruderal species A plant species that tends to be a "pioneer" in the recolonization of a habitat after it is disturbed – for example, by fire, flood, avalanche, or human activities. In many cases, ruderal species are eventually succeeded by other species better suited to long-term survival in the area.

Safe Minimum Standard A form of Precautionary Principle that is often tied to an economic assessment of either a scarce or hard-to-replace resource, where the harm of removing the resource is uncertain. A Safe Minimum Standard rule dictates that some minimum portion of the resource, identified as providing some specified benefit, be saved – provided that the economic cost of doing that is tolerable. Another version dictates that the resource be saved up to the point that the cost of doing so becomes intolerable.

scalar (as in scalar rating/scalar value) A simple physical quantity capable of being represented by a single numerical value. The paradigmatic scalars, from physical science, are such quantities as mass and charge. The values of a scalar form a total ordering. The totality condition of a total ordering says that for any two values, x and y, either $x \leq y$ or $y \geq x$.

set cover problem (computational) A classical NP-complete problem in the theory of computational complexity: Given a family of sets of elements, find the minimal number of sets in this family that still contains all the elements in the union of the entire family. In the context of conservation biology, the sets are places and the elements are species.

spatial granularity A quality that characterizes how finely or coarsely a physical space is divided up. In the broad context of ecology, the size of an area that is taken under consideration. For biodiversity in particular, the size of the area whose biodiversity is assessed.

species Intuitively, a type of organism. In the biological hierarchy of categories of life forms, species is the bedrock category and the bedrock of biology understood from an evolutionary perspective. Surprisingly, there is no single agreed-upon definition and no definition that works well for all organisms. The "biological species" definition – a set of *actually* interbreeding populations or populations *capable* of interbreeding to produce offspring with a like capability (even if, in fact, they do not) – is the definition most widely adopted and cited. This definition focuses on sexual reproductive capability. Other definitions focus on similarities in morphology or other phenotypical traits, phylogeny, or ecological needs and preferences. Bacteria pose particular difficulties because they not only reproduce asexually, but also promiscuously exchange genetic material across species boundaries. A species definition for bacteria typically combines phylogenetic and phenotypical conditions. For example (Rosselló-Mora and Amann 2001, 39): "a monophyletic and genomically coherent cluster of individual organisms that show a high degree of overall similarity in many characteristics, and is diagnosable by a discriminative phenotypic property."

species-area effect or species area relationship (SAR) A positive correlation that is often, but not always, found in terrestrial ecosystems between the size of an area and the number of species encountered within it. Species-area relationships have been explored for both large, contiguous "mainland" areas and for geographically isolated and discontiguous areas such as oceanic or mountaintop islands. The jury is out on whether positive correlations, when they are found, result from a greater land area *simpliciter*, or from the tendency of greater land areas to contain a greater diversity of habitats, which accommodate more different kinds of creatures. The jury is also out on whether a SAR is a viable basis for estimating species richness. That is because (He and Hubbell 2011) a SAR is based on the size of an area sampled to encounter the first individual of a species; a straightforward extrapolation from that number must presume that all individuals of that species are randomly distributed throughout the habitat.

species richness   The number of species in some specified area, which can as small as your bellybutton or as large as the entire planet.

substantial universal (philosophy)   A kind that categorizes an object (a particular, concrete individual). An object is said to instantiate a substantial universal when it satisfies an "is a" relationship to it. See also "instance" and "non-substantial universal".

ta legomena **(philosophy)**   Literally, "what is said". Aristotle uses this phrase in the *Categories* to describe his list of ten distinct kinds, which, he says, serve to classify all things.

taxon   A grouping of organisms at some hierarchical level, such as species, genus, and phylum, among others.

telos **(philosophy)**   Literally, "end" or "goal", or that towards which purpose is directed. This is none other than Aristotle's "final cause", which he explicates in his *Physics* and *Metaphysics*.

trematode   An organism in the class Trematoda – the class of flatworms that are commonly known as flukes. Most have a complex life cycle, which involves a parasitic association with a mollusk or a vertebrate or both.

troglobite, troglobitic   An animal that lives entirely in the darkness of a cave. Troglobitic creatures tend to lack light-adaptive features such as eyes and pigmentation.

trophic structure, tropic level   The set of relationships in a community or ecosystem that define who feeds upon whom. The trophic level of an organism is a measure of how many feeding steps separate it from a primary producer. Trophic structures are often non-hierarchical. An organism can occupy several trophic levels simultaneously, as the result of it or some other organism on which it feeds having multiple trophic relationships.

truth value (philosophy)   In conventional (non-modal, two-valued) logic, the truth or the falsity of a proposition.

ungulate   An animal in one of several orders in the class Mammalia. Most ungulates have hooves – the tips of the toes on which they walk. Most hooved ungulates are either odd-toed Perissodactyla (such as horses, tapirs, and rhinoceroses) or even-toed Artiodactyla (such as pigs, peccaries, hippopotamuses, camels, mouse deer, deer, giraffes, pronghorn, antelopes, sheep, goats, and cattle).

universal (philosophy) A kind or category of kinds. A universal can be either substantial or non-substantial. See "substantial universal" and "non-substantial universal".

vector (epidemiology) An organism, typically an insect or a mollusk, that transmits a pathogen or parasite from one host to another.

virtue ethics (philosophy) The oldest of the three major classes of Western theories of normative ethics, propounded by the ancient Greeks. Modern philosophers placed virtue ethics on the sidelines until the second half of the twentieth century, which witnessed its reinstatement to philosophical respectability. Virtue ethics focuses on the characters of moral agents and their development of character traits – the virtues. One representative version of virtue ethics supposes that, by developing and acting according to the virtues, a person leads, or at least secures some possibility of leading, good life. This orientation differs from that of consequentialist theories, which focus on the consequences of actions. It also differs from deontological theories, which focus on duties and obligations.

zoonosis, zoonotic A disease whose infectious agent normally infects nonhumans but which can also infect humans. A zoonotic disease can be transmitted from a nonhuman host to a human one either directly via contact with the nonhuman host or its excreta, or indirectly via a vector.

zooplankton The class of plankton – free-floating, aquatic organisms – that are relatively small-to-microscopic heterotrophs, and sometimes including the eggs and larvae of non-planktonic animals. This encompasses a very broad range of creatures – such protozoans as foraminiferans, radiolarian, and non-photosynthesizing dinoflagellates; and metazoans, including cnidarians (such as jellies), crustaceans (such as copepods and krill), mollusks of various sorts, and even chordates such as salps and juvenile fish. See also "phytoplankton".

zooprophylaxis The practice of prophylactically introducing incompetent nonhuman hosts of a zoonotic disease into an area in order to suppress the population of pathogens or parasites capable of infecting humans.

References

Links are valid as of March 2012. They are supplied for materials that are freely available online, with the exception of articles from *The New York Times* and the *High Country News*, which require a subscription for full access.

Ackerman, J., J. Bluso-Demers, M. Herzog, C. Robinson-Nilson, and G. Herring. 2010. Impact of salt pond restoration on California Gull displacement and predation on breeding waterbirds: Annual report – December 2010. USGS and San Francisco Bay Bird Observatory. http://www.southbayrestoration.org/documents/technical/Gull%20Annual%20Report%20For%20SBSPRP%202010.pdf.

Ackrill, J.L. 1980. Aristotle on eudaimonia. In *Essays on Aristotle's ethics*, ed. A.E. Rorty, 15–33. Berkeley and Los Angeles: University of California Press.

Aladin, N., and I. Plotnikov. 2004. The Caspian Sea, Lake Basin Management Initiative. http://www.worldlakes.org/uploads/Caspian%20Sea%2028jun04.pdf.

Aldred, J. 2009. Rare 'mountain chicken' frogs airlifted from path of deadly fungus. *The Guardian*, April 21, 2009. http://www.guardian.co.uk/environment/2009/apr/21/wildlife-conservation.

Alter, S.E., E. Rynes, and S.R. Palumbi. 2007. DNA evidence for historic population size and past ecosystem impacts of gray whales. *Proceedings of the National Academy of Science (PNAS)* 104(38): 15162–15167. http://www.pnas.org/content/104/38/15162.full.

Ambrose, S.H. 1998. Late Pleistocene human population bottlenecks, volcanic winter, and differentiation of modern humans. *Journal of Human Evolution* 34(6): 623–651.

Ambrose, S.H. 2003. Did the super-eruption of Toba cause a human population bottleneck? Reply to Gathorne-Hardy and Harcourt-Smith. *Journal of Human Evolution* 45(3): 231–237.

Andersen, J.H., and J. Pawlak. 2006. Nutrients and eutrophication in the Baltic Sea – Effects, causes, solutions. Baltic Sea Parliamentary Conference. Copenhagen: Nordic Council.

Anderson, E.S. 2002. Integration, affirmative action, and strict scrutiny. *New York University Law Review* 77(5): 1195–1271.

Angermeier, P.L., and J.R. Karr. 1994. Biological integrity versus biological diversity as policy directives. *BioScience* 44(10): 690–697.

Antonovics, J., A.D. Bradshaw, and J.R.G. Turner. 1971. Heavy metal tolerance in plants. *Advances in Ecological Research* 7: 1–85.

Aquinas, T. 1264. *Summa contra gentiles*, Trans. V.J. Bourke and ed. J. Kenney (1955–1957). New York: Hanover House. http://dhspriory.org/thomas/ContraGentiles.htm.

Aristotle. 350 B.C.E. a. Categories. In *The Internet Classics Archive*, Trans. E.M. Edghill. http://classics.mit.edu/Aristotle/categories.html.

Aristotle. 350 B.C.E. b. Nichomachean ethics. In *The Internet Classics Archive*, Trans. W.D. Ross. http://classics.mit.edu/Aristotle/nicomachaen.html.

D.S. Maier, *What's So Good About Biodiversity?: A Call for Better Reasoning About Nature's Value*, DOI 10.1007/978-94-007-3991-8, © Springer Science+Business Media B.V. 2012

Arrow, K.J., and A.C. Fisher. 1974. Environmental preservation, uncertainty, and irreversibility. *Quarterly Journal of Economics* 88(2): 312–319.

Arrow, K.J., and R.C. Lind. 1970. Uncertainty and the evaluation of public investment decisions. *American Economic Review* 60: 364–378.

Arrow, K., and H. Raynaud. 1986. *Social choice and multicriterion decision-making.* Cambridge: MIT Press.

Bailey, J.C. (ed.). 1905. *The poems of William Cowper.* London: Methuen.

Belsky, A.J., and D.M. Blumenthal. 1997. Effects of livestock grazing on stand dynamics and soils in upland forests of the interior West. *Conservation Biology* 11(2): 315–327. http://afrsweb. usda.gov/SP2UserFiles/Place/54090000/BlumenthalPDF/1.BelskyandBlumenthal1997-Effect soflivestockgrazingonstanddynamicsandsoils.pdf.

Belsky, A.J., A. Matzke, and S. Uselman. 1999. Survey of livestock influences on stream and riparian ecosystems in the Western United States. *Journal of Soil and Water Conservation* 54: 419–431.

Bendik-Keymer, J. 2010. Species extinction and the vice of thoughtlessness: The importance of spiritual exercises for learning virtue. *Journal of Agricultural and Environmental Ethics* 23(1–2): 61–83.

Bendik-Keymer, J. 2011. A history of unintentional violence: Looking back on marred humanity during the ongoing mass extinction. Forthcoming in *Ethics, Policy & Environment*; references are to a pre-publication draft.

Bendik-Keymer, J. 2012. The sixth mass extinction is caused by us. In *Ethical adaptation to climate change: Human virtues of the future,* ed. A. Thomson and J. Bendik-Keymer. Cambridge: MIT Press.

Bentham, J. 1789. *Introduction to the principles of morals and legislation.* London: Methuen.

Bilger, B. 2009. Swamp things. *The New Yorker,* April 20, 2009, 80–89.

Bishop, R.C. 1986. Resource valuation under uncertainty. In *Advances in applied micro-economics, Risk, uncertainty, and the valuation of benefits and costs,* vol. 4, ed. V.K. Smith, 133–152. Greenwich, CT: JAI Press, Inc.

Bøgh, C., S.E. Clarke, G.E. Walraven, and S.W. Lindsay. 2002. Zooprophylaxis, artefact or reality? A paired-cohort study of the effect of passive zooprophylaxis on malaria in The Gambia. *Transactions of the Royal Society of Tropical Medicine and Hygeine* 96(6): 593–596.

Borchardt, R. 2006. *The passionate gardener.* Kingston, NY: McPherson & Co.

Bowker, G.C. 2006. *Memory practices in the sciences.* Cambridge: The MIT Press.

Brayard, A., G. Escarguel, H. Bucher, C. Monnet, T. Brühwiler, N. Goudemand, T. Galfetti, and J. Guex. 2009. Good genes and good luck: Ammonoid diversity and the end-Permian mass extinction. *Science* 325(5944): 1118–1121.

Britten, N. 2009. Dutch scientists 'grow' meat in laboratory. *The Telegraph,* November 29, 2009. http://www.telegraph.co.uk/foodanddrink/6684854/Scientists-grow-meat-in-laboratory.html.

Bryson, B. 2003. *A short history of nearly everything.* New York: Bantam Dell Publishing Group.

Bryson, G. 2009. State sues US to get beluga whale listing rescinded. *Anchorage Daily News,* January 15, 2009. http://www.adn.com/2009/01/15/654545/state-sues-us-to-get-beluga-whale.html.

Burke, M.K., J.P. Dunham, P. Shahrestani, K.R. Thornton, M.R. Rose, and A.D. Long. 2010. Genome-wide analysis of a long-term evolution experiment with *Drosophila. Nature* 467(7315): 587–590.

Butler, J.L., N.J. Gotelli, and A.M. Ellison. 2008. Linking the brown and the green: Nutrient transformation and fate in the *Sarracenia* microecosystem. *Ecology* 89(4): 898–904.

Caldow, R.W.G., R.A. Stillman, S.E.A. le V. dit Durell, A.D. West, S. McGrorty, J.D. Goss-Custard, P.J. Wood, and J. Humphreys. 2007. Benefits to shorebirds from invasion of a non-native shell-fish. *Proceedings of the Royal Society B* 274: 1449–1455.

Callicott, J.B. 1995. A critique of and an alternative to the wilderness idea. *Wild Earth* 4: 4. In *Environmental ethics,* ed. A. Light and H. Rolston (2003), 437–443. Malden, MA: Blackwell Publishing.

Campbell, L.G., and A.A. Snow. 2009. Can feral weeds evolve from cultivated radish (*Raphanus sativus,* Brassicaceae)? *American Journal of Botany* 96(2): 498–506. http://www.biosci.ohio-state.edu/~asnowlab/Campbellferalweeds09.pdf.

Canter, P.H., H. Thomas, and E. Ernst. 2005. Bringing medicinal plants into cultivation: Opportunities and challenges for biotechnology. *Trends in Biotechnology* 23(4): 180–185. http://cmbi.bjmu.edu.cn/news/report/2004/biotech/14.pdf.

Čapek, K. 2002. *The gardener's year*, Trans. M. Weatherall and R. Weatherall. New York: Modern Library, Random House (First published as *Zahradníkův rok*, 1929).

Caro, T. 2007. The Pleistocene rewilding gambit. *Trends in Ecology & Evolution* 22(6): 281–283. http://www.columbia.edu/~dr2497/PRESS_files/TREE2007.pdf.

Carroll, S.B. 2010. Hybrids may thrive where parents fear to tread. *The New York Times*, September 13, 2010. http://www.nytimes.com/2010/09/14/science/14creatures.html.

Carson, R. 1962. *Silent spring*, ed. Mariner Books (2002). New York: Houghton Mifflin Company.

Chapin III, F.S., E.S. Zavaleta, V.T. Eviner, R.L. Naylor, P.M. Vitousek, H.L. Reynolds, D.U. Hooper, S. Lavorel, O.E. Sala, S.E. Hobbie, M.C. Mack, and S. Díaz. 2000. Consequences of changing biodiversity. *Nature* 405(6783): 234–242.

Chapin III, F.S., P.A. Matson, and H.A. Mooney. 2002. *Principles of terrestrial ecosystem ecology*. New York: Springer.

Cheikh, N., P.W. Miller, and P.W. Kishore. 2000. Role of biotechnology in crop productivity in a changing environment. In *Global change and crop productivity*, ed. K.R. Reddy and H.F. Hodges, 425–436. Wallingford: CAB International.

Chin, R., and B.Y. Lee. 2008. *Principles and practice of clinical trial medicine*. London: Academic Press/Elsevier.

Chivian, E., and A. Bernstein. 2008a. How is biodiversity threatened by human activity? In *Sustaining life: How human health depends on biodiversity*, ed. E. Chivian and A. Bernstein, 28–73. Oxford: Oxford University Press.

Chivian, E., and A. Bernstein. 2008b. Threatened groups of organisms valuable to medicine. In *Sustaining life: How human health depends on biodiversity*, ed. E. Chivian and A. Bernstein, 203–285. Oxford: Oxford University Press.

Chivian, E., and A. Bernstein (eds.). 2008c. *Sustaining life: How human health depends on biodiversity*. Oxford: Oxford University Press.

Chivian, E., A. Bernstein, and J.P. Rosenthal. 2008. Biodiversity and biomedical research. In *Sustaining life: How human health depends on biodiversity*, ed. E. Chivian and A. Bernstein, 162–210. Oxford: Oxford University Press.

Choi, Y.D. 2007. Restoration ecology to the future: A call for new paradigm. *Restoration Ecology* 15(2): 351–353.

Constanza, R., R. d'Arge, R. de Groot, S. Farber, M. Grasso, B. Hannon, K. Limburg, S. Naeem, R.V. O'Neill, J. Paruelo, R.G. Raskin, P. Sutton, and M. van den Belt. 1997. The value of the world's ecosystem services and natural capital. *Nature* 387(6630): 253–260.

Cooper, D.E. 2006. *A philosophy of gardens*. Oxford: Oxford University Press.

Cory, D.C., and B.C. Saliba. 1987. Requiem for option value. *Land Economics* 63(1): 1–10.

Cox, P.A. 2009. Biodiversity and the search for new medicines. In *Biodiversity change and human health*, ed. O.E. Sala, L.A. Meyerson, and C. Parmesan, 269–280. Washington, DC: Island Press.

Czechowicz, K. 2007. Polish geneticists want to recreate the extinct auroch. *Science and Scholarship in Poland*. http://www.naukawpolsce.pap.pl/palio/html.run?_Instance=cms_naukapl.pap.pl&_PageID=1&s=szablon.depesza&dz=szablon.depesza&dep=68335&data=&lang=EN&_CheckSum=-442536143.

Daily, G., and K. Ellison. 2002. *The new economy of nature: The quest to make conservation profitable*. Washington, DC: Island Press.

Darwin, C. 1859. *On the origin of the species by means of natural selection or the preservation of favoured races in the struggle for life*. London: John Murray.

Darimont, C.T., S.M. Carlson, M.T. Kinnison, P.C. Paquet, T.E. Reimchen, and C.C. Wilmers. 2009. Human predators outpace other agents of trait change in the wild. *Proceedings of the National Academy of Science (PNAS)* 106(3): 952–954.

de Saint-Exupéry, A. 1939. *Terre des Hommes* (Published in English translation as *Wind, sand, and stars*). Paris: Gallimard.

de Saint-Exupéry, A. 2000. *The little prince* [Le Petit Prince], Trans. R. Howard. New York: Harcourt.

Dean, C. 2008. The preservation predicament. *The New York Times*, January 29, 2008. http://www.nytimes.com/2008/01/29/science/earth/29habi.html?pagewanted=all.

Diamond, R., S.L. Pimm, M.E. Gilpin, and M. LeCroy. 1989. Rapid evolution of character displacement in myzomelid honeyeaters. *The American Naturalist* 134(5): 675–708.

Díaz, S., J. Fargione, F.S. Chapin III, and D. Tilman. 2006. Biodiversity loss threatens human well-being. *PLoS Biology* 4(8): 1300–1305. http://www.plosbiology.org/article/info: doi/10.1371/journal.pbio.0040277.

Dirzo, R., and P.H. Raven. 2003. Global state of biodiversity and loss. *Annual Review of Environment and Resources* 28: 137–167.

Dobson, A., I. Cattadori, R.D. Holt, R.S. Ostfeld, F. Keesing, K. Krichbaum, J.R. Rohr, S.E. Perkins, and P.J. Hudson. 2006. Sacred cows and sympathetic squirrels: The importance of biological diversity to human health. *PLoS Medicine* 3(6): 714–718. http://www.ecostudies.org/reprints/Dobson_PLOS_2006.pdf.

Donlan, C.J., J. Berger, C.E. Bock, J.H. Bock, D.A. Burney, J.A. Estes, D. Foreman, P.S. Martin, G.W. Roemer, F.A. Smith, M.E. Soulé, and H.W. Greene. 2006. Pleistocene rewilding: An optimistic agenda for twenty-first century conservation. *The American Naturalist* 168(5): 660–681.

Driver, J. 2001. *Uneasy virtue*. Cambridge: Cambridge University Press.

Drucker, A.G., M. Smale, and P. Zambrano. 2005. Valuation and sustainable management of crop and livestock biodiversity: A review of applied economics literature. System-wide Genetic Resources Program (SGRP) of the Consultative Group on International Agricultural Research (CGIAR). Washington, DC: International Food Policy Research Institute (IFPRI); Rome: International Food Policy Research Institute (IFPRI); Nairobi: International Livestock Research Institute (ILRI). http://www.sgrp.cgiar.org/sites/default/files/SGRP%20GR%20Valuation%20_DruckerSmaleZambrano_%20September%2030-final_0.pdf.

Dworkin, R. 1993. *Life's dominion: An argument about abortion, euthanasia, and individual freedom*. New York: Knopf.

Ehrenfeld, D. 1988. Why put a value on biodiversity? In *Biodiversity*, ed. E.O. Wilson, 212–216. Washington, DC: The National Academies Press.

Ehrlich, P.R., and A. Ehrlich. 1981. *Extinction: The causes and consequences of the disappearance of species*. New York: Random House.

Ehrlich, P., and L. Goulder. 2007. Is current consumption excessive?: A general framework and some indications for the United States. *Conservation Biology* 21(5): 1145–1154.

Ehrlich, P.R., and E.O. Wilson. 1991. Biodiversity studies: Science and policy. *Science* 253(5021): 758–762.

Ehrlich, P.R., G.C. Daily, S.C. Daily, N. Myers, and J. Salzman. 1997. No middle way on the environment. *The Atlantic Monthly* 280(6): 98–104. http://www.theatlantic.com/past/docs/issues/97dec/enviro.htm.

Eisner, T. 1982. Chemical ecology and genetic engineering: The prospects for plant protection and the need for plant habitat conservation. Abstract, Symposium on Tropical Biology and Agriculture, Monsanto Co., St. Louis, July 1982.

Elliot, R. 1982. Faking nature. *Inquiry* 25: 81–93. In *Environmental ethics*, ed. A. Light and H. Rolston (2003), 381–389. Malden, MA: Blackwell Publishing.

Ellis, E.C., and N. Ramankutty. 2008. Putting people in the map: Anthropogenic biomes of the world. *Frontiers in Ecology and the Environment* 6(8): 439–447. http://www.ecotope.org/people/ellis/papers/ellis_2008.pdf.

Elton, C. 1958. *The ecology of invasions by animals and plants*. Republished by University of Chicago Press (2000). London: Methuen & Co.

Erickson, W.P., G.D. Johnson, and D.P. Young. 2005. A summary and comparison of bird mortality from anthropogenic causes with an emphasis on collisions. USDA Forest Service General Technical Report PSW-GTR-191, 1029–1042. http://www.fs.fed.us/psw/publications/documents/psw_gtr191/Asilomar/pdfs/1029-1042.pdf.

Erwin, D.H. 1993. *The great Paleozoic crisis: Life and death in the Permian.* New York: Columbia University Press.

European Environment Agency. 2001. Late lessons from early warnings: The precautionary principle 1896–2000. Environmental Issue Report No. 22. Copenhagen: European Environment Agency. http://www.eea.europa.eu/publications/environmental_issue_report_2001_22.

Fackler, M. 2009. Coral transplant surgery prescribed for Japan. *The New York Times*, April 14, 2009. http://www.nytimes.com/2009/04/15/world/asia/15coral.html.

Faith, D.P. 1992. Conservation evaluation and phylogenetic diversity. *Biological Conservation* 61(1): 1–10.

Faith, D.P. 2007. Biodiversity. In *Stanford encyclopedia of philosophy (Fall 2008 Edition)*, ed. E.N. Zalta. http://plato.stanford.edu/archives/fall2008/entries/biodiversity/.

Feinberg, J. 1974. The rights of animals and unborn generations. In *Philosophy and environmental crisis*, ed. W.T. Blackstone, 43–68. Athens, GA: University of Georgia Press.

Fisher, A.C., and W.M. Hanemann. 1986. Option value and the extinction of species. In *Advances in applied micro-economics*, Risk, uncertainty, and the valuation of benefits and costs, vol. 4, ed. V.K. Smith, 169–190. Greenwich, CT: JAI Press, Inc.

Forest, F., R. Grenyer, M. Rouget, T.J. Davies, R.M. Cowling, D.P. Faith, A. Balmford, J.C. Manning, S. Proches, M. van der Bank, G. Reeves, T.A.J. Hedderson, and V. Savolainen. 2007. Preserving the evolutionary potential of floras in biodiversity hotspots. *Nature* 445(7129): 757–760.

Forman, R.T.T., D. Sperling, J.A. Bissonette, A.P. Clevenger, C.D. Cutshall, L. Fahrig, R. France, C.R. Goldman, K. Heanue, J.A. Jones, F.J. Swanson, T. Turrentine, and T.C. Winter. 2003. *Road ecology: Science and solutions.* Washington, DC: Island Press.

Freeman III, A.M. 1984. The sign and size of option value. *Land Economics* 60(1): 1–13.

Freeman III, A.M. 1986. Uncertainty and environmental policy: The role of option and quasi-option value. In *Advances in applied micro-economics*, Risk, uncertainty, and the valuation of benefits and costs, vol. 4, ed. V.K. Smith, 153–167. Greenwich, CT: JAI Press, Inc.

Fridley, J.D., J.J. Stachowicz, S. Naem, D.F. Save, E.W. Seabloom, M.D. Smith, T.J. Stohlgren, D. Tilman, and B. Von Holle. 2007. The invasion paradox: Reconciling pattern and process in species invasions. *Ecology* 88(1): 3–17.

Froment, A. 2009. Biodiversity and health: The place of parasitic and infectious diseases. In *Biodiversity change and human health*, ed. O.E. Sala, L.A. Meyerson, and C. Parmesan, 211–227. Washington, DC: Island Press.

Fromm, E. 1973/1992. *The anatomy of human destructiveness.* New York: Henry Holt.

Gardiner, S.M. 2006. A core precautionary principle. *The Journal of Political Philosophy* 14(1): 33–60.

Gardiner, S.M. 2010a. A perfect moral storm: Climate change, intergenerational ethics, and the problem of moral corruption. In *Climate ethics: Essential readings*, ed. S.M. Gardiner, S. Caney, D. Jamieson, and H. Shue, 87–98. Oxford: Oxford University Press.

Gardiner, S.M. 2010b. Is 'arming the future' with geoengineering really the lesser evil? In *Climate ethics: Essential readings*, ed. S.M. Gardiner, S. Caney, D. Jamieson, and H. Shue, 284–312. Oxford: Oxford University Press.

Gardiner, S.M. 2011. *A perfect moral storm: The ethical tragedy of climate change.* Oxford: Oxford University Press.

Gladwell, M. 2001. The mosquito killer. *The New Yorker*, July 2, 2001, 42–51. http://www.gladwell.com/pdf/malaria.pdf.

Glavin, T. 2006. *The sixth extinction: Journey among the lost and left behind.* New York: Thomas Dunne Books/St. Martin's Press.

Goldman, R.L., H. Tallis, P. Kareiva, and G.C. Daily. 2008. Field evidence that ecosystem service projects support biodiversity and diversify options. *Proceedings of the National Academy of Science (PNAS)* 105(27): 9445–9448. http://www.pnas.org/content/105/27/9445.full.

Gould, S.J. 1996. *Full house.* New York: Harmony Books.

Gould, S.J. 1998. An evolutionary perspective on strengths, fallacies, and confusions in the concept of native plants. *Arnoldia* 58(1): 3–10. http://arnoldia.arboretum.harvard.edu/pdf/articles/483.pdf.

Griffiths, C.J., C.G. Jones, D.M. Hansen, M. Puttoo, R.V. Tatayah, C.B. Muller, and S. Harris. 2010. The use of extant non-indigenous tortoises as a restoration tool to replace extinct ecosystem engineers. *Restoration Ecology* 18(1): 1–7. http://www.torreyaguardians.org/articles/griffiths2010.pdf.

Grifo, F., D. Newman, A.S. Fairfield, B. Bhattacharya, and J.T. Grupenhoff. 1997. The origins of prescription drugs. In *Biodiversity and human health*, ed. F. Grifo and J. Rosenthal, 131–163. Washington, DC: Island Press.

Groc, I. 2010. Beavers sign up to fight effects of climate change. *Discover Magazine*. http://discovermagazine.com/2010/apr/19-beavers-sign-up-fight-effects-climate-change.

Gubler, D.J., P. Reiter, K.L. Ebi, W. Yap, R. Nasci, and J.A. Patz. 2001. Climate variability and change in the United States: Potential impacts on vector- and rodent-borne diseases. *Environmental Health Perspectives* 109(Suppl. 2): 223–233.

Guernier, V., M.E. Hochberg, and J.F. Guégan. 2004. Ecology drives the worldwide distribution of human diseases. *PLoS Biology* 2(6): 740–746. http://www.plosbiology.org/article/info:doi/10.1371/journal.pbio.0020141.

Gurevitch, J., and D.K. Padilla. 2004. Are invasive species a major cause of extinctions? *Trends in Ecology & Evolution* 19(9): 470–474. http://life.bio.sunysb.edu/ee/padillalab/pdfs/Gurevitch%20&%20Padilla%20%28Trends%20in%20Eco&Evo%20-%20Opinion%29%202004.pdf.

Hanemann, W.M. 1984. On reconciling different concepts of option value. Working Paper No. 295, UC Berkeley, Department of Agricultural and Resource Economics, Berkeley. http://www.escholarship.org/uc/item/81w7290x.

Hanemann, W.M. 1989. Information and the concept of option value. *Journal of Environmental Economics and Management* 16: 23–37.

Harris, J.A., R.J. Hobbs, E. Higgs, and J. Aronson. 2006. Ecological restoration and global climate change. *Restoration Ecology* 14(2): 170–176. http://training.fws.gov/EC/Resources/fwca/Climate Change/Ecological restoration and CC.pdf.

Harrison, R.P. 1992. *Forests: The shadow of civilization*. Chicago: The University of Chicago Press.

Harrison, R.P. 2008. *Gardens: An essay on the human condition*. Chicago: The University of Chicago Press.

Harrison, S., J.B. Grace, K.F. Davies, H.D. Safford, and J.H. Viers. 2006. Invasion in a diversity hotspot: Exotic cover and native richness in the Californian serpentine flora. *Ecology* 87(3): 695–703.

Harvey, H. 2008. Back to life. http://www.quaggaproject.org/Publications/VW Quagga article June 2008.pdf.

Hay, J.M., S.D. Sarre, D.M. Lambert, F.W. Allendorf, and C.H. Daugherty. 2010. Genetic diversity and taxonomy: A reassessment of species designation in tuatara (*Sphenodon*: Reptilia). *Conservation Genetics* 11(3): 1063–1081.

Hayes, T., K. Haston, M. Tsui, A. Hoang, C. Haeffele, and A. Vonk. 2003. Atrazine-induced hermaphroditism at 0.1 ppb in American leopard frogs (*Rana pipiens*): Laboratory and field evidence. *Environmental Health Perspectives* 111(4): 568–575.

He, F., and S.P. Hubbell. 2011. Species-area relationships always overestimate extinction rates from habitat loss. *Nature* 473(7347): 368–371.

Hechinger, R.F., and K.D. Lafferty. 2005. Host diversity begets parasite diversity: Bird final hosts and trematodes in snail intermediate hosts. *Proceedings of the Royal Society B* 272(1567): 1059–1066. http://repositories.cdlib.org/postprints/931/.

Helmus, M.R., T.J. Bland, C.K. Williams, and A.R. Ives. 2007. Phylogenetic measures of biodiversity. *The American Naturalist* 169(3): E68–E83. http://udel.edu/~ckwillia/Publications/Helmus%20et%20al.%202007.%20American%20Naturalist.pdf.

Herrel, A., K. Huyghe, B. Vanhooydonck, T. Backeljau, K. Breugelmans, I. Grbac, R. Van Damme, and D.J. Irschick. 2008. Rapid large-scale evolutionary divergence in morphology and performance associated with exploitation of a different dietary resource. *Proceedings of the National Academy of Science (PNAS)* 105(12): 4792–4795. http://www.pnas.org/content/105/12/4792.full.

Himler, A.G., T. Adachi-Hagimori, J.E. Bergen, A. Kozuch, S.E. Kelly, B.E. Tabshnik, E. Ciel, V.E. Duckworth, T.J. Dennehy, E. Zchori-Fein, and M.S. Hunter. 2011. Rapid spread of a bacterial symbiont in an invasive whitefly is driven by fitness benefits and female bias. *Science* 322(6026): 254–256.

Hooper, D.U., F.S. Chapin III, J.J. Ewel, A. Hector, P. Inchausti, S. Lavorel, J.H. Lawton, D.M. Lodge, M. Loreau, S. Naeem, B. Sachmid, H. Setälä, A.J. Symstad, J. Vandermeer, and D.A. Wardle. 2005. Effects of biodiversity on ecosystem functioning: A consensus of current knowledge. *Ecological Monographs* 75(1): 3–35. http://www.cedarcreek.umn.edu/biblio/fulltext/t2038.pdf.

Hughes, J.D. 2009. *An environmental history of the world*, 2nd ed. London: Routledge.

Hunter, D. 2009. Attorney hired for beluga issue. *Anchorage Daily News*, December 31, 2009. http://www.adn.com/news/alaska/wildlife/story/1074409.html.

Hurlbert, S.H. 1971. The nonconcept of species diversity: A critique and alternate parameters. *Ecology* 52(4): 577–586.

Hutton, I., J.P. Parkes, and A.R.E. Sinclair. 2007. Reassembling island ecosystems: The case of Lord Howe Island. *Animal Conservation* 10(1): 22–29.

Jackson, J.B.C. 2006. When ecological pyramids were upside down. In *Whales, whaling, and ocean ecosystems*, ed. J.A. Estes, D.P. DeMaster, D.F. Doak, T.M. Williams, and R.L. Brownell, 27–37. Berkeley: University of California Press.

Jamieson, D. 2002. Ecosystem health: Some preventive medicine. In *Morality's progress: Essays on humans, other animals, and the rest of nature*, 213–224. Oxford: Clarendon Press.

Jamieson, D. 2008. *Ethics and the environment: An introduction*. Cambridge: Cambridge University Press.

Jamieson, D. 2010. Climate change, responsibility, and justice. *Science and Engineering Ethics* 16(3): 431–445.

Janzen, D. 1998. Gardenification of wildland nature and the human footprint. *Science* 279(5355): 1312–1313.

Jenkins, M. 2003. Prospects for biodiversity. *Science*, 302(5648): 1175–1177.

Joling, D. 2010. State sues over beluga endangered listing. The Associated Press, in the *Anchorage Daily News*, June 4, 2010. http://www.adn.com/2010/06/04/1307852/alaska-sues-feds-over-beluga-listing.html.

Jones, C.G., R.S. Ostfeld, M.P. Richard, E.M. Schauber, and J.O. Wolff. 1998. Chain reactions linking acorns to gypsy moth outbreaks and lyme disease risk. *Science* 279(5353): 1023–1026. http://www.ecostudies.org/reprints/Jones_et_al_1998_Science_279_1023-1026.pdf.

Judson, O. 2003. A bug's death. *The wild side*. In *The New York Times*, September 25, 2003. http://www.nytimes.com/2003/09/25/opinion/a-bug-s-death.html?pagewanted=all.

Judson, O. 2008a. A natural selection. *The wild side*. In *The New York Times*, July 22, 2008. http://opinionator.blogs.nytimes.com/2008/07/22/a-natural-selection/.

Judson, O. 2008b. Resurrection science. *The wild side*. In *The New York Times*, November 25, 2008. http://opinionator.blogs.nytimes.com/2008/11/25/resurrection-science/.

Kac, E. 2000. GFP bunny. http://www.ekac.org/gfpbunny.html#gfpbunnyanchor.

Kac, E. 2006. Specimen of secrecy about marvelous discoveries. http://www.ekac.org/specimen.html.

Kamhawi, S., A. Arbagi, S. Adwan, and M. Rida. 1993. Environmental manipulation in the control of a zoonotic cutaneous leishmaniasis focus. *Archives de l'Institut Pasteur de Tunis* 70(3–4): 383–390.

Kareiva, P., and M. Marvier. 2003. Conserving biodiversity coldspots. *American Scientist* 91: 344–351. www.bren.ucsb.edu/academics/courses/297-2S/Readings/Coldspots.pdf.

Karpechenko, G.D. 1928. Polyploid hybrids of *Raphanus sativus* L. X *Brassica oleracea* L. *Zeitschrift für Induktive Abstammungs- und Vererbungslehre* 48: 1–85.

Katz, E. 1992. The big lie: Human restoration of nature. In *Research in Philosophy and Technology*, vol. 12: *Technology and the environment*, ed. F. Ferré, 231–241. Greenwich, CT: JAI Press.

Kawall, J. 2010. The epistemic demands of environmental virtue. *Journal of Agricultural and Environmental Ethics* 23(1–2): 109–128.

Keesing, F., R.D. Holt, and R.S. Ostfeld. 2006. Effects of species diversity on disease risk. *Ecology Letters* 9(4): 485–498. http://www3.interscience.wiley.com/cgi-bin/fulltext/118634073/PDFSTART.

Keesing, F., L.K. Belden, P. Daszak, A. Dobson, C.D. Harvell, R.D. Holt, P. Hudson, A. Jolles, K.E. Jones, C.E. Mitchell, S.S. Myers, T. Bogich, and R.S. Ostfeld. 2010. Impacts of biodiversity on the emergence and transmission of infectious diseases. *Nature* 468(7324): 647–652.

Kellert, S.R. 2005. *Building for life: Designing and understanding the human-nature connection.* Washington, DC: Island Press.

Kellert, S.R. 2009. Biodiversity, quality of life, and evolutionary psychology. In *Biodiversity change and human health*, ed. O.E. Sala, L.A. Meyerson, and C. Parmesan, 99–127. Washington, DC: Island Press.

Khoshbakht, K., and K. Hammer. 2008. How many plant species are cultivated. *Genetic Resources and Crop Evolutions* 55(7): 925–928.

Kifner, J. 1994. Stay-at-Home SWB, 8, into fitness, seeks thrills. *The New York Times*, July 2, 1994. http://www.nytimes.com/1994/07/02/nyregion/about-new-york-stay-at-home-swb-8-into-fitness-seeks-thrills.html.

Kimball, S., and P.M. Schiffman. 2003. Differing effects of cattle grazing on native and alien plants. *Conservation Biology* 17(6): 1681–1693. http://skimball.bio.uci.edu/documents/Kimball&Schiffman.pdf.

Kirchner, J.W., and A. Weil. 2000. Delayed biological recovery from extinctions throughout the fossil record. *Nature* 404(6774): 177–180.

Koh, L.P., P.R. Dunn, N.S. Sodhi, R.K. Colwell, H.C. Proctor, and V.S. Smith. 2004. Species coextinctions and the biodiversity crisis. *Science* 305(5690): 1632–1634. http://vsmith.info/Species-Coextinction.

Krutilla, M.V., C.J. Cicchetti, A.M. Freeman III, and C.S. Russell. 1972. Observations on the economics of irreplaceable assets. In *Environmental quality analysis: Theory and method in the social sciences*, ed. A.V. Kneese and B.T. Bower, 69–112. Baltimore: Resources for the Future/The Johns Hopkins University Press.

Lazaris, A., S. Arcidiacono, Y. Huang, J. Zhou, F. Duguay, N. Chretien, E.A. Welsh, J.W. Soares, and C.N. Karatzas. 2002. Spider silk fibers spun from soluble recombinant silk produced in mammalian cells. *Science* 295(5554): 472–476. For a non-technical summary, see http://news.bbc.co.uk/1/hi/sci/tech/889951.stm.

Leibniz, G.W. 1710. *Essais de Théodicée*, Trans. E.M. Huggard (2005). http://www.gutenberg.org/files/17147/17147-h/17147-h.htm.

Leibniz, G.W. 1714. *Principles of nature and grace, according to reason*, Trans. J. Bennett (2007). http://www.earlymoderntexts.com/pdf/leibprin.pdf.

Leibniz, G.W. 1715. Letter to Christian Wolff, 18 May 1715. In *Philosophical essays,* Trans. G. Ariews and D. Garber (1989), 232–234. Indianapolis, IN: Hackett Publishing Company.

Leopold, A. 1949. *A sand county almanac and sketches here and there.* New York: Oxford University Press.

Levinton, J.S., E. Suatonl, W. Wallace, R. Junkins, B. Kelaher, and B.J. Allen. 2003. Rapid loss of genetically based resistance to metals after the cleanup of a Superfund site. *Proceedings of the National Academy of Science (PNAS)* 100(17): 9889–9891. http://www.pnas.org/content/100/17/9889.full.

Li, W., Z. Shi, M. Yu, W. Ren, C. Smith, J.H. Epstein, H. Wang, G. Crameri, Z. Hu, H. Zhang, J. Zhang, J. McEachern, H. Field, P. Daszak, B.T. Eaton, S. Zhang, and L.F. Wang. 2005. Bats are natural reservoirs of SARS-like coronaviruses. *Science* 310(5748): 676–679.

Light, A. 2000. Ecological restoration and the culture of nature: A pragmatic perspective. In *Restoring nature: Perspectives from the social sciences and humanities*, ed. P.H. Gobster and R.B. Jull, 49–70. Washington, DC: Island Press.

Light, A. 2002. Restoring ecological citizenship. In *Democracy and the claims of nature: Critical perspectives for a new century*, ed. B. Minteer and B.P. Taylor, 153–172. Lanham, MD: Rowman & Littlefield Publishers, Inc.

Light, A. 2003a. The case for a practical pluralism. In *Environmental ethics*, ed. A. Light and H. Rolston, 229–247. Malden, MA: Blackwell Publishing.

Light, A. 2003b. Urban ecological citizenship. *Journal of Social Philosophy* 34(1): 44–63.

Light, A. 2006. Ecological citizenship: The democratic promise of restoration. In *The human metropolis: People and nature in the 21st-century city*, ed. R.H. Platt, 169–181. Amherst: University of Massachusetts Press.

Light, A. 2007a. Restorative relationships: From artifacts to natural systems. Working Paper #2007-05, University of Washington Evans School of Public Affairs, Seattle. http://evans. washington.edu/files/EvansWorkingPaper-2007-05.pdf.

Light, A. 2007b. Does a public environmental philosophy need a convergence hypothesis? Working Paper #2007-06, University of Washington Evans School of Public Affairs, Seattle. http:// evans.washington.edu/files/EvansWorkingPaper-2007-06.pdf.

Liow, L.H., M. Fortelius, E. Bingham, K. Lintulaakso, H. Mannila, L. Flynn, and N.C. Stenseth. 2008. Higher origination and extinction rates in larger mammals. *Proceedings of the National Academy of Science (PNAS)* 105(16): 6097–6102. http://www.pnas.org/content/105/16/6097.full.

Locke, J. 1690. *The second treatise of Civil Government*. Adelaide: The University of Adelaide Library. http://ebooks.adelaide.edu.au/l/locke/john/l81s/index.html.

Lowe, E.J. 2006. *The four-category ontology: A metaphysical foundation for natural science*. Oxford: Oxford University Press.

Lugo, A.E. 1997. The apparent paradox of reestablishing species richness on degraded lands with tree monocultures. *Forestry Ecology and Management* 99(1–2): 9–19.

Lugo, A.E. 2004. The outcome of alien tree invasions in Puerto Rico. *Frontiers in Ecology and the Environment* 2(5): 265–273.

MacArthur, R.H., and E.O. Wilson. 1963. An equilibrium theory of insular zoogeography. *Evolution* 17(4): 373–387.

Maclaurin, J., and K. Sterelny. 2008. *What is biodiversity?* Chicago: University of Chicago Press.

Maklakov, A.A., S. Immler, A. Gonzalez-Voyer, J. Rönn, and N. Kilm. 2011. Brains and the city: Big-brained passerine birds succeed in urban environments. *Biology Letters*, 7(5): 730–732.

Mallozzi, V. 2009. Spend $10,600 on the Yankees – Or for college or a car? *The New York Times*, April 10, 2009. http://www.nytimes.com/2009/04/12/sports/baseball/12cheer.html.

Mann, C.C. 2005. *1491: New revelations of the Americas before Columbus*. New York: Knopf.

Mann, C.C. 2011. *1493: Uncovering the new world Columbus created*. New York: Knopf.

Manson, N.A. 2007. The concept of irreversibility: Its use in the sustainable development and precautionary principle literatures. *The Electronic Journal of Sustainable Development* 1(1): 3–15.

Manwood, J. 1717. *Manwood's treatise of the forest laws*. 4th ed. ed. W. Nelson. London: E. Nutt. http://books.google.com/books?id=2rY1AAAAMAAJ&printsec=frontcover.

Maron, J.L., M. Vila, R. Bommarco, S. Elmendorf, and P. Beardsley. 2004. Rapid evolution of an invasive plant. *Ecological Monographs* 74(2): 261–280.

Marris, E. 2008. Moving on assisted migration. *Nature Reports Climate Change* 2(9): 112–113. http://www.nature.com/climate/2008/0809/full/climate.2008.86.html.

Martin, P., and D. Burney. 1999. Bring back the elephants! *Wild Earth* 9: 57–64. http://findarticles. com/p/articles/mi_m0GER/is_2000_Spring/ai_61426229/.

Marty, J.T. 2005. Effects of cattle grazing on diversity in ephemeral wetlands. *Conservation Biology* 19(5): 1626–1632. www.vernalpools.org/documents/Marty%20Cons%20Bio.pdf.

May, R.M. 1995. Conceptual aspects of the quantification of the extent of biological diversity. In *Biodiversity measurement and estimation*, ed. D.L. Hawksworth, 13–20. London: Chapman & Hall.

Mazmunder, A. 2009. Consequences of aquatic biodiversity for water quality and health. In *Biodiversity change and human health*, ed. O.E. Sala, L.A. Meyerson, and C. Parmesan, 143–157. Washington, DC: Island Press.

McKibben, B. 1989. *The end of nature*. New York: Random House.

McKibben, B. 1995. *Hope, human and wild: True stories of living lightly on the earth*. St. Paul, MN: Hungry Mind Press.

McKibben, B. 2003. *Enough: Staying human in an engineered world*. New York: Henry Holt and Company.

Melillo, J., and O. Sala. 2008. Ecosystem services. In *Sustaining life: How human health depends on biodiversity*, ed. E. Chivian and A. Bernstein, 74–115. Oxford: Oxford University Press.

Meyerson, L.A., O.E. Sala, A. Froment, C.S. Friedman, K. Hund-Rinke, P. Martens, A. Mazmunder, A.N. Purohit, M.B. Thomas, and A. Wilby. 2009. Sustainable allocation of biodiversity to improve human health and well-being. In *Biodiversity change and human health*, ed. O.E. Sala, L.A. Meyerson, and C. Parmesan, 83–96. Washington, DC: Island Press.

Miller, B., G. Ceballos, and R. Reading. 1994. The prairie dog and biotic diversity. *Conservation Biology* 8(3): 677–681. http://www.ecologia.unam.mx/laboratorios/eycfs/faunos/art/GCe/AA39.pdf.

Miller, W., D.I. Drautz, B. Pusey, J. Qi, A.M. Lesk, L.P. Tomsho, M.D. Packard, F. Zhao, A. Sher, A. Tikhonov, B. Raney, N. Patterson, K. Lindblad-Toh, E.S. Lander, J.R. Knight, G.P. Irzyk, K.M. Fredrikson, T.T. Harkins, S. Sheridan, T. Pringle, and S.C. Schuster. 2008. Sequencing the nuclear genome of the extinct woolly mammoth. *Nature* 456(7220): 387–390.

Mitchell, C., C. Ogura, D. Meadows, A. Kane, L. Strommer, S. Fretz, D. Leonard, and A. McClung. 2005. *Hawaii's comprehensive wildlife conservation strategy*. Honolulu: Hawaii Department of Land and Natural Resources. http://www.state.hi.us/dlnr/dofaw/cwcs/process_strategy.htm.

Molyneux, D.H., R.S. Ostfeld, A. Bernstein, and E. Chivian. 2008. Ecosystem disturbance, biodiversity loss, and human infectious disease. In *Sustaining life: How human health depends on biodiversity*, ed. E. Chivian and A. Bernstein, 286–323. Oxford: Oxford University Press.

Moore, G.E. 1903. *Principia ethica*. Cambridge: Cambridge University Press.

Muir, J. 1911. *My first summer in the sierra*. Boston: Houghton Mifflin Company.

Munshi-South, J., and K. Kharchenko. 2010. Rapid pervasive genetic differentiation of urban white-footed mouse (*Peromyscus leucopus*) populations in New York City. *Molecular Biology* 19(19): 4242–4254.

Myers, N. 1996a. The biodiversity crisis and the future of evolution. *The Environmentalist* 16: 37–47.

Myers, N. 1996b. Environmental services of biodiversity. *Proceedings of the National Academy of Science (PNAS)* 93(7): 2764–2769. http://www.pnas.org/content/93/7/2764.full.

Myers, N., R.A. Mittermeier, C.G. Mittermeier, G.A.B. da Fonseca, and J. Kent. 2000. Biodiversity hotspots for conservation priorities. *Nature* 403(6772): 853–858. http://www.nicholas.duke.edu/people/faculty/myers/myersn.etal.2000.pdf.

Nash, M. 2009. Bring in the cows. *High Country News* 41(9): 8–9, 23. http://www.hcn.org/issues/41.9/bring-in-the-cows/.

Nehring, K., and C. Puppe. 2002. A theory of diversity. *Econometrica* 70(3): 1155–1198.

Newman, D.J., J. Kilama, A. Bernstein, and E. Chivian. 2008. Medicines from nature. In *Sustaining life: How human health depends on biodiversity*, ed. E. Chivian and A. Bernstein, 116–161. Oxford: Oxford University Press.

Niklas, K.J., B.H. Tiffney, and A.H. Knoll. 1983. Patterns in vascular land plant diversification. *Nature* 303(5918): 614–616.

Normile, D. 2009. Bringing coral reefs back from the living dead. *Science* 325(5940): 559–561.

Norton, B. 1987. *Why preserve natural variety?* Princeton: Princeton University Press.

Norton, B. 2001. Conservation biology and environmental values, *Can there be a universal earth ethic?* In *Protecting biological diversity, roles and responsibilities*, ed. C. Potvin, M. Kraenzel, and G. Seutin, 71–102. Montreal: McGill-Queens University Press.

Norton, B.G. 2005. *Sustainability: A philosophy of adaptive ecosystem management*. Chicago: Chicago University Press.

Norton, B. 2006. Toward a policy-relevant definition of biodiversity. In *The endangered species act at thirty*, vol. 2, ed. J.M. Scott, D.D. Goble, and F.W. Davis, 49–58. Washington, DC: Island Press.

Norton, B.G., and A.C. Steinemann. 2001. Environmental values and adaptive management. *Environmental Values* 10(4): 473–506.

Obst, M., P. Funch, and G. Giribet. 2005. Hidden diversity and host specificity in cycliophorans: A phylogeographic analysis along the North Atlantic and Mediterranean Sea. *Molecular Biology* 14(14): 4427–4440.

Oliveira-Santos, L.G.R., and F.A.S. Fernandez. 2010. Pleistocene rewilding, Frankenstein ecosystems, and an alternative conservation agenda. *Conservation Biology* 24(1): 4–5.

O'Neill, J. 1992. The varieties of intrinsic value. *The Monist* 75(2): 119–137. In *Environmental ethics*, ed. A. Light and H. Rolston (2003), 131–142. Malden, MA: Blackwell Publishing.

Padilla, F.M., and F.I. Pugnaire. 2006. The role of nurse plants in the restoration of degraded environments. *Frontiers in Ecology and the Environment* 4(4): 196–202.

Palmer, M., E. Bernhardt, E. Chornesky, S. Collins, A. Dobson, C. Duke, B. Gold, R. Jacobson, S. Kingsland, R. Kranz, M. Mappin, M.L. Martinez, F. Micheli, J. Morse, M. Pace, M. Pascual, S. Palumbi, O.J. Reichman, A. Simons, A. Townsend, and M. Turner. 2004. Ecology for a crowded planet. *Science* 304(5675): 1251–1252.

Parry, M.L., O.F. Canziani, J.P. Palutikof, P.J. van der Linden, and C.E. Hanson, eds. 2007. *Contribution of Working Group II to the Fourth Assessment Report (AR4) of the Intergovernmental Panel on Climate Change (IPCC)*. Cambridge: Cambridge University Press. http://www.ipcc-wg2.gov/publications/AR4/index.html.

Pelzer, K.D., and N. Currin. 2009. *Zoonotic diseases of cattle publication, 400–460*. Blacksburg: Virginia Polytechnic Institute and State University. http://pubs.ext.vt.edu/400/400-460/400-460_pdf.pdf.

Pettit, P. 2008. Freedom and probability: A comment on Goodin and Jackson. *Philosophy & Public Affairs* 36(2): 206–220.

Pettit, P. 2011. The instability of Berlin's freedom as non-interference. *Ethics* 121(4): 693–716.

Plummer, M.L., and R.C. Hartman. 1986. Option value: A general approach. *Economic Inquiry* 24(3): 455–471.

Pope John Paul II. 1999. Post-synodal apostolic exhortation Ecclesia in America of the Holy Father John Paul II to the bishops, priests and deacons, men and women religious, and all the lay faithful on the encounter with the living Jesus Christ: The way to conversion, communion and solidarity in America. http://www.vatican.va/holy_father/john_paul_ii/apost_exhortations/documents/hf_jp-ii_exh_22011999_ecclesia-in-america_en.html.

Powell, L.F. 1978. In *Regents of the University of California v. Bakke* (No. 7811), 438 U.S. 265. http://www4.law.cornell.edu/supct/html/historics/USSC_CR_0438_0265_ZO.html.

Pritchard, J.K., J.K. Pickrell, and G. Coop. 2010. The genetics of human adaptation: Hard sweeps, soft sweeps, and polygenic adaptation. *Current Biology* 20(4): R208–R215.

Purvis, A., and A. Hector. 2000. Getting the measure of biodiversity. *Nature* 405(6783): 212–219.

Rahel, F.J. 2002. Homogenization of freshwater faunas. *Annual Review of Ecology and Systematics* 33: 291–315.

Rapport, D.J., P. Daszak, A. Froment, J.F. Guegan, K.D. Lafferty, A. Larigauderie, A. Mazumder, and A. Winding. 2009. The impact of anthropogenic stress at global and regional scales on biodiversity and human health. In *Biodiversity change and human health*, ed. O.E. Sala, L.A. Meyerson, and C. Parmesan, 41–60. Washington, DC: Island Press.

Rawls, J. 1971. *A theory of justice*. Cambridge: The Belknap Press of Harvard University Press.

Raz, J. 1986. *The morality of freedom*. Oxford: Oxford University Press.

Reiners, W.A., and J.A. Lockwood. 2010. *Philosophical foundations for the practices of ecology*. Cambridge: Cambridge University Press.

Reise, K., S. Olenin, and D.W. Thieltges. 2006. Are aliens threatening European aquatic coastal ecosystems? *Helgoland Marine Research* 60(2): 77–83.

Revkin, A. 2009. U.S. curbs use of species act in protecting polar bear. *The New York Times*, May 8, 2009. http://www.nytimes.com/2009/05/09/science/earth/09bear.html.

Richards, D. 2005. Biodiversity performance measures – Their role in Rio Tinto. Presentation for Setting Targets – Delivering Outcomes: Biodiversity Performance Measures Workshop. Earthwatch Institute, London. http://www.businessandbiodiversity.org/ppt/Presentation 4 - Dave Richards Rio Tinto.ppt.

Rick, T.C., and J.M. Erlandson. 2009. Coastal exploitation. *Science* 325(5943): 952–953.

Ricketts, T.H., G.C. Daily, P.R. Ehrlich, and C.D. Michener. 2004. Economic value of tropical forest to coffee production. *Proceedings of the National Academy of Science (PNAS)* 101(34): 12579–12582. http://www.pnas.org/content/101/34/12579.full.

Ricotta, C. 2005. Through the jungle of biological diversity. *Acta Biotheoretica* 53: 29–38.

Ricotta, C. 2007. A semantic taxonomy for diversity measures. *Acta Biotheoretica* 55: 23–33.

Rodriguez, L.F. 2006. Can invasive species facilitate native species? Evidence of how, when, and why these impacts occur. *Biological Invasions* 8(4): 927–939.

Roemer, G.W., and R.K. Wayne. 2003. Conservation in conflict: The tale of two endangered species. *Conservation Biology* 17(5): 1251–1260.

Roemer, G.W., T.J. Coonan, D.K. Garcelon, J. Bascompte, and L. Laughrin. 2001. Feral pigs facilitate hyperpredation by golden eagles and indirectly cause the decline of the island fox. *Animal Conservation* 4: 307–318.

Rolston, H. 1988. *Environmental ethics: Duties to and values in the natural world*. Philadelphia: Temple University Press.

Rolston, H. 1994. Value in nature and the nature of value. In *Philosophy and natural environment*. Royal institute of Philosophy, Suppl., ed. R. Attfield and A. Belsey, 13–30. Cambridge: Cambridge University Press. http://lamar.colostate.edu/~rolston/value-n.pdf.

Rominger, E.M. 2007. Culling mountain lions to protect ungulate populations – Some lives are more sacred than others. *Transactions of the 72nd North American Wildlife and Natural Resources Conference*, Washington, DC, 186–193. http://www.wildlifemanagementinstitute.org/PDF/2-Culling Mountain Lions....pdf.

Rosner, H. 2010. One tough sucker. *High Country News* 42(10): 12–18. http://www.hcn.org/issues/42.10/one-tough-sucker.

Ross, T. 2011. Alaska's 'abundance management'. *High Country News* 43(3): 10–16. http://www.hcn.org/issues/43.3/palin-politics-and-alaka-predator-control.

Ross, P.I., M.G. Jalkotzy, and M. Festa-Bianchet. 1997. Cougar predation on bighorn sheep in southwestern Alberta during winter. *Canadian Journal of Zoology* 75(5): 771–775.

Rosselló-Mora, R., and R. Amann. 2001. The species concept for prokaryotes. *FEMS Microbiology Reviews* 25(1): 39–67.

Rothschild, B.M., L.D. Martin, G. Lev, H. Bercovier, G.K. Bar-Gal, C. Greenblatt, H. Donoghue, M. Spigelman, and D. Brittain. 2001. *Mycobacterium tuberculosis* complex DNA from an extinct bison dated 17,000 years before the present. *Clinical Infectious Diseases* 33(3): 305–311.

Rothstein, E. 2008. The forest premeditated: Illusions of wildness in a botanical garden. *The New York Times*, June 16, 2008. http://www.nytimes.com/2008/06/16/arts/16conn.html?pagewanted=all.

Sagoff, M. 1997. Do we consume too much? *The Atlantic Monthly* 279(6): 80–96. http://www.theatlantic.com/past/docs/issues/97jun/consume.htm.

Sagoff, M. 1999. What's wrong with exotic species? *Philosophy and Public Policy Quarterly*, 19(4). https://scholar.vt.edu/access/content/user/hullrb/PUBLIC/sagoffexoticspecies.pdf. (A shorter version of Sagoff (2003)).

Sagoff, M. 2003. Native to a place, or what's wrong with exotic species? In *Values at sea: Ethics for the marine environment*, ed. D.G. Dallmeyer, 93–110. Athens: University of Georgia Press. (A longer version of Sagoff (1999)).

Sala, O.E., F.S. Chapin III, J.J. Armesto, E. Berlow, J. Bloomfield, R. Dirzo, E. Huber-Sanwald, L.F. Huenneke, R.B. Jackson, A. Kinzig, R. Leemans, D.M. Lodge, H.A. Mooney, M. Oesterheld, N.L. Poff, M.T. Sykes, B.H. Walker, M. Walker, and D.H. Wall. 2000. Global biodiversity scenarios for the year 2100. *Science* 287(5459): 1770–1774.

Sala, O.E., L.A. Meyerson, and C. Parmesan (eds.). 2009. *Biodiversity change and human health*. Washington, DC: Island Press.

Salvo, M. 2009. Western wildlife under hoof: Public lands livestock grazing threatens iconic species. Chandler, AZ: WildEarth Guardians. http://www.wildearthguardians.org/site/DocServer/report-WWUH_4_30_highres.pdf?docID=662.

Sanderson, E.W., M. Jaiteh, M.A. Levy, K.H. Redford, A.V. Wannebo, and G. Woolmer. 2002. The human footprint and the last of the wild. *BioScience* 52(10): 891–904. http://sedac.ciesin.columbia.edu/wildareas/documents/human_footprint_Sanderson_etal2002.pdf.

Sandler, R. 2007. *Character and environment: A virtue-oriented approach to environmental ethics*. New York: Columbia University Press.

Sandy, E. 2010. Italian scientists turn back time, revive and ancient cattle breed. *Institute for International Journalism* (E.W. Scripps School of Journalism at Ohio University). http://scrippsiij.blogspot.com/2010/02/scientists-turn-ack-time-revive-ancient.html.

Sarkar, S. 1999. Wilderness preservation and biodiversity conservation – Keeping divergent goals distinct. *BioScience* 49(5): 405–412. http://uts.cc.utexas.edu/~consbio/Cons/Sarkar-Wilderness.pdf.

Sarkar, S. 2002. Defining 'biodiversity'; assessing biodiversity. *The Monist* 85(1): 131–155. http://uts.cc.utexas.edu/~consbio/Cons/Sarkar.Monist.pdf.

Sarkar, S. 2005. *Biodiversity and environmental philosophy: An introduction.* Cambridge: Cambridge University Press.

Sarkar, S. 2006. Ecological diversity and biodiversity. *Acta Biotheoretica* 54: 133–140.

Satz, D. 2010. *Why some things should not be for sale.* Oxford: Oxford University Press.

Saul, A. 2003. Zooprophylaxis or zoopotentiation: The outcome of introducing animals on vector transmission is highly dependent on mosquito mortality while searching. *Malaria Journal* 2: 32. http://malariajournal.com/content/2/1/32.

Sax, D.F., and S. Gaines. 2003. Species diversity: From global decreases to local increases. *Trends in Ecology & Evolution* 18(11): 561–566.

Sax, D.F., and S. Gaines. 2008. Species invasions and extinction: The future of native biodiversity on islands. *Proceedings of the National Academy of Science (PNAS)* 105(Suppl. 1): 11490–11497. http://www.pnas.org/content/105/suppl.1/11490.full.

Sax, D.F., J.J. Stachowicz, J.H. Brown, J.F. Bruno, M.N. Dawson, S.D. Gaines, R.K. Grosberg, A. Hastings, R.D. Holt, M.M. Mayfield, M.I. O'Connor, and W.R. Rice. 2007. Ecological and evolutionary insights from species invasions. *Trends in Ecology & Evolution* 22(9): 465–471.

Scarry, E. 1999. *On beauty and being just.* Princeton: Princeton University Press.

Schippmann, U., D.J. Leaman, and A.B. Cunningham. 2002. Impact of cultivation and gathering of medicinal plants on biodiversity: Global trends and issues. In *Biodiversity and the ecosystem approach in agriculture, forestry and fisheries.* Rome: Food and Agricultural Organization of the United Nations (FAO). http://www.fao.org/DOCREP/005/AA010E/AA010E00.HTM.

Schmidtz, D. 2008a. Choosing ends. In *Person, polis, planet: Essays in applied philosophy*, 3–61. Oxford: Oxford University Press; A revision of Schmidtz, D. 1994. Choosing ends. *Ethics* 104(2): 226–251.

Schmidtz, D. 2008b. Reasons for altruism. In *Person, polis, planet: Essays in applied philosophy*, 62–77. Oxford: Oxford University Press.

Sepkoski, J.J. 1982. A compendium of fossil marine families. *Milwaukee Public Museum Contributions in Biology and Geology* 51: 1–125.

Sexton, J.P., J.K. McKay, and A. Sala. 2002. Plasticity and genetic diversity may allow saltcedar to invade cold climates in North America. *Ecological Applications* 12(6): 1652–1660.

Shanks, P. 2010. Two more 'Lazarus' projects. *Biopolitical times.* http://www.biopoliticaltimes.org/article.php?id=5057.

Singer, P. 1979. Not for humans only: The place of nonhumans in environmental issues. In *Ethics and problems of the 21st century*, ed. K.E. Goodpaster and K.M. Syre, 191–206. Notre Dame: University of Notre Dame Press.

Skevington, S.M. 2009. Qualify of life, biodiversity, and health: Observations and applications. In *Biodiversity change and human health*, ed. O.E. Sala, L.A. Meyerson, and C. Parmesan, 129–140. Washington, DC: Island Press.

Slotte, T., H. Huang, M. Lascoux, and A. Ceplitis. 2008. Polyploid speciation did not confer instant reproductive isolation in *Capsella* (Brassicaceae). *Molecular Biology and Evolution* 25(7): 1472–1481. http://mbe.oxfordjournals.org/cgi/reprint/25/7/1472.

Smith, S.Q., and T.H. Jones. 2004. Tracking the cryptic pumiliotoxins. *Proceedings of the National Academy of Science (PNAS)* 101(21): 7841–7842. http://www.pnas.org/content/101/21/7841.full.

Snyder, W.E., and E.W. Evans. 2006. Ecological effects of invasive arthropod generalist predators. *Annual Review of Ecology, Evolution, and Systematics* 37: 95–122.

Snyder, W.E., and A.R. Ives. 2001. Generalist predators disrupt biological control by a specialist parasitoid. *Ecology* 83(2): 705–716.

Soluck, D.A., and J.S. Richardson. 1997. The role of stoneflies in enhancing growth of trout: A test of the importance of predator–predator facilitation within a stream community. *Oikos* 80: 214–219.

Sommer, L. 2011. Gulls threaten South Bay salt pond restoration work. KQED *Quest*. http://science.kqed.org/quest/audio/gulls-threaten-south-bay-salt-pond-restoration-work/.

Soulé, M. 1985. What is conservation biology? *BioScience* 35: 727–734.

Soulé, M. 1990. The onslaught of alien species, and other challenges in the coming decades. *Conservation Biology* 4(3): 233–239.

Spicer, J. 2006. *Biodiversity: A beginner's guide*. Oxford: Oneworld Publications.

Sterba, J. 2001. *Three challenges to ethics*. New York: Oxford University Press.

Stiling, P. 2002. *Ecology, theories and applications*, 4th ed. Upper Saddle River, NJ: Prentice Hall.

Stohlgren, T.J., D.T. Barnett, and J.T. Kartesz. 2003. The rich get richer: Patterns of plant invasions in the United States. *Frontiers in Ecology and the Environment* 1(1): 11–14.

Strand, G. 2008. *Inventing Niagara: Beauty, power and lies*. New York: Simon & Schuster.

Takacs, D. 1996. *The idea of biodiversity, philosophies of paradise*. Baltimore: The Johns Hopkins University Press.

Tanaka, M.M., J.R. Kendal, and K.N. Laland. 2009. From traditional medicine to witchcraft: Why medical treatments are not always efficacious. *PLos ONE* 4(4). http://www.plosone.org/article/info:doi/10.1371/journal.pone.0005192.

Taylor, P. 1981. The ethics of respect for nature. *Environmental Ethics* 3(3): 197–218. http://www.umweltethik.at/download.php?id=391.

Taylor, P. 1989. *Respect for nature*. 2nd printing with corrections. Princeton: Princeton University Press.

Taylor, K. 2009. Voyage of the dammed. *High Country News* 41(10): 12–13, 24. http://www.hcn.org/issues/41.10/voyage-of-the-dammed.

The Nature Conservancy. 2011. Super sucker saves reefs: Underwater vacuum removes invasive algae. http://www.nature.org/ourinitiatives/regions/northamerica/unitedstates/hawaii/explore/super-sucker.xml.

The Nature Conservancy International Leadership Council. 2008. Biodiversity offsets meeting agenda, June 19–20, 2008. http://www.nature.org/aboutus/workingwithcompanies/ilc_agenda_june.pdf.

Thomas, M.B., K.D. Lafferty, and C.S. Friedman. 2009. Biodiversity and disease. In *Biodiversity change and human health*, ed. O.E. Sala, L.A. Meyerson, and C. Parmesan, 229–243. Washington, DC: Island Press.

Thoreau, H.D. 1864. *The maine woods*. Boston: Ticknor and Fields.

Tilman, D., P.B. Rich, J. Knops, D. Sedin, T. Mielke, and C. Lehman. 2001. Diversity and productivity in a long-term grassland experiment. *Science* 294(5543): 843–845.

Tingley, R., B.L. Phillips, and R. Shine. 2011. Establishment success of introduced amphibians increases in the presence of congeneric species. *The American Naturalist* 177(3): 382–388.

Townsend, A.R., L.A. Martinelli, and R.W. Howarth. 2009. The global nitrogen cycle, biodiversity, and human health. In *Biodiversity change and human health*, ed. O.E. Sala, L.A. Meyerson, and C. Parmesan, 159–178. Washington, DC: Island Press.

U.S. Fish and Wildlife Service. 2009. San Clemente Loggerhead Shrike *Lanius ludovicianus mearnsi* | 5-year review: Summary and evaluation. Carlsbad, CA: U.S. Fish and Wildlife Service, Carlsbad Fish and Wildlife Office. http://www.fws.gov/ecos/ajax/docs/five_year_review/doc2631.pdf.

van Dam, J.A., H.A. Aziz, M.A.A. Sierra, F.J. Hilgen, L.W. van den Hoek Ostende, L.J. Ourens, P. Mein, A.J. van der Meulen, and P. Palaez-Campomanes. 2006. Long-period astronomical forcing of mammal turnover. *Nature* 443(7112): 687–692.

van Kooten, G.C., and E.H. Bulte. 2000. *The economics of nature: Managing biological assets*. Malden, MA: 83–93: Blackwell.

Vartanyan, S.L., V.E. Garutt, and A.V. Sher. 1993. Holocene dwarf mammoths from Wrangel Island in the Siberian Arctic. *Nature* 362(6418): 337–340.

Vellend, M., L.J. Harmon, J.L. Lockwood, M.M. Mayfield, A.R. Hughes, J.P. Wares, and D.F. Sax. 2007. Effects of exotic species on evolutionary diversification. *Trends in Ecology & Evolution* 22(9): 481–488.

Vining, A.R., and D.L. Weimer. 1998. Passive use benefits: Existence, option, and quasi-option value. In *Handbook of public finance*, ed. F. Thompson and M.T. Green, 319–345. New York: Marcel Dekker, Inc.

Vitousek, P.M., and L.R. Walker. 1989. Biological invasion by *Myrica Faya* in Hawai'i: Plant demography, nitrogen fixation, ecosystem effects. *Ecological Monographs* 59(3): 247–265.

Vitousek, P.M., H.A. Mooney, J. Lubchenco, and J. Melillo. 1997. Human domination of earth's ecosystems. *Science* 277(5325): 494–499.

Wade, N. 2010a. Scientists cite fastest case of human evolution. *The New York Times*, July 1, 2010. http://www.nytimes.com/2010/07/02/science/02tibet.html.

Wade, N. 2010b. Adventures in very recent evolution. *The New York Times*, July 19, 2010. http://www.nytimes.com/2010/07/20/science/20adapt.html?pagewanted=all.

Wade, N. 2010c. Natural selection cuts broad swath through fruit fly genome. *The New York Times*, September 20, 2010. http://www.nytimes.com/2010/09/21/science/21gene.html.

Ward, D., and D. Brownlee. 2000. *Rare earth*. New York : Copernicus Books, an imprint of Springer.

Wardle, D.A., K.I. Bonner, G.M. Barker, G.W. Yeates, K.S. Nicholson, R.D. Bardgett, R.N. Watson, and A. Ghani. 1999. Plant removals in perennial grassland: Vegetation dynamics, decomposers, soil biodiversity, and ecosystem properties. *Ecological Monographs* 29(4): 535–568.

Warwick, R.M., and P.J. Somerfield. 2008. All animals are equal, but some animals are more equal than others. *Journal of Experimental Marine Biology and Ecology* 366(1–2): 184–186.

Weinberger, D. 2007. *Everything is miscellaneous: The power of the new digital disorder*. New York: Henry Holt and Company.

Weisbrod, B.A. 1964. Collective consumption properties of individual consumption goods. *Quarterly Journal of Economics* 78(3): 471–477.

Weisman, A. 2007. *The world without us*. New York: Thomas Dunne Books, St. Martin's Press.

Whittaker, R.H. 1972. Evolution and measurement of species diversity. *Taxon* 21: 213–251.

Wigglesworth, M. 1871. God's controversy with New England. *Proceedings of the Massachusetts Historical Society* 12: 83–93. http://digitalcommons.unl.edu/etas/36/.

Wilby, A., C. Mitchell, D. Blumenthal, P. Daszak, C.S. Friedman, P. Jutro, A. Mazumder, A. Prieur-Richard, M. Desprez-Loustau, M. Sharma, and M.B. Thomas. 2009. Biodiversity, food provision, and human health. In *Biodiversity change and human health*, ed. O.E. Sala, L.A. Meyerson, and C. Parmesan, 13–39. Washington, DC: Island Press.

Wilcove, D.S., D. Rothstein, J. Dubow, A. Phillips, and E. Losos. 1998. Quantifying threats to imperiled species in the United States: Assessing the relative importance of habitat destruction, alien species, pollution, overexploitation, and disease. *BioScience* 48(8): 607–615.

Williams, B. 1995. *Making sense of humanity and other philosophical papers*. Cambridge: Cambridge University Press.

Wilson, E.O. 1984. *Biophilia*. Cambridge, MA: Harvard University Press.

Wilson, E.O. 1987. The little things that run the world (The importance and conservation of invertebrates). *Conservation Biology* 1(4): 344–346.

Wilson, E.O. 1992. *The diversity of life*. Cambridge: Harvard University Press.

Wilson, E.O. 1996. Biophilia and the environmental ethic. In *In search of nature*, E.O. Wilson, 163–179. Washington, DC: Island Press.

Wilson, E.O. 2002. *The future of life*. New York: Knopf.

Wolman, D. 2010. Grass is singing over the tumbled graves. *High Country News* 42(9): 12–18. http://www.hcn.org/issues/42.9/accidental-wilderness (The title of the online version differs from the print version's.).

Wood, P.M. 2000. *Biodiversity and democracy: Rethinking society and nature*. Vancouver: UBC Press.

Wood, T.E., N. Takebayashi, M.S. Barker, I. Mayrose, P.B. Greenspoon, and L.H. Riesebert. 2009. The frequency of polyploid speciation in vascular plants. *Proceedings of the National Academy of Science (PNAS)* 106(33): 13875–13879. http://www.pnas.org/content/106/33/13875.full.

World Health Organization. 2009. *World health statistics*. Geneva: WHO. http://www.who.int/entity/whosis/whostat/EN_WHS09_Full.pdf.

Yi X, Y. Liang, E. Huerta-Sanchez, X. Jin, Z.X.P. Cuo, J.E. Pool, X. Xu, H. Jiang, N. Vinckenbosch, T.S. Korneliussen, H. Zheng, T. Liu, W. He, K. Li, R. Luo, X. Nie, H. Wu, M. Zhao, H. Cao, J. Zou, Y. Shan, S. Li, Q. Yang, Asan, P. Ni, G. Tian, J. Xu, X. Liu, T. Jiang, R. Wu, G. Zhou, M. Tang, J. Qin, T. Wang, S. Feng, G. Li, Huasang, J. Luosang, W. Wang, F. Chen, Y. Wang, X. Zheng, Z. Li, Z. Bianba, G. Yang, X. Wang, S. Tang, G. Gao, Y. Chen, Z. Luo, L. Gusang, Z. Cao, Q. Zhang, W. Ouyang, X. Ren, H. Liang, H. Zheng, Y. Huang, J. Li, L. Bolund, K. Kristiansen, Y. Li, Y. Zhang, X. Zhang, R. Li, S. Li, H. Yang, R. Nielsen, Jun Wang, and Jian Wang. 2010. Sequencing of 50 human exomes reveals adaptation to high altitude. *Science* 329(5987): 75–78.

Index

Aladin, Nicolai
  on effects of invading organisms,
    486–488
Alien species, 9. *See also* Exotic species;
    Invasive species; Native species
  as affecting status of prey and predator,
    505
  contribution to biodiversity, 181, 318,
    485–489
  in fallacies of accident, 48–49
  as natives in alien environment, 320–322,
    333–334, 405
  vs. native species, 81, 380,
    489–492, 505
  relevance to biodiversity, 213
  relevance to disease, 216
  as serving conservation, 380, 504
  as urban guests, 501
Ambrose, Stanley, 138, 138n5, 229
Andersen, Jesper
  on hypoxia, 182
Angermeier, Paul
  on categories of biodiversity,
    89, 107
  and "ecological hierarchy" of categories,
    89n28
  and ecosystem integrity, 123
Anthropocentrism, 18–20. *See also* Value
  and anthropogenic value, 19–20
  and appropriate fit, 451, 466–467
  glossary entry, 516
  and value monism/pluralism, 20
  and virtue ethics, 32
Anthropomorphization of nature, 395–396,
    428, 438
  and appropriate fit, 426, 473–475, 497
  and virtue ethics, 477

Appropriate fit. *See also* Biophilia; Existence
    value; Edenic state of nature;
    Gardens; Landscape art; Modal
    robustness; Parks; Respect for
    nature; Steroidal beavers, humans
    as; Virtue ethics; Wilderness
  absence of scientific basis, 430–434
  as for friendship, 433–434
  vs. anthropocentrism, 466–467
  as attractive alternative, 351
  and Białowieża Puszcza, 503
  vs. biophilia, 470
  and Chernobyl, 504–505
  and "core" of natural value, 427
  vs. Edenic state of nature, 468–470
  elements of, 351, 422–423
  vs. existence value, 475–476
  and gardens, 460–463
  and Hanford Site, 503–504
  implications of
    for abiotic parts of nature, 496–497
    for alien species, 485–492
    for conservation, 501–509
    for conundrum of how much
      biodiversity, 495–496
    for economic valuation, 492–494
    for historical shifts in attitudes towards
      nature, 497–500
    for nature within human domains,
      500–501
  inaptness of social, political analogies in
    support of, 437–438
  as integration in the world as given, 424–426
  as integration into the world at large, 452
  vs. intrinsic value theory, 476–477
  and Korean Demilitarized Zone, 503
  and landscape art, 465

D.S. Maier, *What's So Good About Biodiversity?: A Call for Better*
Reasoning About Nature's Value, DOI 10.1007/978-94-007-3991-8,
© Springer Science+Business Media B.V. 2012